17th Russian Youth Conference on Physics and Astronomy (PhysicA.SPb/2014)

Journal of Physics: Conference Series Volume 661

St. Petersburg, Russia
28 – 30 October 2014

Editors:

Nikita S. Averkiev
Sergey A. Poniaev
Grigorii S. Sokolovskii

ISBN: 978-1-5108-1734-0
ISSN: 1742-6588

Printed from e-media with permission by:

Curran Associates, Inc.
57 Morehouse Lane
Red Hook, NY 12571

CURRAN ASSOCIATES INC. proceedings.com

Some format issues inherent in the e-media version may also appear in this print version.

Printed by Curran Associates, Inc. (2016)

Institute of Physics
Dirac House, Temple Back
Bristol BS1 6BE UK

Phone: 44 1 17 929 7481
Fax: 44 1 17 920 0979

techtracking@iop.org

Additional copies of this publication are available from:

Curran Associates, Inc.
57 Morehouse Lane
Red Hook, NY 12571 USA
Phone: 845-758-0400
Fax: 845-758-2634
Email: curran@proceedings.com
Web: www.proceedings.com

Table of contents

Volume 661

17th Russian Youth Conference on Physics and Astronomy (PhysicA.SPb/2014)
28–30 October 2014, St Petersburg, Russia

Accepted papers received: 6 November 2015
Published online: 2 December 2015

Preface

Papers

Astronomy and astrophysics

Biophysics

Impurities and defects in solid state

Mathematical physics and numerical methods

Nanostructured and thin-film materials

Optics and spectroscopy

Physics of ferroics

Physics of quantum structures

Plasma physics, hydro- and aero-dynamics

Other topics of physics

XVII Russian Youth Conference on Physics and Astronomy (PhysicA.SPb)

The seventeenth Russian Youth Conference on Physics and Astronomy (PhysicA.SPb) was held from 28–30 October 2014 in Saint Petersburg, Russia. The Conference continues the tradition of Saint Petersburg Seminars on Physics and Astronomy originating from the mid-1990s. Since then PhysicA.SPb maintains both the scientific and educational quality of contributions delivered to the young audience. This is the main feature of the Conference that makes it possible to combine the whole spectrum of modern Physics and Astronomy within one event.

PhysicA.SPb/2014 has brought together more than 200 students, young scientists and their professor colleagues from many universities and research institutes across the whole of Russia as well as from Belarus, Ukraine, Finland, the Netherlands, France and Germany. Oral and poster presentations were combined into the well-defined sections of Astronomy and Astrophysics, Optics and spectroscopy, Physics of ferroics, Nanostructured and thin-film materials, Mathematical physics and numerical methods, Biophysics, Plasma physics, hydro- and aero-dynamics, and Physics of quantum structures.

This volume of *Journal of Physics: Conference Series* presents the extended contributions from participants of PhysicA.SPb/2014 that were peer-reviewed by expert referees through processes administered by the Presiders of the Organising and Programme Committees to the best professional and scientific standards.

The Editors: **Nikita S. Averkiev, Sergey A. Poniaev and Grigorii S. Sokolovskii**

Journal of Physics: Conference Series **661** (2015) 011002 doi:10.1088/1742-6596/661/1/011002

Peer review statement

All papers published in this volume of *Journal of Physics: Conference Series* have been peer reviewed through processes administered by the proceedings Editors. Reviews were conducted by expert referees to the professional and scientific standards expected of a proceedings journal published by IOP Publishing.

Search for the optical counterparts of the γ-ray pulsars J2055+2539, J2043+2740, J1957+5033

D M Beronya[1], Yu A Shibanov[1,2], D A Zyuzin[2] and V N Komarova[3]

[1] St. Petersburg Polytechnic University, Politekhnicheskaya 29, St. Petersburg, 195251, Russia
[2] Ioffe Institute, Politekhnicheskaya 26, St. Petersburg, 194021, Russia
[3] Special Astrophysical Observatory, Nizhnij Arkhyz, 369167, Russia

E-mail: daria.beronya@gmail.com

Abstract. First optical observations of γ-ray pulsars J2055+2539, J2043+5740, J1957+5033 have been obtained with the BTA 6m optical telescope. We found no reliable counterpart candidates to these pulsars. However, upper limits on optical fluxes were derived. Comparison of PSR J2055+2539 and PSR J2043+2740 optical limits and their X-ray spectral data likely suggests the presence of spectral breaks between the optical and X-rays in the multiwavelength emission of these pulsars. X-ray spectral data for PSR J1957+5033 are not yet available. In spite of rather deep Hα-observations of PSR J2055+2539, a bow shock nebula, which was expected for the pulsar, has not been detected. Probably, the density of the interstellar medium surrounding the pulsar is not enough for a bright Hα bow shock to be formed. We also find the further deeper observations of pulsars J2055+2539 and J2043+5740 are reasonable, while a large gamma-ray position error ellipse for J1957+5033 is crowded by background sources and its coordinates should be updated using X-ray observations before any further optical study.

1. Introduction

Rapidly rotating, highly magnetized neutron stars are able to accelerate charged particles to extreme energies (of the order of some TeVs) unattainable in terrestrial accelerators. As a result, they are able to produce an emission in the γ-ray band.

Before the launch of the Fermi Large Area Telescope only 6 pulsars were known as γ-ray emitters. All of these have been identified also at optical wavelengths. The second Fermi LAT catalogs was published in 2013 and included 117 γ-ray pulsars unveiled in three years of observations. Only for 38 of these pulsars did optical observations carry out [1].

Optical investigations of pulsars enable to provide more accurate distance estimations. They are also necessary to obtain multiwavelength spectra, that is important for creating appropriate models of pulsar radiation.

In this work, we focus on searching for optical counterpart candidates to pulsars J2055+2539, J2043+2740 and J1957+5033. Certain parameters of pulsars are summarized in Table 1. These are equatorial α_{2000}, δ_{2000} and galactic l, b coordinates, rotation periods P, period derivatives $\dot{P}$, characteristic ages τ, surface magnetic fields B and spin down energy loss rates $\dot{E}$.

PSR J2043+2740 was discovered in the radio in the Arecibo millisecond-pulsar survey at 430 MHz in 1994 [2]. In 2010 Abdo et al. published the results of the first six month of *Fermi* data analysis, containing γ-ray high-confidence pulsation parameters for this pulsar [3].

Table 1. Parameters of pulsars J2055+2539, J2043+2740, J1957+5033.

PSR	$\alpha_{2000}, \delta_{2000}$	l, b	P, s	$\dot{P}$, 10^{-15}	τ, Myr	B, 10^{12} G	$\dot{E}$, 10^{33} erg s^{-1}
J2055+2539	$20^h55^m48\overset{s}{.}8$ $+25°40'02''$	70°.69 -12°.52	0.32	4.08 ± 0.11	1.24	1.16	4.9
J2043+2740	$20^h43^m43\overset{s}{.}5$ $+27°40'56''$	70°.61 -9°.15	0.1	1.27 ± 0.01	1.2	0.354	5.6
J1957+5033	$19^h57^m38\overset{s}{.}9$ $+50°33'18''$	84°.58 11°.01	0.38	7.08 ± 0.05	0.839	1.65	5.3

PSR J2055+2539 and PSR J1957+5033 were discovered in γ-rays in a blind frequency search using 11 months of data acquired by the *Fermi* LAT [4].

The observations in X-rays were performed for all of these pulsars. PSR J2043+2740 was observed in 2002 with *XMM-Newton* for a total on-source time of 17 ks and detected in the 0.3 − 10 keV energy band at high level of significance. PSR J2055+2539 was also observed with *XMM-Newton* (0.2−12 keV) using 26 and 136 ks of data taken in 2009 and 2013, respectively. These observations allowed to determine its position and revealed a bright comet-like trail behind the pulsar. Of particular interest in this case is H_α-imaging, which enables pulsar bow shocks to be detectable. PSR J1957+5033 was observed with *Chandra* in 2014, but the data are not yet available.

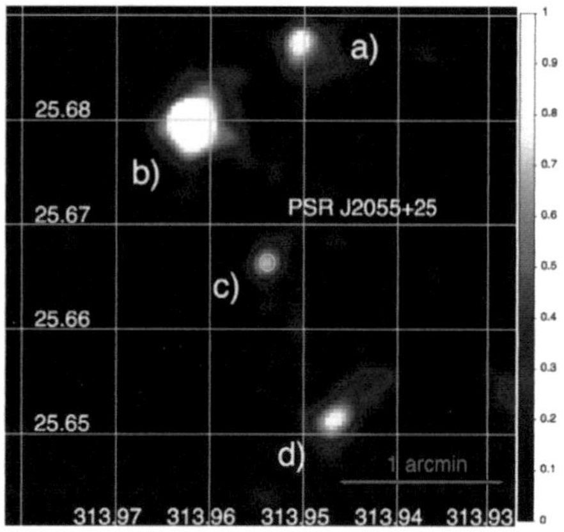

a) b)

Figure 1. XMM-Newton X-ray archival images of the regions around PSR J2055+2539 using data obtained in 2009 (a) ([4], ObsID 0605470401) and 2013 (b) (ObsID 0724090101). The color scale is in counts per square arcsecond. The green ellipse and the red x mark represent the best pulsar location derived from pulsar timing.

2. Optical observations

The fields of the pulsars J2055+2539, J2043+2740 and J1957+5033 were observed on August 3, 2013 with the multi-mode focal reducer SCORPIO of the BTA, a 6 meter aperture optical telescope at the Special Astrophysical Observatory. Data binning of 2×2 was applied resulting to the pixel scale of $0\overset{''}{.}357$. The observing conditions were photometric with seeing varied from 2.0 to 2.8. Parameters of filters we used are shown in Table 2. The online version of the article is supplemented with an attached file including the summary of the BTA observations.

Data reduction was performed in a standard way by means of `IRAF` tools. `Imgalign` task from `ctio IRAF` package was used in order to align images for each filter before combining them. Stacking of images was performed applying `avsigclip` algorithm with no scaling or weighting. Astrometrical calibration was made with `ccmap/ccsetwcs IRAF` tasks using the positions of about 10 bright, but non-saturated field stars from USNO-B1.0 catalogs, having the nominal accuracy of $0\farcs24$. After removing the stars with large residuals 1σ rms-errors were about $0\farcs07$–$0\farcs1$ in RA and $0\farcs05$–$0\farcs08$ in Dec for the fields of the three pulsars. For our pulsars 1σ *Fermi* position uncertainty is typically grater that $1''$ and below we neglect much smaller astrometric referencing uncertainties of the optical images. For the photometric calibrations in the FN657 and SED707 filters we used the GD248 photometric standard star (Oke, 1990), taken in the same night as the targets. Since no photometric standard were observed in VR$_c$ bands, the photometric calibration in these filters was performed using stars from astrometric catalogs.

Table 2. Parameters of optical filters.

Filter	T_{max}, %	λ_c, $\AA$	FWHM, $\AA$	Function
V	49	5470	790	pulsar detection in the optical continuum
R$_c$	54	6620	1500	
SED707	77	7040	207	
FN657	73	6555	75	pulsar bow shock detection in the H$_\alpha$

2.1. PSR J2055+2539 observations

The pulsar field was observed in the V, R$_c$, FN657 and SED707 bands with total exposures of 120s, 360s, 1800s and 300s, respectively (Figures 2, 3).

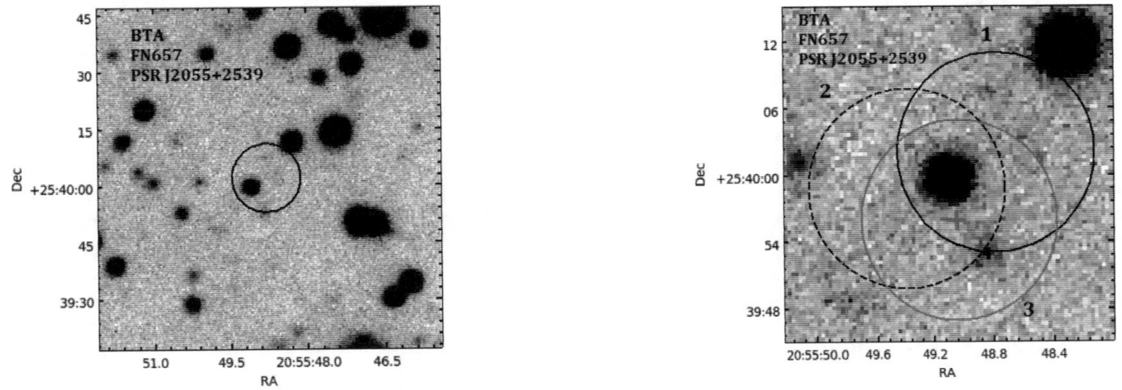

a) b)

Figure 2. a) $1\farcm5 \times 1\farcm5$ image fragment of the PSR J2055+2539 field. b) Intersection of 3σ uncertainty ellipses of the *Fermi* (1) and *XMM-Newton* (2, 3 - according to the data of 2009 and 2013, respectively) pulsar positions marked on the $0\farcm5 \times 0\farcm5$ image fragment of the pulsar field.

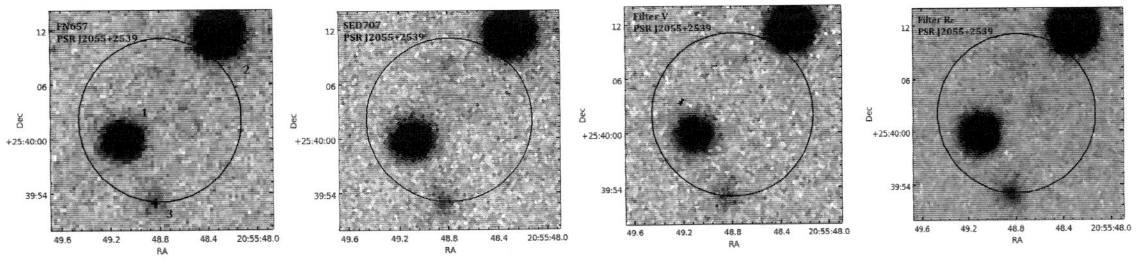

Figure 3. $0\farcm4 \times 0\farcm4$ PSR J2055+2539 vicinity in the four bands.

There are three objects within or just off the uncertainty ellipse. Objects 1 and 2 are obviously background stars, which can be found in the astrometrical catalogs. Taking into account errors their magnitudes obtained with aperture photometry coincided with ones of the catalogs. Source 3 is fainter and absent in catalogues. Further studies are necessary to reveal its nature. Our results of aperture photometry for sources 1,2,3 are presented in Table 3.

Table 3. The results of aperture photometry of the three objects visible in Figure 3.

Object	α_{2000}	δ_{2000}	FN657	SED707	V	R_c
1	20:55:49.1	+25:39:59.7	19.1±0.2	19.24±0.11	20.13±0.56	19.18±0.28
2	20:55:48.3	+25:40:11.6	18.0±0.2	18.15±0.11	18.77±0.56	18.18±0.28
3	20:55:48.9	+25:39:52.9	21.64±0.21	21.83±0.13	22.48±0.56	21.64±0.29

Based on the image in the FN657 band it is not obviously possible to conclude the detection of H_α bow shock emission. Apparently, the density of the interstellar medium surrounding this high Galactic altitude pulsar is not enough for a bright H_α bow shock to be formed.

2.2. PSR J1957+5033 observations

The pulsar field was observed in the V, R_c, FN657 and SED707 bands with total exposures of 120s, 60s, 1200s and 600s, respectively (Figure 4).

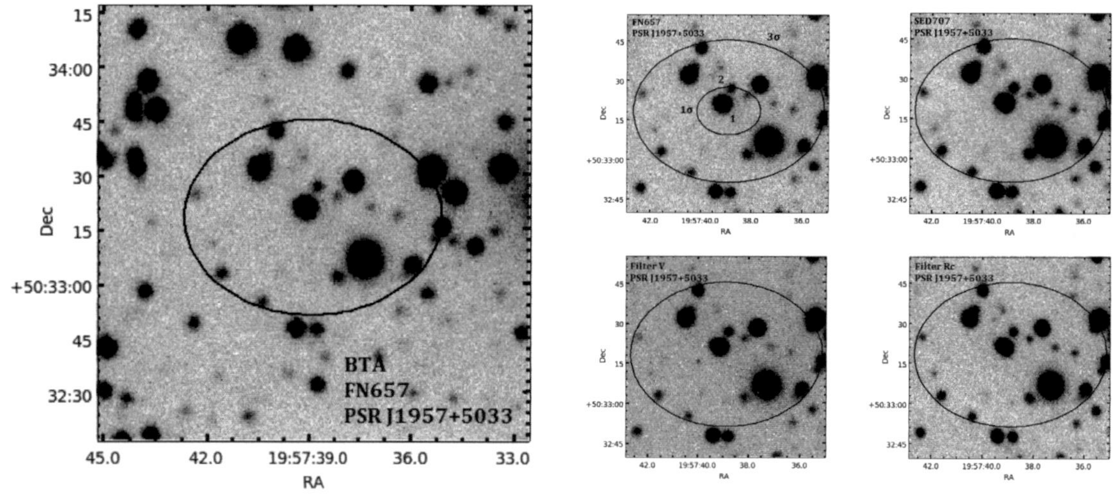

a) b)

Figure 4. a) $2' \times 2'$ image fragment of the PSR J1957+5033 field. b) $1'.25 \times 1'.25$ pulsar vicinity in the four bands. 1σ and 3σ pulsar *Fermi* position error ellipses are shown.

Two stars are seen within 1σ *Fermi* error ellipse. They are marked as 1 and 2 in Figure 4. Their parameters can be found in the astrometrical catalogs (Table 4).

Table 4. Parameters of stars 1 and 2 marked in Figure 4, taken from astrometrical catalogs.

Object	$\alpha_{2000}, \delta_{2000}$	B USNO-A2.0	R USNO-A2.0	B Nomad	R Nomad	V GSC 2.3	R GSC 2.2
1	19:57:39.135 +50:33:20.6	17	16.8	17.25	16.58	17.11	16.54±0.24
2	19:57:38.786 +50:33:26.9	–	–	19.56	19.68	–	–

The identification of PSR J1957+5033 is complicated by large *Fermi* positional errors and many stellar objects within 3σ uncertainty ellipse. However, X-ray observation made by the ACIS-S instrument on the *Chandra* telescope, having a spatial resolution of $0\rlap{.}''5$, will enable to narrow down the region of the expected pulsar position.

2.3. PSR J2043+2740 observations

The pulsar field was observed in the R_c, FN657 and SED707 bands with total exposures of 180s, 600s and 300s, respectively (Figures 5). An extended source is detected on the border of the 3σ pulsar ellipse (Figure 5). The object is not included in any astrometrical catalog. We performed aperture photometry on the source. Its magnitude is R_c=21.01±0.09. It is also visible in FN657 and SED707 bands, while its relation to the pulsar remains unclear.

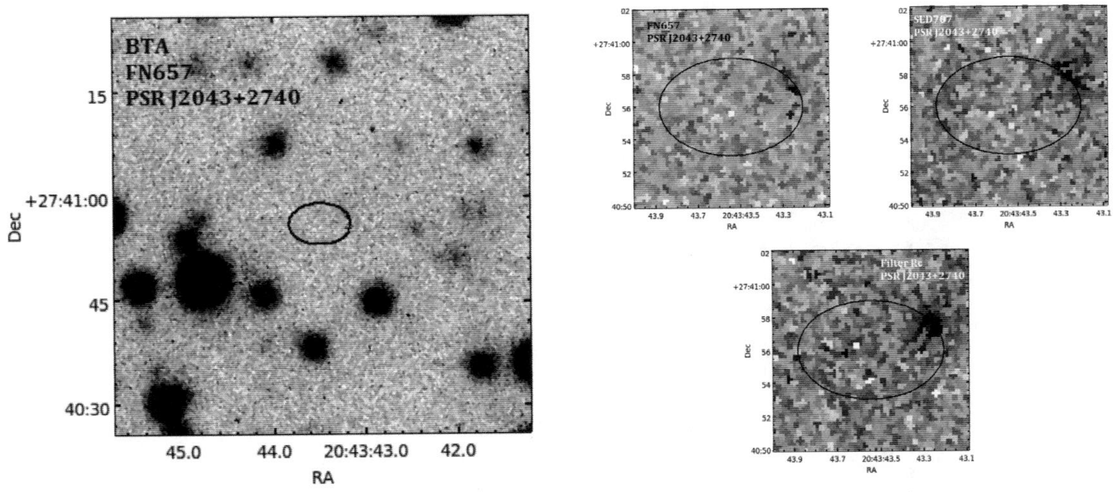

a) b)

Figure 5. a) $1' \times 1'$ image fragment of the PSR J2043+2740 field. b) $0\rlap{.}'2 \times 0\rlap{.}'2$ pulsar vicinity in the three bands.

3. The upper limits on the optical flux

We have not detected any plausible pulsar optical counterpart or signature of the H_α bow shock for all three pulsars observed with the BTA. 3σ detection limits estimated by a standard way are given in Table 5. Interstellar extinctions for each pulsar were inferred from X-ray data, the total Galactic extinction and HI column density for given directions, and pulsar distance estimates.

Table 5. 3σ magnitude upper limits and 3σ extinction corrected flux upper limits.

PSR	V (F_{lim}, μJy)	R_c (F_{lim}, μJy)	FN657 (F_{lim}, μJy)	SED707 (F_{lim}, μJy)
J2055+2539	22.78 (5.354)	23.22 (2.57)	22.75 (4.935)	22.77 (4.591)
J1957+5033	23.07 (2.931)	22.61 (3.512)	22.66 (4.13)	23.22 (2.408)
J2043+2740	–	21.67 (11.718)	21.87 (12.141)	22.44 (6.773)

4. Multiwavelength observations of pulsars J2055+2539, J1957+5033, J2043+2740

In order to compare the obtained flux upper limits with available information about X-ray and γ-ray spectra we have carried out multiwavelength spectral analysis for each pulsar. We used the results of *XMM-Newton* observations of pulsars J2043+2740 [5] and J2055+2539 (ObsID 0724090101) for the X-rays, and the *Fermi* online catalog, containing the γ-ray spectral parameters [1]. The X-ray and γ-ray spectra are best fitted by a pure power law and an exponentially cut-off power law, respectively. The parameters of fits are listed in the attached file. The unabsorbed multiwavelength spectra are shown in Figure 6.

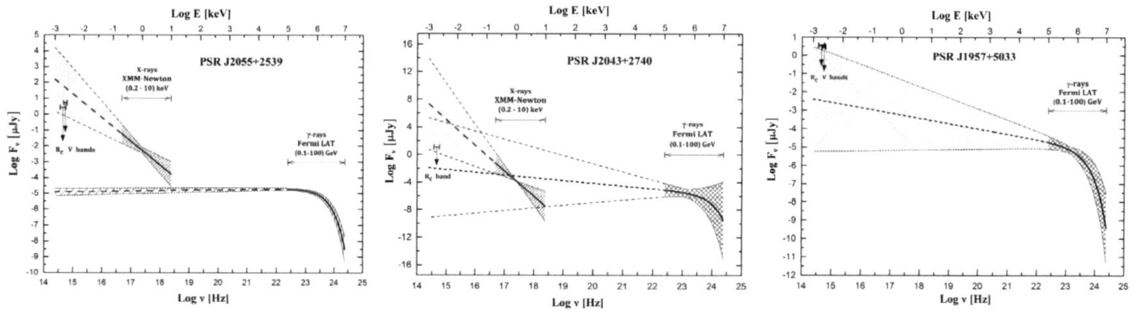

Figure 6. Multiwavelength spectra of pulsars J2055+2539, J2043+2740, J1957+5033. The detection ranges are marked by horizontal bars. The optical detection limits are marked by horizontal bars with arrows. Solid lines show the best-fit spectral models to the *XMM-Newton* and *Fermi* data, extrapolated towards the optical range (dashed line). The shaded regions correspond to 1σ uncertainties of the fits.

5. Conclusions

We presented the results of the first optical observations of the γ-ray pulsars J2055+2539, J2043+5740, J1957+5033, taken with the BTA. Although the pulsars are undetected at our rather short exposures, the derived optical flux upper limits appear to be informative. In case of PSR J2055+2539 and PSR J2043+5740, an inspection of multiwavelength spectra likely suggests a break between X-ray nonthermal power law component and the optical range. In [6], it is assumed that each pulsars nonthermal spectrum can be empirically described by a single power law stretching across many orders of magnitude in frequency, with a break or cut-off at GeV energies. As it is seen from Figure 6, the BTA upper limits are 3-5 orders of magnitude higher than γ-ray extrapolation toward to the optical range, what makes the suggested dependence to be consistent with our data. In spite of an 30 minute total exposure of the combined H_α image of PSR J2055+2539, the expected bow shock has not been detected. Probably, this high-altitude pulsar is surrounded with the low-density interstellar medium, which cannot produce a detectable bow shock. We also find the further deeper optical observations of pulsars J2055+2539 and J2043+5740 are reasonable, while J1957+5033 position should be improved based on Chandra observations.

Acknowledgments

DAZ was supported by RFBR, research project No. 14-02-31600 mol_a.

References

[1] Abdo A A 2013 *ApJS* **208** 17 (*Preprint* arXiv:1305.4385v3)
[2] Thorsett S E *et al* 1994, IAU circ.6012
[3] Abdo A A *et al* 2010 *ApJS* **187** 460 (*Preprint* arXiv:1006.2134v1)
[4] Saz Parkinson P M *et al* 2010 *ApJ* **725** 571 (*Preprint* arXiv:1006.2134v1)
[5] Becker W *et al* 2004 *ApJ* **615** 908
[6] Durant M *et al* 2012 Multiwavelength spectroscopy of PSR B0656+14 (*Preprint* arXiv:1109.1984v3)

Statistical approach to thermal evolution of neutron stars

M V Beznogov[1] **and D G Yakovlev**[2]

[1] St. Petersburg Academic University, 8/3 Khlopina Street, St. Petersburg 194021, Russia
[2] Ioffe Physical Technical Institute, 26 Politekhnicheskaya, St. Petersburg 194021, Russia

E-mail: mikavb89@gmail.com

Abstract. Studying thermal evolution of neutron stars (NSs) is one of a few ways to investigate the properties of superdense matter in their cores. We study the cooling of isolated NSs (INSs) and deep crustal heating of transiently accreting NSs in X-ray transients (XRTs, binary systems with low-mass companions). Currently, nearly 50 of such NSs are observed, and one can apply statistical methods to analyze the whole dataset. We propose a method for such analysis based on thermal evolution theory for individual stars and on averaging the results over NS mass distributions. We calculate the distributions of INSs and accreting NSs (ANSs) in XRTs over cooling and heating diagrams respectively. Comparing theoretical and observational distributions one can infer information on physical properties of superdense matter and on mass distributions of INSs and ANSs.

1. Introduction

We study neutron stars of two types. First, they are *isolated middle-aged* ($10^2 - 10^6$ yr) NSs born hot is supernova explosions but cooling down mostly via neutrino emission from their superdense cores. They are mainly isothermal inside; a noticeable temperature gradient is available only in thin outer heat blanketing envelopes (e.g., [1, 2]).

Second, we analyze old ($t \gtrsim 10^8 - 10^9$ yr) *transiently accreting quasi-stationary NSs in compact binaries with low-mass companions*; such systems are called XRTs. These NSs accrete matter from time to time (in the active states of XRTs) from their companions. The accreted matter is compressed under the weight of newly accreted material which is accompanied by nuclear transformations and deep crustal heating [3, 4] with characteristic energy release of 1–2 MeV per accreted nucleon. The accretion episodes are supposed to be neither long nor intense to destroy the internal thermal equilibrium. Nevertheless, the deep crustal heating should be sufficiently strong to keep the NSs warm and power observable thermal emission of ANSs during quiescent states of XRTs [5]. The mean NS heating rate is determined by the mass accretion rate $\langle \dot{M} \rangle$ averaged over characteristic cooling times of NSs ($\gtrsim 10^3$ yr).

INSs are usually studied by calculating their *cooling curves* (time dependence of their effective surface temperature $T_s(t)$ or (equivalently) thermal surface luminosity $L_\gamma(t)$). The curves are calculated under different assumptions on the neutrino emission in the stellar core, and then compared with the data.

ANSs in XRTs are investigated by calculating their *heating curves*, which give average T_s or L_γ for ANSs in quiescent states as a function of $\langle \dot{M} \rangle$ (which is also compared with observations).

It is important, that the cooling and heating curves have much in common (e.g., [6, 7]) and allow one to study fundamental physics of NSs. The most important cooling/heating regulators to be tested are:

(i) A level of neutrino luminosity of NSs;

(ii) NS mass M and equation of state (EOS) of superdense matter in the stellar core which regulate the neutrino emission level;

(iii) Composition of the NS heat blanketing envelope which determines the relation between the internal and surface temperatures.

Observations of INSs and ANSs are rapidly progressing. Let us utilize the accumulated statistics and develop a statistical theory of NS evolution. It will take into account that cooling and heating curves can strongly depend on M. Then one can introduce the probability to find a source in different places of cooling/heating diagrams by averaging these curves over mass distributions of INSs and ANSs. Naturally, these mass distributions can be different. Comparing theoretical and observational distributions over diagrams one can study not only individual cooling regulators mentioned above but also the mass distributions of NSs. A more extended version of this study is published elsewhere [8]. Note that statistical studies of populations of INSs have been performed earlier in several publications (e.g., [9]) but they have been quite different.

2. Statistical theory. Broadening of direct Urca threshold

Let us present the simplest version of the statistical theory. It is based on *ordinary* theory of NS cooling and heating, e.g., [10, 11]. For illustration, consider NSs with nucleonic cores and one phenomenological EOS in the core described in [12]. The authors of [12] denoted it as HHJ. The maximum NS mass for this EOS is $M_{\max} = 2.16\,M_\odot$ (to be consistent with mass measurements $M \approx 2M_\odot$ of two massive NSs [13, 14]). In our case the radius of the most massive stable NS is $R = 10.84$ km and the central density $\rho_c = 2.45 \times 10^{15}$ g cm^{-3}. The powerful direct Urca process of neutrino emission opens at $M > M_D = 1.72\,M_\odot$. At lower M this process is forbidden, while at higher M it is allowed in the central kernel of a star (at densities $\rho > \rho_D = 1.00 \times 10^{15}$ g cm^{-3}); then NSs undergo very rapid neutrino cooling.

We calculate thermal evolution of INSs and ANSs using our generally relativistic cooling code [15] on a dense grid of masses M, from $1.1\,M_\odot$ to $2.1\,M_\odot$. The cooling curves of INSs are obtained by direct running the code.

The heating curves of ANSs are calculated as stationary solutions of the heat balance equation (e.g., [16]),

$$L_h^\infty = L_\nu^\infty + L_\gamma^\infty, \tag{1}$$

where L_h^∞ is the averaged deep crustal heating power (redshifted for a distant observer and determined by the time-averaged mass accretion rate $\langle \dot{M} \rangle$). The interior of the star is assumed to be isothermal while the internal temperature is related to T_s by the heat blanketing solution (e.g., [2]). Calculated cooling curves will be plotted on the $T_s^\infty - t$ diagram, while heating curves will be plotted on the $L_\gamma^\infty - \langle \dot{M} \rangle$ diagram. These will be ordinary cooling and heating curves which have been extensively studied by the theory. As a rule, the highest cooling or heating curve refers to the low-mass NSs (with slow neutrino emission) while the lowest curve belongs to the maximum-mass star with highest neutrino cooling rate. The space between the highest and lowest cooling curves is filled by the curves for NSs of different M but this filling can be very non-uniform (e.g., [17]).

Now we put forward the statistical theory of NS evolution. All NSs will be assumed to have the same internal structure (EOS, neutrino emission properties) but, naturally, they have different parameters such as mass, the amount of light elements in heat-blanketing envelopes, magnetic fields, rotation, etc. Instead of deterministic cooling/heating curves in appropriate diagrams we introduce the probability to find a star in different places of a diagram. These probabilities are obtained by averaging the cooling/heating curves over statistical distributions of variable parameters such as distributions of masses M and masses $\Delta M_{\rm le}$ of light elements in the heat-blanketing envelopes. Then the cooling/heating, that initially followed specific trajectories, is replaced by statistical probabilities to find NSs at different stages of their evolution.

In the first approximation it is reasonable to neglect the effects of magnetic fields and rotation on thermal states of NSs. To study INSs and ANSs we introduce the mass distribution functions for these sources, $f_{\rm i}(M)$ and $f_{\rm a}(M)$. These functions are naturally different; ANSs should be overall more massive than INSs.

Such distribution functions are taken in the form

$$\begin{aligned} f_{\rm i}(M) &= \frac{1}{N_{\rm i}}\frac{1}{\sqrt{2\pi}\,\sigma_{\rm i}}\exp\left(-\frac{(M-\mu_{\rm i})^2}{2\sigma_{\rm i}^2}\right), \\ f_{\rm a}(M) &= \frac{1}{N_{\rm a}}\frac{1}{\sqrt{2\pi}\,M\sigma_{\rm a}}\exp\left(-\frac{(\ln[M/{\rm M}_\odot]-\mu_{\rm a})^2}{2\sigma_{\rm a}^2}\right), \end{aligned} \tag{2}$$

where $\sigma_{\rm i,a}$ and $\mu_{\rm i,a}$ are the parameters of the distributions; $N_{\rm i,a}$ are normalization factors. These distributions are used in the mass range from $1.1\,{\rm M}_\odot$ to $2.1\,{\rm M}_\odot$; for $M < 1.1\,{\rm M}_\odot$ and $M > 2.1\,{\rm M}_\odot$ they are artificially set to zero. Note that for the normal distribution function $f_{\rm i}(M)$, $\mu_{\rm i}$ is the most probable mass. However, for the log-normal distribution $f_{\rm a}(M)$ the most probable mass is ${\rm M}_\odot \exp\left(\mu_{\rm a}-\sigma_{\rm a}^2\right)$. We expect that ANSs should be overall heavier than INSs as a result of accretion.

The heat transparency of the blanketing envelope is determined by the mass $\Delta M_{\rm le}$ of light (accreted) elements (mainly, hydrogen and helium) in these envelopes. The higher $\Delta M_{\rm le}$, the larger thermal conductivity in the envelope, and the higher $T_{\rm s}$ for a given internal temperature of the star (e.g., [2]). However, $\Delta M_{\rm le}$ cannot exceed $\Delta M_{\rm le\,max} \approx 10^{-7}\,M$ because at formally larger $\Delta M_{\rm le}$ the light elements at the bottom of the heat blanketing envelope transform into heavier ones. We will consider $\Delta M_{\rm le} \le \Delta M_{\rm le\,max}$ as a random quantity characterized by a distribution $f_{\rm acc}(\Delta M_{\rm le})$

$$f_{\rm acc}(\Delta M_{\rm le}) = {\rm const} \quad {\rm at} \quad \Delta M_{\rm le} \le 10^{-7} M. \tag{3}$$

In fact, $f_{\rm acc}(\Delta M_{\rm le})$ is uncertain (especially for ANSs); we take (3) as a first approximation.

Our cooling code [15] allows us to study the effects of superfluidity on evolution of NSs. To reduce the number of variable parameters, we employ a semi-phenomenological approach. In particular, we will broad out artificially a step-like density dependence of the neutrino emissivity $Q_{\rm D}$ provided by the direct Urca process [18]. In the absence of superfluidity the direct Urca process switches on sharply with increasing density, from $Q_{\rm D}=0$ at $\rho<\rho_{\rm D}$ to finite $Q_{\rm D}$ at $\rho \ge \rho_{\rm D}$. Moreover, in our model HHJ EOS, NS cores consist of neutrons with admixture of protons, electrons and muons, and we have two direct Urca processes, electronic and muonic ones (e.g., [19]). Accordingly, we have two density thresholds for the onset of the these processes (and the emissivities of both processes – if open – are the same). The density threshold for the muonic process is always higher than for the electronic one. Accordingly, when we increase M (or $\rho_{\rm c}$) the electronic direct Urca always switches on first. It sharply (by 6–7 orders of magnitude) increases the neutrino luminosity of the NS, and becomes the leading one. The switch on of the muonic process with further increase of M is relatively unimportant (although included in the calculations). A sharp step-like onset of the direct Urca process is known to be incompatible with observations. One needs to broaden its threshold. We will describe this broadening on a phenomenological level by multiplying the direct Urca neutrino emissivities by a broadening factor b. For instance, for the electronic direct Urca we take

$$Q_{\rm D} = Q_{\rm D0}\, b(x), \quad b(x) = 0.5\,[1+{\rm erf}\,(x)]\,, \tag{4}$$

where Q_{D0} is the threshold emissivity, $x = (\rho - \rho_D)/(\alpha\rho_D)$, erf$(x)$ is the error function, so that $b(x) \to 0$ at $x \to -\infty$ and $b(x) \to 1$ at $x \to \infty$, and α is a parameter assumed to be small, $\alpha \ll 1$. This parameter determines a narrow range of densities $|\rho - \rho_D| \sim \alpha\rho_D$ in which the direct Urca process gains its strength. Similar broadening is introduced for the muonic direct Urca process but it is almost unimportant.

For example, the broadening of the direct Urca threshold can be provided by proton superfluidity (e.g., [19]). This superfluidity is characterized by the proton critical temperature $T_{cp}(\rho)$ (e.g., [20]). The critical temperatures are very model dependent. A scatter of theoretical $T_{cp}(\rho)$ is large, so that we do not rely on specific theoretical models but consider $T_{cp}(\rho)$ on phenomenological level. Proton superfluidity is expected to be strong in the outer core of the NS (with $T_{cp}(\rho) \gtrsim 3 \times 10^9$ K) but becomes weaker or fully disappears in the inner core, at a few nuclear matter densities. As long as it is strong, it greatly suppresses the direct Urca process (even if it is formally allowed) by the presence of a large gap in the energy spectrum of protons. When proton superfluidity weakens with growing ρ, the superfluid suppression is removed and the direct Urca becomes very powerful. It switches on after exceeding some threshold density, but not very sharply, as if the threshold is broadened.

3. Results and conclusions

Some results are presented in Figs. 1 and 2 (more details can be found in [8]). The probabilities are calculated by averaging over neutron star masses in accordance with (2) and over the amount of light elements in the heat blanketing envelopes (3). The probability distributions are plotted by grayscaling (in relative units). White regions refer to zero or very low probability. The direct Urca threshold is broadened in accordance with (4) at $\alpha = 0.1$. We see that the theoretical distributions describe the data reasonably well.

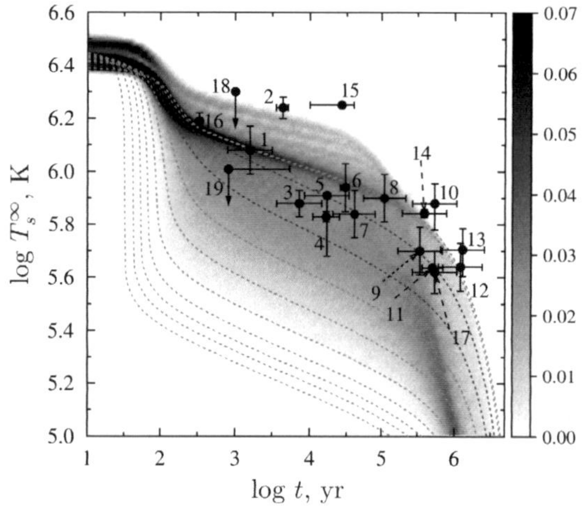

Figure 1: Probability to find an INS in different places of the $T_s^\infty - t$ plane compared with observations. Dotted lines show 11 "reference" cooling curves for stars with iron envelopes and masses $M = 1.1, 1.2, \ldots, 2.1\,M_\odot$ (see text for details).

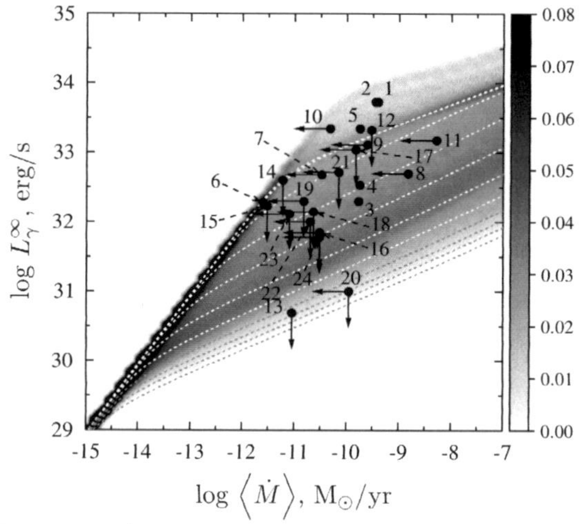

Figure 2: Same as in Fig. 1 but for ANSs in the $L_\gamma^\infty - \langle\dot{M}\rangle$ plane.

The data for INSs include (1) PSR J1119–6127,(2) RX J0822–4300, (3) PSR J1357–6429, (4) Vela, (5) PSR B1706–44, (6) PSR J0538+2817, (7) B2334+61, (8) PSR B0656+14, (9) Geminga, (10) PSR B1055–52, (11) RX J1856.4–3754, (12) PSR J2043+2740, (13) RX

J0720.4–3125, (14) PSR J1741–2054, (15) XMMU J1731–347, (16) Cas A NS, (17) PSR J0357+3205 (18) Crab, (19) PSR J0205+6449.

XRTs in Fig. 2 are (1) Aql X-1,(2) 4U 1608–522, (3) MXB 1659–29, (4) NGC 6440 X-1, (5) RX J1709–2639, (6) IGR 00291+5934, (7) Cen X-4, (8) KS 1731–260, (9) 1M 1716–315, (10) 4U 1730–22, (11) 4U 2129+47, (12) Terzan 5, (13) SAX J1808.4–3658, (14) XTE J1751–305, (15) XTE J1814–338, (16) EXO 1747–214, (17) Terzan 1, (18) XTE 2123–058, (19) SAX J1810.8–2609, (20) 1H 1905+000, (21) 2S 1803–45, (22) XTE J0929–314, (23) XTE J1807–294, (24) NGC 6440 X-2. The references to observational data can be found in [8].

Our results are preliminary and illustrative. We have taken one EOS of nucleon matter in the NS core. Moreover, we have introduced the broadening of the direct Urca threshold and distribution functions over NS masses phenomenologically. Nevertheless, we have varied the broadening of the direct Urca threshold [the parameter α in (4)], and typical mass ranges of INSs and ANSs. Using such 'trial and error' procedure we have obtained a reasonable agreement with observations of NSs of both types for $\alpha = 0.1$, $\mu_\mathrm{i} = 1.4$, $\sigma_\mathrm{i} = 0.15$, $\mu_\mathrm{a} = 0.47$, and $\sigma_\mathrm{a} = 0.17$. Lower α would result in two distinctly different populations of NSs of each type (INSs and ANSs), warmer ($M \leq M_\mathrm{D}$) and colder ($M > M_\mathrm{D}$) ones, in disagreement with the data. At $\alpha \approx 0.1$ these populations merge in one population (for NSs of each type) which qualitatively describes the data. Larger α result in overabundance of colder stars, in contradiction with the data [8]. The selected mass distribution functions (2) and the obtained values of their parameters do not contradict observations and theoretical expectations [21] but they are not unique. The most probable mass for ANSs is $1.55\,\mathrm{M}_\odot$, whereas for INSs it is $1.4\,\mathrm{M}_\odot$. Thus, ANSs are overall heavier than INSs as it was expected.

The advantage of our approach is that it explains *all the data*. The explanation essentially requires (i) the presence of the direct Urca process in the inner cores of massive NSs (to interpret the coldest ANSs); (ii) a moderate broadening of the direct Urca threshold ($\alpha \approx 0.1$) to merge the populations of warm and colder NSs of each type into one (observable) population; and (iii) higher typical masses of ANSs [to explain the very cold accreting source 13 (SAX 1808.4–3658) and the absence of very cold middle-aged INSs]. In this case the averaging over the masses of light elements in (3) plays relatively minor role but is helpful to explain the existence of warmer isolated and accreting NSs. Nevertheless, these stars can be explained [8] by assuming the presence of strong proton superfluidity in NSs with $M < M_\mathrm{D}$. This superfluidity suppresses the modified Urca process, which is the major process of neutrino emission in low-mass NSs. Such stars will become slower neutrino coolers, and hence warmer sources (e.g., [11], and references therein). The required broadening of the direct Urca threshold can be produced by weakening of proton superfluidity in the massive stars. Therefore, the obtained explanation, in physical terms, can be reached by assuming the presence of proton superfluidity in neutron star cores. This superfluidity should be strong in low-mass stars but weakens in high-mass ones whose neutrino emission is greatly enhanced by the direct Urca process (as discussed above).

The present scenario is different from the minimal cooling model [22, 23]. The latter model assumes that the enhanced neutrino cooling is produced by the neutrino emission due to the triplet-state pairing of neutrons. This enhancement is much weaker than that due to the direct Urca process; it cannot explain the observations of the coldest XRTs (such as SAX 1808.4–3658).

The disadvantage of our analysis is that it should be elaborated. For instance, one can try other EOSs of superdense matter. We expect that the results will be similar but rescaled with respect to the values of M_D for new EOSs. One can also try different mass distributions of INSs and ANSs. In addition, the distribution of light elements in the heat blanketing envelopes is not entirely arbitrary but is regulated by diffusion of ions. Another issue would be to include the effects of rotation and magnetic fields, and also the effects of nucleon superfluidity.

Acknowledgments
The authors acknowledge partial support by the State Program "Leading Scientific Schools of Russian Federation," grant NSh 294.2014.2. The work of MB has also been partly supported by the Dynasty Foundation and Foundation for Support of Education and Science (Alferov's Foundation), and the work of DY by Russian Foundation for Basic Research (grants 14-02-00868-a and 13-02-12017-ofi-M).

References

[1] Gudmundsson E H, Pethick C J and Epstein R I 1983 *Astrophys. J.* **272** 286–300
[2] Potekhin A Y, Chabrier G and Yakovlev D G 1997 *Astron. Astrophys.* **323** 415–428
[3] Haensel P and Zdunik J L 1990 *Astron. Astrophys.* **227** 431–436
[4] Haensel P and Zdunik J L 2007 *Astrophys. Space Sci.* **308** 363–369
[5] Brown E F, Bildsten L and Rutledge R E 1998 *Astrophys. J. Lett.* **504** L95–L98
[6] Yakovlev D G and Haensel P 2003 *Astron. Astrophys.* **407** 259–263
[7] Yakovlev D G, Levenfish K P and Haensel P 2003 *Astron. Astrophys.* **407** 265–271
[8] Beznogov M V and Yakovlev D G 2015 *MNRAS* **447** 1598–1609
[9] Popov S, Grigorian H, Turolla R and Blaschke D 2006 *Astron. Astrophys.* **448** 327–334
[10] Page D, Lattimer J M, Prakash M and Steiner A W 2009 *Astrophys. J.* **707** 1131–1140
[11] Yakovlev D G and Pethick C J 2004 *Annu. Rev. Astron. Astrophys.* **42** 169–210
[12] Kaminker A D, Kaurov A A, Potekhin A Y and Yakovlev D G 2014 *MNRAS* **442** 3484–3494
[13] Demorest P B, Pennucci T, Ransom S M, Roberts M S E and Hessels J W T 2010 *Nature* **467** 1081–1083
[14] Antoniadis J, Freire P C C, Wex N, Tauris T M, Lynch R S, van Kerkwijk M H, Kramer M, Bassa C, Dhillon V S, Driebe T, Hessels J W T, Kaspi V M, Kondratiev V I, Langer N, Marsh T R, McLaughlin M A, Pennucci T T, Ransom S M, Stairs I H, van Leeuwen J, Verbiest J P W and Whelan D G 2013 *Science* **340** 448
[15] Gnedin O Y, Yakovlev D G and Potekhin A Y 2001 *MNRAS* **324** 725–736
[16] Haensel P and Zdunik J L 2003 *Astron. Astrophys.* **404** L33–L36
[17] Gusakov M E, Kaminker A D, Yakovlev D G and Gnedin O Y 2005 *MNRAS* **363** 555–562
[18] Lattimer J M, Pethick C J, Prakash M and Haensel P 1991 *Phys. Rev. Lett.* **66** 2701–2704
[19] Yakovlev D G, Kaminker A D, Gnedin O Y and Haensel P 2001 *Phys. Rep.* **354** 1–155
[20] Lombardo U and Schulze H J 2001 *Physics of Neutron Star Interiors* (*Lecture Notes in Physics, Berlin Springer Verlag* vol 578) ed Blaschke D, Glendenning N K and Sedrakian A p 30
[21] Kiziltan B, Kottas A, De Yoreo M and Thorsett S E 2013 *Astrophys. J.* **778** 66
[22] Page D, Lattimer J M, Prakash M and Steiner A W 2004 *Astrophys. J. Suppl.* **155** 623–650
[23] Gusakov M E, Kaminker A D, Yakovlev D G and Gnedin O Y 2004 *Astron. Astrophys.* **423** 1063–1071

Journal of Physics: Conference Series **661** (2015) 012003 doi:10.1088/1742-6596/661/1/012003

Auto– and cross–correlation analysis of the QSOs radio wave intensity

S A Demin[1,2], O Yu Panischev[1,2], Yu A Nefedyev[1,2]
[1] Kazan Federal University, Institute of Physics, Kazan, 420008 Russia
[2] Kazan Federal University, Engelhardt Astronomical observatory, Zelenodolski region, st. Observatory, 422526 Russia

E-mail: serge_demin@mail.ru

Abstract. We discuss here the Flicker-Noise Spectroscopy approach to studying astrophysical systems, for example the radio wave intensity of quasi-stellar object (QSO) 1641+399 and BL Lacertae (BL Lac) 0215+015 in different frequency ranges. The presented method allows to parameterize the study dynamics using a short set of characteristics. The considering sources have a significant differences in manifesting the non-stationary effects, dynamical intermittency and synchronization. The radio wave intensity dynamics of the BL Lac 0215+015 is characterized by well-defined set of natural frequencies, persistent behavior with low effects of non-stationarity and high level of frequency-phase synchronization. For dynamics of the QSO 1641+399 reverse occurs including the asymmetrical structure of cross-correlator. Our findings show that using the flicker-noise spectroscopy approach to studying astrophysical objects allows to carry out the more detail analysis of their behavior and evolution.

1. Introduction

The main factors determining the evolution of accretive astrophysical systems are nonlinearity, non-stationarity, dynamic intermittency and also the collective effects [1]. The disk accretion of the astrophysical objects is largely determined by processes in high-temperature plasma. The variety of accretive objects and the high number effects realized in plasma of accretion disk, lead to necessity of using all the accessible analytical methods for the further understanding the properties of evolution of these systems [2].

In this work we use the Flicker-Noise Spectroscopy (FNS) [3] which is a phenomenological approach to studying the auto- and cross-correlation dynamics in time series. We study the dynamics of radio wave intensity of the QSOs and BL Lacs in frequency bands 2 and 8 GHz [4]. Experimental data have been provided by N. Tanizuka (Laboratory for Complex Systems Analysis, Osaka Prefecture University). The data registration has been carried out in period from 1979 to 1988 years (3 309 days).

QSOs and BL Lacs are the most energetic and distant members of a class of objects called active galactic nuclei (AGN). QSOs and BL Lacs are extremely luminous and were first identified as being high redshift sources of electromagnetic energy, including radio waves and visible light, that appeared to be similar to stars, rather than extended sources similar to galaxies. The QSO's spectra contain very broad emission lines, unlike any known from stars, hence the name "quasi-stellar". The BL Lacs – another type of extragalactic radio sources, as distinct from QSOs, are characterized by continuous spectrum. According to the most common hypothesizes, QSOs and BL Lacs are the active nuclei of distant galaxies, in which a super massive black hole absorbs a matter from the gas-dust disk. At moving the layers of falling matter the collective effects and resonance phenomena arise. Active galactic nuclei are some of the most intrigue phenomena in the Universe because of a variety of their peculiar properties: a colossal energy release, a nonthermal character of radiation and its high and rapid variability in whole electromagnetic spectrum, from radio to gamma band. Studying these sources is an important in solving a high number of significant physical problems (for example, variations of fundamental constants over cosmic time [5]).

The prospectivity of using the FNS to study the signals generated by the accretive astrophysical objects is determined by introducing the information quantifiers characterizing the dynamic components in different frequency ranges. The FNS advantage consists in extracting information from the series of distinct irregularities (spikes, jumps, discontinuities of derivatives of various orders) by analyzing the behavior of time, spatial and power dynamic variables at each existential level of the hierarchical organization of the system. Thus, the most valuable information is obtained by analyzing the power spectra and the difference moments (''structural functions'') of various orders. It is necessary to point out, that the difference moments are formed exclusively by irregularities of a jump type. On the other hand, the power spectra are formed by the contributions of two types of irregularities: peaks and jumps. As distinct from capabilities of linear and non-linear methods, which are used in studying the radio wave intensity of QSOs and BL Lacs (Hurst exponent, fractal and correlation dimensions, phase space maps) [2], FNS allows to discover the system – specific "resonances" and their interferential contributions at lower frequencies, chaotic "random walk" ("irregularity-jump") components at larger frequencies, and chaotic "irregularity-spike" (inertial) components in the highest frequency range. The introduced cross-correlation expression allows to describe the collective phenomena in distributed systems [3].

2. Principles of flicker-noise spectroscopy and basic relations

In FNS, introduced parameters for signal $V(t)$, where t is time, are related to the autocorrelation function [3]

$$\psi(\tau)=\langle V(t)V(t+\tau)\rangle_{T-\tau}, \qquad \langle(\ldots)\rangle_{T-\tau}=\frac{1}{T-\tau}\int_0^{T-\tau}(\ldots)\mathrm{d}t, \tag{1}$$

where τ is the time lag parameter ($0\le\tau\le T_M$) and T_M is the upper bound of τ ($T_M\le T/2$). The averaging over interval $T-\tau$ implies that all the characteristics that can be extracted by analyzing functions $\psi(\tau)$ should be regarded as the average values on this interval. To extract the information contained in $\psi(\tau)$ ($\langle V(t)\rangle = 0$ is assumed), the following transforms, or ''projections'', of this function are analyzed: cosine transforms (power spectrum estimates) $S(f)$, where f is the frequency,

$$S(f)=2\int_0^{T_M}\langle V(t)V(t+t_1)\rangle_{T-\tau}\cos(2\pi f t_1)\mathrm{d}t_1 \tag{2}$$

and its difference moments (Kolmogorov transient structure functions) of the second order $\Phi^{(2)}(\tau)$:

$$\Phi^{(2)}(\tau)=\left\langle[V(t)-V(t+\tau)]^2\right\rangle_{T-\tau}. \tag{3}$$

The information contents of $S(f)$ and $\Phi^{(2)}(\tau)$ are generally different, and the parameters for both functions are needed to solve parameterization problems. By considering the intermittent character of signals under study, interpolation expressions for the stochastic components $S_s(f)$ and $\Phi_s^{(2)}(\tau)$ of $S(f)$ and $\Phi^{(2)}(\tau)$, respectively, were derived using the theory of generalized functions [6]. It was shown that the stochastic components of structure functions $\Phi^{(2)}(\tau)$ are formed only by jump-like irregularities (''random walks''), and stochastic components of functions $S(f)$, which characterize the ''energy side'' of the process, are formed by spike-like (inertial) and jump-like irregularities [7].

In FNS parameterization, the original signal $V(t)$ is separated into three components: system-specific ''resonances'' and their interferential contributions at lower frequencies, stochastic jump-like (''random walk'') component at larger frequencies, and stochastic spike-like (inertial) component in the highest frequency range. For simplicity, we will further refer to jump-like and spike-like irregularities as ''jumps'' and ''spikes'', respectively.

Let us write the basic interpolation expressions for stochastic components. The parameters characterizing the dynamic correlations on every level of the evolution hierarchy are assumed to be the same. Consider the simplest case, in which there is only one characteristic scale in the sequences of spikes and jumps:

$$\Phi_s^{(2)}(\tau)\approx 2\sigma^2\cdot\left[1-\Gamma^{-1}(H_1)\cdot\Gamma(H_1,\tau/T_1)\right]^2, \Gamma(s,x)=\int_x^{\infty}\exp(-t)\cdot t^{s-1}\mathrm{d}t, \quad \Gamma(s)=\Gamma(s,0), \tag{4}$$

where $\Gamma(s)$ и $\Gamma(s, x)$ are the complete and incomplete gamma functions, respectively ($x\ge 0$ и $s>0$); σ is the standard deviation of the measured dynamic variable with dimension; H_1 – is the Hurst exponent, which describes the rate at which the dynamic variable ''forgets'' its values on the time intervals that are less than the correlation time T_1.

The interpolating function for power spectrum component formed by spikes can be written as

$$S_s(f) \approx \frac{S_s(0)}{1+(2\pi f T_0)^n}, \tag{5}$$

here $S_s(0)$, T_0 и n are the phenomenological parameters introduced to describe the power spectra of experimental data. As a measure of chaotic signal component formed by spikes and jumps in high frequency range, it is useful to choose $S_s(T_0^{-1})$, determined as ''spikiness'' factor.

The dynamics of complex systems includes both stochastic components, i.e., spikes and jumps, and system-specific slowly varying regular components associated with a set of frequencies. These frequencies correspond to internal and external resonances and their interferences. It should be noted that the whole set of resonance and interferential frequencies may get rearranged during the evolution of an open system. All the specific frequencies and their interferential contributions, which manifest themselves as oscillations in the $S(f)$, will be further called ''resonant''. It is assumed that $S(f)$ can be presented as a linear superposition of stochastic component $S_s(f)$ and resonant component $S_r(f)$:

$$S(f) = S_s(f) + S_r(f). \tag{6}$$

Here, we assume that the resonant components are statistically stationary (they depend only on time lag τ). This allows us to estimate $\psi_r(\tau)$ as an ''incomplete'' cosine transform of $S_r(f)$ by applying the Wiener–Khinchin Theorem:

$$\psi_r(\tau) \approx 2\int_{1/T}^{f_{LF}} S_r(f)\cos(2\pi f\tau)\mathrm{d}f, \tag{7}$$

where $f_{max} = 0.5f_d$, f_d is the sampling frequency. It should be noted that (7) is an approximation applied to a finite discrete time series assuming the wide-sense stationarity of the resonant signal component. The resonant component $\Phi_r^{(2)}(\tau)$ in this case is found by

$$\Phi_r^{(2)}(\tau) = 2[\psi_r(0) - \psi_r(\tau)]. \tag{8}$$

The stochastic component of $\Phi^{(2)}(\tau)$ can then be estimated as

$$\Phi_s^{(2)}(\tau) = \Phi^{(2)}(\tau) - \Phi_r^{(2)}(\tau). \tag{9}$$

These equations allow one to sequentially separate out resonant and stochastic components of structure functions and power spectrum estimates for experimental time series and perform the parameterization of the components.

The information about the dynamics of correlations in variables $V_i(t)$ and $V_j(t)$, measured at different points i and j, can be extracted by analyzing the temporal variations of various cross-correlation functions. Here, we will use the simplest ''two-point'' correlation expression characterizing the links between $V_i(t)$ and $V_j(t)$ [3, 8]:

$$q_{ij}(\tau,\theta_{ij}) = \left\langle \left[\frac{V_i(t)-V_i(t+\tau)}{\sqrt{2}\sigma_i}\right]\left[\frac{V_j(t+\theta_{ij})-V_j(t+\theta_{ij}+\tau)}{\sqrt{2}\sigma_j}\right]\right\rangle_{T-\tau-|\theta_{ij}|},$$
$$\sigma_i(\tau) = \left\{\left\langle [V_i(t)-V_i(t+\tau)]^2\right\rangle_{T-\tau-|\theta_{ij}|}\right\}^{\frac{1}{2}}, \tag{10}$$

where τ is the ''lag'' time ($\tau > 0$), θ_{ij} is the ''time shift'' parameter. The cross-correlation expression $q_{ij}(\tau,\theta_{ij})$ is a function of temporal parameters τ and θ_{ij}, which can be represented as a three-dimensional plot. Of most interest for the analysis are the intervals of τ and θ_{ij} where the cross-correlation function q_{ij} approaches positive unity (maximum level of positive correlations) or negative unity (maximum level of negative correlations). The value of θ_{ij} corresponding to maximum values of cross-correlation $q_{ij}(\tau,\theta_{ij})$ characterizes the cause-and-effect relation (''flow direction'') between signals $V_i(t)$ and $V_j(t)$. When $\theta_{ij} > 0$, the flow moves from point i to point j, when $\theta_{ij} < 0$, from j to i. When the distance between points i and j is fixed, the value of θ_{ij} can be used to estimate the rate of information transfer between these two points. The magnitude and behavior of the two-parameter expression (10) may significantly depend on the value of selected averaging interval T and upper-bound values of τ and θ_{ij}, which we will refer to as τ_{max} and θ_{max}. From the statistical reliability point of view, we set a constraint of $\tau_{max} + |\theta_{max}| \leq T/2$. Analysis of cross-correlators allow to study the collective phenomena in distributed system [8, 9].

3. FNS analysis of QSO and BL Lac radio wave intensity

3.1. FNS parameterization of the radio wave intensity

To demonstrate the FNS approach capabilities we analyze the spectral flux density of the radio wave intensity from two sources: QSO 0215+015 and BL Lac 1641+399 [4]. In figure 1 and in table 1 we show the FNS-parameterization results. Power spectrum of 0215+015 has the peaks corresponding by quasi-periodic processes, dominating in studied dynamics. As distinct from 0215+015, the 1641+399's radio wave intensity is characterized by peaks absence that means the strong dynamic intermittence in disk accretion.

The different moment of the second order for 0215+015 is in satisfactory agreements with the FNS interpolation curve and demonstrates the distinct quasi-oscillating structure reflecting a strong influence of the jumps on studied dynamics. At the same time the influence of these irregularities on 1641+399 dynamics is lower.

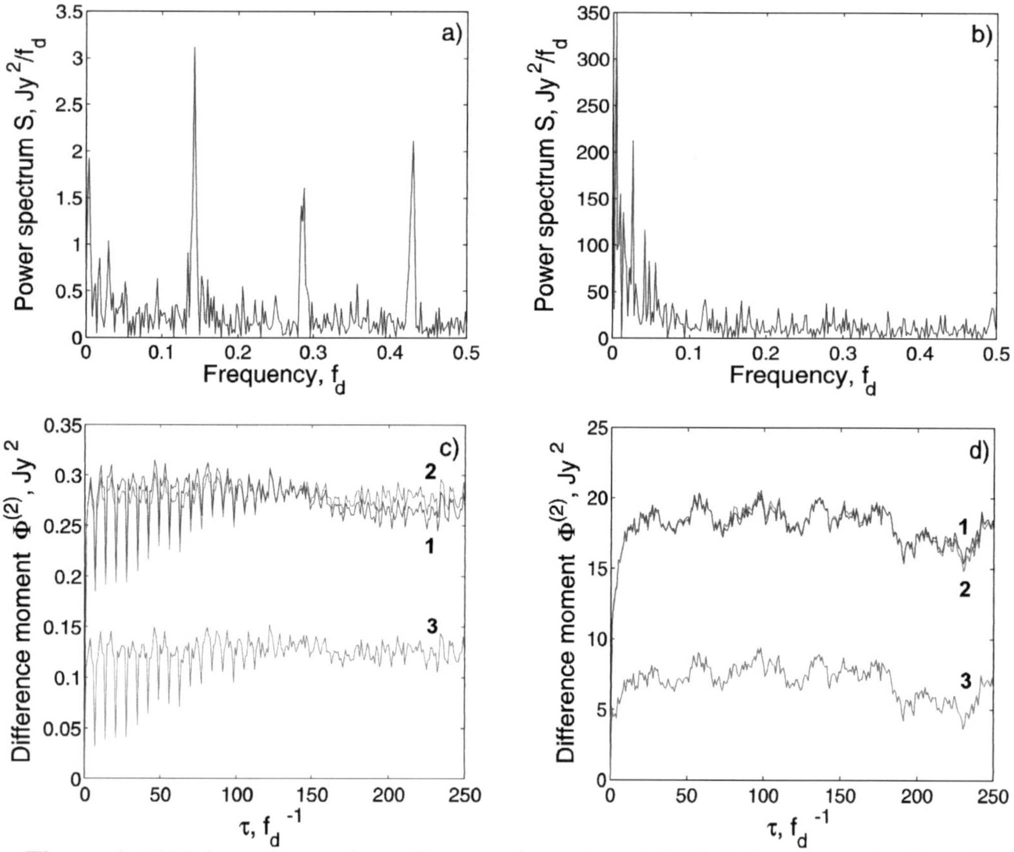

Figure 1. FNS-dependences for radio wave intensity of 0215+015 (a, c) and 1641+399 (b, d) in frequency band 2 GHz. Power spectrum $S(f)$ (a, b) and difference moment of the second order $\Phi^{(2)}(\tau)$ (c, d): calculated on the basis of the experimental data (1, blue curve); interpolation FNS-dependence (2, green curve) and the resonant component $\Phi_r^{(2)}(\tau)$ (3, red curve).

Analysis of findings in table 1 allow us to conclude that the values of ''spikiness'' factor reflects the strong intermittence effects in dynamics of 1641+399 and low manifestation of these effects in activity of 0215+015. H_1 parameter shows the high trend stability in accretion processes of 0215+015 and the low trend stability in 1641+399.

Table 1. The FNS-parameters for signals of spectral flux density of the studied radio wave intensity.

Source	Frequency	σ (rel. un.)	H_1	$S_s(T_{01}^{-1})$ (rel.un.)$^2 \cdot f_d^{-1}$	n
0215+015	*2 GHz*	0.28	1.82	0.61	0.34
	8 GHz	0.42	0.70	1.52	0.33
1641+399	*2 GHz*	2.36	0.20	36.24	0.80
	8 GHz	2.41	0.3	133.84	0.76

3.2. Cross-correlations and frequency phase synchronization in spectral flux density

Figure 2 demonstrates 3D cross-correlators for radio wave intensity of the QSO and BL Lac. For the most visualisation of the cross-correlators we choose the following values of parameters: T = 1 000 days; $\tau_{\max}$ = 0.4 T; θ_{ij} = 0.1 T. Cross-correlators for 0215+015 radio wave intensity (figure 2a) has a clear oscillating structure and the specific frequency set. It means that the QSO radio wave intensity is characterized by the strong frequency-phase synchronization for signals in bands 2 and 8 GHz. The corresponding dependence for BL Lac 1641+399 (figure 2b) shows the large scale diffuse structure with no the characterized frequency set. It means that the synchronization here is low manifesting in comparison of 0215+015.

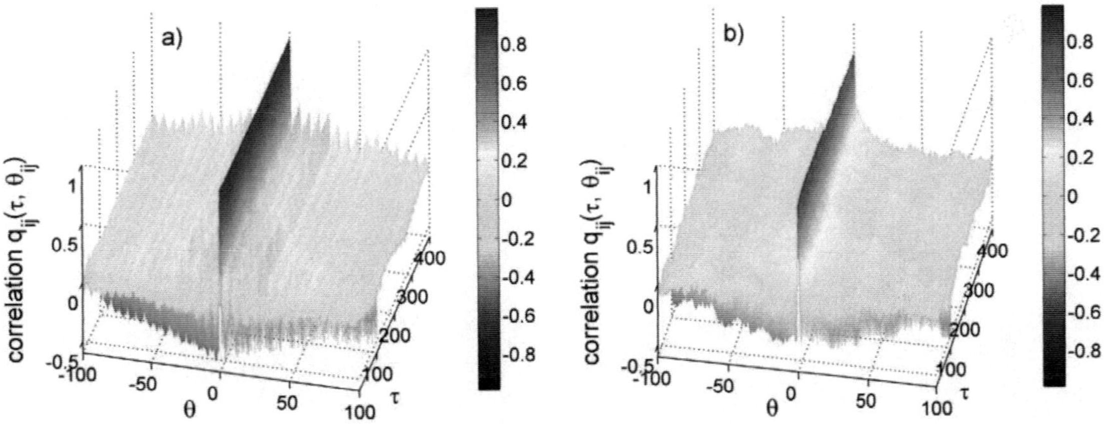

Figure 2. 3D cross-correlation dependences $q_{ij}(\tau,\theta_{ij})$ for radio wave intensity of 0215+015 (a) and 1641+399 (b) in frequency bands 2 and 8 GHz.

4. Conclusion

In this paper we have presented the findings of analysing the radio wave intensity of the two sources by means of the Flicker-Noise Spectroscopy. The basis of FNS is the introducing the describing relations for the different types of irregularities reflecting the resonant and chaotic components of the dynamics. It allows to parameterize a dynamics by using a small set of parameters. The two-point FNS cross-correlators allow to describe the synchronization and collective effects in QSO and BL Lac radio wave intensity.

Acknowledgments

The authors are grateful to Prof. Timashev S.F. (Karpov Institute of Physical Chemistry, Moscow) for help in discussion of results. This work was supported by the Russian Foundation for Basic Research, projects no. 12-02-97000-r_povolgye_a and 15-02-01638-a.

References

[1] Demin S A, Panischev O Yu and Nefedyev Yu A 2014 *NPCS* **17(2)** 177

[2] Khan M R and Tanizuka N 2001 Comparative Time Series Analysis of the Radio Wave Intensity of the QSOs observed by VLBI (KSP) and GBI *CRL TDC Symp.* (Kashima: Kashima Space Research Center) pp 4–7
[3] Timashev S F 2007 *Flicke-Noise Spectroscopy: Information in Chaotic Signals* (Moscow: Fizmatlit) [in Russian]
[4] Tanizuka N and Khan M R 2003 Dynamical structure of quasar radio wave intensity fluctuations from daily to yearly period *Noise and Fluctuations – ICNF-2003* ed J Sikula (Brno: Brno University of Technology) pp 813–816
[5] Petitjean P, Srianand R, Chand H, Ivanchik A, Noterdaeme P and Gupta N 2009 *Space Sci. Rev.* **148** 289
[6] Vladimirov V S 1971 *Equations of Mathematical Physics* (New York: Marcel Dekker)
[7] Schuster H G 1984 *Deterministic Chaos: An Introduction* (Weinheim: Physik-Verlag)
[8] Timashev S F, Polyakov Yu S, Yulmetyev R M, Demin S A, Panischev O Yu, Shimojo S and Bhattacharya J 2010 *Laser Phys.* **20** 604
[9] Panischev O Yu, Demin S A and Bhattacharya J 2010 *Physica A* **389** 4958

2D axisymmetric model of particle acceleration in colliding shock flows system

P.E. Gladilin[1], A.M. Bykov[1], S.M. Osipov[1] and V.I. Romanskiy[1]
[1]Ioffe Physical-Technical Institute of the Russian Academy of Sciences, Saint-Petersburg, Russia,
E-mail: `peter.gladilin@gmail.com`

Abstract. We present the 2D axisymmetric model of particle acceleration at colliding shocks from supernova remnant and stellar wind from the nearby star. The model is the expansion of the previously developed plane-parallel model and takes into account three three-dimensional structure of the stellar wind and the supernova remnant shock. Numerical and analytical calculations provides the energetic and spatial distributions of the particles accelerated by colliding shock flows system. The presented model can be used in calculations of the emission spectra of different stellar associations and star clusters with colliding shock flows.

1. Introduction

The interaction of the supernova remnant (SNR) with the wind of a nearby massive star or the star cluster is the short-lived phenomenon. It should occur in young massive star clusters and OB-associations. This interaction leads to localized sources of X-ray and γ-ray emission embedded in the diffuse emission of the SNR. In the paper [1] there was noted that the emission arising in the SNR/stellar wind interaction is a substantial fraction of the total X-ray luminosity several thousand years after the SNR-stellar wind interaction. The simulations developed in [1] predict the presence of X-ray bright rings in that kind of sources. In the present paper we investigate the properties of these systems as the particle accelerator and the source of X-ray and γ-ray emission but a few hundreds years before the beginning of the shocks interaction.

We model a class of particle accelerators associated with the close approaching of a young SNR shock and a fast shocked stellar wind. Some basic features of such systems were noticed in our previous papers [2, 3]. In figure 1 there is presented the schematic view of the system. The modeled stage starts a few hundred years before the SNR shock collides with the wind termination shock when the distance between the shocks D_0 becomes less than 1 pc. At this stage the maximal energy particles accelerated via diffusive shock acceleration (DSA) mechanism at the SNR shock reach the fast wind termination shock and begin to scatter back by magnetic fluctuations carried by the fast stellar wind. Thus, the high energy particles that have mean free path $\Lambda(p)$ larger than D_0 start to be accelerated by converging fast flows. This is the most favorable circumstance for the efficient Fermi acceleration.

Numerical calculations based on the non-linear time-dependent model of particle acceleration in Colliding Shock Flows (CSF) system [2, 3] showed that these systems have a set of important properties. Maximal energies of the particles accelerated in these systems can be by the order of magnitude higher than in the case of an isolated SNR shock [4]. These sources can make a

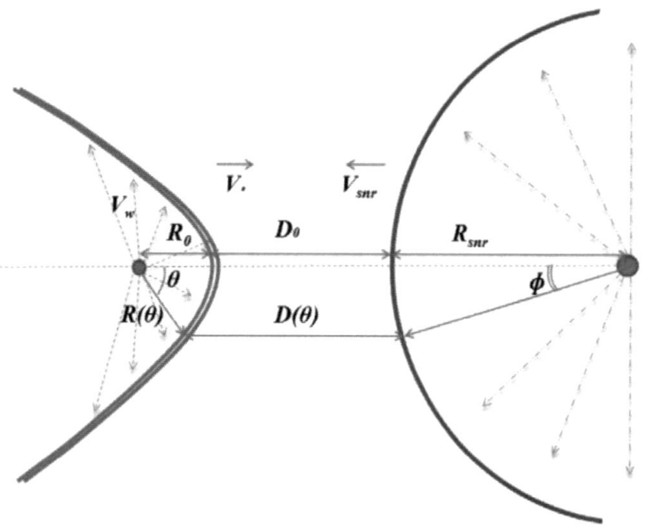

Figure 1. The schematic view of the modeled system. SNR shock approaches the bowshock of the nearby star.

significant contribution to the total flux of galactic cosmic rays (CR) in the high energy range at $10^{14} - 10^{17}$ eV [3]. The specific property of CSF system is that due to high efficiency of acceleration the particles have very hard spectral energy distribution (SED) with the spectral index $\gamma = 1$ during the time of efficient acceleration ($\tau_a \geq 300$ yr). The existing numerical codes provided the model spectra of X-ray and γ-ray emission from these sources [3, 5], that have a set of properties of a "dark accelerators" - non-identified sources of gamma-emission.

The non-linear plane-parallel model of particle acceleration at colliding shock flows did not account for the three-dimensional geometry of the flows. The main goal of this paper is to generalize the model for an arbitrary choice of flow velocities and imply the 3D model expansion with taking into account the shapes of the stellar wind bowshock and the SNR's shell. According to this new model we have calculated new SED of the particles accelerated in the source.

2. The geometry of the model

The shape of the stellar wind bowshock in the thin shell limit can be described as follows [6]:

$$R(\theta) = \sqrt{\frac{\dot{m}V_w}{4\pi\rho_a V_*^2}} \cdot \frac{\sqrt{3(1-\theta ctg\theta)}}{sin\theta} \qquad (1)$$

where θ - is the angle between the symmetric axis and selected direction at the bowshock, $\dot{m}$ - is the mass loss rate of the star, V_w - the velocity of the stellar wind, ρ_a- is the density of the ambient medium, V_* - is the star velocity. The star is assumed to be situated in the origin ($R = 0$).

Assuming that the shape of the SNR shock is the sphere with the radius R_{snr} the distance between the bowshock and the SNR shock at arbitrary time moment t can be expressed as follows (see figure 1 for the illustration):

$$D(\theta) = D_0 + R_{snr}(1 - cos\phi) + \\ +R_0\left(1 - \frac{R(\theta)cos\theta}{R_0}\right) - (V_* + V_{snr}cos\phi)(t - t_0) \quad (2)$$

where D_0 - is the distance between the shocks at $t = t_0$ and $\theta = 0$, R_{snr} - is the radius of the SNR, $R_0 = R(\theta = 0) = \sqrt{\frac{\dot{m}V_w}{4\pi\rho_a V_*^2}}$ - is the radius of the stop point of the stellar wind, V_{snr} - is the velocity of the SNR shock, $\phi = \arctan\left(\frac{R(\theta)\sin\theta}{R_{snr}}\right)$. We have assumed above that the star moves parallel to the axis but in general case V_* should be replaced by $V_* * cos\psi$, where ψ - is the angle between the axis and the direction of star motion.

Equation (2) allows to calculate the distance between colliding shocks for any distance from the axis (different θ's).

3. Non-symmetric non-linear model of particle acceleration at CSF systems

Recently we have developed the semi-analytical model of particle acceleration in the symmetric CSF system (with equal velocities of colliding flows) [3]. Here we present a generalized case of this model with arbitrary choice of flows velocities. For detailed discussion of the previous model see [3].

Consider a model describing a population of high energy CR particles with $\Lambda(p) > D_0$ in a vicinity of two approaching shocks with $R_{\rm snr} \gg D_0$ and $R_{\rm sw} \gg D_0$. This model is appropriate for high energy particles (with $E_p \geq 1$ TeV) that are able to cross the distance between the shocks without scattering.

To derive the distribution function at the shock in the case of two colliding shocks we employ a steady-state diffusion-convection equation with the CR particle injection rate $Q(x,p)$:

$$u(x)\frac{\partial f(x,p)}{\partial x} - \frac{\partial}{\partial x}\left[D(x,p)\frac{\partial}{\partial x}f(x,p)\right] = \\ = \frac{p}{3}\frac{du(x)}{dx}\frac{\partial f(x,p)}{\partial p} + Q(x,p)\delta(x), \quad (3)$$

where $D(x,p)$ is the diffusion coefficient, $u(x)$ is the fluid velocity.

Consider points $x = -x_{0_1}$ and $x = x_{0_2}$ to be coordinates of the free escape boundaries (FEB) at the upwards of both the shocks. At these points particles escape from the accelerator into interstellar medium. The shocks collide at $x = 0$. Integrating Eq. 3 from $x = -x_{0_1}$ to $x = x_{0_2}$ and introducing

$$u_{p1} = u_1 - \frac{1}{f_0(p)}\int_{-x_{0_1}}^{0} dx(du/dx)f(x,p)\ , \\ u_{p2} = u_2 - \frac{1}{f_0(p)}\int_{0}^{x_{0_2}} dx(du/dx)f(x,p), \quad (4)$$

where u_i is the fluid velocities immediately upstream (at $x = 0^-$ and $x = 0^+$, $i = 1,2$), one can evaluate the CR distribution function on the shock:

$$f_0(p) = \frac{3\eta\rho_a(u_1 + u_2)}{8\pi m_p(u_{p1}(p) + u_{p2}(p))}\frac{1}{p^3} \\ -\frac{3}{(u_{p1}(p) + u_{p2}(p))}\int_{p_0}^{p}\frac{\phi_{esc}\,p'^2dp'}{p^3}, \quad (5)$$

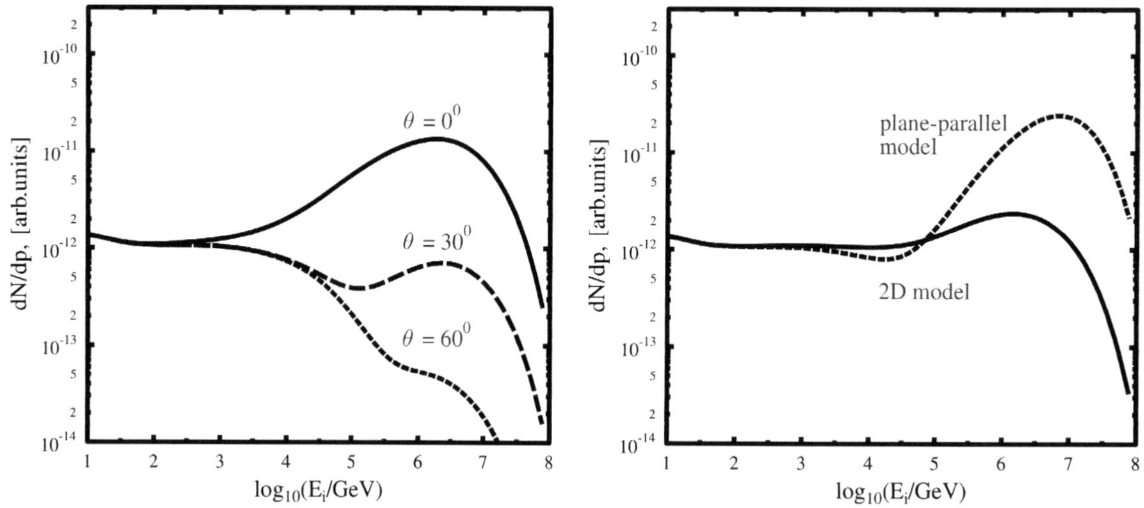

Figure 2. Left: calculated spectra for the different distances from the axis (i.e. for different choices of θ - 0^0, 30^0, 60^0) at time $t - t_0 = 200$ yrs; **right:** average spectrum of the particles *(solid line)* at time $t - t_0 = 200$ yrs comparing to the previously calculated spectrum in the plane-parallel case in model [3] *(dashed line)*

where $\phi_{esc}(p) = -\left(D_1 \frac{\partial f}{\partial x}|_{x01} + D_2 \frac{\partial f}{\partial x}|_{x02}\right)$, ρ_a - is the ambient density, η - is the particle injection rate, m_p - mass of proton. The source point $Q(x,p)$ in the diffusion-convection equation (3) was chosen as follows:

$$Q(x,p) = \frac{\eta n(u_1 + u_2)/2}{4\pi p_{inj}^2} \cdot \delta(p - p_{inj})\delta(x) \tag{6}$$

where p_{inj} - is the injection momentum, $n = \rho_a/m_p$. The function u_p is instrumental to account for the nonlinear modification of the flow due to backreaction of the accelerated particles and it is included into the non-linear calculations (for details see e.g. [7, 8]).

Eq. (5) represents the momentum distribution function for the MHD flow with two colliding shocks with arbitrary chosen velocities. Note, that for every momentum p function $f_0(p)$ is proportional to p^{-3} with correction factor $u_p(p)$. The first term in Eq. (5) reflects CR injection and the second term is due to the escaping flux. The second term is most important for the CR spectral shape at the highest energies and rules out the particle escape and the source contribution to the galactic cosmic rays flux.

4. Calculations

Taking into account expressions (2) and (5) we have calculated SEDs via non-linear time-dependent model [3] of particle acceleration at the CSF system. For each $D(\theta)$ (for $0 < \theta < 90^0$) there were calculated corresponding projections $V_w \cdot cos\theta$ and $V_* \cdot cos\phi$, which are effective velocities of oppositely directed flows for the chosen off-axis distance. For the whole set of pairs of these effective velocities we have calculated SEDs of the accelerated particles using non-linear model described in Section 3.

In figure 2 (left) we present the set of calculated SEDs for the different distances from the axis (i.e. for different choices of θ). Model parameters are the following: SNR velocity $V_{snr} = 5000$ km/s, $V_* = 100$ km/s, $V_{sw} = 2000$ km/s, $D_0 = 0.5$ pc. One can easily see the differences in the spectra: the lower the distance between the shocks (small θ)the better seen a "bump" from the particles accelerated in CSF system at the high energy range. That means that the particles begin to "feel" the second shock and started to be efficiently accelerated by the two-shocks system.

In figure 2 (right) is shown the calculated spectrum of the particles to be compared to the SED for the plane-parallel case in [3]. The presented spectrum is the average of the SEDs calculated for the angles $-90^0 < \theta < 90^0$. Spectra in figure 2 are for the time $t - t_0 = 200$ yrs, pointing the moment when shocks have almost collided. Accounting for 2D geometry results in significant decrease of p_{max} and intensity, comparing with it in the plane-parallel case. The effect of CSF is efficient only when the shocks are close enough ($D_0 << R_{snr}, Rsw, dD(\theta)/dR_{curv} << 1$). When one of these condition breaks ($\theta \geq 60^0$) particle experiences only first order Fermi acceleration at the isolated shock and the spectrum tends to the theoretical test-particle limit $dN/dp \propto 1/p^2$ that is typical for the DSA mechanism.

We have splitted the simulation area into 3 arc surfaces: $0^0 - 30^0$, $30^0 - 60^0$, $60^0 - 90^0$, which have produced accelerated particle densities with corresponding weights: 0.84, 0.15 and 0.01. The central region of interaction makes crucial contribution to the overall spectrum while intensity of the further off-axis regions is weak and can be neglected for $\theta \geq 60^0$ with $\approx 1\%$ error in the total spectrum.

5. Conclusion

We have presented the 2D non-linear axisymmetric model of particle acceleration at colliding shocks from supernova remnant and stellar wind from the nearby star. Comparing with the previously developed model we have apply the dissymmetry of the system. This model allows to calculate particle SEDs for arbitrary choice of the distance between colliding shocks. Computer simulations show that the spectra calculated by 2D model have lower intensity and lower p_{max} than it is for the plane-parallel case.

However, the larger velocity of the stellar wind (larger R_0) and the SNR size, the lower the differences in the spectra between plane-parallel and 2D model cases. But accounting for the three-dimensional effects is crucial in the case of relatively small R_0 or R_{snr}, i.e. for the young SNRs and stars with low V_{sw}.

If we assume that SN explodes near the center of the cluster, the ejecta will interact with not just one, but with large number of O-type stars. Assuming that each interaction region leads to a luminosity of 10^{35} erg/s and that there are 100 nearby O-type stars, a luminosity of 10^{37} erg/s is predicted from the whole cluster. This luminosity is well enough to contribute to the galactic CR flux at the high energy range.

6. Acknowledgments

Gladilin P.E., Osipov S.M. and Romanskiy V.I. are supported by RFBR grant mol-a-14-02-31721 for the young scientists.

Journal of Physics: Conference Series **661** (2015) 012004 doi:10.1088/1742-6596/661/1/012004

References

[1] Velázquez P F, Koenigsberger G and Raga A C 2003 *ApJ* **584** 284–292 (*Preprint* arXiv:astro-ph/0211491)
[2] Bykov A M, Gladilin P E and Osipov S M 2011 *Memories of Italian Astronomical Society* **82** 800 (*Preprint* 1111.2587)
[3] Bykov A M, Gladilin P E and Osipov S M 2013 *MNRAS* **429** 2755–2762 (*Preprint* 1212.1556)
[4] Gladilin P E, Bykov A M and Osipov S M 2014 *Journal of Physics Conference Series* **572** 012003
[5] Bykov A M, Ellison D C, Gladilin P E and Osipov S M 2012 *American Institute of Physics Conference Series* (*American Institute of Physics Conference Series* vol 1505) ed Aharonian F A, Hofmann W and Rieger F M pp 46–55 (*Preprint* 1212.1985)
[6] Wilkin F P 1996 *ApJl* **459** L31
[7] Amato E and Blasi P 2005 *MNRAS* **364** 76–80 (*Preprint* arXiv:astro-ph/0509673)
[8] Caprioli D, Amato E and Blasi P 2010 *Astroparticle Physics* **33** 307–311 (*Preprint* 0912.2714)

Journal of Physics: Conference Series **661** (2015) 012005 doi:10.1088/1742-6596/661/1/012005

A new estimation of $HD/2H_2$ at high redshift using the spectrum of the quasar J 2123−0050

V V Klimenko[1,2], S A Balashev[1,2], A V Ivanchik[1,2], D A Varshalovich[1,2]

[1] Ioffe Institute, Polytekhnicheskaya 26, 194021 Saint-Petersburg, Russia
[2] Peter the Great St.-Petersburg Polytechnic University, Polytekhnicheskaya 29, 195251 Saint-Petersburg, Russia

E-mail: slava.klimenko@gmail.com

Abstract. We present a new analysis of the quasar spectrum J 2123−0050 obtained using VLT/UVES. The H_2/HD absorption system at $z = 2.059$ was analysed. This system consists of two subsystems with $z_A = 2.05933$ and $z_B = 2.05955$. The HD lines have been detected only in subsystem A with the column density of $\log N = 13.87 \pm 0.06$. We have determined the column density of H_2 in this subsystem, $\log N = 17.93 \pm 0.01$, which is about three times larger than estimation derived early from analyses of quasar spectrum obtained using KECK/HIRES [1]. The derived ratio $HD/2H_2 = (4.28 \pm 0.60) \times 10^{-5}$ is the largest in quasar spectra, nevertheless it coincides with the primordial deuterium abundance within 2σ error. Additionally, we have found some evidence in the partial covering effect for the H_2 system.

1. Introduction

The physical conditions in the cold dense phase of the neutral interstellar medium (ISM) can be examined by analysis of molecular absorptions in spectra of background sources (e.g. stars in local galaxies or quasars and gamma-ray bursts at high redshift). H_2 and HD are widespread molecules in the ISM. The populations of different rotational levels of H_2 and HD molecules and the relative abundance of HD molecules ($HD/2H_2$) are sensitive to physical conditions of the medium: kinetic temperature, volume density, intensity of UV background radiation, cosmic ray ionization rate, dust and metal content and others (e.g. [2]).

Firstly, H_2 and HD molecules have been detected in the diffuse Galactic ISM in UV absorption ($\lambda \sim 1000$ Å) [3]. Due to atmospheric extinction, these absorption lines can be detected only on spacecraft observatories (such as Copernicus, FUSE, HST). However, because of the cosmological redshift, H_2 and HD molecules in galaxies at high redshift ($z > 2$) can be study in visible-band observations with ground-based telescopes (such as the 8.2-metre Very Large Telescope (VLT) and 10-metre KECK telescope).

To date, since the first detection in 1985 [4], only 27 H_2 absorption systems at high redshift have been identified in quasar spectra, because it is difficult task even for large telescopes. Since the first detection of HD molecules in 2001 [5], HD molecules have been detected in 9 out of 27 systems (e.g. [6]). The result of the analysis of the H_2/HD system at $z = 2.06$ in the quasar spectrum J 2123−005 obtained using the High Resolution Echelle Spectrometer (HIRES) at KECK [1] stands out among other measurements for clouds at high redshift. The measured ratio, $HD/2H_2 = (7.94^{+2.32}_{-1.46}) \times 10^{-5}$ [1], is more than two times larger than the primordial deuterium abundance $(D/H)_{Pr} = (3.05 \pm 0.22) \times 10^{-5}$ (e.g. [7]), whereas all other measurements are consistent or less than the primordial value.

Additionally, this H_2 system has been investigated by [8] to test possible cosmological variation of the proton-to-electron mass ratio, $\mu = m_p/m_e$. However, the results of the analyses are different in the number of the H_2 components and the derived total column densities.

Here we present a new independent analysis of this H_2/HD system at $z_{abs} = 2.059$ in the spectrum of the quasar J 2123−0050 obtained using the Ultraviolet and Visible Echelle Spectrograph (UVES) at VLT (Program ID 081.A-0242, PI: Ubachs). The signal-to-noise ratio in this spectrum is nearly twice as large as one in the spectrum obtained using KECK/HIRES, that allows us to measure H_2 and HD column densities with higher accuracy.

2. Observational Data

The spectrum of J 2123-0050 was obtained using UVES/VLT. This quasar was observed in the wavelength 3024-3884 Å and 4584-9645 Å in 2008 (Program ID 081.A-0242, PI: Ubachs). The data are in free access from the UVES archive. 1D spectra of individual exposure was taken from the Advanced UVES archive. All the spectra are corrected for the motion of the observatory around the barycenter of the solar system and converted to vacuum wavelengths. We interpolate these spectra into an common wavelength array and generate the weighted mean combined spectrum using the inverse square of error as the weight. The total exposure time is 11.3-h, which allowed a signal-to-noise ratio to be achieved about 30 at the resolution of 49600 for the part of the spectrum containing H_2 and HD absorption lines.

3. The analysis of the H_2/HD system

Molecular hydrogen is detected in several rotational levels, up to J = 5, in two components at $z_A = 2.05932(5)$ and $z_B = 2.05955(2)$. Since the velocity shift between A and B components ($\simeq$ 22 km/s) almost three times exceeds the full width half maximum of UVES instrumental function (FWHM$\simeq$ 6 km/s) and Doppler parameters of H_2 lines ($b \sim 1-5$ km/s), therefore the individual components are resolved even for saturated H_2 lines of the low J = 0, 1 levels. We use a single-component model for the analysis of each of the subsystems A and B.

3.1. Partial coverage of the H_2 absorption lines

It is known that, when not taking into account the partial coverage effects, the column density of an absorber can be underestimate by a factor up to two orders of magnitude [9, 10]. To date, the partial coverage effect has been detected in three of 27 known H_2 systems in quasar spectra [10, 11, 12]. The detail duscussion and manifestation of the partial coverage effect are presented in [10, 11]. Here we present the fourth case of the partial coverage effect.

We demonstrate that the H_2 cloud does not cover the emission region of QSO J 2123−0050 completely. A residual flux at the bottom of saturated H_2 absorption lines is detected at the level 1-3 per cent of the continuum. The comparison of the residual fluxes at the bottom of H_2 lines and saturated Lyα forest lines ($\Delta\lambda > 0.5$ Å, i.e. $\delta v > 40$ km/s) is shown in Fig.1. The residual flux was calculated as the median value of the flux in pixels at the bottom of a line, which have the fluxes within $f_c \pm 1\sigma_i$ (where f_c is the flux at the centre of the line).

A non-zero residual flux can be also the result of a convolution saturated H_2 lines with instrumental function of the UVES. In order to find out the cause of the detected residual flux we use the following procedure. We fit the saturated H_2 lines at J = 0 and J = 1 using four model: (i) without a source of a residual flux, (ii) with a source of a residual flux in continuum, (iii) with a source of a residual flux in the emission lines (Lyβ and O VI), and (iv) with two sources of a residual flux in continuum and in emission lines simultaneously. For each model we have determined a value of the corrected Akaike Information Criterion (AICC, [13]), which allows for selecting a preferred model. Note, that the difference of criteria, ΔAICC = 10, implies

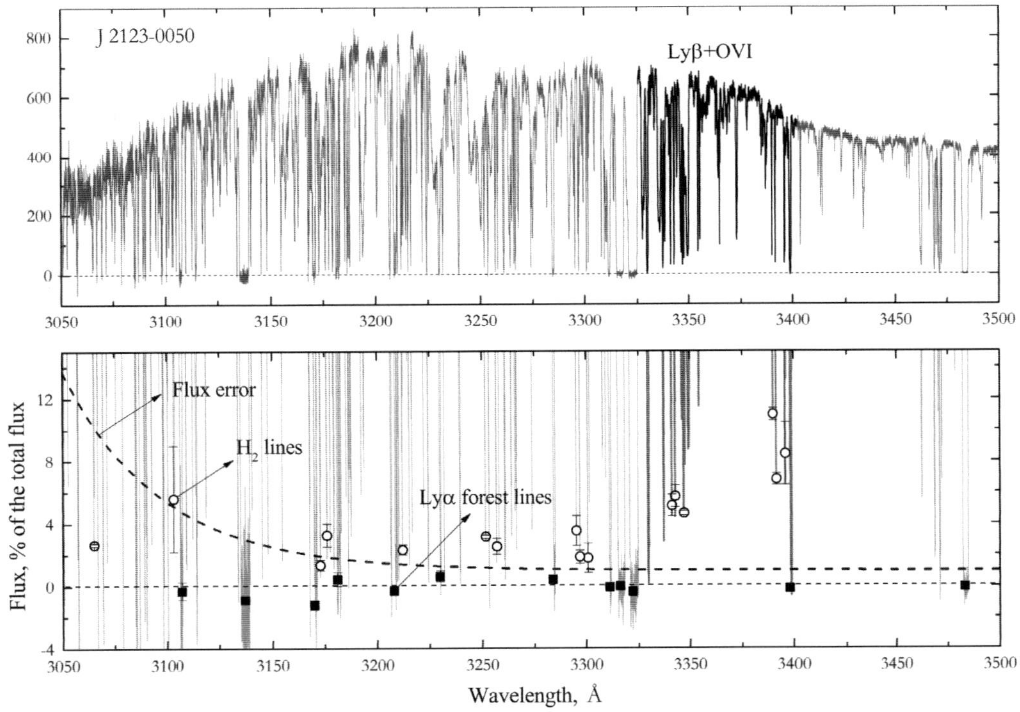

Figure 1. Top panel: A part of J 2123−0050 quasar spectrum (VLT/UVES) containing H_2 absorption lines. Emission lines are named. Bottom panel: The residual fluxes at the bottom of H_2 (open circles) and Lyα forest (black squares) absorption lines. The dashed line shows the mean error for pixels at the bottom of the absorption lines.

Table 1. Best-fitting results of the analyses for different models of the H_2 system. The columns present statistical estimations, residual fluxes (RF) and logarithm of column densities

Model	χ^2_{red}	AICC	ΔAICC	$\mathrm{RF_{Cont}}$ (%)	$\mathrm{RF_{Emiss}}$ (%)	log N(J = 0)	log N(J = 1)
i	1.11	374.7	0	no	no	17.36 ± 0.02	17.79 ± 0.01
ii	1.05	363.9	-10.8	1.1 ± 0.4	no	17.38 ± 0.02	17.80 ± 0.01
iii	0.94	330.0	-44.7	no	2.7 ± 0.5	17.35 ± 0.02	17.78 ± 0.01
iv	0.91	321.3	-53.4	1.1 ± 0.3	3.4 ± 0.5	17.37 ± 0.02	17.79 ± 0.01

strong evidence for the model with the lower value of the criterion. The results of the analysis are shown in Table 1. Following the above discussion we used model IV to describe the H_2 absorption system in the spectrum of J 2123−0050.

3.2. H_2 column densities

We have performed a Voigt-profile fitting of the H_2 system using model IV with two additional sources of the residual flux. The flux detected by an observer is divided into three parts. The main source is covered by both H_2 clouds (A and B). Two additional sources are not covered by H_2 clouds and produce a non-zero residual flux at the bottom of the H_2 lines. One additional source produces the uniform residual flux in all H_2 lines, another one produces only in the H_2 absorption lines located on top of the quasar emission lines (Lyβ and O VI). The best-fitting

Table 2. Best fitting results of the analysis of the H_2 system at $z = 2.06$ towards J 2123−005.

subsystem A $z_A = 2.0593245$				subsystem B $z_B = 2.0595532$			
	J	$\log N_A$	b_A, km/s		J	$\log N_B$	b_B, km/s
H_2	0	17.37 ± 0.02	1.71 ± 0.04	H_2	0	14.00 ± 0.02	5.18 ± 0.51
	1	17.79 ± 0.01	2.07 ± 0.05		1	14.84 ± 0.01	5.24 ± 0.11
	2	15.99 ± 0.03	2.97 ± 0.05		2	14.41 ± 0.01	5.02 ± 0.18
	3	15.54 ± 0.02	3.56 ± 0.06		3	14.50 ± 0.01	4.71 ± 0.30
	4	14.16 ± 0.01	4.64 ± 0.23		4	13.84 ± 0.03	7.86 ± 0.57
	5	13.75 ± 0.03	5.10 ± 0.51		5	13.50 ± 0.06	7.36 ± 0.80
HD	0	13.87 ± 0.06	2.0 ± 0.6	HD	0	undetected	

Table 3. Comparison of the results of the H_2/HD system analyses.

This work				Tumlinson et al. [1]			
	J	$\log N$	b, km/s		J	$\log N$	b, km/s
H_2	0	17.37 ± 0.02	1.71 ± 0.04	H_2	0	16.86 ± 0.24	1.5 ± 0.2
	1	17.79 ± 0.01	2.07 ± 0.05		1	17.55 ± 0.10	1.5 ± 0.2
HD	0	13.87 ± 0.06	2.0 ± 0.6	HD	0	13.84 ± 0.20	1.9 ± 0.5
$HD/2H_2 = (4.28 \pm 0.60) \times 10^{-5}$				$HD/2H_2 = (7.94^{+2.32}_{-1.46}) \times 10^{-5}$			

parameters are present in Table 2. The best fit of the H_2 lines at $J = 0, 1$ are shown in Fig. 2. The total H_2 column densities are $\log N^A_{H_2 tot} = 17.93 \pm 0.01$ and $\log N^B_{H_2 tot} = 15.16 \pm 0.01$. Using the H_2 ortho-to-para ratio for the component A we determine the kinetic temperature of cloud A, $T_A = 138.6 \pm 2.5$ K. As the result of lower H_2 column density in cloud B, the H_2 lines are optically-thin, and the populations of the H_2 rotational levels are mainly caused by UV pumping.

4. HD column density

In subsystem A, we have detected HD lines, corresponding to $J = 0$ level. The laboratory wavelengths of HD transitions were taken from [14], the oscillator strengths were taken from the calculations by [15]. For estimating parameters we use the Markov Chain Monte Carlo (MCMC) method. We have determined the HD column density, $\log N(\mathrm{HD}) = 13.87 \pm 0.05$ taking into account the partial coverage effect. But it should be note that for lines with a small value of optical depths the partial coverage does not affect on column densities determination. The best fitting profiles of HD lines are shown in Fig. 3. The derived $HD/2H_2$ ratio in subsystem A, $(4.28 \pm 0.60) \times 10^{-5}$, is the largest of such values obtained in quasar spectra. Nevertheless, it coincides (within 2σ) with the value of the primordial deuterium abundance $(D/H)_{Pr} = (3.05 \pm 0.22) \times 10^{-5}$ (e.g. [7]).

5. Conclusion

The comparison of our result with the result obtained in the work [1] is present in Table 3. Since the signal-to-noise ratio of the spectrum obtained using UVES/VLT about three times larger than ones in the spectrum obtained using HIRE/SKECK, our estimates of H_2 and HD column densities have the higher accuracy. We have found that the $HD/2H_2$ ratio in subsystem A is

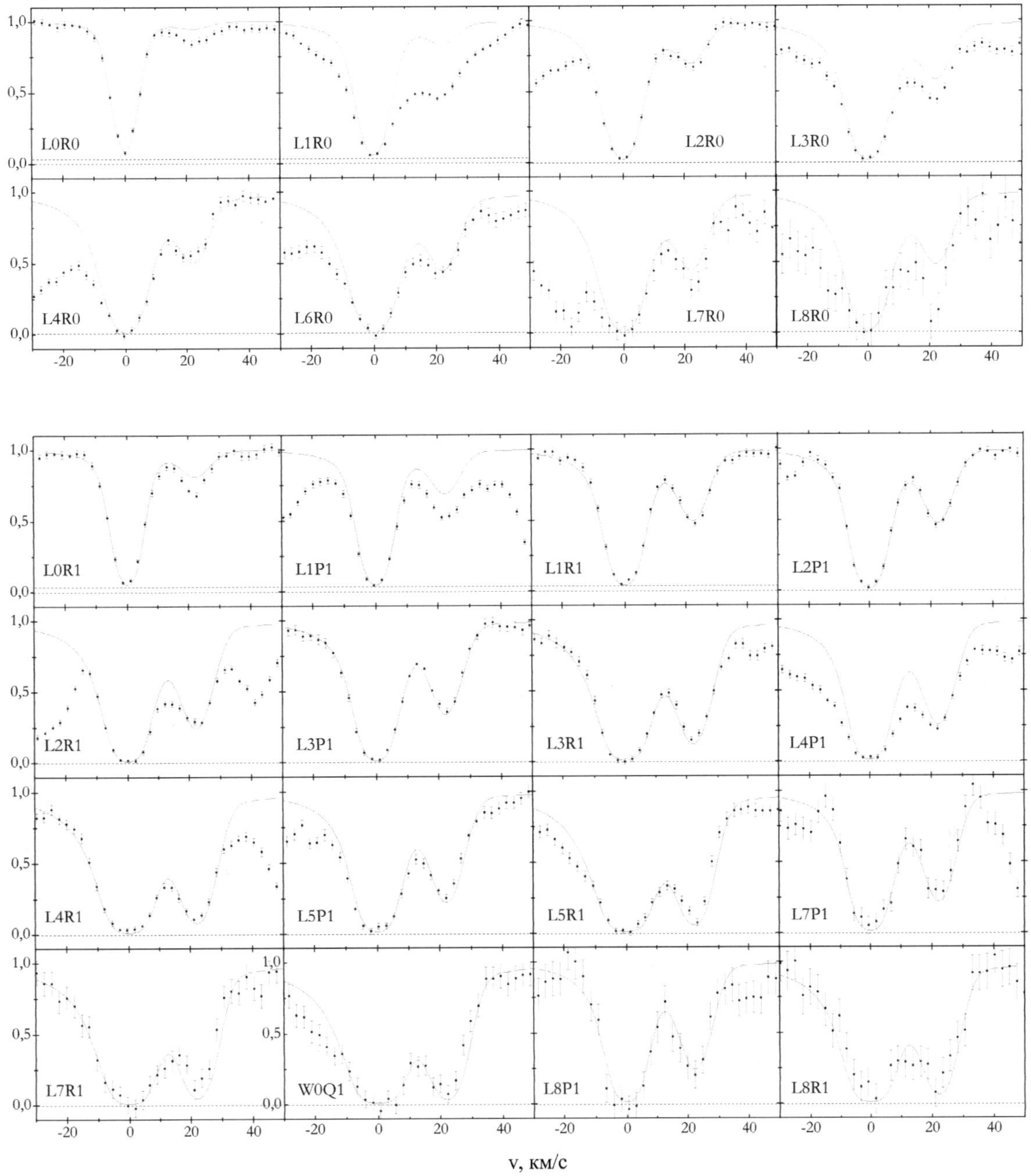

Figure 2. The synthetic spectrum of the H_2 absorption lines at $z_{abs} = 2.059$ (solid line) in the observed quasar spectrum of J 2123−0050 (UVES/VLT). The H_2 lines associated with J = 0 and J = 1 levels are shown in the top and bottom panels, respectively.

more than twice as lower as the value obtained by [1]. This is caused mainly by quite different estimates of the H_2 column densities at J = 0.

Probably, it might be explained by complexity of the analysis of the H_2 lines for J = 0. All these lines in the spectrum of J 2123−005 are blended by the Lyα forest lines (except for the

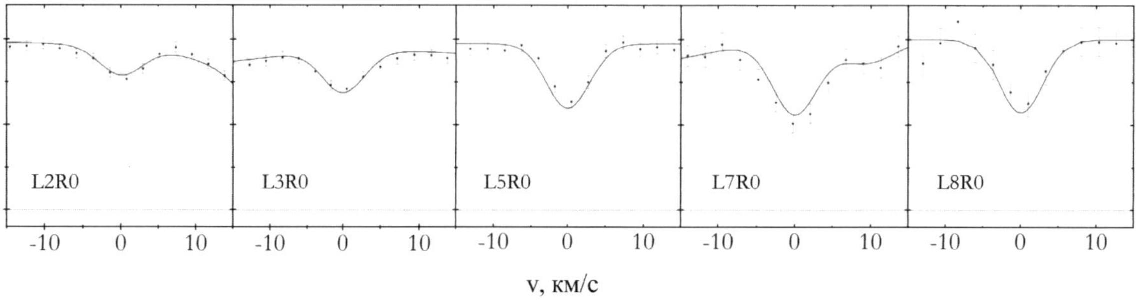

Figure 3. The synthetic spectrum of the HD absorption lines at $z_{abs} = 2.059$ (solid line) in the observed spectrum of J 2123−0050 (UVES/VLT).

L0R0 line, see Fig.2). Therefore, to determine the H_2 column density from $J = 0$ level, it is necessary to correct continuum level near H_2 lines by the inclusion of the Lyα forest lines. We have found that, in some cases the Lyα forest lines located at the wings of the H_2 lines, but sometimes the Lyα forest line located near the center of the H_2 lines. Insufficient or excessive use of the Lyα blends in the H_2 system analysis might cause over- or underestimation of the determined column density of H_2, respectively, especially for moderate signal-to-noise ratios.

5.1. Acknowledgments

This work is based on observations carried out at the European Southern Observatory (ESO) under program ID 81.A-0242 (PI: Ubachs), with UVES spectrograph installed at the Kueyen UT2 on Cerro Paranal, Chile. The work has been supported by the Russian Scientific Foundation (14-12-00955).

References

[1] Tumlinson J, Malec A L, Carswell R F, Murphy M T, Buning R, Milutinovic N, Ellison S L, Prochaska J X, Jorgenson R A, Ubachs W and Wolfe A M 2010 *Astr. Journal Lett.* **718** L156–L160 (*Preprint* `1007.0030`)
[2] Liszt H S 2015 *"Atrophys. Journal"* **799** 66 (*Preprint* `1411.5055`)
[3] Spitzer Jr L and Jenkins E B 1975 *"Ann. Rev. Astron. and Astrophys."* **13** 133–164
[4] Levshakov S A and Varshalovich D A 1985 *"MNRAS"* **212** 517–521
[5] Varshalovich D A, Ivanchik A V, Petitjean P, Srianand R and Ledoux C 2001 *Astronomy Letters* **27** 683–685 (*Preprint* `astro-ph/0107310`)
[6] Ivanchik A V, Balashev S A, Varshalovich D A and Klimenko V V 2015 *Astron. Rep.* **59** 100
[7] Olive K A, Petitjean P, Vangioni E and Silk J 2012 *"MNRAS"* **426** 1427–1435 (*Preprint* `1203.5701`)
[8] Malec A L, Buning R, Murphy M T, Milutinovic N, Ellison S L, Prochaska J X, Kaper L, Tumlinson J, Carswell R F and Ubachs W 2010 *"Highlights of Astronomy"* **15** 321–321 (*Preprint* `1001.4080`)
[9] Ivanchik A V, Petitjean P, Balashev S A, Srianand R, Varshalovich D A, Ledoux C and Noterdaeme P 2010 *"MNRAS"* **404** 1583–1590 (*Preprint* `1002.2107`)
[10] Klimenko V V, Balashev S A, Ivanchik A V, Ledoux C, Noterdaeme P, Petitjean P, Srianand R and Varshalovich D A 2015 *ArXiv e-prints* (*Preprint* `1501.07210`)
[11] Balashev S A, Petitjean P, Ivanchik A V, Ledoux C, Srianand R, Noterdaeme P and Varshalovich D A 2011 *"MNRAS"* **418** 357–369 (*Preprint* `1107.4808`)
[12] Albornoz Vásquez D, Rahmani H, Noterdaeme P, Petitjean P, Srianand R and Ledoux C 2014 *"A&A"* **562** A88 (*Preprint* `1310.8569`)
[13] Sugiura N 1978 *Commun. Stat. A-Theor.* **7** 13–26
[14] Ivanov T I, Dickenson G D, Roudjane M, de Olivera N, Joyeux D, Nahon L, Tchang-Brillet W Ü L and Ubachs W 2010 *"Molecular Physics"* **104** 771–786
[15] Abgrall H and Roueff E 2006 *"A&A"* **445** 361–372

Impulsive event probability in the radiocarbon record in VIII – XI centuries

E G Kovylova and A N Konstantinov
St. Petersburg State Polytechnical University, 29 Polytekhnicheskaya St., St. Petersburg, Russia

E-mail: fomalhaut.hm@gmail.com

Abstract. The abundance of cosmogenic isotopes in natural samples is the main source of information about past variations of cosmic ray intensity, in the solar activity and in the strength of the geomagnetic field. Sharp increases could originate from powerful impulsive events such as solar flares, gamma-rays from supernova explosions and gamma-ray bursts. A significant increase in the radiocarbon record has been detected recently in tree rings around AD 775 [1]. Both large solar proton event (SPE) [2] and gamma-ray burst (GRB) in our Galaxy [3] are favored as a source. However, either of the explanations faces difficulties of low event rate because of detection of a similar peak around AD993 [4], [5]. What is more, we know other similar result [6]. We carried out a statistical analysis of these three data sets. It is shown that AD 775 event differs fundamentally from AD 993 and AD 1006 events, because the last two can be explained without the assumption of the impulsive event.

1. Data analysis technique

Record of $\Delta^{14}C$ contains many signals corresponding to variations of known sources. Therefore, the main goal is to filter the data and estimate the probability of the presence of an independent statistically significant signal.

Allowance for the influence of the magnetic field of the Earth, and long-term variations in solar activity, is carried out by subtraction of calibration curve IntCal09 from data [7]. If a statistically significant jump is detected, the correlation coefficient between the data and a profile corresponding to the case of impulsive event is estimated. The profile is taken from [3]. It should be pointed out, that authors of [3] had to modify the exchange rates of carbon cycle model to obtain the desired profile.

Since we do not have information about solar activity during that period, the modulating effects of solar activity on the Galactic Cosmic Ray (GCR) flux are taken into account as follows: a sine wave is chosen using the least squares method. Then its amplitude and period are compared with the measured values [8].

Data obtained as a result of the subtraction of the calibration curve and the best fit sine wave, are analysed by a normality test to determine if statistically significant signal is present.

2. Results

Analysis of the data around AD 993 (figure 1) showed a large value of the correlation coefficient between the data and a profile corresponding to the case of impulsive event ($r_1 = 0.82$). Best fit sine wave period was equal to $T = 14 \pm 1.1$ years, the amplitude $A = 3.8 \pm 1$ ‰, which is consistent with the observational data [8]. Thus, the shape of the curve can also be explained by the influence of solar modulation, namely the 11 year cycle (r_2 = 0.57). Consequently, the analysis result does not let us conclusively assert that the impulsive event had been registered around AD 993.

A similar conclusion was obtained by considering a record of $\Delta^{14}C$ around AD 1006 (figure 2) (r_1 = 0.71, r_2 = 0.62). The best fit sine wave parameters are T = 14.2 ± 1 years, A = 3.7 ± 1.3 ‰.

Joint analysis of the data around AD 993 and AD 1006 demonstrated the coordination in the localization of sine wave and its parameters within the error limits(T = 14 years, A = 3.9 ± 0.8 ‰). This result is a powerful argument for explaining the shape of the curve $\Delta^{14}C$ by the influence of solar modulation, especially due to the fact that these two data sets were obtained at different times in different laboratories using different samples. The result does not contradict the indirect data for solar activity in the past [9].

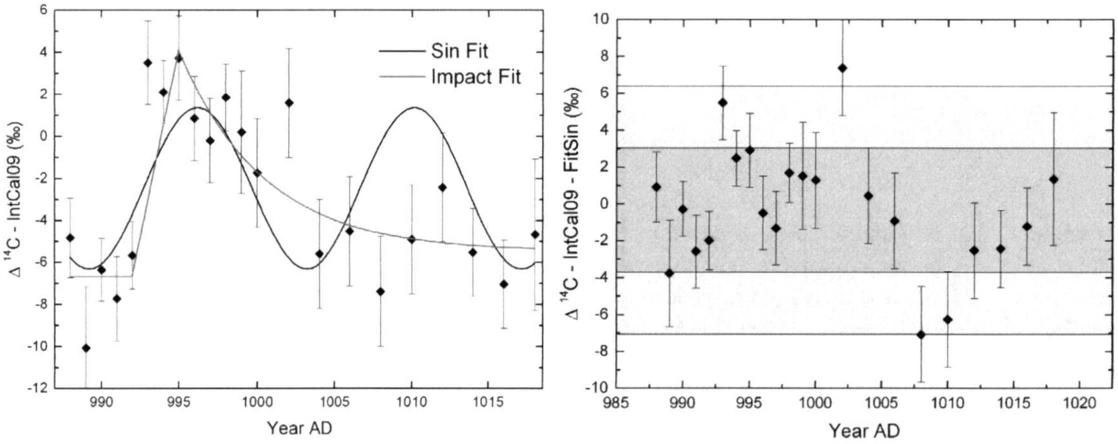

Figure 1: Data around AD 993. (Left) data obtained as a result of the subtraction of the calibration curve; (right) data obtained as a result of the subtraction of the calibration curve and the best fit sine wave.

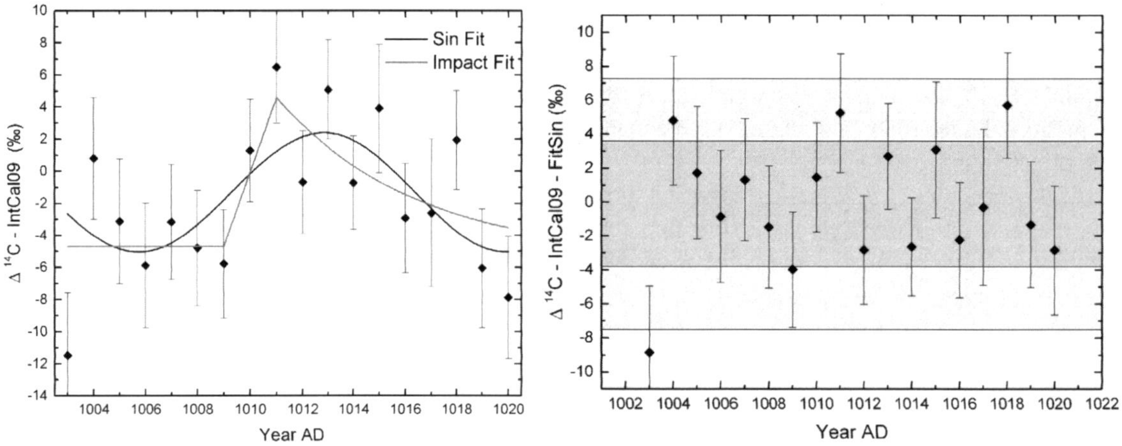

Figure 2: Data around AD 1006. (Left) data obtained as a result of the subtraction of the calibration curve; (right) data obtained as a result of the subtraction of the calibration curve and the best fit sine wave.

The correlation coefficient between the data around AD 775 (figure 3) and a profile corresponding to the case of impulsive event is extremely high (r_1 = 0.93). When we try to localize the wave of 11 year cycle the sine wave was obtained with parameters T = 22.6 ± 2 years, A = 5.3 ± 1.2 ‰. These values do not allow us to treat it as a reflection of the 11 year solar modulation.

As a result of approximation with allowance for variability of solar cycle parameters a curve which is in good agreement with the experimental data was obtained (r_2 = 0.82). Its parameters are in reasonable agreement with 11 year cycle characteristics (A_1= 5.6 ± 1.2 ‰, T_1 = 16 ± 1.8 years, A_2= 1.6 ± 0.5 ‰, T_2 = 13.4 ± 1.8 years). Data obtained as a result of the subtraction of the calibration curve and modified sine wave are not a noise, but do contain a statistically significant signal with high probability.

It can therefore be concluded that AD 775 event can not be explained as a result of coaction of known permanent sources.

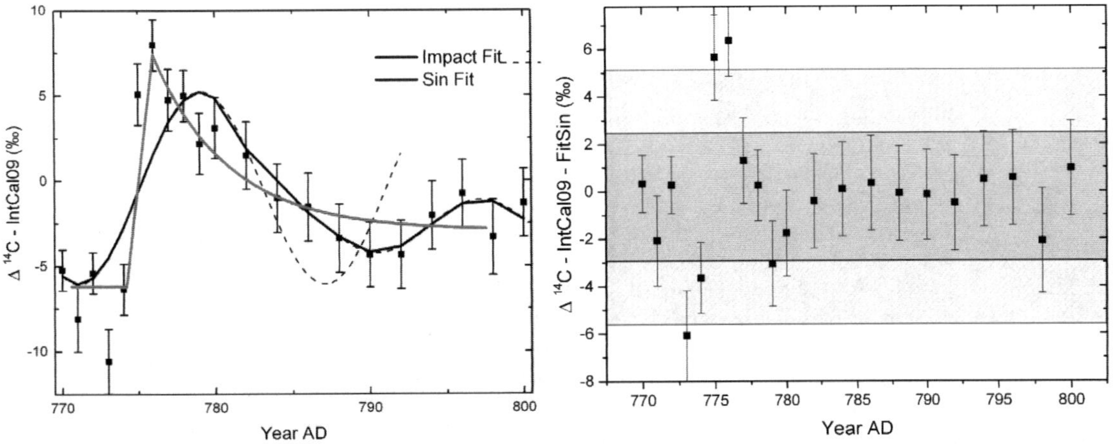

Figure 3: Data around AD 993. (Left) data obtained as a result of the subtraction of the calibration curve; (right) data obtained as a result of the subtraction of the calibration curve and modified sine wave.

3. Conclusions

Based on the analysis of high-resolution records of $\Delta^{14}C$ in tree rings in periods of 770-800, 988-1018 and 1003-1020 AD we came to the following conclusions:

1. AD 993 and AD 1006 events can be explained as a result of coaction of known permanent sources that affect the ^{14}C production rate, and can not be definitely interpreted as a response to an impulsive event.
2. AD 775 event differs fundamentally from AD 993 and AD 1006 events, because it can not be explained without the assumption of the impulsive event.
3. The main objection to the two most likely causes of events 775 is removed.

References

[1] Miyake, F. et al., 2012, Nature 486, 240 – 242
[2] Usoskin, I. G. et al., 2013, Astronomy and Astrophysics, 552, L3
[3] Pavlov, A. K. et al., 2013, MNRAS 435, 2878
[4] Miyake, F. et al., 2013, Nature Communications 4, 1748
[5] Neuhauser, R., Hambaryan, V. V., 2014, Astron. Nachr., 88, 789 - 803
[6] Damon, P. E. et al., 1995, Radiocarbon 37, 599 - 604
[7] Reimer, P. J. et al., 2009, Radiocarbon 51, 1111 – 1150
[8] Stuiver, M. et al., 1998, Radiocarbon 40,1127 – 1151
[9] Schove, D. J., 1955, Journal of Geophysical Research 60, №2, 127

Journal of Physics: Conference Series **661** (2015) 012007 doi:10.1088/1742-6596/661/1/012007

Numerical calculations of cosmic ray cascade in the Earth's atmosphere using different particle interaction models

A V Nesterenok[1] and V O Naidenov
Ioffe Institute, 26 Polytechnicheskaya St., 194021, Saint Petersburg, Russia

E-mail: [1] alex-n10@yandex.ru

Abstract. The interaction of primary cosmic rays with the Earth's atmosphere is investigated using the simulation toolkit GEANT4. Two reference lists of physical processes – QGSP_BIC_HP and FTFP_BERT_HP – are used in the simulations of cosmic ray cascade in the atmosphere. The cosmic ray neutron fluxes are calculated for mean level of solar activity, high geomagnetic latitudes and sea level. The calculated fluxes are compared with the published results of other analogous simulations and with experimental data.

1. Introduction

The Earth's atmosphere is permanently bombarded by energetic particles of cosmic rays. The cosmic ray particles interact by electromagnetic and nuclear processes with the atoms of the Earth's atmosphere and produce secondary particles. Some of the particles produced in the cascade can reach the Earth's surface and induce nuclear reactions there. It is essential to know the temporal and spatial variation of cosmic ray particle fluxes in the atmosphere for many geophysical and engineering problems such as production of cosmogenic nuclides, radiation dosimetry and single event upsets in the microelectronics.

To quantify the effect of cosmic rays on the Earth's environment it is important to know precisely the flux of cosmic ray shower particles in function of position, atmospheric depth, and time. A number of studies have been devoted to the calculation of cosmic ray particle spectra in the atmosphere [1-5]. A common approach to the treatment of atmospheric particle cascade is the Monte Carlo calculations. In this method, the trajectories of all the particles in the cascade are simulated until they are absorbed by nuclei, stopped or escape from the atmosphere. Monte Carlo calculations are needed to account accurately for decay and energy loss processes, and for the energy dependence of the cross sections. Several simulation codes such as CORSIKA, PHARMA, FLUKA and GEANT4, were developed or used for the calculation of particle fluxes in the atmosphere [1, 2, 6, 7]. The simulation codes give an opportunity to employ different particle interaction models and cross sections in the simulations. The simulation toolkit GEANT4 was used in the numerical calculations of cosmic ray cascade in the Ref. [3]. The interaction models and cross sections were chosen in the code in the Ref. [3] to match available experimental data on particle fluxes. The aim of this study is to do simulations using the same model of the atmosphere and the ground as in the Ref. [3] but with different list of particle interaction models and cross sections.

2. The calculational model

Particle fluxes in matter were calculated using simulation toolkit Geant4 version 10.0 [7]. The model of the target (the Earth's atmosphere and the ground), the energy and type of incident particles and particle interaction models are set by special commands in the code. A detailed description of the physical model and calculation of particle fluxes is given in the Ref. [3]. Here, a brief overview is given.

2.1. The simulation of particle interactions

Two reference lists of physical processes were used in the calculations – QGSP_BIC_HP and FTFP_BERT_HP – that differ in description of hadron-nucleus interactions [8]. In the FTFP_BERT_HP list, the fritiof precompound model is employed for describing inelastic scattering of hadrons and mesons with nuclei at high energies > 4 GeV, and the Bertini intranuclear cascade model is employed at energies below this energy limit. In the QGSP_BIC_HP list, the quark-gluon string precompound model is used to describe inelastic scattering of nucleons and mesons with nuclei at high energies, the binary cascade is used for describing nucleon-nucleus interactions at medium energies. High precision neutron models are used for simulating neutron-nucleus interactions at neutron energies below 20 MeV in both physics lists. These models are based on the G4NDL data library that comes largely from the ENDF-B VII and JENDL libraries.

The list of physical processes used in the numerical calculations in the Ref. [3] was similar to the FTFP_BERT_HP reference list, but the JENDL/HE cross section data were employed in the simulation of neutron-nucleus interactions at neutron energies > 20 MeV.

2.2. Model of the Earth's atmosphere and surface

The Earth was modelled as a sphere of a radius of 6371 km. The elemental composition of the surface was assumed to be an average composition of the upper continental crust – in weight percents, 48.3% O, 31.1% Si, 8.2% Al, 3.6% Fe, 2.6% Ca, 2.4% Na, 2.3% K, 1.5% Mg [9]. The surface density was taken to be 2.65 g/cm^2. The atmosphere was considered as a spherical shell of 100 km thickness. The chemical composition of the air (by mass) is nitrogen 75.5%, oxygen 23.2% and argon 1.3%. The total thickness of the atmosphere was taken to be equal to the atmospheric depth of 1034 g/cm^2 at sea level. The atmosphere was divided into concentric subshells with a thickness of 15 g/cm^2 and constant air density within one subshell. Additional division was done near air-ground interface. The dependences of air density and temperature on altitude were taken from the COSPAR reference atmosphere data CIRA-86 (http://ccmc.gsfc.nasa.gov/modelweb/atmos/cospar1.html). The data were averaged over the latitude.

2.3. Calculation of particle fluxes

The energy range of the spectrum of primary particles was split into intervals. The number of energy intervals was taken to be 16 in the energy ranges 1–10, 10–10^2 GeV/nucleon and was taken to be 8 in the energy ranges 0.1–1, 10^2–10^3, 10^3–10^4 GeV/nucleon. Within each interval the primary energy was sampled at random for energies up to 10 GeV/nucleon and was sampled according to power spectrum with the exponent -2.7 for higher energies. The primary particle distribution over the nadir angle θ was adopted to be proportional to $\cos\theta\sin\theta$ with the nadir angle being sampled in the range 0–90°. The simulation of particle cascade in matter initiated by the primary particle of a given type was performed with results being stored separately for each energy interval of the primary. Thus, the yield functions of secondary particle fluxes corresponding to the primary of a given type and energy were computed. We considered two most abundant species of galactic cosmic rays – protons and α-particles. We considered primary particles with energies up to 10 TeV for protons and up to 3 TeV/nucleon for α-particles.

The contribution of all Z > 2 nuclei to particle cascade in matter was determined by applying a scaling factor f to the results, obtained for α-particles. This scaling factor is the ratio of nucleon number densities of all Z > 2 nuclei to α-particles in the galactic cosmic rays. The value of $f = 1.4$ was used [3].

The total number of primary particles for which the cascade simulations were performed is about 7 million for protons and 2 million for α-particles. Simulations were performed on the supercomputer at the Saint-Petersburg Branch of the Joint Supercomputer Center of the Russian Academy of Sciences (http://scc.ioffe.ru/).

2.4. The spectra of primary cosmic rays near the Earth's orbit

Primary cosmic rays at energies above several hundred MeV per nucleon are mostly of galactic origin, about 90% of particles being protons and 10% helium nuclei. The particle fraction of heavier nuclei does not exceed 1%. At energies below few GeV per nucleon the flux of galactic cosmic rays in space at the Earth's orbit significantly depends on the level of solar activity. In the force field model the level of the solar modulation of particle spectra is described by the modulation potential, which characterizes the energy losses of cosmic rays in the heliosphere. The differential energy spectrum of cosmic ray nuclei of type i in the force field approximation is given by [10]:

$$J_i^{\text{diff}}(E) = J_{\text{LIS},i}(E+\Phi_i)\frac{E\left(E+2m_pc^2\right)}{\left(E+\Phi_i\right)\left(E+\Phi_i+2m_pc^2\right)},$$

where $J_{\text{LIS},i}(E)$ gives the local interstellar spectrum of nuclei i, E is nucleus kinetic energy per nucleon, m_pc^2 is proton's rest-mass energy, $\Phi_i = (eZ_i/A_i)\varphi$, Z_i and A_i are nucleus charge and mass numbers, respectively, e is the elementary charge, and φ is the modulation potential which is related to solar activity. We took the parameterization of the local interstellar spectrum from the Ref. [11]:

$$J_{\text{LIS},i}(E) = C_i\frac{p(E)^{-2.78}}{1+0.487p(E)^{-2.51}}, \qquad p(E) = \left[E\left(E+2m_pc^2\right)\right]^{1/2},$$

where E is expressed in GeV per nucleon, C_i is the normalization factor; C_p = 1.9·10^4 (m^2 s sr GeV)$^{-1}$ for protons, and C_{He} = 9.5·10^2 (m^2 s sr GeV/nucleon)$^{-1}$ for helium nuclei. The value of modulation potential of 0.65 GV was adopted in this study as an average value for the present epoch [11]. The power functions for the primary particle spectra were employed for energies higher than 100 GeV. The parameters of the spectra approximation for protons and helium nuclei were taken according to the results of ATIC2 experiment [12]. The calculated yield functions of secondary particle fluxes were convoluted with a given energy spectrum of the primary cosmic rays.

The geomagnetic field inhibits low energy charged primaries from penetrating the atmosphere for equatorial and mid-latitude positions. The cosmic ray fluxes in the atmosphere are unaffected by the geomagnetic field at high geomagnetic latitudes $\lambda > 60°$ [4].

3. Results and discussion

Figure 1 shows calculated neutron angle-integrated differential fluxes at the atmosphere-ground interface at sea level at high geomagnetic latitudes. The modulation potential value of 0.65 GV was taken in the calculations. For comparison, we present the calculated and measured neutron angle-integrated differential fluxes from the Refs. [2, 3, 13]. The data from the Ref. [2] are the results of Monte Carlo simulations using the PHITS code coupled with the nuclear data library JENDL/HE. The spectrum was calculated for semi-infinite atmosphere without considering air-ground interface. The data from the Ref. [3] are the results of numerical calculations using the simulation toolkit GEANT4. The atmosphere and the ground models in the Ref. [3] were the same as in the present calculations but the different particle interaction models and cross sections were used. The data from the Ref. [13] are the results of neutron spectrum measurements scaled to sea level high geomagnetic latitudes and mean level of solar modulation.

The high energy neutrons are the main carrier of the nuclear cascade and come predominantly from vertical direction [3]. At high neutron energies, there is disagreement between the neutron differential fluxes calculated in this work and the flux calculated in the Ref. [3] despite the fact, that the GEANT4

Journal of Physics: Conference Series **661** (2015) 012007 doi:10.1088/1742-6596/661/1/012007

code system was used in both simulations. The discrepancy in the spectra may be suggested as arising from different particle interaction models and cross section data used in the simulations. In particular, the JENDL/HE cross section data were used in the Ref. [3] that can be the main reason of disagreement between calculated fluxes. The discrepancy between the simulation data reaches the factor 2 and is much larger than the estimated error of neutron flux measurements using the neutron monitors – about 15% at energies above 150 MeV and lower at lower neutron energies [13]. There is a good agreement between calculated neutron fluxes from the Refs. [2, 3] and neutron spectrum measurements [13] at high neutron energies E > 10 MeV.

Through elastic and inelastic scattering the neutron energy is reduced and, after many collisions neutrons thermalize to energies corresponding to the ambient gas temperature. A significant fraction of high energy neutrons reaches epithermal and thermal energies [14]. The fluxes of low energy neutrons E < 1 MeV are isotropic [3]. The differences between calculated low energy neutron spectra in this work and in the Ref. [3] can be attributed to the difference between calculated high energy neutron fluxes.

Ground level low energy neutron flux depends strongly on the landscape geometry and the chemical composition of the ground [5]. The cross sections of scattering and capture of low energy neutrons differ substantially for different chemical elements. The different models of the atmosphere-ground interface were considered in the Refs. [2, 3], that can be the reason of discrepancy between calculated fluxes at low neutron energies.

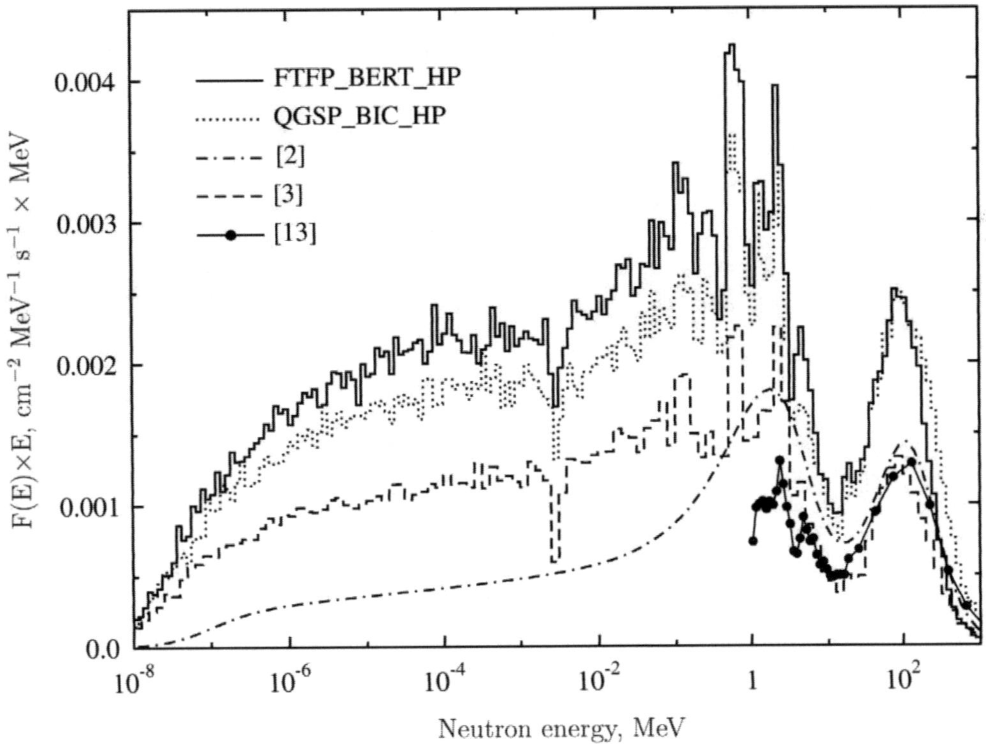

Figure 1. The neutron angle-integrated differential flux in the atmosphere. The neutron flux values multiplied by neutron energy are plotted in the vertical axis with a logarithmic scale on the horizontal axis.

Journal of Physics: Conference Series **661** (2015) 012007 doi:10.1088/1742-6596/661/1/012007

4. Conclusions

The cosmic ray particle cascade in the atmosphere was calculated using the GEANT4 simulation toolkit. Different particle interaction models and cross sections were used. A comparison is made between calculated neutron fluxes in this work and published results of analogous simulations. The results strongly depend on particle interaction models used in the simulations. The discrepancy between neutron fluxes calculated using the different physical models is much larger than the estimated error of neutron monitor measurements.

References

[1] Heck D, Knapp J, Capdevielle J N, Schatz G and Thouw T 1998 *Forschungszentrum Karlsruhe Report FZKA 6019* (Karlsruhe: Forschungszentrum Karlsruhe GmbH)
[2] Sato T, Yasuda H, Niita K, Endo A and Sihver L 2008 *Radiation Research* **170** 244–59
[3] Nesterenok A 2013 *Nucl. Instrum. Methods Phys. Res.* B **295** 99–106
[4] Nesterenok A and Naidenov V 2012 *Nucl. Instrum. Methods Phys. Res.* B **270** 12–8
[5] Masarik J, Kim K J and Reedy R C 2007 *Nucl. Instrum. Methods Phys. Res.* B **259** 642–5
[6] Battistoni G, Cerutti F, Fassò A, Ferrari A, Muraro S, Ranft J, Roesler S and Sala P R 2007 *AIP Conf. Proc. (Batavia, Illinois, USA, 6-8 September 2006)* vol 896, ed M Albrow and R Raja (Melville, NY: American Institute of Physics) pp 31–49
[7] Allison J et al 2006 *IEEE Transactions Nucl. Sci.* **53** 270–8
[8] GEANT4 collaboration 2013 Physics Reference Manual (version: geant4 10.0) available from *http://geant4.cern.ch/support/index.shtml* (accessed 18.11.14)
[9] Rudnick R L and Gao S 2003 Composition of the continental crust *in Treatise on Geochemistry* vol 3, ed R L Rudnick et al (Elsevier) pp 1–64
[10] Gleeson L J and Axford W I 1968 *Astrophys. J.* **154** 1011–26
[11] Usoskin I G, Bazilevskaya G A and Kovaltsov G A 2011 *J. Geophys. Res.* **116** A02104
[12] Panov A D et al 2009 *Bull. Russ. Acad. Sci.: Phys.* **73** 564–7
[13] Gordon M S, Goldhagen P, Rodbell K P, Zabel T H, Tang H H K, Clem J M and Bailey P 2004 *IEEE Transanctions Nucl. Sci.* **51** 3427–34
[14] Liu B and Phillips F M 1994 *Water Resources Research* **30** 3115–25

An effective potential for electron-nucleus scattering in neutrino-pair bremsstrahlung in neutron star crust

D D Ofengeim[1,2], A D Kaminker[1] and D G Yakovlev[1]

[1] Ioffe Institute, Politechnicheskaya 26, 194021, St. Petersburg, Russia
[2] St. Petersburg Academic University, 8/3 Khlopina Street, St. Petersburg 194021, Russia

E-mail: ddofengeim@gmail.com

Abstract. We derive an analytic approximation for the emissivity of neutrino-pair bremsstrahlung (NPB) due to scattering of electrons by atomic nuclei in a neutron star (NS) crust of any realistic composition. The emissivity is expressed through generalized Coulomb logarithm by introducing an effective potential of electron-nucleus scattering. In addition, we study the conditions at which NPB in the crust is affected by strong magnetic fields and outline the main effects of the fields on neutrino emission in NSs. The results can be used for modelling of many phenomena in NSs, such as cooling of young isolated NSs, thermal relaxation of accreting NSs with overheated crust in soft X-ray transients and evolution of magnetars.

1. Introduction

It is well known that studies of thermal evolution of NSs allow one to explore the physics of superdense matter in NS interiors [1]. The thermal evolution is mainly regulated by neutrino emission from various NS layers [2]. Here we focus on the important neutrino emission mechanism in the NS crust, which is the emission of neutrino pairs (of any flavors) in collisions of electrons (e) with atomic nuclei (A, Z), called also NPB,

$$\mathrm{e} + (A, Z) \rightarrow \mathrm{e} + (A, Z) + \nu + \widetilde{\nu}. \tag{1}$$

It has been extensively studied as reviewed in [2].

We employ the formalism of [3] which includes the most complete collection of physical effects important in NPB and allows one to calculate the emissivity of the process Q [erg s^{-1} cm^{-3}] for any composition of the NS crust. To use these results in modeling of NS phenomena, it is convenient to have an analytic fit of Q. The authors of [3] fitted Q for the ground-state crust. Here we calculate Q for different compositions and obtain the required fit. In addition, we investigate the conditions at which the NPB is affected by magnetic fields. More extended version of this investigation is published elsewhere [4].

2. Formalism

Let us analyze the NPB in the crust of an NS. Under the crust we mean the envelope of the star which contains atomic nuclei; it extends [1] to the density $\rho \approx 1.5 \times 10^{14}$ g cm^{-3}. The nuclei are crystallized or form Coulomb liquid. At any value of ρ the matter is assumed to contain spherical nuclei of one species. The nuclei are immersed in the sea of electrons, and at densities higher than the neutron drip density $\rho_{\rm ND} \approx (4-6)\times 10^{11}$ g cm^{-3} [2], also in the sea of free neutrons.

Following [3] we consider the case of ultrarelativistic and strongly degenerate electrons. Then $p_{\rm F} \gg m_{\rm e}c$, that is $\rho \gg 10^6$ g cm^{-3}, where $p_{\rm F} = \hbar(3\pi^2 n_{\rm e})^{1/3}$ is the electron Fermi momentum, $m_{\rm e}$ the electron rest-mass, and $n_{\rm e}$ is the electron number density. The electrons are degenerate at $cp_{\rm F} \gg k_{\rm B}T$, T being the temperature and $k_{\rm B}$ the Boltzmann constant. The electrons form nearly ideal Fermi gas and the nuclei (ions) are fully ionized. The electric neutrality of the matter implies $n_{\rm e} = Zn_{\rm i}$, where $n_{\rm i}$ is the number density of the nuclei, and Z is the nucleus charge number. Let us introduce also $A_{\rm tot}$, the total number of nucleons per one nucleus, and $A = A_{\rm nuc}$, the total number of nucleons confined in one nucleus. In the outer crust ($\rho < \rho_{\rm ND}$) we have $A_{\rm tot} = A$, while in the inner crust ($\rho \geq \rho_{\rm ND}$) one gets $A_{\rm tot} > A$. The mass density of the matter in the crust is $\rho \approx m_{\rm u}A_{\rm tot}n_{\rm i}$, where $m_{\rm u}$ is the atomic mass unit.

Let us introduce the dimensionless parameters,

$$x = \frac{p_{\rm F}}{m_{\rm e}c}, \quad \Gamma = \frac{Z^2e^2}{k_{\rm B}Ta}, \quad \tau = \frac{T}{T_{\rm pi}}, \quad \xi = \frac{R_{\rm p}p_{\rm F}}{\hbar}, \tag{2}$$

where $a = (4\pi n_{\rm i}/3)^{-1/3}$ is the ion sphere (spherical Wigner-Seitz cell) radius; $T_{\rm pi} = \hbar\omega_{\rm pi}/k_{\rm B}$ is the ion plasma temperature determined by the ion plasma frequency $\omega_{\rm pi} = \sqrt{4\pi Z^2e^2n_{\rm i}/(Am_{\rm u})}$; $R_{\rm p}$ is the proton core radius of the nucleus. Thus, x is the relativity parameter of degenerate electrons ($x \gg 1$ in our case); Γ is the Coulomb coupling parameter ($\Gamma \lesssim 1$ for ion gas; $\Gamma \gtrsim 1$ for ion liquid or crystal; crystallization occurs at $\Gamma = 175$); τ determines the importance of quantum effects in ion motion; ξ specifies the importance of finite sizes of the nuclei with $\xi \to 0$ for point-like nuclei; see [1] for details.

According to [3], the NPB emissivity is

$$Q = 5.362\times 10^{15} Z^2 (n_{\rm i}/10^{34}\ {\rm g\ cm^{-3}})(T/10^9\ {\rm K})^6 L R_{\rm NB}\ {\rm erg\,cm^{-3}\ s^{-1}}. \tag{3}$$

Here $R_{\rm NB} = 1 + 0.00554Z + 0.0000737Z^2$ is an approximate non-Born correction. Furthermore, L is the generalized Coulomb logarithm to be determined; $L = L_{\rm liq}$ in the liquid phase; $L = L_{\rm ph} + L_{\rm sl}$ in the solid phase, where $L_{\rm ph}$ is the contribution of electron-phonon scattering, and $L_{\rm sl}$ is the contribution of Bragg diffraction of electrons on the static lattice. For liquid ions,

$$L_{\rm liq} = \int_0^1 S_{\rm liq}(q)|V(q)|^2 \mathcal{R}_{\rm c}(y) y^3\,{\rm d}y, \tag{4}$$

where $\hbar q$ is the electron momentum transfer in a reaction event and $y = \hbar q/(2p_{\rm F})$. The function $S_{\rm liq}(q)$ is the ion structure factor in the Coulomb liquid [5]; $\mathcal{R}_{\rm c}(y) = 1 + 2y^2(\ln y)/(1-y^2)$ comes from the squared matrix element; $V(q) = F(q)/(y^2+y_0^2)$ is the Fourier transform of the Coulomb potential screened by electron polarization (included into y_0 [3]); $F(q)$ is the nuclear form factor which takes into account proton charge distribution within the nucleus [3].

Let us consider this charge distribution as uniform. Then $F(q) = F(u) = 3(\sin u - u\cos u)/u^3$, $u = 2\xi y$. At $\rho \lesssim 10^{12}$ g cm^{-3} one typically has $R_{\rm p} \ll a$, and finite sizes of atomic nuclei are unimportant ($\xi \to 0$, $F(q) = 1$). At $10^{12} \lesssim \rho \lesssim 10^{13}$ g cm^{-3} the approximation of uniformly charged proton core works well but at higher ρ it breaks down [1]. However, in this case one can take realistic proton charge distribution, calculate its root-mean-square (rms) value and use this value instead of $R_{\rm p}$ in the formulas obtained formally for the uniform charge distribution [6].

The expression for $L_{\rm ph}$ is

$$L_{\rm ph} = \int_{(4Z)^{-1/3}}^1 S_{\rm ph}(q)|V(q)|^2 \mathcal{R}_{\rm c}(y) y^3\,{\rm d}y. \tag{5}$$

An appropriate effective structure factor $S_{\rm ph}$ is obtained in [7] and takes into account multiphonon processes, $S_{\rm ph} = \left[\exp(w_1 y^2) - 1\right]\exp(-wy^2)$,

$$w_1 = \left(12\pi^2\right)^{1/3} \frac{Z^{2/3}u_{-2}}{\Gamma} \frac{b\tau}{\sqrt{(b\tau)^2 + u_{-2}^2 \exp(-7.6\tau)}}, \tag{6}$$

$$w = \left(12\pi^2\right)^{1/3} \frac{Z^{2/3}u_{-2}}{\Gamma} \left(1 + \frac{u_{-1}}{2u_{-2}} \frac{\exp(-9.1\tau)}{\tau}\right), \tag{7}$$

where $b = 231$, $u_{-1} = 2.798$, and $u_{-2} = 12.972$. The crystal is assumed to have the body centered cubic structure, but the results are almost insensitive to the lattice type [3]. Finally,

$$L_{\rm sl} = \frac{1}{12Z} \sum_{\boldsymbol{K}\neq 0} (1-y^2)y^2 |V(K)|^2 I(t_V, y)\exp(-wy^2), \tag{8}$$

where $\boldsymbol{K}$ is a reciprocal lattice vector, $y = \hbar|\boldsymbol{K}|/(2p_{\rm F})$,

$$\frac{1}{t_V} = \frac{\Gamma}{Z}\left(\frac{4}{3\pi Z}\right)^{2/3} \sqrt{1-y^2}\,|V(K)|\exp(-wy^2). \tag{9}$$

The function $I(t_V, y)$ was analyzed in [3]. It accounts for the electron band structure effects [8].

Note that the approach [3] neglects the quantum effects of ion motion in the liquid state [$S_{\rm liq}(q)$ in (4) is classical, independent of τ]. Note also that in our case L is explicitly independent of x. We will calculate and fit the Coulomb logarithm which can be presented as a function of four arguments, $L = L(Z, \Gamma, \tau, \xi)$.

3. Coulomb logarithm and effective potential

Our fit of L will use the concept of effective potential for the electron-nucleus scattering in NPB. Similar effective potentials have been introduced by A. Y. Potekhin to fit Coulomb logarithms [6] for electric and thermal conductivities of electrons due to electron-nucleus scattering.

In our case $V_{\rm eff}$ is introduced as

$$L = \int_0^1 |V_{\rm eff}(y)|^2 \mathcal{R}_{\rm c}(y) y^3\,{\rm d}y, \quad |V_{\rm eff}|^2 = S_{\rm eff}(y)\frac{|F(2\xi y)|^2}{y^4}, \tag{10}$$

where $S_{\rm eff}$ is the effective structure factor which we present in the form $S_{\rm eff} = \left[\exp(w_1^* y^2) - 1\right]\exp(-w^* y^2)$. Then w_1^* and w^* can be taken similar to (6) and (7),

$$w_1^* = B\frac{b\tau_*}{\sqrt{(b\tau_*)^2 + u_{-2}^2 \exp\left[-7.6\left(\tau_* + 18/\Gamma_*\right)\right]}}, \tag{11}$$

$$w^* = B\left[1 + \frac{u_{-1}}{2u_{-2}\tau_*}\exp\left(-9.1\left\{\tau_* + \frac{216}{\Gamma_*}\right\}\right)\right], \tag{12}$$

$$B = \frac{3}{5}\frac{(12\pi^2)^{1/3}u_{-2}Z^{4/5}}{1 + (\xi/Z)\Gamma_*/(200+\Gamma_*)}\left(\Gamma_*^2 + \frac{1037}{(1+\Gamma_*/204)^4}\frac{\Gamma_*^{(1+0.15\xi)/2}}{Z^{0.10}}\right)^{-1/2}. \tag{13}$$

Here τ_* and Γ_* are rescaled τ and Γ, respectively. They are related to original Γ, τ, ξ and Z as

$$\tau_* = 0.095\left(\frac{2\tau}{0.095\sqrt{\tau^2+4}}\right)^{1/\Lambda}, \quad \Lambda = 1 + \frac{(Z/35.5)^2}{1+\Gamma/(223Z)}, \tag{14}$$

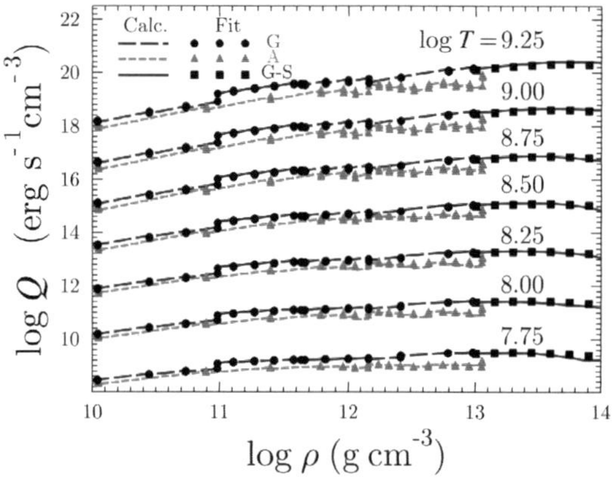

Figure 1: Density dependence of NPB emissivity Q for ground-state (G) and accreted (A) crusts of neutron star at seven temperatures, $T = 10^{7.75}$, 10^8, $\ldots 10^{9.25}$ K. Lines are numerical values, while symbols are our fits. At $\rho \gtrsim 10^{13}$ g cm^{-3} the smooth composition model of ground-state crust (G-S) is used.

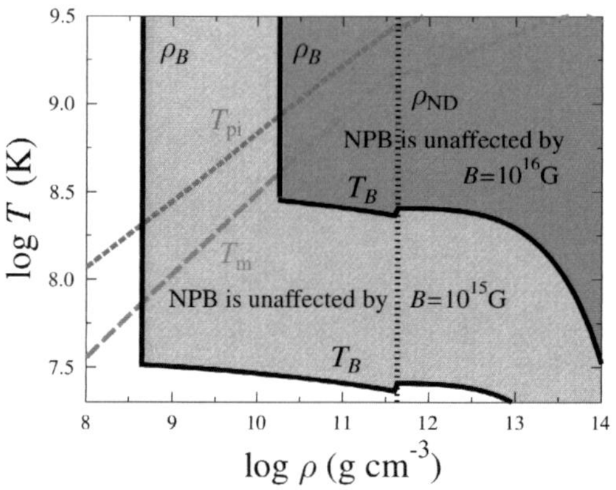

Figure 2: Density-temperature diagram for the ground-state NS crust. The densely shaded region is expected to be weakly affected by the magnetic field $B = 10^{16}$ G, while slightly shaded region by the field $B = 10^{15}$ G (see text for details).

$$\Gamma_* = \Gamma \exp\left\{ -\frac{\tau_*^{0.053} \, (Z/11.3)^{4/9}}{1 + \tau_*{}^G \, [\Gamma/(19.3 Z^{1.7})]^H} \right\}, \tag{15}$$

with $G = 0.5 + 0.002 Z\xi$ and $H = [1.52 + 0.9(1 - \tau_*)]/(1 + 0.5\xi)$.

For an analytic integration of (10) we approximate $\mathcal{R}_c$ as $\mathcal{R}_c(y) \approx 1 - y$ (with the maximum relative error of 0.08 at $y = 0.3$ over $0 \le y \le 1$) and $|F(u)|^2 \approx \exp(-\alpha u^2)$ with $\alpha = 0.23$ (maximum absolute error of 0.032 at $u = 1.6$ for any $u \ge 0$). Then

$$L = f\left(w^* + 4\alpha\xi^2\right) - f\left(w^* - w_1^* + 4\alpha\xi^2\right), \quad f(x) = \frac{1}{2}\left[\sqrt{\frac{\pi}{x}}\,\mathrm{erf}(\sqrt{x}) + \ln(x) - \mathrm{Ei}(-x)\right], \tag{16}$$

erf(x) and Ei(x) being the error function and exponential integral, respectively; $f(x) = 0.71139 + x/6$ as $x \to 0$. Note that $L_{\rm ph}$ roughly reproduces the total L [4]; thus $L_{\rm ph}$ is basic for our fit. Modified w^* and w_1^* accurately reproduce $L_{\rm liq}$, while Γ_* and τ_* tune the fit of $L_{\rm sl}$.

The fit contains a number of coefficients determined by comparing with the calculations. The calculations has been done on the wide grid of the parameters: Z=6,...,50 (10 points), $A = 1.9Z, \ldots, 3Z$ (10 points), $R_{\rm p} = (0.0, \ldots, 0.2) \times a$ (10 points), $\log n_{\rm i}[\mathrm{cm}^{-3}] = 29, \ldots, 34$ (15 points), and $\log T[\mathrm{K}] = 8, \ldots, 9$ (15 points). We have excluded some points as described in [4]. After that the fit has been done on 172550 grid points. The rms relative fit error of L is about 8%, with the maximum error $\sim$ 50% at $Z = 11$, $A = 21$, $n_{\rm i} = 10^{34}$ cm^{-3}, $T = 10^8$ K, and $R_{\rm p} = 0.022\,a$, which is sufficient for applications.

Fig. 1 displays Q versus ρ in the NS crust at seven temperatures, $T = 10^{7.75}$, 10^8, $\ldots 10^{9.25}$ K. We use two models of the crust, the ground-state and accreted ones [1, 9]. The numerical values of Q for the ground-state crust are shown by the long-dashed lines, and for the accreted crust by the short-dashed lines. The fits are plotted by filled dots and triangles, respectively. The accreted crust is composed of lighter nuclei with lower Z leading to weaker NPB. Slight jumps of the calculated and fitted Q values at some densities are associated with the change of nuclides in dense matter with growing ρ [1]. The fits reproduce calculated Q values quite well.

At $\rho \gtrsim 10^{13}$ g cm^{-3} the accreted crust is almost indistiguishable from the ground-state one, and we do not plot the data for the accreted crust at higher densities. Note that the NPB for the dense ground-state crust starts to depend on the proton density profiles within the nuclei (see above). To demonstrate the quality of our fits in this regime we use the smooth-composition model of the ground-state crust at $\rho \gtrsim 10^{13}$ g cm^{-3} (the solid lines), calculate the rms proton core radii and use them in the fits (squares). Again, the fit quality remains good.

4. Effects of magnetic fields on NPB

Many NSs possess strong magnetic fields which can affect the NPB. Accurate consideration of NPB in a B-field is is still not done. Here we formulate the conditions at which B-fields can modify the NPB (through electrons and atomic nuclei).

A field $\boldsymbol{B}$ changes the motion of electrons because of Landau quantization of electron states; e.g. [1]. In our case of strongly degenerate relativistic electrons the importance of magnetic effects is mostly determined by the characteristic density $\rho_B \approx 7045(A_{\rm tot}/Z)(B/10^{12}\ {\rm G})^{3/2}$ g cm^{-3}. At $\rho \lesssim \rho_B$ the electrons occupy the ground Landau level while at $\rho \gg \rho_B$ they populate many Landau levels. Accordingly, we expect that the field $\boldsymbol{B}$ strongly affects the NPB at $\rho \lesssim \rho_B$.

Strong magnetic fields modify vibration properties (phonon modes) of Coulomb crystals and, hence, influence the NPB. Phonon modes of magnetized Coulomb crystals have been studied in a number of works, e.g. [10, 11] and references therein. Contributions of various modes to physical properties of matter are affected by the fields in certain ranges of ρ and T (see Fig. 6 of [11], where B-fields modify phonon heat capacity). Magnetic field effects are mostly pronounced at $T \ll T_B = \hbar\omega_B/k_{\rm B}$, where $\omega_B = ZeB/(A_{\rm nuc}m_{\rm u}c)$ is the ion cyclotron frequency. It is natural to expect that at $T \gtrsim T_B$ the NPB is not affected by B-fields through the ion motion.

Fig. 2 shows possible ranges of ρ and T in the NS crust where magnetic fields can influence the NPB for the smooth composition model of the ground-state matter. We expect that the densely shaded region is weakly affected by the magnetic $B = 10^{16}$ G, whereas the slightly shaded region is weakly affected by the field $B = 10^{15}$ G. Any of these two regions is bounded by the density ρ_B and by the temperature T_B. Hence, the fields $B \lesssim 10^{15}$ G have practically no effects on the NPB in the NS crust under formulated conditions (see above), while higher fields may affect the NPB, especially at low densities. In addition, in Fig. 2 we plot the neutron drip density $\rho_{\rm ND} \approx 4.3 \times 10^{11}$ g cm^{-3}, the melting temperature $T_{\rm m}$ of the crystal and the ion plasma temperature $T_{\rm pi}$ for the ground state matter neglecting the effects of magnetic fields.

5. Neutrinos and magnetic fields

Based on [2] let us outline general effects of B-fields on neutrino emission of NSs. There are many neutrino emission mechanisms there. In addition, the B-fields introduce new, specific mechanisms, like synchrotron emission of neutrino-pairs by electrons. Some effects of B-fields are almost unexplored (e.g., on plasmon decay into neutrino pairs). The strongest effects of B-fields occur in the low-density matter, first of all in the outer NS crust, while deeper layers are affected by magnetic fields (like $B \lesssim 10^{14}$ G) much weaker.

Higher B influence the neutrino emission in deeper layers, in the inner crust and the NS core. To modify the strongest neutrino mechanism, the direct Urca process (which can be open at high densities in the inner NS core), one typically needs $B \sim 10^{16}$ G. At this B charged particles in the core populate many Landau levels, and the main B-field effect will be to broaden the direct Urca threshold [12]. One needs superstrong $B \gtrsim 10^{18}$ G for the particles in the NS core to occupy one or a few Landau levels, but such fields are close to the maximum fields which NSs can contain [1]. The direct Urca process in the extreme case of only one populated Landau level is well studied [12]. If B is lower and more Landau levels are occupied, the direct Urca emissivity will generally tend to the $B = 0$ case but the broadening of the direct Urca threshold will be strong. In addition, neutrino emission in NS cores is strongly affected by superfluidity and superconductivity of NS matter; such effects are superimposed by the effects of B-fields which further complicates studies of neutrino emission of NSs.

6. Conclusions

Using the formalism [3], which includes rich spectrum of physical effects, we have derived a universal fit for the NPB emissivity Q in an NS crust. We have expressed the Coulomb logarithm L which determines Q using the method of effective potential $V_{\rm eff}$.

Our fit is valid for any realistic composition of the NS crust, in wide ranges of parameters, particularly, at densities from about 10^8 g cm^{-3} to the crust bottom and at temperatures from about a few times of 10^7 K to a few times of 10^9 K. It is valid for the nuclei with $6 \lesssim Z \lesssim 50$ and $1.9Z \lesssim A_{\rm nuc} \lesssim 3Z$. The fit is convenient for using in computer codes which simulate thermal evolution of NSs. The fit assumes the presence of nuclei of one type at any values of ρ and T. For a multicomponent plasma one can employ the approach of mean nucleus (mean ion), e.g. [1]. Our fit neglects the effects of B-fields but we have studied its applicability in magnetized NSs and we have also outlined general effects of magnetic fields on neutrino emission processes in NSs.

As the leading neutrino process in a NS crust, NPB is important for modelling transient phenomena in warm NSs, particularly, thermal relaxation in young (age 10–100 yr) isolated NSs [13, 6] and in accreting NSs with overheated crust in soft X-ray transients (e.g. [14, 15]). The NPB can also be important in X-ray superbursts in accreting NSs (e.g. [16]), and in thermal evolution of magnetars [17], as well as in massive cooling white dwarfs.

Acknowledgments

This work was supported by the Russian Science Foundation, grant 14-12-00316. Authors are also grateful to D A Baiko for his helpful comments.

References

[1] Haensel P, Potekhin A Y and Yakovlev D G 2007 *Neutron stars 1: Equation of State and Structure* (New York: Springer)
[2] Yakovlev D G, Kaminker A D, Gnedin O Y and Haensel P 2001 *Phys. Rep.* **354** 1
[3] Kaminker A D, Pethick C J, Potekhin A Y, Thorsson V and Yakovlev D G 1999 *Astron. Astrophys.* **343** 1009
[4] Ofengeim D D, Kaminker A D and Yakovlev D G 2014 *EPL* **108** 31002
[5] Young D A, Corey M E and DeWitt H E 1991 *Phys. Rev.* A **44** 6508
[6] Gnedin O Y, Yakovlev D G and Potekhin A Y 2001 *MNRAS* **324** 725
[7] Baiko D A, Kaminker A D, Potekhin A Y and Yakovlev D G 1998 *Phys. Rev. Lett.* **81** 5556
[8] Pethick C J and Thorsson V 1997 *Phys. Rev.* D **56** 5748
[9] Haensel P and Zdunik J L 1990 *Astron. Astrophys.* **229** 117
[10] Baiko D A 2009 *Phys. Rev.* E **80** 046405
[11] Baiko D A and Yakovlev D G 2013 *MNRAS* **433** 2018
[12] Baiko D A and Yakovlev D G 1999 *Astron. Astrophys.* **342** 192
[13] Lattimer J M, van Riper K A, Prakash M and Prakash M 1994 *Astrophys. J.* **425** 802
[14] Degenaar N et al. 2013 *Astrophys. J.* **775** 48
[15] Degenaar N, Wijnands R and Miller J M 2013 *Astrophys. J.* **767** L31
[16] Gupta S, Brown E F, Schatz H, Möller P and Kratz K-L 2007 *Astrophys. J.* **662** 1188
[17] Mereghetti S 2013 *Braz. J. Phys.* **43** 356
[18] Yakovlev D G, Kaminker A D and Gnedin O Y 2001 *Astron. Astrophys.* **379** L5

Solar Cycle Characteristics and Their Relationship with Dynamo Theory

P A Otkidychev[1], **H Popova**[2], **V Popov**[3]

[1]Mountain Astronomical Station of the Pulkovo Observatory, Kislovodsk, Russia
[2]Moscow State University, Moscow, Russia
[3]Financial University, Moscow, Russia

E-mail: [1]`otkid@list.ru`, [2]`popovaelp@mail.ru`, [3]`masterlu@mail.ru`

Abstract.
We try to establish the correlation between different parameters of "butterfly-diagrams" derived from the analysis of solar observational data for the 12–23 solar activity cycles and the values in the models of α-Ω-dynamo using RGO – NASA/Marshall data set. We have ascertained that there is a linear relationship between S and BT/L for all the investigated cycles, where S is the mean area of the sunspots (umbrae), B is the mean magnetic field strength, T is duration of a cycle and L is the mean latitude of the sunspots in a cycle.

1. Introduction

It is generally accepted that the 11-year cycle of solar activity is connected with the action of the magnetic dynamo, the mechanism of which is based on the mutual work of the α-effect and the differential rotation [1]. It is assumed that Sun magnetic field has two components: the poloidal and the toroidal. The toroidal magnetic field originates from the poloidal one owing to the action of the differential rotation, which is situated inside the convective zone of the Sun. The reverse process of the transformation of the toroidal magnetic field into the poloidal one is realized as a result of the violation of the mirror symmetry of convection in a rotating body. The Coriolis force acting on the arising and diverging (descending and converging) vortices leads to the prevalence of right vortices in the northern hemisphere and left vortices in the southern one. The electromotive force, appearing as a result of the action of the Faraday's electromagnetic induction, after the averaging by velocity pulsations gain the component which is parallel to the mean magnetic field. It is this component that make close the self-excitation circuit in the Parker dynamo.

The 11-year cycle of solar activity is connected with its spot-producing activity. At the beginning of the solar activity cycle the spots appear at the mean latitudes and during approximately 11 years approach to the equator. Such latitudinal-temporal sunspot distribution looks like butterflies and is called butterfly-diagram or Maunder butterflies.

According to conceptions of dynamo theory, the butterfly appearance is connected with the motion of the toroidal component of the magnetic field (dynamo-wave) from higher latitudes to the equator in each hemisphere. The latitudinal-temporal diagram of the level lines of the magnetic field toroidal component obtained even using simplest dynamo models, qualitatively reproduces observational data of the butterfly-diagram. The form of the butterfly-diagram

Table 1. Cycle durations

	Total	Northern "wings"	Southern "wings"
16	08.1923–08.1933	08.1923–08.1933 (0+0=0)	08.1923–08.1933 (0+0=0)
17	09.1933–01.1944	09.1933–01.1944 (3+0=3)	10.1933–12.1943 (7+17=24)
18	02.1944–03.1954	07.1944–03.1954 (19+0=19)	02.1944–03.1954 (1+0=1)
19	04.1954–09.1964	05.1954–09.1964 (1+0=1)	06.1954–09.1964 (6+0=6)
20	10.1964–05.1976	10.1964–05.1976 (0+0=0)	04.1965–05.1976 (37+0=37)
21	06.1976–08.1986	06.1976–08.1986 (13+0=13)	06.1976–08.1986 (0+0=0)
22	09.1986–04.1996	09.1986–04.1996 (5+0=5)	10.1986–04.1996 (12+0=12)
23	05.1996–12.2008	05.1996–09.2008 (0+24=24)	05.1996–12.2008 (17+0=17)

obtained in models is in essential dependence from the controlling model parameters. In [2], [6], [7], [8] it was shown numerically and analytically that meridional circulation directed against the dynamo-wave propagation can essentially slow down its motion. Besides, intensive meridional circulation "blows away" the dynamo-wave to the poles. In [4] it was shown that the turbulent diffusivity coefficient also has ability to influence the duration of solar activity cycle. Also there was shown that the α-effect latitudinal profile has influence upon the form of the Maunder butterfly.

Thus, the characteristics of the theoretical latitudinal-temporal diagrams for the magnetic fields of the Sun and other stars significantly depend on the controlling parameters in the dynamo models. On the other hand, the attempt to reproduce the observational latitudinal-temporal distributions for the stars magnetic fields by the choice of the appropriate values and the controlling processes depending on the coordinates and time could give the information about the physics of the investigated process. However, it is should be remembered that the physics of the process will be limited by the bounds of the chosen model.

2. Observations

In this paper we use Greenwich data (RGO – NASA/Marshall) of monthly and daily areas and center coordinates of the sunspot groups during the period 1923–2008, which entirely embraces cycles 16–23 of the solar activity, as well as for the umbrae during the period 1923–1976 (cycles 16–20 respectively). On the basis of this data one forms the butterfly-diagrams for each cycle, separately for the northern and the southern solar hemispheres. For each "wing" of the Maunder butterfly we applied linear approximation.

This linear approximation may be invalidated since at the beginning of a cycle the spots "from the previous cycle" (i.e. the spots with low latitudes) continue to occur for some time. Similarly, at the end of a cycle the spots "from the following cycle" (i.e. the spot with high latitudes) start to arise. Though the relative number of groups with such spots is very small (in relation to the total number of groups in a cycle, which is of the order of twenty thousand), their presence would bring distortion in the linear approximation and would lead to an error in finding the "wing" tilt angles. Therefore, we have deleted the groups that belong to "adjacent" cycles, i.e. ones with low latitudes in the beginning of the cycle and with high latitudes in the end of the cycle. In Table 1 it is shown: total cycle duration T according to NGDC/NOAA data; cycle duration for a northern "wing" separately; the number of groups deleted in the beginning and in the end of a northern "wing"; the same information for a southern "wing".

As can be seen from the above, the number of the deleted groups is negligible, so their contribution to the mean latitude of the cycle is insignificant. Also it is noteworthy to note

Table 2. The basic parameters of cycles 16–23

	S_{spot}, mh	S_{umb}, mh	B, Gs	T, years	L_{mean}, °	L_{north}, °	L_{south}, °
16	707	129	2695	10.03	14.7	14.95	14.42
17	957	178	2400	10.41	15.3	15.23	15.31
18	1185	193	2555	10.14	15.4	15.48	15.28
19	1424	233	2190	10.47	17.3	17.30	17.33
20	846	129	2350	11.65	14.7	15.30	14.19
21	1242	—	2230	10.23	15.0	14.87	15.17
22	1174	—	2720	9.66	17.1	17.71	16.46
23	796	—	2520	12.60	14.9	14.92	14.95

that the northern and southern "wings" pass to the following cycle not simultaneously but independently. Besides, the tangents of the tilt angles (found from the linear approximation) for the northern and the southern "wings" are not equal to each other in a general case, which also indicates the independence of the spot latitudes in the opposite hemispheres.

The mean latitude of each "wing" (separately for the north and for the south) was calculated as the sum of products of daily groups latitudes (by the absolute value) by daily groups areas, divided by the total sum of group areas. It is of interest that the mean latitudes of the North and the South are almost equal for each cycle. The mean cycle latitude L was calculated as the arithmetical mean of the North and the South.

Further, we have calculated the mean cycle sunspot (umbra) area S as the mean value of monthly average sunspot (umbra) areas for cycles 16–23 (16–20 in the case of umbrae) in the millions of the solar hemisphere (mh).

At last we used the data of the mean magnetic field strength B for cycles 16–23 [10], which was calculated by A.G. Tlatov on the basis of Mount Wilson Observatory data for the sunspots with areas more than 100 mh (actually, the vast majority of the magnetic flux is concentrated in the spots of such area).

The data for S, B and L are summarized in Table 2.

3. Results

The analysis shows that there exists an almost direct dependence ($R = -0.87$) between S and BT/L for all the studied cycles. This dependence can be defined by the following relation: $S = 2539 - 0.89BT/L$, $R = 0.87$, $\sigma = 132$ (Figure 1, (a)).

In case of umbra area the dependence is even more linear ($R = -0.95$), but here we have only five cycles for which both the umbra areas and the magnetic field were calculated (and hence, only five points to plot) (Figure 1, (b)). In the case of umbrae the dependence is following: $S = 500 - 0.20BT/L$, $R = 0.95$, $\sigma = 16$. This is understandable, since it is in the umbrae the greater part of the magnetic field is concentrated.

Let us analyze the obtained result. With other things being equal the more powerful cycle (i.e. the cycle with greater S) corresponds either to the smaller value of the magnetic field B, or to the smaller cycle duration T, or to the greater latitude L. It allows to assume that each cycle involves approximately the same amount of magnetic energy, however, the ways of its realization in each cycle are various. So, the increased sunspot activity in the general case leads to the fact that the spots will be generated at higher latitudes, but the magnetic field strength in the spots will be smaller, and the cycle will be transient.

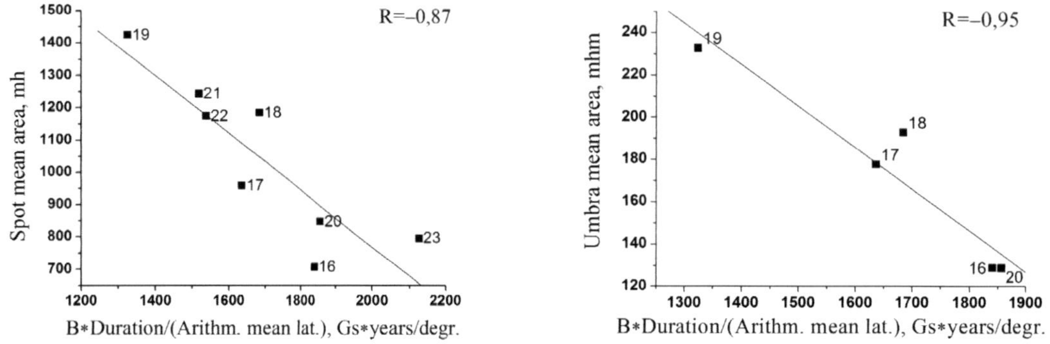

(a) (b)

Figure 1. The dependence between the mean magnetic field strength, the mean spot (a) and umbra (b) area, the duration and the arithmetical mean latitude for cycles 16–23

Indeed, all four values are changed in combination, since there is no dependence between any pair of individual values. The only exception is the pair of values (S, L) for which there is a weak dependence with the correlation coefficient $R = 0.74$ (Figure 2). However, this dependence becomes much better, if instead of the mean sunspot area S we take the relative number of groups containing very large spots. In [9] there was calculated the ratio of groups containing spots with an area of more than 800 mh for cycles 19–23 according to the data of Kislovodsk Mountain Solar Station. In Figure 3 there is dependence between this ratio and the mean latitude L ($R = 0.87$). Considering the fact that the spots of large and huge size contains the essential part of the magnetic flux, we can suggest that the meridional circulation, which is responsible for the latitudinal "butterfly wings" shift, is associated with the total energy of the magnetic field.

The mean cycle spot area generally is proportional to the tangent of the "wing" tilt angle, although there is no direct relationship. There is no dependence between the magnetic field strength and the tangent of the tilt angle.

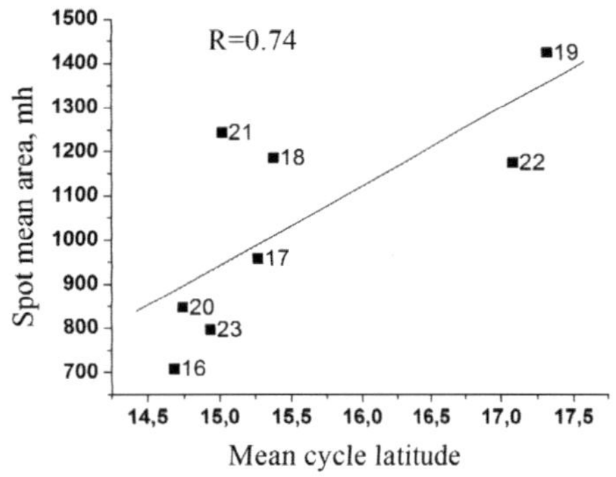

Figure 2. The dependence between the mean sunspot area and the arithmetical mean sunspot latitude for cycles 16–23

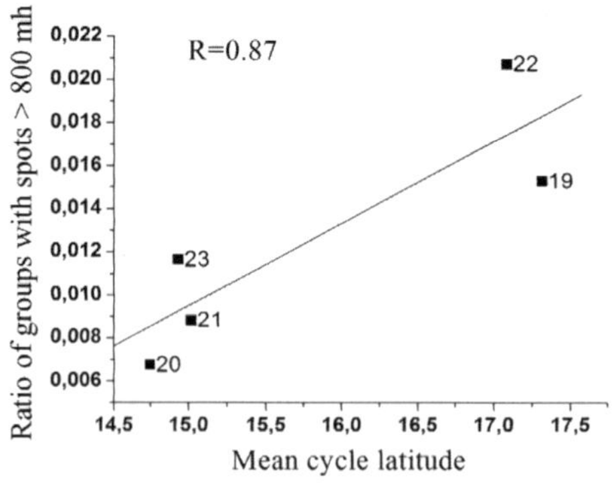

Figure 3. The dependence between the ratio of groups with spots of area more than 800 mh and the arithmetical mean sunspot latitude for cycles 19–23

4. Simulations

Here we use a simple dynamo model which is a straightforward generalization of the initial Parker [5] migratory dynamo. The governing dynamo equations are derived by averaging the magnetic field over the radius within the thin shell, where the dynamo mechanism operates, and neglecting terms describing the effects of curvature near the pole. In this case the dynamo equations have the form:

$$\frac{\partial A}{\partial t} = \frac{\alpha_0 B}{1+\xi^2 B^2} + \beta \frac{\partial^2 A}{\partial \theta^2} - V\frac{\partial A}{\partial \theta}, \tag{1}$$

$$\frac{\partial B}{\partial t} = -D\cos\theta \frac{\partial A}{\partial \theta} + \beta \frac{\partial^2 B}{\partial \theta^2} - \frac{\partial (VB)}{\partial \theta}. \tag{2}$$

Here B represents the toroidal magnetic field, A is the azimuthal component of the vector potential of the poloidal magnetic field, and θ is the latitude measured from the equator. The factor $\cos\theta$ is the decrease in the differential rotation at higher latitudes [3]. The second equation neglects the small contribution of the α-effect, i.e. we use so-called α-Ω-approximation. Curvature effects are absent in the diffusion terms. In Eq. 1–2 the parameter D is the dimensionless dynamo-number (a measure of the intensity of dynamo action), and β is the turbulent diffusivity. We used a simple scheme for the stabilization of the magnetic field growth, namely, the algebraic quenching of the helicity. This scheme assumes that $\alpha = \alpha_0(\theta)/(1+\xi^2B^2) \approx \alpha_0(\theta)(1-\xi^2B^2)$, where $\alpha_0(\theta) = \sin\theta$ is the helicity in the unmagnetized medium and $B_0 = \xi^{-1}$ is the magnetic field for which the α-effect is considerably suppressed.

For reasons of symmetry ($\alpha(-\theta) = -\alpha(\theta)$), the equations 1–2 may be considered only for one hemisphere (for example, the northern one) with the conditions of the antisymmetry (the case of dipole symmetry) or the symmetry (the case of quadrupole symmetry) at the equator. In this paper we restrict ourselves to the consideration of dipole symmetry with the simplest kinematic value of the helicity in the unmagnetized medium $\alpha_0 = \sin\theta$. As boundary condition at the poles we use the following: $A(-\pi/2,t) = B(-\pi/2,t) = A(\pi/2,t) = B(\pi/2,t) = 0$, because we find only the solutions with the dipole symmetry.

Following [6], we consider latitudinal profile of the meridional circulation: $V(\theta) = v\sin 2\theta$. Since in our model the latitude is measured from the equator, the value with a positive sign corresponds to the meridional circulation directed against the dynamo-wave propagation.

Since the parameters of the Mounder butterflies vary from cycle to cycle, we modify them for each cycle but keep them constant during each cycle. Any periodic dependence of their change over time is not revealed. Maybe, this is because the number of cycles studied is small.

According to such simplest dynamo theory the synchronous latitudinal butterfly motion which is due to that the meridional circulation has the same module in both hemispheres. Both the meridional circulation and the turbulent diffusion have an influence upon the butterfly tilt angle. Thus, the fact that the cycles with less duration have a bigger tilt angle is reproduced in the model by increasing the meridional circulation or by the decrease of the turbulent diffusion coefficient.

The increase of the meridional circulation always leads to the butterfly shift to upper latitudes and to the increase of the cycle duration. However, the field amplitude initially decreases and then grows with the further increase of the meridional circulation. When the turbulent diffusion coefficient is decreasing the cycle duration is growing, the field amplitude is increasing, and the butterfly is not shifted by the latitude. Increasing of the dynamo-number module leads to an increase in the magnetic field amplitude but latitude and duration of the cycle does not change. From the dynamo theory it follows that in the observed cycles these parameters should be modified together. Therefore, for example, there is no evident dependence between the mean butterfly latitude and the cycle duration, since according to the dynamo theory the cycle duration would be directly proportional to the mean butterfly latitude in the case that only the meridional circulation should be changed.

The average value of the latitude of a butterfly is located about 15°. The maximum deviation from the mean latitude for the observed cycles is 18%. The maximum deviation of the duration of the 11-year cycle is approximately 11%, the deviation from the mean amplitude of the field is 13%.

To solve Eq. 1–2 we used numerical method of lines in Mathcad 11. One can consider the typical values of the parameters for obtaining 11-year solar cycle: $D \approx -10000$, amplitude of the meridional circulation ≈ 0.5 in model units (≈ 2 meters per second), and $\beta \approx 1$. In this case the normalized cycle duration is approximately equal 1 diffusive unit. The observed deviation from the mean latitude can be obtained by changing amplitude of the meridional circulation $\Delta v \approx 0.1$ in model units (which gives a deviation of 20% from its mean value) or $\Delta\beta \approx 0.25$ (deviation from the mean value of 25%). The observed deviation from the mean duration can be obtained by changing the amplitude of the meridional circulation to $\Delta v \approx 1.5$ in model units (deviation from the mean value is 300%) or $\Delta\beta \approx 0.05$ (deviation from the mean value is 5%). In order to obtain observed deviation from the mean amplitude of the toroidal magnetic field one needs to change D ($\Delta D \approx 2000$, or 20%), or v ($\Delta v \approx 2.5$, or 500%), or β ($\Delta\beta \approx 0.01$, or 1%).

5. Conclusion

The linear dependence $f(BT/L, S)$ from cycle to cycle possibly is realized due to the fact that the equal quantity of energy is spent on each cycle, but it is redistributed in different ways in each cycle, therefore we can observe the change of the other cycle characteristics. It is the dynamo coefficient that is responsible for the magnetic field generation in the dynamo models. Hence it is quite probable that it does not change from cycle to cycle.

According to the numerical simulations changes of butterfly latitude are most sensitive to changes in the amplitude of the meridional circulation, while changes of cycle duration and amplitude of the magnetic field are most sensitive to changes in the coefficient of turbulent diffusion.

Since the clear periodicity in changes in the characteristics of Mounder butterflies were not found, it can be assumed that the changes of control parameters (dynamo-number, coefficient of turbulent diffusion, and meridional circulation) have some stochastic nature.

Acknowledgments

The work is supported by the grants of Russian Foundation for Basic Research 12-02-00170, 12-02-00884.

References

[1] Brandenburg A, Subramanian K 2005 *Physics Reports*, **417**, pp 1–209
[2] Dikpati M, Gilman P A 2001 *Astrophys.J.*, **559**, pp 428–442
[3] Kuzanyan K M, Sokoloff D D 1995 *Geophys. Astrophys. Fluid Dynam.*, **81**, p 113
[4] Munoz-Jaramillo A, Nandy D and Martens P C H 2011 *The Astrophysical Journal Letters*, **727**, pp 23–29
[5] Parker E N 1955 *Astrophys.J.*, **122**, pp 293–314
[6] Popova E P 2009 *Astronomy Reports*, **53**, pp 863–868
[7] Popova H P, Reshetnyak, M Yu, and Sokoloff D 2008 *Astronomy Reports*, **52**, pp 157–163
[8] Popova H, Sokoloff D 2008 *Astron.Nachr.*, **329**, 7, pp 766–768
[9] Otkidychev P A, Skorbezh N N 2013 *Solar Activity Cycle: The Observal and Theoretical Aspects* (St. Petersburg: Pulkovo) pp 187–190
[10] Tlatov A G 2012 *Solar Activity Cycle: The Observal and Theoretical Aspects* (St. Petersburg: Pulkovo) pp 133–136

Journal of Physics: Conference Series **661** (2015) 012010 doi:10.1088/1742-6596/661/1/012010

Model of dynamical structures in synchrotron images of pulsar wind nebulae

A E Petrov[1,2] **and A M Bykov**[2,1]

[1] Department of Cosmic Research, Saint-Petersburg State Polytechnical University, Polytekhnicheskaya st. 29, St. Petersburg 195251, Russia

[2] Laboratory for High Energy Astrophysics, Ioffe Physical Technical Institute, Polytekhnicheskaya st. 26, St. Petersburg 194021, Russia

E-mail: alexey.e.petrov@gmail.com

Abstract. The relativistic pulsar winds are forming the pulsar wind nebulae (PWN) — the unique extended sources of non-thermal radiation detected in all bands of the electromagnetic spectrum. High angular resolution imaging of the PWN with modern orbital and ground-based telescopes makes possible to study the highly non-equilibrium processes in the pulsar wind plasma. Physical interpretation of the observed dynamic structures in the nebulae requires modeling of the relativistic pulsar wind. The main mechanism of emission of the magnetized relativistic pair plasma in the PWN is the synchrotron radiation, and the observed dynamical structures may be related with a propagation of perturbations of the magnetic field. A kinetic approach to highly non-equilibrium relativistic pair plasma allows us to evaluate the structure of the perturbation of the magnetic field propagating transverse to the mean quasi-stationary magnetic field. We present synchrotron images with the dynamic structures in the PWN simulated in the relativistic pair plasma with the strong scattering of pairs by the stochastic magnetic field fluctuation.

1. Introduction

The pulsar wind nebulae (PWN), formed by the relativistic pulsar winds, present one of the most interesting classes of objects for the high energy astrophysics. The emission of these nebulae is detected across the electromagnetic spectrum. The conversion mechanism, responsible for the highly efficient transformation of the energy of a rotating neutron star into the electromagnetic energy of a PWN is one of the actual problems. High angular resolution observations of the PWN, possible thanks to modern orbital and ground-based telescopes, allow studying highly non-equilibrium processes in the pulsar wind plasma. Such observations revealed the presence of dynamical structures, like jets and wisps, in the Crab nebula[1, 2]. The origin of the high energy gamma-ray flares, discovered in 2011 by Fermi and AGILE telescopes, is one of the last hot issues[2–4].

The study of the origin of the observed dynamical structures can help us to answer the question about their possible relationship with the gamma-ray flares. It also may be useful for understanding the physics of the processes, involved in the energy conversion mechanism in the relativistic pair plasma of the PWN. To interpret these structures, the modeling of the relativistic pulsar wind is required.

The main mechanism of emission of the magnetized relativistic pair plasma in the PWN is the synchrotron radiation. Thus, a propagation of perturbations of the magnetic field may be responsible for the observed dynamical structures.

Applying a kinetic approach to highly non-equilibrium relativistic pair plasma, we can evaluate the structure of the perturbation of the magnetic field propagating transverse to the mean quasi-stationary magnetic field. We describe the scheme of obtaining the evolution equation for the magnetic perturbation propagating in the relativistic pair plasma with the strong scattering of pairs by the stochastic magnetic field fluctuation. We discuss some solutions of this equation and present the synchrotron images of their evolution.

2. The evolution equation

2.1. The kinetic model

We consider a local weakly-nonlinear perturbation of the magnetic field, propagating transverse to the mean quasi-stationary magnetic field B_0 in highly non-equilibrium relativistic pair plasma. Due to locality, one-dimensional analysis is applicable. We assume that $\mathbf{B}_0 \parallel Oz$, and that the perturbation, considered as a linearly polarized wave with electric field $\mathbf{E} \parallel y$ and magnetic field $\mathbf{b} \parallel z$, propagates along the Ox axis. The 'collisional' regime of the strong scattering of pairs by the stochastic magnetic field fluctuation is supposed. We introduce the typical frequency of such 'collisions' (scatterings) ν and the gyrofrequency $\Omega = eB_0/mc\gamma$. Here e is the positron charge, m – the mass of a particle, c – the light velocity, γ – the Lorentz-factor of a particle. The typical frequencies of the processes under consideration ω are supposed to be small in comparison with ν. The frequency of scatterings by the magnetic field fluctuation may not exceed the gyrofrequency. We write for the frequencies:

$$\omega \ll \nu \leq \Omega, \quad \nu = a\Omega \tag{1}$$

We use the kinetic approach and seek the distribution functions of positrons and electrons in the form of $f_\alpha = f_0 + \tilde{f}_\alpha$, where f_0 is the isotropic part of the distribution function, $\alpha = p, e$. The subscripts 'p' and 'e' refer to positrons and electrons, respectively. The collision operator is taken in the form of the relaxation time approximation

$$-\nu\left(f - f_0\right) = -\nu\tilde{f}.$$

The system of kinetic equations for the components of pair plasma takes the form:

$$\partial_t f_p + v_x\partial_x f_p - \Omega\cdot\hat{\mathbf{O}}f_p + e\left(E_y - \frac{1}{c}v_x b\right)\partial_{p_y} f_p + \frac{e}{c}v_y b\partial_{p_x} f_p = -\nu_1\tilde{f}_p + \nu_2\tilde{f}_e \tag{2}$$

$$\partial_t f_e + v_x\partial_x f_e + \Omega\cdot\hat{\mathbf{O}}f_e - e\left(E_y - \frac{1}{c}v_x b\right)\partial_{p_y} f_p - \frac{e}{c}v_y b\partial_{p_x} f_p = -\nu_2\tilde{f}_e + \nu_1\tilde{f}_p. \tag{3}$$

Here $\hat{\mathbf{O}} = [\mathbf{p}\times\partial_{\mathbf{p}}]$. The collisional operators are taken in the symmetric with respect to the sorts of the particles form.

2.2. The derivation of the evolution equation

In this subsection we give a brief review of the derivation of the evolution equation, performed in [5].

The solution of the kinetic system (2)—(3) is obtained in two stages. At the first stage, the equations are linearized (the terms with nonlinear combinations of fields are neglected) and solved using the Fourier transforms. This allows to obtain the dispersion equation and the terms $\delta f^1_{p,e}$, $\delta f^2_{p,e}$, $\delta f^3_{p,e}$ of the linear variation of the distribution function.

At the second stage, we solve the system using the successive approximation method, taking into account the nonlinear terms. In the first approximation again only linear terms are left. The equations are solved by building the Green functions. In the second approximation the nonlinear terms are considered and the same procedure is performed. This allows to obtain the nonlinear variation of the distribution function $\delta f^4_{p,e}$.

The current density, defined by the variations of the distribution functions, could be derived as $j_y = \sum\limits_{i,\alpha} e_\alpha \int v_y \delta f^i_\alpha d^3p$. This result should be substituted to the Maxwell equation for $rot\mathbf{B}$. Thus, we get the evolution equation, which takes the form of the Korteweg—de Vries—Burgers equation:

$$\partial_\tau h + \partial_\xi h + \partial^3_\xi h + \lambda h \partial_\xi h = \chi \partial^2_\xi h \tag{4}$$

Here $h = b/B_0$; τ,ξ — the dimensionless temporal and spatial coordinates. The coefficients of this equation are expressed via the integrals of the form of $\sum\limits_\alpha \int v_y \delta f^i_\alpha d^3p$.

2.3. The initial value problem

The equation (4) has an analytical solution only in important special cases. One of such cases is realized when $\nu \ll \Omega$: in this limit one can neglect the term $\sim \partial^2_\xi h$ describing the dissipation. Thus, the equation will take the Korteweg—de Vries form. One of the Korteweg—de Vries equation solutions is a soliton — a long lived solitary wave, propagating without changing of its shape:

$$h(\xi, \tau) = \frac{h_0}{\cosh^2 \left[w^{-1} \left(\xi - \left(1 + 4w^{-2}\right) \tau \right) \right]}. \tag{5}$$

This solution has an important parameter — the soliton width:

$$w = \sqrt{12/\lambda h_0} \tag{6}$$

We emphasize that the derivation of the evolution equation is performed for an arbitrary isotropic part of the distribution function f_0. In the further modeling of the synchrotron images of the local magnetic field structures in the relativistic pair plasma of the Crab nebula we take f_0 in the form of a broken power-law. We choose the parameters of f_0 resembling the values which could be locally estimated from the spectral energy distribution of the average emission of the Crab nebula, presented in Figure 2 of [2]. We take the breaking Lorentz-factor $\gamma_b = 10^6$, and the spectral indices $p_1 = 1.35$ and $p_2 = 2.15$ at low and high energies, respectively. The induction of the mean quasi-stationary magnetic field is taken, according to observations, to be equal to 200 μG. The concentration of particles is estimated from the assumption $P \approx B_0^2/8\pi$.

In Figure 1 we present the results of the numerical solution of the initial value problem for the equation (4) for the case of significant damping. The amplitude and the spatial scale of the initial perturbation of the magnetic field are matching the solitary peaks emerging in the simulation of the cyclotron instability by Spitkovsky & Arons [6] (see their Figure 2). We use the Gaussian form for the initial perturbation, and take the amplitude $b/B_0 = 0.35$ and the standard deviation to be equal to $1.4 \cdot 10^{16}$ cm. We suppose it to propagate with the velocity $u = c/3$.

In Figure 1(a) we present the results for the limit $\nu = \Omega$. One can see that the amplitude of the propagating perturbation decreases, and its width increases. In addition, the perturbation decelerates. This behaviour is similar to the specific relations between amplitudes, widths and velocities of the soliton solutions. The decreasing of the amplitude is quite slow: ~10% from the initial value on the timescale of a month. The perturbation remains to be a relatively narrow peak (width of the order of a few units of 10^{16} cm) for tens of days. On longer timescales a 'tail' emerges behind the peak likely due to damping.

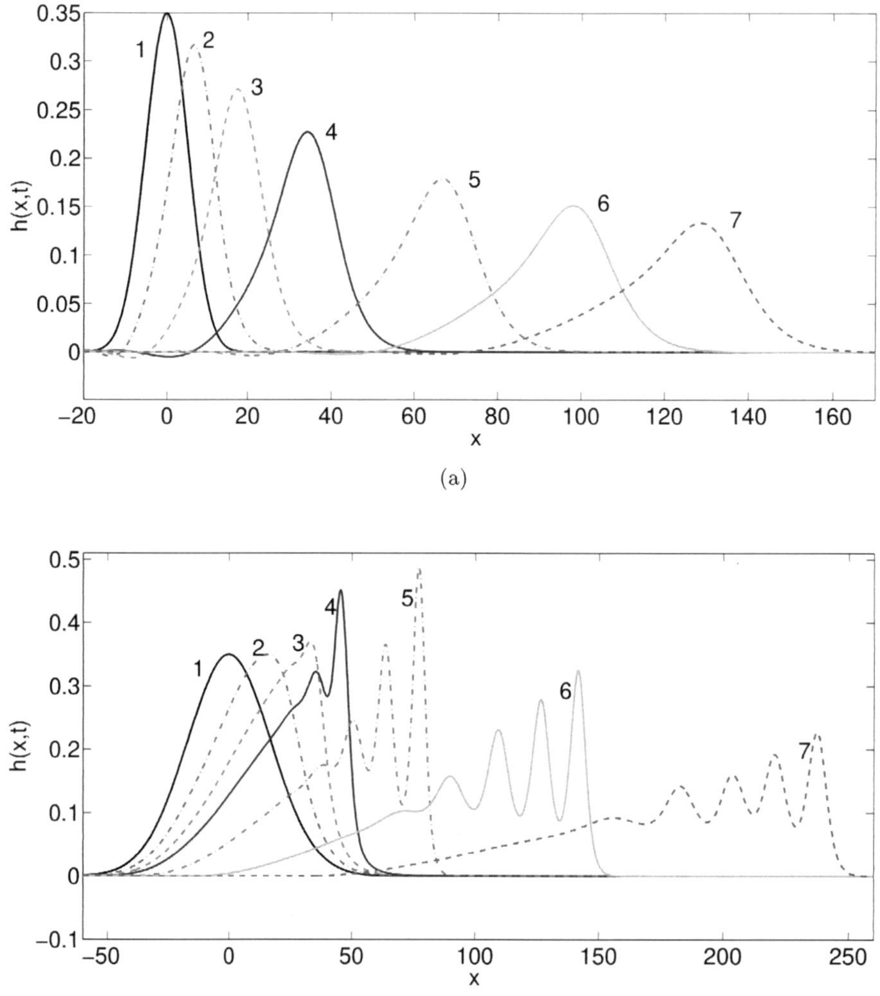

(a)

(b)

Figure 1. The evolution of the initial perturbation (Gaussian with the amplitude $b/B_0 = 0.35$ and the standard deviation equal to $1.4 \cdot 10^{16}$ cm) in case of (a): $\nu = \Omega$; (b): $\nu = 0.1\Omega$. The perturbation at various moments of time is presented. The list of the corresponding moments of time is shown in Table 1. The variable $x = \xi - U\tau$, where τ,ξ — the dimensionless time and coordinate, U — the dimensionless velocity of the initial perturbation. The unit length is equal to $2.8 \cdot 10^{15}$ cm in (a) and $8.1 \cdot 10^{14}$ cm in (b).

The results shown in Figure 1(b) correspond to the frequency $\nu = 0.1\Omega$. Here the dispersion term $\sim \partial_\xi^3 h$ dominates over the damping term, and one can see the formation of narrow peaks. The spatial scale of these peaks is of the order of a few units of 10^{15} cm. This process is similar to the decay of the initial perturbation into solitons, which emerges in the collisionless case. In the same time, one can see clear manifestations of the damping.

The angular resolution of the Hubble Space Telescope $\sim 0.1''$ allows us to resolve an object of a size larger than $3 \cdot 10^{15}$ cm in the Crab nebula. The localized perturbations of the magnetic field result in synchrotron features, which could be detected as bright wisps and filaments.

Table 1. The time moments of the evolution, shown in Figure 1, in units of 10^7 sec

Figure	t_1	t_2	t_3	t_4	t_5	t_6	t_7
1(a)	0.0	0.37	1.03	2.06	4.11	6.17	8.23
1(b)	0.0	0.091	0.181	0.272	0.543	1.21	2.42

3. The synchrotron images

The main mechanism of emission of the magnetized relativistic pair plasma of the pulsar wind nebula is the synchrotron radiation. Thus, the propagating perturbations of the magnetic field could be responsible for the observed dynamical structures in the nebulae.

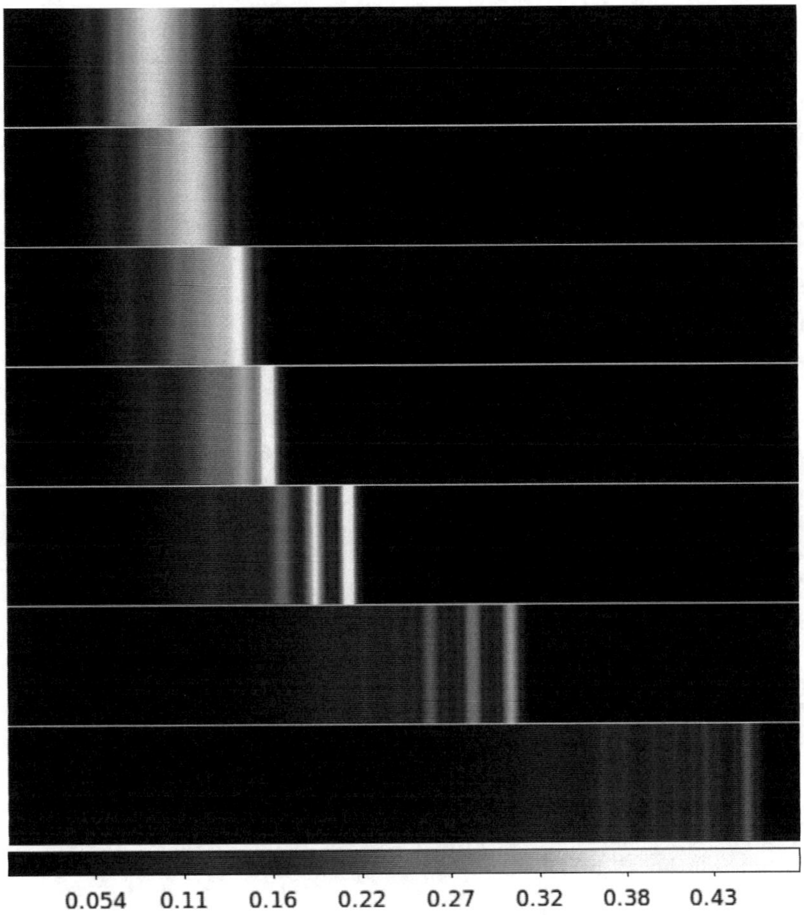

Figure 2. The synchrotron images of the evolution of the initial perturbation presented in Figure 1(b): the spatial distribution of the intensity (normalized to the background value). The images correspond to the moments of time in the bottom row of Table 1. The size of the simulated box is $2.6 \cdot 10^{17}$ cm. The frame moves with the velocity of the initial perturbation. The wavelength of emission $\lambda = 555$ nm (corresponding photon energy $E = 2.23$ eV).

The solution of the initial value problem for the evolution equation (4) is a spatial distribution of the magnetic field induction. One has to evaluate the Stokes parameters for this distribution to get the coordinate distribution of the intensity of emission and the parameters describing the polarization.

The synchrotron images of the evolution of the initial perturbation shown in Figure 1(b) are presented in Figure 2. The simulated box moves with the velocity of the initial perturbation.

One can see the formation of thin bright stripes, which correspond to narrow peaks, shown in Figure 1(b). These stripes have widths $\sim 10^{15}$ cm. For the leading stripes the significant enhancement of the calculated emission intensity over the background level holds on a timescale of tens of days.

4. Conclusions

In this paper we discuss the modeling of the dynamical structures in the highly-nonequilibrium relativistic pair plasma of the pulsar wind nebulae. We describe a kinetic model of a propagation of a weakly-nonlinear perturbation of the magnetic field transverse to the mean quasi-stationary magnetic field. The regime of the strong scattering of pairs by the stochastic magnetic field fluctuation is considered. The locality of the perturbation is supposed, thus, the one-dimensional analysis is performed. We review the procedure of derivation of the evolution equation from the kinetic equations for the pair plasma components. The obtained evolution equation has the form of the Korteweg—de Vries—Burgers (KdVB) equation.

We solve the initial value problem for the KdVB equation numerically, taking a localized perturbation of the magnetic field as the initial condition. We show that for the limit $\nu = \Omega$ the perturbation retains the character of a single peak for a long time, but its amplitude decreases due to dissipation. The perturbation decelerates and broadens, and a long 'tail' emerges behind it likely due to damping. For the case of $\nu = 0.1\Omega$, when the dispersion dominates over the dissipation, the changes in the shape are more substantial: the groups of solitary peaks are formed. They have the spatial scale of the order of a few units of 10^{15} cm. For this case we evaluate the synchrotron emission intensity at the obtained magnetic field distribution and build the images, representing the coordinate distribution of the intensity. In these images one can see bright thin stripes, corresponding to the peaks mentioned above. The widths of these stripes are of the order of the scale which could be resolved by the Hubble Space Telescope in the Crab nebula ($3 \cdot 10^{15}$ cm). For the leading stripes the significant enhancement of the calculated synchrotron emission intensity over the background level holds on a timescale of tens of days.

Thus, the perturbations described by the evolution equation (4) can be responsible for the observed dynamical structures, like wisps in the Crab nebula. Also, it can be shown that the characteristic spatial scales of the derived structures are sensitive to the pressure of the pair plasma. This can open up a prospective of an observational constraining of the distribution function parameters and studying of the relativistic pairs gas properties.

References

[1] Hester J J 2008 *Annu. rev. astron. astrophys.* **46** 127-55
[2] Bühler R and Blandford R 2014 *Rep. Prog. Phys.* **77** 066901
[3] Arons J 2012 *Space Sci. Rev.* **173** 341-67
[4] Bykov A M, Pavlov G G, Artemyev A V and Uvarov Yu A 2012 *MNRAS* **421** L67-71
[5] Petrov A E and Bykov A M 2014 *J. Phys.: Conf. Ser.* **572** 012005
[6] Spitkovsky A and Arons J 2004 *ApJ* **603** 669-81

Optical observations of PSR J2021+3651 with the GTC

E. B. Ryspaeva[1], A. Yu. Kirichenko[1,2], Yu. A. Shibanov[1,2], P. S. Shternin[1,2], A. A. Danilenko[2], D. A. Zyuzin[2], M. Durant[3], O. Yu. Kargaltsev[4], G. G. Pavlov[5], A. Cabrera-Lavers[6,7]

[1] St. Petersburg Polytechnic University, Politekhnicheskaya 29, St. Petersburg, 195251, Russia
[2] Ioffe Institute, Politekhnicheskaya 26, St. Petersburg, 194021, Russia
[3] Department of Medical Biophysics, Sunnybrook Hospital M6 623, 2075 Bayview Avenue, Toronto M4N 3M5, Canada
[4] Department of Physics, The George Washington University, Washington, DC 20052, USA
[5] Department of Astronomy & Astrophysics, Pennsylvania State University, 525 Davey Lab, University Park, PA 16802, USA
[6] Instituto de Astrofisica de Canarias, E-38205 La Laguna, Tenerife, Spain
[7] Universidad de La Laguna, Dept. Astrofisica, E-38206 La Laguna, Tenerife, Spain

E-mail: `e.ryspaeva@yandex.ru`

Abstract. We analyzed first optical observations of 17 kyr-old PSR J2021+3651 and its pulsar wind nebula obtained with the 10.4-m Gran Telescopio Canarias (GTC) telescope in the Sloan r' band. In addition, we reanalyzed archival X-ray data obtained with *Chandra*. The pulsar and the nebula were not detected in the optical down to 27.20 and 24.85 magnitude 3σ limits, respectively. Using the optical and X-ray data we conclude that PSR J2021+3651, like the Vela pulsar, is a very inefficient nonthermal emitter in the optical and X-rays, while its gamma-ray efficiency is consistent with an average efficiency for all gamma-ray pulsars of similar ages.

1. Introduction

The pulsar PSR J2021+3651 with a period of 103.7 ms was discovered in the radio with the 305-m Arecibo telescope [1]. With a characteristic age $\tau_c \approx 17$ kyr and a spin-down luminosity $\dot{E} \approx 3.4 \times 10^{36}$ erg s^{-1}, PSR J2021+3651 is one of the youngest and most energetic rotation-powered pulsars known. The canonical dipole magnetic field estimate from the pulsar period and its derivative is $B \approx 3.2 \times 10^{12}$ G. PSR J2021+3651 was identified in X-rays with *Chandra* [2]. The extended pulsar wind nebula (PWN) G75.2+0.1 was also revealed in this range. Its brightest internal part, within $\sim 30''$ of the pulsar, has a torus-like morphology with axial jets. This PWN was dubbed the Dragonfly Nebula owing to its specific spatial shape [3]. In addition, PSR J2021+3651 was detected in gamma-rays with the *Fermi* observatory [4]. The distance D to the pulsar is very uncertain and varies from 1 to 12 kpc, depending on the method applied to derive this parameter [3, 5].

Thus the pulsar was extensively studied in various spectral domains, but it has never been observed in the optical. In this work first optical observations of the PSR J2021+3651 are analyzed together with the archival X-ray data.

Journal of Physics: Conference Series **661** (2015) 012011 doi:10.1088/1742-6596/661/1/012011

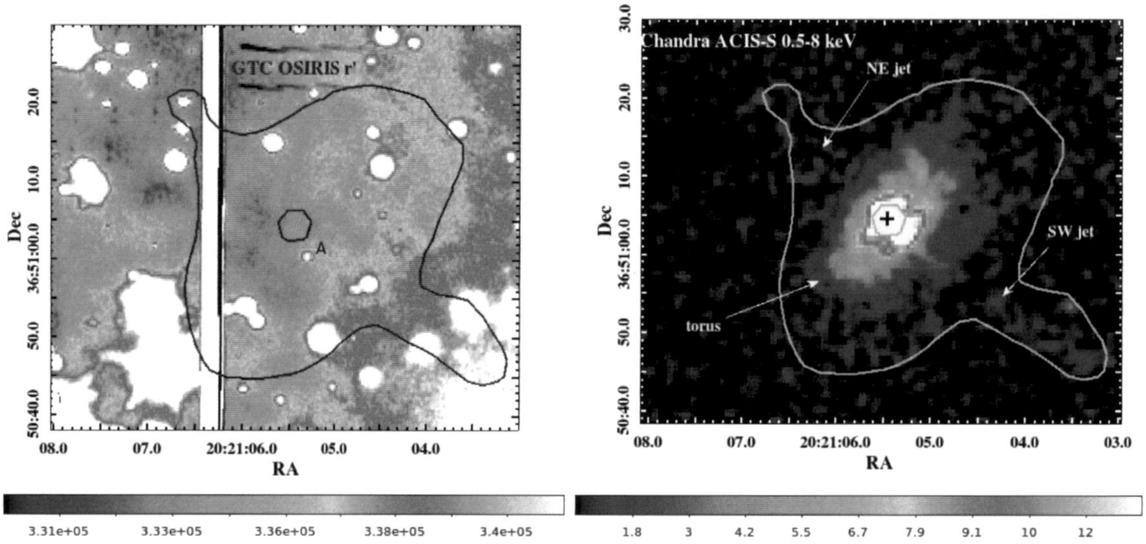

Figure 1. *Left*: $\sim 50'' \times 50''$ GTC/OSIRIS optical Sloan r' image centered at the PSR J2021+3651 X-ray position. The closest to the pulsar source "A" is marked. *Right*: *Chandra*/ACIS-S X-ray image of the same field obtained by merging all available archival data. The pulsar position is marked by "+" in the X-ray image. Contours indicating the X-ray PWN boundary and the pulsar location are shown in both images. Colorbars show flux intensity in counts pixel^{-1}.

2. GTC data

PSR J2021+3651 was observed with the Optical System for Imaging and low-intermediate Resolution Integrated Spectroscopy (OSIRIS) at the GTC in Sloan r' band in 2011 September 28 [5]. Observing conditions were photometric with a mean seeing of about 0.″9. Sixteen images with exposures of 158 s each were obtained. Standard data reduction was performed with `IRAF` tools. For astrometric referencing we used the positions of stars from the USNO-B1 astrometric catalog. The resulting 1σ referencing uncertainty is $\approx$ 0.″26. For photometric calibration we used the G158-100 Sloan standard observed at the same night as our target. The atmospheric extinction for the Sloan r'-band of 0.10$\pm$0.01 mag airmass^{-1} was taken from the OSIRIS user manual. The resulting magnitude zero-point for our data is $29^{\rm m}_{.}13 \pm 0^{\rm m}_{.}02$.

3. Results

3.1. Optical counterpart

Figure 1 shows optical (left panel) and X-rays (right panel) images of the pulsar field. In X-rays the pulsar and PWN are clearly seen. The latter demonstrates a torus-like structure with two axial SW and NE jets [3]. Neither pulsar, nor PWN are detected in the optical. The nearest detected optical source "A" with $r' = 24.40^{+0.04}_{-0.04}$ is located at about 4.″8 or at $\approx 6\sigma$ from the pulsar X-ray position and thus is an unrelated object. We estimated optical flux upper limits for the pulsar and PWN using a standard method. For the pulsar, we used a mean background deviation within a circular aperture with 4 pixel (1″) radius centered at the pulsar position. We accounted for an aperture correction of $0^{\rm m}_{.}1$ obtained by us using bright background stars. The resulting 3σ upper limit on the pulsar flux density is ≤ 0.04 μJy ($r' \geq 27.20$). For the nebula, we used an elliptical aperture, centered at the pulsar, with semi-axes of 6.″2 and 10.″6 and a position angle of 137°, which encapsulates most of the X-ray PWN equatorial torus emission. The 3σ upper limit on the spatially integrated PWN flux density is ≤ 0.36 μJy ($r' \geq 24.85$). An average upper limit for the surface flux density is $\leq 1.74\times10^{-3}$ μJy arcsec^{-2} ($r' \geq 30.63$ mag arcsec^{-2}).

3.2. X-ray data

For comparison of the optical upper limits with the X-ray spectra of the pulsar and PWN, we reanalyzed archival X-ray data, which were obtained with *Chandra* in the ACIS-S configuration in December 2006 and in February 2003 (obsIDs 3901, 7603, 8502) with 114 ks exposure in total. The pulsar equatorial coordinates RA=20:21:05.46 and Dec=36:51:04.52 were obtained in [3] and verified by us using the X-ray image in Figure 1. We extracted pulsar and PWN spectra using the `CIAO 4.5` *specextract* task. For extraction of the PSR J2021+3651 spectrum, we used a circle region with a radius of $0\farcs74$, centered at the pulsar. For extraction of the PWN spectrum, we used the same elliptical region as used in the Section 3.1. We fitted all spectra in 0.3–10 keV range using `XSPEC` v.12.8.1 tools.

The pulsar and PWN spectra were fitted simultaneously, using the BB+PL, and NSMAX+PL spectral models for the pulsar, and the PL model for the PWN. The PL is a power law model describing nonthermal emission spectral component. It contains two unknown parameters, the normalization K and the photon index Γ. The BB and NSMAX are the blackbody and magnetized hydrogen atmosphere models, respectively, describing the thermal emission from a neutron star (NS) surface. We also used the model of the photoelectric absorption PHABS in a product with each of these models to account for the interstellar absorption towards the pulsar. It is described by the hydrogen column density N_H, which was set as a global fit parameter. Standard interstellar medium abundances and absorption cross-sections were applied. In what follows, we do not indicate this multiplicative component explicitly for brevity. The simplest BB thermal emission model contains two unknown parameters – the temperature T and the normalization R^2/D^2, where R is the effective emitting area radius. The NSMAX model describes a more realistic case of the NS hydrogen atmosphere, hence its parametrization is more complicated. In addition to T and normalization, it depends on the surface gravity via the gravitational redshift parameter $1+z = (1-2.952M_{NS}/R_{NS})^{-0.5}$, where M_{NS} and R_{NS} are the NS mass and circumferential radius in the units of Solar mass and km, respectively. We fixed $1+z$=1.21, which corresponds to reasonable NS parameters $M_{NS} = 1.4\,M_{\odot}$ and $R_{NS} = 13$ km. We also selected the specific NSMAX model with a surface magnetic field of 4×10^{12} G, which is close to $B \approx 3.2 \times 10^{12}$ G derived from PSR J2021+3651 spindown measurements.

The PWN is bright enough and contaminates the spectrum of the pulsar. To model the contribution of the PWN to the spectrum extracted from the pulsar aperture, we added a second PL component of the PWN origin to the pulsar spectral model. The second PL photon index was tied with the PWN photon index, and its normalization was tied with the 5% fraction of the total PWN normalization. The latter value was estimated by Van Etten [3] and independently confirmed by us via modeling of *Chandra*/ACIS point spread function and analyzing the PWN spatial brightness profiles.

The described spectral fitting procedure can constrain only the R/D ratio. To obtain R and D separately one needs to account for an additional information. We used the fact that N_H increases with the distance to a Galactic source. To estimate the $N_H - D$ relation for the pulsar line of sight we employed the so-called "red-clump" stars method [6]. This relation was used to constrain D from the inferred N_H fit range, and then to constrain R from D and the thermal component normalization. The best spectral fit results for the pulsar are presented in Table 1. For the torus part of the Dragonfly the best-fit photon spectral index is Γ=$1.4^{+0.1}_{-0.1}$ and the unabsorbed flux in 2–10 keV range is Log $F_X = -12.2^{+0.1}_{-0.1}$ erg cm^{-2} s^{-1}.

Both applied models equally well describe the pulsar spectrum, according to the reduced χ^2 values about unity, see Table 1. The BB+PL model leads to a temperature of 155^{+14}_{-14} eV. It matches well to $T = 160^{+20}_{-20}$ eV and 150^{+20}_{-20} eV obtained by Van Etten et al. [3] and Hessels et al. [2], respectively, who used the same spectral models, but fixed N_H at a certain value. The distance of $1.8^{+1.7}_{-1.4}$ kpc is now mainly determined by N_H and the N_H–D relation [5]. As it is typical for pulsars, which X-ray spectral data can be equally well fitted by the blackbody

Table 1. The best-fit parameters of the pulsar X-ray spectrum described by the BB+PL and NSMAX+PL models. The reduced χ^2 values in the last column obtained for 469 degrees of freedom. All errors correspond to 90% credible intervals.

Model	N_H 10^{21} cm^{-2}	Γ	K 10^{-5} photons keV^{-1} cm^{-2} s^{-1}	T eV	R km	D kpc	χ^2
BB+PL	$5.8^{+0.5}_{-0.5}$	$1.8^{+0.6}_{-0.6}$	$1.0^{+1.0}_{-0.6}$	155^{+14}_{-14}	$1.3^{+1.5}_{-1.0}$	$1.8^{+1.5}_{-1.4}$	0.98
NSMAX+PL	$6.0^{+0.5}_{-0.5}$	$1.3^{+0.7}_{-0.8}$	$0.5^{+0.7}_{-0.4}$	63^{+9}_{-8}	$12.0^{+19.5}_{-9.6}$	$1.8^{+1.7}_{-1.4}$	1.00

and NS atmosphere models, in former case R is apparently a factor of 10 smaller and T is a factor of 2.5 larger than those for the latter model. If the hydrogen atmosphere model is used, $R = 12.0^{+19.5}_{-9.6}$ km implies that emission can come from the bulk of the NS surface with the effective surface temperature, redshifted for distant observer, $T = 63^{+9}_{-8}$ eV, close to that of the Vela pulsar [7]. For the BB model, the thermal emission can be interpreted as coming from a small area with a radius of $1.3^{+1.5}_{-1.0}$ km, which is compatible with a canonical pulsar hot polar cap radius of ~ 0.6 km for a 100-ms pulsar [8].

4. Multiwavelength spectrum of the pulsar

The best-fit N_H values obtained from the X-ray spectral analysis suggest a total interstellar extinction towards the Dragonfly in the V band $A_V \approx 3.3$ using the standard $N_H - A_V$ relation [9]. This results in the extinction $A_{r'} \approx 2.8$ in the r' band using a standard extinction law with R_V=3.1 [10]. Based on this, upper limits on the dereddened flux densities for the pulsar and PWN in the r' band are about 0.57 μJy and 4.85 μJy, respectively. In Figure 2 we compare the derived pulsar upper limit with extrapolation of the unabsorbed X-ray spectrum as fitted by the BB+PL model. The solid line in Figure 2 shows the total best-fit model, including the contribution of PWN nonthermal photons to the spectrum extracted from the pulsar aperture. The dashed line shows solely the PL component of the pulsar. As expected, the PWN contribution is substantial only in the high-energy tail. For completeness we also show the extrapolation of the *Fermi* best-fit gamma-ray spectrum of the pulsar [4].

For all rotational powered pulsars detected in the optical and X-rays the nonthermal emission component dominates in the optical and usually shows a break between the optical and X-rays with significant spectral flattening in the optical [11, 12, 13, 14]. As seen in Figure 2, the pulsar optical flux upper limit does not exceed the extrapolation of the best-fit X-ray spectral model. This does not exclude the presence of the break for PSR J2021+3651, although the extrapolation is still rather uncertain and the optical limit is not deep enough. The same conclusions holds in the case when NSMAX+PL model is applied.

The optical upper limit for the PWN exceeds significantly the long wavelength extrapolation of its X-rays spectrum and does not constrain the PWN spectral parameters [5].

The pulsar's 0.1–100 GeV γ-ray luminosity L_γ and efficiency $\eta_\gamma = L_\gamma/\dot{E}$ derived in the 2nd Fermi Pulsar Catalog [4] using the distance of 10 kpc from [2] appear to be unreasonably high [2]. In contrast, for the distance D=$1.8^{+1.7}_{-1.4}$ kpc inferred from our analysis, Log $L_\gamma = 35.3^{+0.6}_{-1.3}$ erg s^{-1} and Log $\eta_\gamma = -1.2^{+0.6}_{-1.3}$ become consistent with the average values of the respective distributions

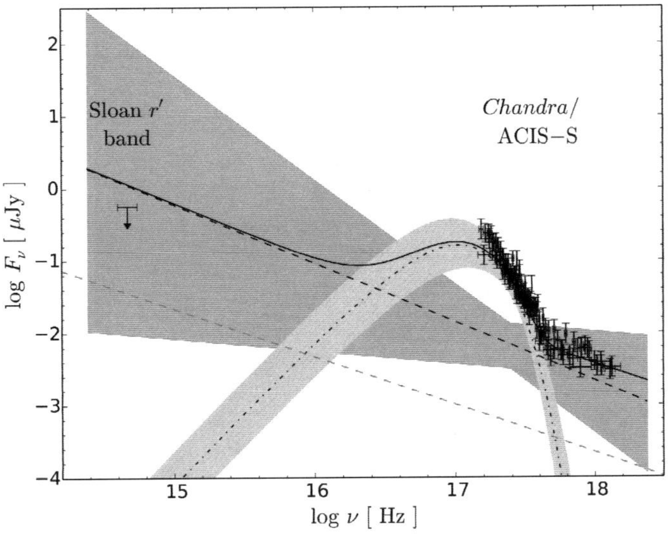

Figure 2. Unabsorbed spectrum of PSR J2021+3651. The solid line is the best-fit model for the *Chandra* X-ray spectrum which includes BB and PL pulsar spectral components. The best-fit model is extrapolated towards the optical. The GTC dereddened 3σ flux upper limit in the r' band is shown by the bar with the arrow. Dash-dotted and dashed lines with light- and dark-gray regions are the BB and PL pulsar spectral and PWN components with their 90% uncertainties, respectively. The difference between the solid and dashed lines is clearly visible at the high-energy tail and reflects the PWN contribution. The gray dash-dotted line shows the extrapolation of the *Fermi* gamma-ray best-fit spectrum.

for γ-ray pulsars with similar $\dot{E}$ and/or characteristic ages [4].

The X-ray luminosity and efficiency are Log $L_X = 31.1^{+1.3}_{-2.0}$ erg s^{-1} and Log $\eta_X = -5.4^{+1.3}_{-2.0}$, respectively, assuming the distance range derived in this work. For the 90% distance interval upper boundary of 3.5 kpc, upper limits on the optical luminosity and efficiency translated to the V band are Log $L_V \leq 29.9$ erg s^{-1} and Log $\eta_V \leq -6.7$, respectively. Both efficiencies are comparable to those of the Vela pulsar, which is known to be very inefficient nonthermal emitter in these ranges as compared to other younger and older pulsars. This makes PSR J2021+3651 a new member of the small sample of Vela-like pulsars forming a puzzling minimum in the efficiency-age dependence near a 10 kyr age [15].

5. Conclusions

We analyzed deepest up-to-date optical images of the PSR J2021+3651 field down to $r' = 27.2$. We did not detect the pulsar optical counterpart, but obtained important constrains on its optical-X-ray spectral energy distribution. From the X-ray spectral analysis and new constraints on the distance to a pulsar we estimated its luminosities and efficiencies in the optical, X-ray, and gamma-ray spectral domains. Based on that we concluded that PSR J2021+3651, like the Vela pulsar, is a very inefficient nonthermal emitter in the optical and X-rays, but its gamma-ray efficiency is consistent with an average efficiency for all pulsars of similar ages.

Acknowledgements

The work was supported by the Russian Science Foundation, grant 14-12-00316. The scientific results reported in this article are partially based on data obtained from the Chandra Data Archive, observations made by the Chandra X-ray Observatory.

References

[1] Roberts M. S. E., Hessels J. W. T., Ransom S. M. *et al* 2002, *The Astrophysical Journal* **577** L19.

[2] Hessels, J. W. *et al* 2004, *The Astrophysical Journal* **612** 389
[3] Van Etten, A., Romani, R., Ng, C.-Y., 2008, *The Astrophysical Journal* **680** 1417
[4] Abdo, A. A. *et al* 2013, *The Astrophysical Journal* **208** 17
[5] Kirichenko, A. *et al* 2014, *The Astrophysical Journal* (submitted)
[6] Durant, M., & van Kerkwijk, M. H., 2006, *The Astrophysical Journal* **650** 1070
[7] Pavlov, G. G., Zavlin, V. E., Sanwal, D. *et al* 2001, *The Astrophysical Journal* **552** L129
[8] Sturrock, P. A. 1971, *The Astrophysical Journal* **164** 529
[9] Predehl, P., Schmitt, J. H. M., 1995, A&A, 293, 889
[10] Cardelli, J. A., Clayton, G. C., & Mathis, J. S. 1989, *The Astrophysical Journal* **345** 245
[11] Mignani, R. P., Pavlov, G. G., & Kargaltsev, O. 2010, *The Astrophysical Journal* **720** 1635
[12] Danilenko, A. A., Zyuzin, D. A., Shibanov, Y. A. *et al* 2011, MNRAS, 415, 867
[13] Durant, M., Kargaltsev, O., & Pavlov, G. G. 2011, *The Astrophysical Journal* **743** 38
[14] Shibanov, Y. A., Zharikov, S. V., Komarova, V. N. *et al* 2006, A&A, 448, 313
[15] Zharikov, S., Shibanov, Y., & Komarova, V. 2006, *Advances in Space Research* **37** 1979

Propeller regime of the accretion onto young stars in the ballistic approach

S G Shulman[1,2], **V P Grinin**[1,2] **and I S Potravnov**[2]

[1] St.-Petersburg State University, 7-9 Universitetskaya Nab., St.-Petersburg, 199034, Russia
[2] Pulkovo Astronomical Observatory of RAS, 65/1 Pulkovskoye chaussee, St.-Petersburg, 196140, Russia

E-mail: sgshulman@rambler.ru

Abstract. The form of the outflows generated during the interaction of the accretion disc with the tilted stellar magnetosphere in the propeller regime is studied in the ballistic approach. The problem is solved analytically. Different ways of the flux expansion are considered. The results are applied to the description of the narrow absorption details observed in the spectra of young stars in a blue wing of the sodium D Na I resonance lines.

1. Introduction

In the spectra of some young stars (RZ Psc [1], [2], MWC 480 [3] and others) unusual discrete absorption components are seen in the blue wing of the sodium D Na I lines (Figure 1). The number, shape and radial velocities of these components vary from night to night.

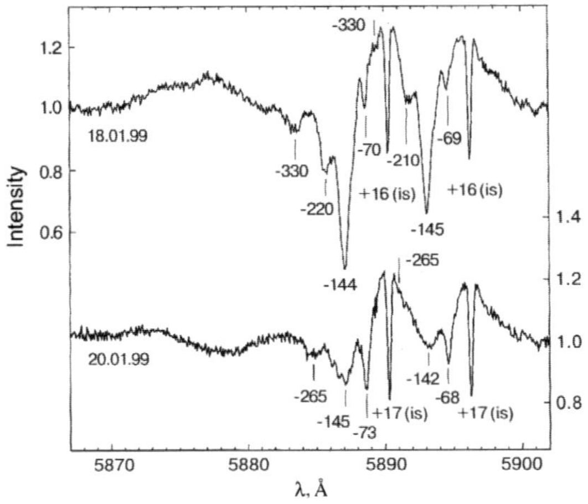

Figure 1. Two spectra of MWC 480 [3] in the neighborhood of the sodium resonance doublet. The radial velocities of the individual components of the lines are indicated. The strong changes in the helium 5876 Å line are seen in the spectrum.

We assume [4] that absorption components may be a result of the interaction of an accretion disc with a magnetosphere in the propeller regime. The possibility of this regime was first noticed by Illarionov and Sunyaev [6] during the study of an accretion onto neutron stars. It is realized when the effective radius of the magnetosphere exceeds a corotation one. In this case the rotating magnetosphere of the star rejects the falling matter. As a result the accretion flow onto the star may be rather weak. We suppose that the outflowing matter may cause the narrow absorption components at the intersection of the light of sight with the different parts of the gas stream.

Taking into consideration the expression for the radius of the Alfven surface [7] we can draw a conclusion that a decrease in the mass accretion which is usual during the young stars evolution may lead the accretion process to turn into the propeller regime (the radius of the Alfven surface $r_A \sim \dot{M}^{-2/7}$, where $\dot{M}$ is the mass accretion rate).

2. Form of the outflows

In this work we consider a case when a dipole magnetic field of the star is misaligned with the rotation axis of the star (and the disc). In this case the magnetically accelerated matter forms two spiral outflows which originate from the disc-magnetosphere boundary [8]. The outflowing matter may be ejected to an infinite distance. However, returning into the accretion disc is also conceivable.

We suppose that the matter is accelerated to a velocity $V_0 = \beta V_I$ (where V_I is the circular velocity of the ejection point and β is parameter of the model) and ejected forming with the disc plane an angle α. We assume that the direction of the outflow in the ejection point is orthogonal to the radius-vector. At the same time the region where the matter is launched from the disc is rotating with the angular velocity of the magnetosphere and the star Ω. As a result rotating spiral tracery is formed. The form of these outflows can be described analytically in the ballistic approach [9].

Let us introduce a system of Cartesian rotating coordinates centered at the star's center. The X axis is directed to the ejection point. The Y axis lies in the disc plane. Y and Z axes are directed to the outflow direction. In the case of $\beta < \sqrt{2}$ when the matter returns into disc the expression for the outflows form is:

$$\begin{cases} x &= \frac{\beta r_c}{\sqrt{2-\beta^2}} \sin E \cos\alpha \sin\theta + \frac{r_c}{2-\beta^2}\left(\cos E - \left(\beta^2 - 1\right)\right)\cos\theta \\ y &= \frac{\beta r_c}{\sqrt{2-\beta^2}} \sin E \cos\alpha \cos\theta - \frac{r_c}{2-\beta^2}\left(\cos E - \left(\beta^2 - 1\right)\right)\sin\theta \\ z &= \frac{\beta r_c}{\sqrt{2-\beta^2}} \sin E \sin\alpha \end{cases}, \tag{1}$$

where

$$\theta = \Omega \frac{E - \left(\beta^2 - 1\right)\sin E}{\varkappa}\left(\frac{r_c}{2-\beta^2}\right)^{\frac{3}{2}}. \tag{2}$$

Here r_c is the corotation radius, $\varkappa = \sqrt{GM}$ the gravitation parameter which is equal to root square from the gravitation constant multiplied by the star's mass, $E \in [0,\, 2\pi]$ is the eccentric anomaly.

In the case of $\beta > \sqrt{2}$ when the matter is ejected to an infinite distance the expression is slightly different:

$$\begin{cases} x &= \frac{\beta r_c}{\sqrt{\beta^2-2}} \sinh H \cos\alpha \sin\theta + \frac{r_c}{\beta^2-2}\left(\left(\beta^2 - 1\right) - \cosh H\right)\cos\theta \\ y &= \frac{\beta r_c}{\sqrt{\beta^2-2}} \sinh H \cos\alpha \cos\theta - \frac{r_c}{\beta^2-2}\left(\left(\beta^2 - 1\right) - \cosh H\right)\sin\theta \\ z &= \frac{\beta r_c}{\sqrt{\beta^2-2}} \sinh H \sin\alpha \end{cases}, \tag{3}$$

where

$$\theta = \Omega \frac{\left(\beta^2 - 1\right)\sinh H - H}{\varkappa}\left(\frac{r_c}{\beta^2-2}\right)^{\frac{3}{2}}. \tag{4}$$

$H \in [0,\, +\mathrm{inf})$ is an analog of the eccentric anomaly for the hyperbola.

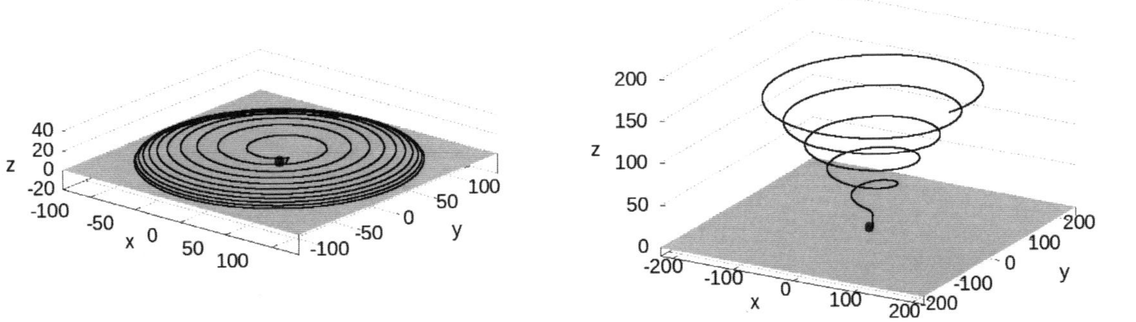

Figure 2. Forms of the outflows in the cases of $\beta < \sqrt{2}$ (left panel) and $\beta > \sqrt{2}$ (right panel).

2.1. Asymptotic solution

When the ejected matter does not return onto the disc ($\beta > \sqrt{2}$) there is an asymptotic form of the formula (3). We pay attention to the circumstance that $\sinh H$ and $\cosh H$ grows fast with an increase in H. Moreover, when H is large $\sinh H \approx \cosh H$. After simple transformations, we obtain that at large H a spiral turns out on the cone with the star on its top:

$$\frac{x^2 + y^2}{1 + \beta^2 (\beta^2 - 2) \cos^2 \alpha} - \frac{z^2}{\beta^2 (\beta^2 - 2) \sin^2 \alpha} = 0. \tag{5}$$

Hence, when the observer looks through the cone's side there might be few discrete components from different spiral's turns.

3. Outflow expansion

For the modeling of the outflow expansion two different ways have been used. In the first way we assume that the cross section of the stream expands isotropically with the sound speed:

$$c = \sqrt{\frac{\gamma \mathcal{R}}{\mu}} \sqrt{T}, \tag{6}$$

here γ is the adiabatic index (5/3 for a monatomic gase), $\mathcal{R}$ the molar gas constant, μ the molar mass, T the temperature.

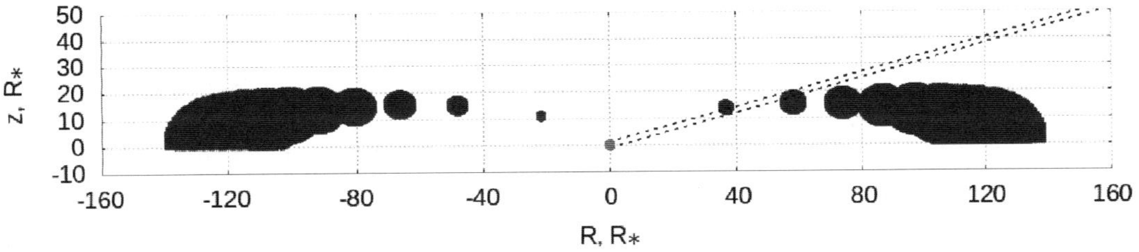

Figure 3. Cross section of the outflow in the case of the expansion with the sound velocity for $\beta < \sqrt{2}$. R_* is the radius of the star. Dotted lines show the channel from the star to the observer.

In the second case we modeled trajectories of the particles thrown with some velocity dispersion. For the size of this dispersion we took the root mean square speed of the matter in the disc:

$$V_T = \sqrt{\frac{3\mathcal{R}}{\mu}}\sqrt{T}. \tag{7}$$

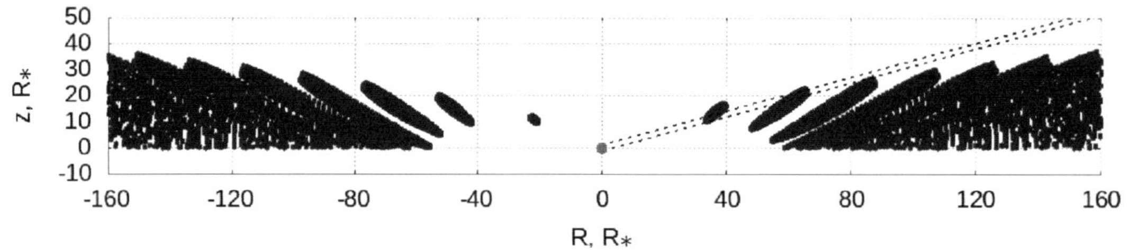

Figure 4. The same as on the figure 3 for the expansion caused by the dispersion in the velocities.

As can be seen from figures 3–5 both methods show close results in the beginning. Later there are some differences. We suppose that the both ways are important and take part in the real expansion. The expansion with the sound speed may be more important in the beginning of the flow whereas the dispersion in the velocities should take larger part in the extrinsic spiral turns. The mechanism of the expansion influences the components shape. It should be studied in detail with more observation data.

Each crossing of the channel from the star to the observer with the outflow gives one absorption component.

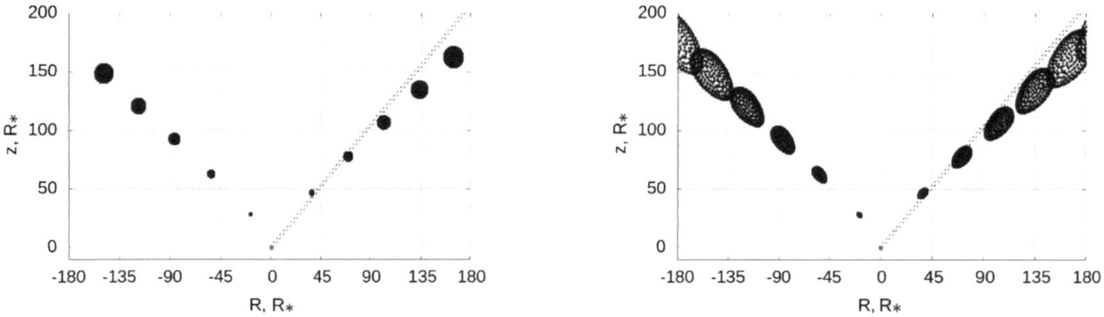

Figure 5. The same as on the figures 3–4 in the case $\beta > \sqrt{2}$.

4. Radial velocities

Radial velocities of the components depend on the radial velocity of the outflow which can be easily obtained from the integrals of motion:

$$v_r = \varkappa\sqrt{\frac{2}{r} + \frac{\beta^2 - 2}{R_0} - \frac{\beta^2 R_0}{r^2}}, \tag{8}$$

here r is the distance from the star. This allows us to get radial velocities from the model and compare them with the observations. When $\beta < \sqrt{2}$ the radial velocity decreases until zero in the point where the matter contacts the disc. Otherwise the minimum velocity is positive.

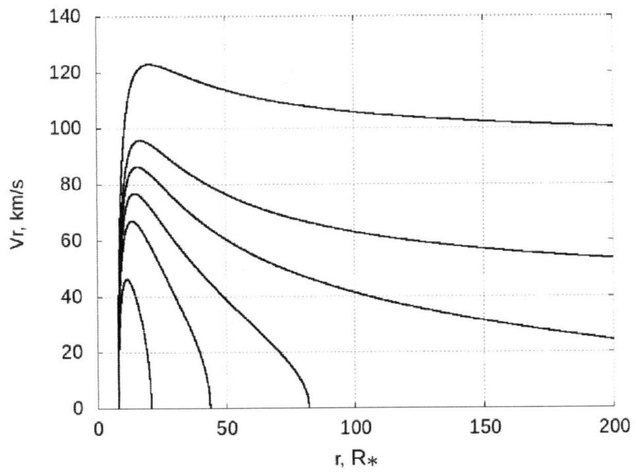

Figure 6. Radial velocities for the outflows with different meaning of β. The parameter β decreases from the upper line ($\beta = 1.6$) to the bottom line ($\beta = 1.2$).

5. Conclusion

The model considered above demonstrates a possibility to explain the narrow discrete absorption components of the sodium resonance doublet lines in the spectra of some young stars. The more detailed discussion of the accretion onto the young stars in the propeller regime and the derivation of the equations presented above will be given in papers [4] and [9].

In this model we does not consider an acceleration of the matter. Therefore, it does not provide the full description of the process, but makes further researches of this spectral peculiarity well grounded. The more detail model can predict the number and radial velocities of the components. To make more definite conclusions about the nature of the components we need the spectral monitoring of these stars. After this monitoring we'll have a possibility to specify the model and compare it with the observations.

The analytical results may be also used for the description of some other problems such as outflow modeling from any rotating source. It might be disc winds or jets from binary systems or active galactic nucleus.

Acknowledgments

We thank the referee, D. P. Barsukov, for the useful comments that improved the paper.

References

[1] Potravnov I S, Grinin V P and Ilyin I V 2013 *Astrophysics* **56** 453–60
[2] Potravnov I S, Grinin V P, Ilyin I V and Shakovskoy D N 2014 *Astron. Astrophys.* **563** A139
[3] Kozlova O V, Grinin V P and Chuntonov G A 2003 *Astrophysics* **46** 265–81
[4] Grinin V P, Potravnov I S, Ilyin I V and Shulman S G 2015 *Astron. Lett.* (in press)
[5] Grinin V P, Potravnov I S and Musaev F A 2010 *Astron. Astrophys.* **524** A8
[6] Illarionov A F and Sunyaev R A 1975 *Astron. Astrophys.* **39** 185–95
[7] Lamb F K, Pethick C J and Pines D 1973 *Astrophys. J.* **184** 271-90
[8] Romanova M M, Ustuugova G V, Koldoba A V and Lovelace R V E 2009 *Mon. Not. R. Astron. Soc.* **399** 1802–28
[9] Shulman S G 2015 *Astrophysics* (in press)
[10] Subbotin M F 1968 *Introduction to the Theoretical Astronomy* (Moscow: Nauka) (in Russian)

Journal of Physics: Conference Series **661** (2015) 012013 doi:10.1088/1742-6596/661/1/012013

Measurements of the Cosmic Microwave Background Temperature at high redshift by analysis of CO excitation

A I Sobolev, A V Ivanchik, D A Varshalovich, S A Balashev

Department of Theoretical Astrophysics, Ioffe Physical Technical Institute, Saint Petersburg, Russia

E-mail: Asoboll@gmail.com

Abstract. The linear relation between the temperature of the Cosmic Microwave Background and the cosmological redshift is a fundamental prediction of the standard hot Big Bang model. Measurements of the CMB temperature in the past can be done by means of observing radiative excitation of atomic and molecular species like C and CO. It can be detected in quasar spectra with z $\sim$ 2. We study excitation of CO rotational levels both by the CMB radiation and collisions. Corrections of determination of the CMB temperature due to collisional excitation by the analysis of CO molecule rotation level populations are proposed.

1. Introduction

The Cosmic Microwave Background (CMB) is the thermal radiation left over from the early stages of the Universe. Its existence is a fundamental prediction of the hot Big Bang theory. In the standard cosmological model, the CMB undergo adiabatic expansion and the CMB black-body temperature has to be the linear function of redshift: $T_{\rm CMB}(z) = T^0_{\rm CMB} \cdot (1+z)$, where $T^0_{\rm CMB} = 2.7255 \pm 0.0006$K is the local CMB temperature [1]. However this relation can become nonlinear, e.g. $T_{\rm CMB}(z) = T^0_{\rm CMB} \cdot (1+z)^{1-\beta}$ [2], if local position invariance is violated [3] or if the number of photons is not conserved. However, for the latter, it was shown [4] that is extremely difficult to fulfill "adiabatic" conditions, without producing the spectral distortions, which are constrained by FIRAS/COBE at the level $\delta I/I \sim 10^{-4}$ [5].

In that sense it is necessary to directly measure the evolution of $T_{\rm CMB}$. At low redshifts ($z \lesssim 0.6$) there is a method to determine $T_{\rm CMB}$ by using Sunyaev-Zeldovich effect [6], that is seen as CMB spectral distortions towards the galaxy clusters.

At higher redshifts ($z \sim 2$) $T_{\rm CMB}$ can be measured by analysis of the excitation of different atoms or molecules, which is measured using absorption lines in high redshift quasar spectra. To date populations of fine structure levels of C I and rotational levels of CO are used.

In case of dominant excitation by the CMB radiation, the excitation temperature of this species is equal to the CMB temperature. The most stringent constraints on $T_{\rm CMB}(z)$ were obtained using CO rotational levels [3]. For the lowest vibrational level of the ground electronic state of CO the rotational transition energies between adjacent $J-1$ and J rotational levels are $E_J = 5.54 \cdot J$K. Such energies are close to the CMB temperature, thus the rotational states of CO molecules are subject to the strong radiative excitation by CMB. However the presence of CO molecules requires screening from UV radiation, since CO is easily photodissociated by UV photons. Therefore CO molecules are usually located in the center of a dense gas regions, which are refer to the translucent and dense molecular clouds. In such a dense medium the collisional excitation of CO molecules can be significant. In this work we examine the impact of the collisional excitation of CO on the CO excitation temperature that can be derived from the populations of the CO rotational levels.

1.1. $T_{\rm CMB}$ measurements based on CO molecules observations

Both absorption and emission lines of CO molecules are observed at high redshift. The study of CO absorption lines in quasar spectra allows one to separate a few components usually seen in CO profiles (which are shifted due to relative motions) and to measure accurately column densities of each rotational level for each component. Although CO molecule have a permanent dipole momentum and thereby have strong interaction with the radiation, the CO molecules were detected at high redshift in quasar spectra only recently in 2008-2011 [3, 8–10]. This is because of the complex CO chemistry: CO can be formed only in the medium that is well screened from the ambient UV radiation field, which destroys CO by photodestruction concerned with transitions in the resonance lines. The screening can be produced by the dust, H_2, and by the CO selfshielding that happens only in the deepest layer of interstellar clouds. At the same time such screening is preventing CO molecule from being detected since it absorb quasar spectrum as well. $T_{\rm CMB}$ measurements preformed by using analysis of CO absorption lines with high redshifts are summarized in [3]. These observations cover redshifts from z = 1.73 to z = 2.69 and were obtained neglecting the collisional excitation of CO molecules. Increase of the observation precision will require taking into account the collisional excitation.

2. Excitation of CO rotational levels by radiation and collisions

The radiative pumping of CO rotational levels via electronic transitions is negligible for the typical conditions of interstellar medium. Therefore the radiative excitation of CO rotational levels is dominated by CO pure rotational transitions. The main contribution of radiation at frequencies corresponding to these transitions (especially at high redshifts) is given by CMB radiation, which spectral energy distribution $u(\nu)$ is well established to be blackbody. Therefore probabilities of rotational transitions between adjacent $(i+1)^{\rm th}$ and $i^{\rm th}$ states depend only on $T_{\rm CMB}$ and can be described as:

$$W^{CMB}_{i,i+1} = B_{i,i+1}u(\nu_{i,i+1}); \qquad W^{CMB}_{i+1,i} = A_{i+1,i} + B_{i+1,i}u(\nu_{i,i+1}) \tag{1}$$

where $\nu_{i,i+1}$, $A_{i+1,i}$ and $B_{i+1,i}$ are the frequency and Einstein coefficients of the $i+1 \rightarrow i$ transition, respectively. For our calculations we used these data taken from HITRAN[1] database.

In case of a diffuse and translucent interstellar cloud it is sufficient to consider collisional excitation of rotational levels of CO molecules only by atomic hydrogen, molecular hydrogen and helium. To describe collisional transitions one needs to know its probabilities, number densities of all species that interact with CO molecules and the kinetic temperature of the gas, $T_{\rm Kin}$. Probabilities of collisional transitions are experimentally measured. For collisions of CO with atomic and molecular hydrogen we used data obtained by Yang et al. [11,12]. For collisions with helium we used data obtained by Cecchi-Pestellini et al. [13]. Since primordial helium to hydrogen ratio is well defined and can be taken equal to 0.083 [14], only three parameters determine collisional excitation of CO. It is convenient to take it as following: the total number density of particles, n, the hydrogen molecular fraction, $f = 2n({\rm H_2})/(2n({\rm H_2}) + n({\rm H}))$ (which specifies molecular hydrogen abundance), and the kinetic temperature of the gas, $T_{\rm Kin}$. One can note that parahydrogen and orthohydrogen contribute differently in collisional excitation of CO

[1] `http://hitran.iao.ru/`

molecules and should be treated as two species. However in the considered selfshielded regions of H_2 medium their ratio is defined by $T_{\rm Kin}$ and thereby the number of parameters remains the same.

The population of the CO molecule rotational levels can be expressed by number densities, N_i, of each CO level. In stationary condition one can write the detailed balance for $i^{\rm th}$ rotational level.

$$\sum_{i\neq j} W_{ji}N_j - \sum_{j\neq i} W_{ij}N_i = 0, \tag{2}$$

where W_{ij} is a probability of rotational transition from the state i to the state j that can be expressed as follows:

$$W_{ij}^{\rm tot}(T_{\rm CMB}, T_{\rm Kin}, n, f) = \left\{\frac{f}{2}\left[\alpha_{para}(T_{\rm Kin})W_{ij}^{H_{2para}}(T_{\rm Kin}) + (1-\alpha_{para}(T_{\rm Kin}))W_{ij}^{H_{2ortho}}(T_{\rm Kin})\right]\right.$$
$$\left. + (1-f)W_{ij}^{H}(T_{\rm Kin}) + 0.083\cdot W_{ij}^{He}(T_{\rm Kin})\right\}\frac{n}{1.083 - f/2} + W_{ij}^{CMB}(T_{\rm CMB}) \tag{3}$$

where α_{para} is a fraction of parahydrogen. $W_{ij}^{H_{2para}}$, $W_{ij}^{H_{2ortho}}$, W_{ij}^{H} and W_{ij}^{He} are the probabilities of transitions from i to j rotational level of CO by collisions with para-H_2, ortho-H_2, H and He, respectively. A stationary solution of eq. 2, $N^{\rm CO}$ (the vector describing the relative population of rotational levels of CO), is an eigenvector in the null space of the following matrix:

$$\tilde{W}_{ij} = W_{ji}^{\rm tot} - \delta_{ij}\sum_k W_{ik}^{\rm tot} \tag{4}$$

To summarize, $N^{\rm CO}$ is defined by radiative and collisional transitions, which are determined by medium parameters $T_{\rm CMB}$, $T_{\rm Kin}$, n, and f.

3. Results

The figure 1 shows the dependence of relative population of second and first excited rotational states compared to the population of the ground state of CO molecule on the total number density of gas. For this calculation we fixed $T_{\rm Kin} = 100\,{\rm K}$, which is the typical kinetic temperature, measured in H_2-bearing clouds at high redshifts, and $T_{\rm CMB} = 9.54\,{\rm K}$ (that corresponds to the typical redshift of absorption systems, $z \approx 2.5$). We present the result of calculation in two limits: all hydrogen in the molecular phase, $f = 1$, (shown by red lines) and all hydrogen in the atomic phase, $f = 0$ (shown by blue lines). It can be seen that collisional excitation of CO by molecular hydrogen is more significant than excitation by atomic hydrogen in astrophysical conditions [15, 16].

We have checked that our calculations satisfy the limiting cases: at low and high number density the relative population approaches the equilibrium distribution with the temperature equal to $T_{\rm CMB}$ and $T_{\rm kin}$ temperature, respectively.

$$N_J^{eq}(T) = (2J+1)exp(-\frac{E_J}{kT}) \tag{5}$$

Also we compare calculated relative populations of CO levels with the outputs from Radex [17] in case of collisions with molecular hydrogen only. In general, population distributions are consistent. However differences are growing with the level number. We measured average relative difference at the set of typical parameters: $T_{\rm CMB} = 9.54\,{\rm K}$, $T_{\rm Kin} = 100\,{\rm K}$, and $n = 200\,{\rm cm}^{-3}$. For the first three levels (J=0,1,2), the value is about $2\cdot 10^{-4}$. For the first 6 levels, the value

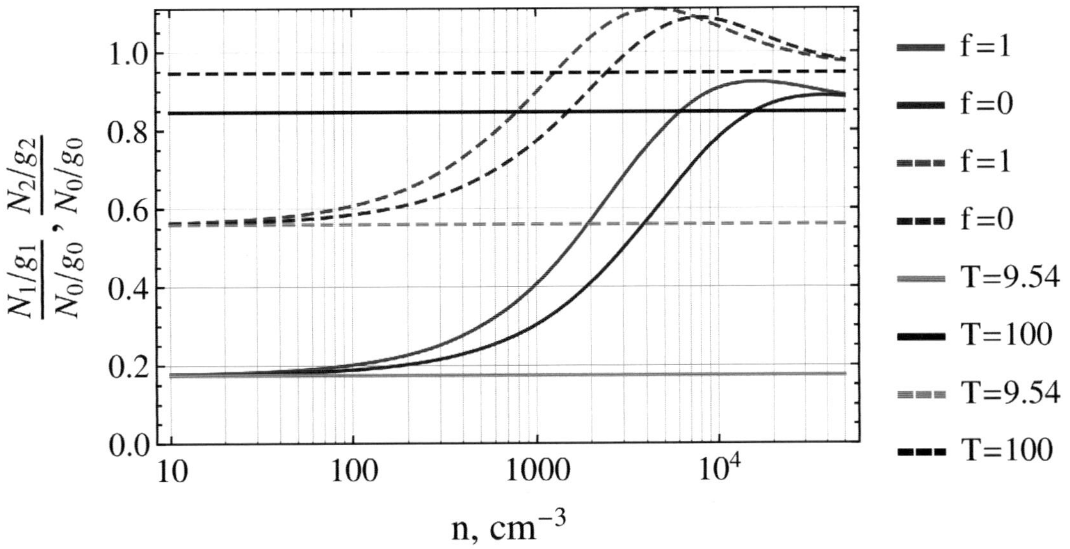

Figure 1. Dependence of CO levels relative populations on the total particle concentration. Red and blue lines show cases of molecular and atomic hydrogen, respectively. Dashed lines correspond to the first excited states, and solid lines correspond to the second excited states. Limiting values with equilibrium CMB temperature and kinetic temperature represented by green and black lines.

is about $2 \cdot 10^{-3}$. And for 16 levels, average relative difference is about 10^{-2}. This is probably due to different constants used in calculations.

It is worth mentioning that at high number densities $n \gtrsim 2000\,\mathrm{cm}^{-3}$ inverse population of CO rotational levels is obtained , i.e. $N_1/N_0 > g_1/g_0 = 3$. The inverse population of the first rotational level occurs since the transition from the second to the first excited state at such number densities is mainly due to radiative transition, while all other transitions are dominated by collisional transitions. Thus the population of the first excited state of CO molecules is increased relative to the equilibrium state with the temperature equal to the kinetic gas temperature. Such behavior of the CO excitation temperature well agree with calculations performed by Varshalovich & Khersonskii in 1978 [18] for the local T_{CMB}.

The figure 2 shows the excitation diagram of CO rotational levels for the set of typical parameters: $T_{\mathrm{CMB}} = 9.54\,\mathrm{K}$, $T_{\mathrm{Kin}} = 100\,\mathrm{K}$, and $n = 200\,\mathrm{cm}^{-3}$. Although, excitation temperatures differ for each transition, the determined excitation temperature describes rotational level populations relatively good. At higher levels the intensity of spontaneous emission drops below the intensity of collisional transitions. This leads to the change in the population distribution behavior starting from 6th excited rotational level.

At low number densities (up to few hundreds per cubic centimeter) our calculation demonstrates (see Fig. 1) that the excitation temperature differs little from the CMB temperature. It means that collisional excitation leads to the correction for measurement of T_{CMB} from the observed excitation temperature of CO. It is convenient to formulate this correction for T_{CMB} in an approximate way.

Figure 3 shows the dependence of estimated excitation temperature of CO molecules on the number density calculated at fixed $T_{\mathrm{Kin}} = 100\,\mathrm{K}$ and $T_{\mathrm{CMB}} = 9.54\,\mathrm{K}$. We have determined the excitation temperature of CO as averaged excitation temperature of the first two rotational

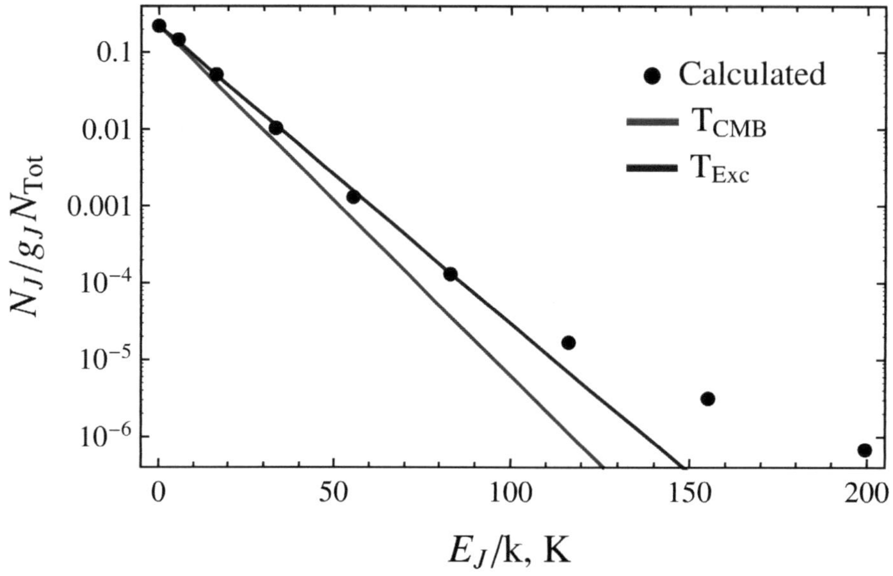

Figure 2. Excitation diagram of the first 9 rotational levels of CO for fixed $T_{CMB} = 9.54\,K$, $T_{Kin} = 100\,K$, and $n = 200\,cm^{-3}$. Blue line corresponds to the determined excitation temperature and red line corresponds to the CMB temperature.

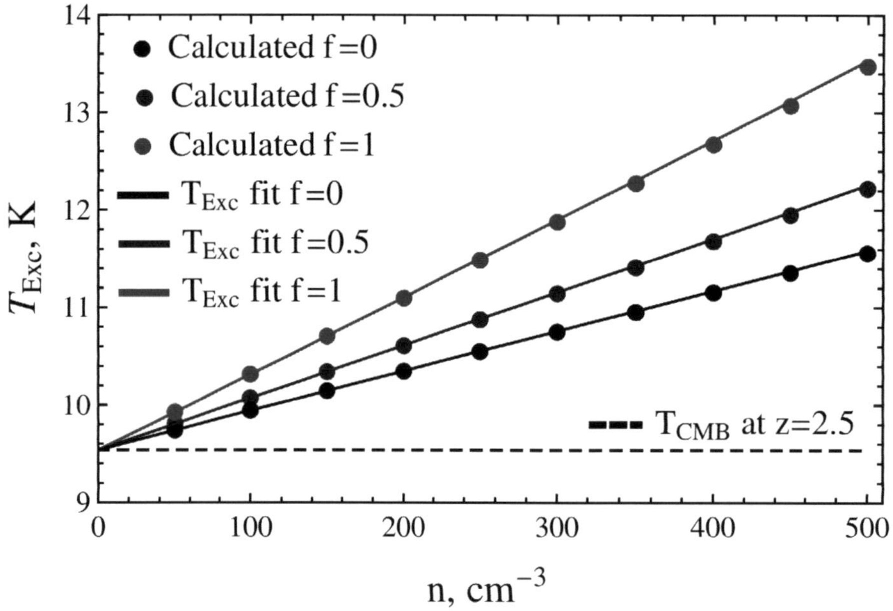

Figure 3. Dependence of the excitation temperature of CO molecules on total number density of the medium, for fixed $T_{Kin} = 100\,K$ and $T_{CMB} = 9.54\,K$.

levels. To estimate the excitation temperature from modeled populations we have used unweighted χ^2 criteria. As can be seen T_{Exc} dependence on the particle number density for different molecular fractions of hydrogen is very close to linear one. Usually in observations

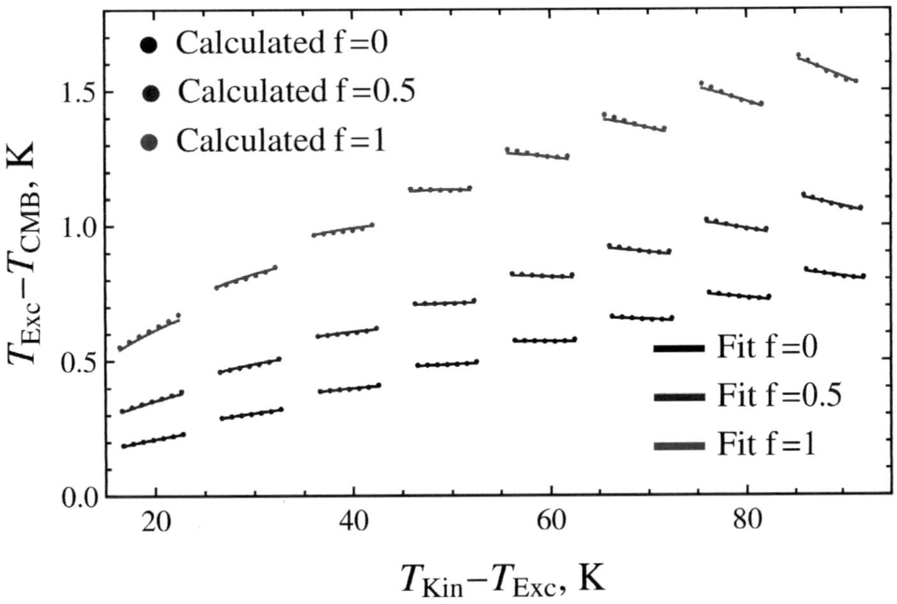

Figure 4. Dependence of the estimated correction for the CMB temperature on the kinetic gas temperature for different hydrogen molecular fractions. Figure presents approximation (lines) and calculated values (dots).

populations of the exited states of CO is obtained and $T_{\rm Exc}$ is derived. Thus correction for $T_{\rm CMB}$ taking into account collisional excitation of CO molecules can be expressed as linear function of the number density:

$$T_{\rm CMB} = T_{\rm Exc} - n \cdot \Theta(f, T_{\rm Kin}, T_{\rm Exc}) \tag{6}$$

We found that in first order it depends only on the difference between the kinetic and the CMB temperatures. We proposed a fit function as following:

$$\Theta(f, T_{\rm Kin}, T_{\rm Exc}) = C_1(f)(1-\exp(-C_2(f)(T_{\rm Kin}-T_{\rm Exc})))+C_3(f)(1+C_4(f)T_{\rm Exc})(T_{\rm Kin}-T_{\rm Exc}) \tag{7}$$

The approximation constants were obtained by comparison of the considered approximation with the calculated $T_{\rm Exc}$ on the grid of $T_{\rm CMB}$ and $T_{\rm Kin}$ parameters: $T_{\rm CMB}$ was varied from 7 K to 13 K with 1 K step, $T_{\rm Kin}$ was varied from 30 K to 100 K with 10 K step, and n from 0 cm^{-3} to 500 cm^{-3} with 50 cm^{-3} step. The figure 4 shows the approximations for the excitation temperature. The approximation constants for different molecular fractions are presented in the table 1.

$$T_{\rm CMB} = T_{\rm Exc} - n \cdot \{C_1(f)(1-\exp(-C_2(f)(T_{\rm Kin}-T_{\rm Exc}))) + C_3(f)(1+C_4(f)T_{\rm Exc})(T_{\rm Kin}-T_{\rm Exc})\} \tag{8}$$

Eq. (8) gives the estimated CMB temperature taking into account collisional excitation of CO molecules for the range of the total number densities.

Table 1. The approximation constants for Θ (7) as a function of the molecular fraction

f	C_1, K · cm^3	C_2, K^{-1}	C_3, cm^3	C_4, K^{-1}	rms error	max error
0	$1.241 \cdot 10^{-3}$	$1.784 \cdot 10^{-2}$	$2.693 \cdot 10^{-5}$	$2.633 \cdot 10^{-2}$	$7.4 \cdot 10^{-3}$	$2.9 \cdot 10^{-2}$
0.5	$2.186 \cdot 10^{-3}$	$3.195 \cdot 10^{-2}$	$2.734 \cdot 10^{-5}$	$3.363 \cdot 10^{-2}$	$9.4 \cdot 10^{-3}$	$3.4 \cdot 10^{-2}$
1	$4.581 \cdot 10^{-3}$	$3.542 \cdot 10^{-2}$	$2.404 \cdot 10^{-5}$	$5.499 \cdot 10^{-2}$	$1.2 \cdot 10^{-2}$	$4.8 \cdot 10^{-2}$

4. Conclusions

We have calculated the population distribution of the CO molecule rotational states taking into account radiative and collisional excitations. The impact of collisional excitation by H, H_2, He on the level populations of CO molecules has been analyzed. We have confirmed that collisional excitation by atomic hydrogen is less significant than by molecular hydrogen in typical astrophysical conditions. Also it was shown that the appearance of an inverse population in CO molecules by a combination of collisional and radiative excitation is possible. Correction due to collisional excitation for determination of the Cosmic Microwave Background temperature at high redshifts by observations of populations of CO rotational levels has been calculated. We have derived approximation for this correction and its dependence on physical conditions in the interstellar medium (number density, kinetic temperature and molecular fraction).

References

[1] Fixsen D J 2009 *APJ* **707** 916-20
[2] Lima J A S, Silva A I and Viegas S M 2000 *MNRAS*, **312** 747-52
[3] Noterdaeme P, Petitjean P, Srianand R, Ledoux C and Lopez S 2011 *A&A* **526** L7
[4] Chluba J 2014 *MNRAS*, **443**, 1881
[5] Mather J C et al. 1994 *APJ* **420**, 439-44
[6] Fabbri R, Melchiorri F and Natale V 1978 *AP&SS* **59** 223-36
[7] Rawle T D, Egami E, Bussmann R S, Gurwell M, Ivison R J, Boone F Combes F, et al., 2014, *APJ*, 783, 59
[8] Srianand R, Noterdaeme P, Ledoux C and Petitjea P 2008 *A&A* **482** L39-42
[9] Noterdaeme P, Ledoux C, Srianand R, Petitjean P and Lopez 2009 *A&A* **503** 765-70
[10] Noterdaeme P, Petitjean P, Ledoux C, Lopez S, Srianand R and Vergani S D 2010 *A&A* **523** A80
[11] Yang B, Stancil P C, Balakrishnan N and Forrey R C 2010 *APJ* **718** 1062-9
[12] Yang B, Stancil P C, Balakrishnan N, Forrey R C and Bowman J M 2013 *APJ* **771** 49
[13] Cecchi-Pestellini C, Bodo E, Balakrishnan N and Dalgarno A 2002 *APJ* **571** 1015-20
[14] Planck collaboration 2014 *A&A* **571** A16
[15] Green S and Thaddeus P 1976 *APJ* **205** 766-85
[16] Shepler B C, Yang B H, Dhilip Kumar T J, Stancil P C, Bowman J M, Balakrishnan N, Zhang P, Bodo E and Dalgarno A 2007 *A&A* **475** L15-8
[17] Van der Tak F F S, Black J H, Schoier F L, Jansen D J and van Dishoeck E F 2007 *A&A* **468** 627-35
[18] Varshalovich D A and Khersonskii V K 1978 *SvA*, **22** 192-4

Selenocentric reference coordinates net in the dynamic system

N Y Varaksina [1,2], Y A Nefedyev[1,2], K O Churkin[2], R R Zabbarova[2] and S A Demin[1,2]

[1] Kazan Federal University, Institute of Physics, Kazan, 420008 Russia
[2] Kazan Federal University, Engelhardt Astronomical observatory, Zelenodolski region, st. Observatory, 422526 Russia

E-mail: vnu_357@mail.ru

Abstract. In this work, the selenocentric dynamic reference net was developed for the first time in the field of selendesy in order to address problems with space navigation. Three tasks were addressed in this research: a) the analysis of the mathematical model of the orthogonal coordinate transformation accuracy; b) the identification of the basic dynamic reference system objects with ones that are contained in reducing catalogues; and c) the extension of the base points net of the basic dynamic reference system. The result was a dynamic coordinate system summary that contains 1 162 objects. The correlation analysis of this net was carried out and was found to coincide with modern dynamic coordinate systems that have been obtained. This selenocentric reference catalogue covers the full visible area of the Moon.

1. Introduction

Currently, Moon is a subject of research in many space experiments in the fields of astronomy and planetary science. The modern space missions “Clementine” [1], “Lunar Reconnaissance Orbiter” (LRO) [2] and “KAGUYA (SELENE)” [3] swiftly and radically changed attempts at the study of the moon [4]. Today, there are prospects in the industrial and robotic exploration of the moon for creating long-term manned lunar bases. Development of these space technologies imposes special requirements on the results of coordinate-time support. These requirements include the development of reference systems, establishing mutual orientation of dynamic and inertial coordinate systems and the investigation of the dynamics and geometry of the celestial bodies. This fully applies to the dynamic and geometric parameters of the moon that refer to its centre of mass. However, despite the outstanding results in this direction derived from observations of space missions “Apollo”, “Zond”, “LRO” and “Kaguya” the task of creating a global selenocentric network has still not been completely solved.

2. Formulation of the problem

Modern reference selenodetic networks are not equally accurate for the different coordinate axes, and they may even have an ellipsoidal distribution of errors. It should be noted that there are three main aspects to the problem [4].

First, Doyle et al. [5] carried out a detailed analysis of the internal precision of the Apollo system. Based on this analysis, we can formulate the following conclusions. To transform the topographical origin of Apollo three ALSEP stations were used. Since the mean square errors of the transformation were less than m and a measurement error was about m, we can assume that the points near and between the three ALSEP stations have errors of positions less than m. The offset from the location of the ALSEP increases the error up to ±400 m, and this is a big part of the observed area. The errors of the point positions lying near the borders of the studied areas can reach ±400 m and even exceed ±1000 m. When the mission LRO have been received, very accurate altitude data but horizontal coordinates of the lunar objects have the dubious binding to the dynamical coordinate system too. Despite the fact that the spacecraft can scan the lunar surface with accuracy down to a centimeter (for example LRO, Kaguya, Clementine etc.), we obtain a very precise physical lunar surface but inaccurate dimensioning of the dynamic coordinate system.

The accurate satellite binding of lunar objects can only be carried out using one of two methods:

1) simultaneous scanning of the lunar surface with the exact definition of the coordinates of the laser impingement point and the binding of the orbiter to the stars, or
2) installation of lighthouses on the entire lunar surface. At the present time, the technical capabilities for both of these options are unavailable.

Second, at the present time, all data for the lunar selenography can be divided into two types. On one hand, the first data type obtained by the lunar surface laser scanning from onboard satellites describes the lunar relief well, but it does not give the coordinates for the reference objects on the moon. Other data types obtained on the basis of directly binding lunar objects to the stars gives the exact coordinates of reference objects, but they do not describe the lunar relief with sufficient accuracy. All of these systems have a different frame of reference and orientation of the axes. It is well known that data obtained from all of the space missions relate to a quasi-dynamic frame of reference in which the centre of origin is the lunar centre mass (LCM), but their axes do not coincide with the axes of inertia on the moon.

Third, most modern selenodetic catalogues have a quasi-dynamic system of coordinates. Either they have a centre of origin of coordinates that does not coincide with the LCM or the axes do not coincide with the axes of inertia on the moon. At the present time, it should also be noted that there is no reference dynamic selenocentric coordinate system covering enough area on the lunar surface based on satellite observations. In addition, despite the accuracy of the lunar physical relief obtained by space missions, the reference surface of this relief has a rather vague figure. Therefore in most articles describing satellite topographic data, much attention is paid to the high accuracy of the obtained physical relief and the question of the surface reference is glossed over. So far, there is no method for constructing the lunar photogrammetric topographic satellite map based on the integration of the thousands of separate pictures of the lunar surface to a single system on an absolute basis. Thus, we cannot say that satellite topographic data allows for the production of full-value topographic models with a complete reference surface.

3. The method of the making of the dynamical selenocentric reference coordinate system DSC

The dynamical selenocentric catalogue (DSC) on the surface of the Moon was based on large-scaled Moon and stars pictures taken by new unique method of separate plates. In contrast to the methods of taking pictures of the Moon without stars, in this method we have an absolute orientation definition, 0-point of coordinate system and its scale. Choosing the Moon craters included to the bearing net DSC we used following criterions. At first, we chose round shaped craters. Secondly, these craters have not very big size. In third, chosen craters must be well-observed. In the fourth, the net craters must be in others selenodesic catalogue objects lists and satisfy IAU recommendations. Required parameters are got from 2 amendment equation of this type:

$$\mathbf{A}\times\boldsymbol{\theta}+\boldsymbol{\varepsilon}=\mathbf{Z}, \tag{1}$$

$\mathbf{A}(A_{ij})$ – structured transition matrix, $\boldsymbol{\theta}(\Delta\xi,\Delta\eta,\Delta\zeta)$ – column vector of required parameters, $\boldsymbol{\varepsilon}$ – column vector of random observational errors, $\mathbf{Z}$ – column vector of observations.

The determination of the required parameters $\hat{\boldsymbol{\theta}}(\Delta\hat{\xi},\Delta\hat{\eta},\Delta\hat{\zeta})$ is:

$$\hat{\boldsymbol{\theta}}=(\mathbf{A}^T\mathbf{P}\mathbf{A})^{-1}(\mathbf{A}^T\mathbf{P}\mathbf{Z}), \tag{2}$$

and their errors are determined by the covariance matrix

$$\mathbf{D}(\hat{\boldsymbol{\theta}})=\frac{\mathbf{V}^T\mathbf{P}\mathbf{V}}{2m-3}(\mathbf{A}^T\mathbf{P}\mathbf{A})^{-1}, \tag{3}$$

$\mathbf{V}$ – vector of residual deviations.

The analysis of the DSC catalogue showed it fully satisfy following criterions: includes a sufficient number of control points for discovering the Moon figure and implementation of the exact binding to them; contains objects with coordinates referred to the Moon's ephemeris center of mass and covering quite big part of the Moon surface; presented points' accuracy reaches ± 35 meters in planning coordinates and in average ± 60 meters in heights.

The target of the research is to improve the accuracy of the condensation and expansion selenocentric control system based on optimal coordinate transformations. Elements of the matrix and the displacement vector can be obtained from common points for DSC and converted into a system in the particular directory.

The main problem with extending the DSC (system $\mathbf{X}$) is a precise determination of the orientation matrix elements and displacement vector of the origin of coordinates during the transition from the $\mathbf{Y}$ coordinate system to another system by common points:

$$\mathbf{X}=\mathbf{A}\mathbf{Y}+\mathbf{X}_0 . \qquad (4)$$

$\mathbf{A}$ – the transition matrix, $\mathbf{X}_0$ – the origin of the coordinates displacement vector of the $\mathbf{Y}$ to the zero point of the $\mathbf{X}$ coordinate system. The exact solution of the problem becomes more important in the coordinates extrapolation. In our case, it is especially important as the objects of the reverse side of the moon are outside the set of control points.

Achieving the highest possible accuracy of transformation coordinates (TC) is provided by the following tasks:

a) Justifying the applicability of the adaptive regression modelling (ARM-approach), which is provided for the formation of the model (4);

– Assessing the quality of the model TC;

– Diagnosing compliance with the terms of the computational scheme MNE (in particular, Gauss-Markov conditions);

– A numerical adjustment to a material breach of the condition;

b) Methods of the ARM-approach application development, which includes:

– Criteria for the accuracy of the transition;

– A set of competing mathematical models of the TC;

– The corresponding set of methods for structural and parametric identification;

– The scenario data, providing an assessment of the prognostic properties of the model diagnostics of the conditions and adaptation in violation to provide the required optimal estimates of the properties (consistency, unbiasedness and efficiency).

In the ARM-approach [6], it is postulated that the matrix model structure of TC (5) is unknown for each pair of catalogues. In general form, for example, the first equation (4) can be rewritten in the matrix of the regression equation form:

$$\mathbf{Y}=\mathbf{X}\boldsymbol{\beta}+\boldsymbol{\varepsilon}, \qquad (5)$$

adding the error term ε and considering that the first row of the matrix is the $\boldsymbol{\beta}$ vector. Obviously, in structural identification, the parameters of the equation are simultaneously estimated, namely the vector $\boldsymbol{\beta}$ elements for the simple case (5).

Because of the errors in the coordinates determination in both systems and the possible multicollinearity (interdependence) of the estimates, the matrix $\mathbf{A}$ often does not satisfy the orthogonality conditions of the transition from $\mathbf{Y}$ to $\mathbf{X}$, written as:

$$\mathbf{ATA}=\mathbf{E}, \det \mathbf{A}=1 \qquad (6)$$

In this regard, the basic deterministic transformation should be considered to be the equation (4), considered together with the conditions (6). This task, with accuracy up to the centres of the $\mathbf{Y}$ and $\mathbf{X}$ systems' divergence and the scale factor problem, is solved by a numerical method of optimization.

At this stage, as a model of approximation, the transformation was applied to model (4) without the condition (6). Later, in the concentration of the fundamental DSC, algebraic polynomials of the second and third degree, as well as the two-component description where the first component is a deterministic model (4) and (6), and the second includes marked polynomials to describe the residue after the first component, will be used. The optimal model will be derived from the ARM approach.

4. Summary and conclusions

For the moon, there are several dynamical coordinate systems, and among them, the most informative is the Dynamical Selenocentric Catalogue (DSC), which was built at the Engelhardt Astronomical Observatory (EAO). Plane coordinates in the catalogue were obtained by direct binding of lunar objects to the stars, as well as altitude data that was adjusted based on the data LRO mission. Since its creation, lunar objects that are directly tied to the stars can be considered DSC-made in a dynamic coordinate system.

The analysis of the DSC catalogue showed that it fully satisfied the following criteria: the inclusion of a sufficient number of control points for discovering the moon's figure, the implementation of the exact binding to those points, containing objects with coordinates that referred to the moon's ephemeris centre of mass and covering a large part of the moon's surface, and presented points that have an accuracy that reaches metres in planning coordinates and on average meters in heights. In this work, direct binding to DSC was used to create the union lunar dynamic network on the lunar surface, covering the best lunar sphere and containing sufficient reference points for navigation tasks.

Acknowledgments

This work was funded by the subsidy allocated to Kazan Federal University for the state assignment in the sphere of scientific activities and within the of a research projects RFBR N 13-02-00792-a.

References

[1] Smith D E, Zuber M T, Neumann G A and Lemoine F G 1997 *J. Geophys. Res.* **102** 1591

[2] Kirk R L, Archinal B A, Gaddis L R and Rosiek M R 2008 Cartography for lunar exploration: 2008 status and mission plans *The International Archives of Photogrammetry, Remote Sensing and Spatial Information Sciences* (Beijing) pp. 1473–1490

[3] Araki H, Tazawa S, Noda H, Ishihara Y, Goossens S, Sasaki S, Kawano N, Kamiya I, Otake H, Oberst J and Shum C 2009 *Science* **323** 897

[4] Nefedyev Y, Valeev S, Mikeev R, Varaksina N and Andreev A 2012 *Advanced in Space Research* **50** 1564

[5] Doyle F J, Elassal A A and Lucas J R 1977 Selenocentric geodetic reference systems *NOAA Tech. Rep. NOS 70, NGS 5* (Washington, D.C.: Depart. of Commerce)

[6] Valeev S G 1986 *Earth, Moon and Planets* **35** 1

Lorentzian' analysis of the accuracy of modern catalogues of stellar positions

N Y Varaksina[1,2], Y A Nefedyev[1,2], K O Churkin[2], R R Zabbarova[2] and S A Demin[1,2]
[1] Kazan Federal University, Institute of Physics, Kazan, 420008 Russia
[2] Kazan Federal University, Engelhardt Astronomical observatory, Zelenodolski region, st. Observatory, 422526 Russia

E-mail: vnu_357@mail.ru

Abstract. There is a new approach for the estimation of the position accuracy and proper motions of the stars in astrometric catalogues by comparison of the stars' positions in the researched and Hipparcos catalogues in different periods, but under a standard equinox. To verify this method was carried out the analysis of the star positions and proper motions UCAC2, PPM, ACRS, Tycho-2, ACT, TRC, FON and Tycho catalogues. As a result of this study was obtained that the accuracy of positions and proper motions of the stars in Tycho-2 and UCAC2 catalogues are approximately equal. The results of the comparison are represented graphically.

1. Introduction

The Hipparcos catalogue provides highly accurate data on the positions (about 1 mas) and proper motions (about 1 mas/yr) of the stars [1]. According to the IAU resolution, this catalogue is accepted as the astrometrical standard. Therefore Hipparcos is used for comparison to estimate the accuracy of other astrometrical catalogues.

This work offers a new graphic method for estimation of the position accuracy and proper motions of the stars in astrometrical catalogues. This method is based on comparison of the stars' positions in different periods and *J*.2000.0 equinox with their positions as given in the Hipparcos catalogue in the same periods and *J*.2000.0 equinox. The number n of the absolute value of coordinates' differences is more than the given value Δ concerning the general number N of stars, is determined for each period in the researched and Hipparcos catalogues and as a percentage. The sum of the absolute values of coordinates' differences is taken as 100%, which is equal to the general number of stars compared in both catalogues. The years for which the differences of coordinates are calculated are put on the abscissa axis and the percentages are put on the y-axis.

The use of a new method can be justified by the expediency of comprehensive research of the catalogues' accuracy and by the use of a visual representation of the results obtained by the given method. The method is illustrated by examples of research on the PPM [2, 3], ACRS [4], Tycho-2 [5], ACT [6], TRC [7], FON [8], Tycho [1], UCAC2 [9] catalogues.

2. Graphic method of comparison of positions' accuracy and proper motions of stars in astrometrical catalogues

The epoch of the modern PPM and Tycho-2 catalogues is $T(0)=J.2000.0$ and for the Hipparcos and Tycho catalogues is $T(0)=J.1991.25$. Obviously, for an epoch $T(j)$ the coordinates of stars of the Hipparcos catalogue will be

$$\alpha\{H\}=[\alpha\{1991.25\}+\mu\alpha\times\{T(j)-1991.25\}],\delta\{H\}=[\delta\{1991.25\}]+\mu\delta\times\{T(j)-1991.25\}, \qquad (1)$$

and for the researched catalogue

$$\alpha\{cat\}=[\alpha\{T(0)\}]+\mu\alpha\times\{T(j)-T(0)\}],\delta\{H\}=[\delta\{1991.25\}+\mu\delta\times(T(j)-1991.25)]. \qquad (2)$$

Consider the set of common stars in the researched and Hipparcos catalogues $i=1,2,3....N$. Form the differences of coordinates for all stars for an epoch $T(j)$.

$$\Delta(\alpha)\times\cos(\delta)=\{\alpha(cat)-\alpha(H)\}\times\cos(\delta),\ \Delta(\delta)=\{\delta(cat)-\delta(H)\}. \quad (3)$$

Take the absolute values of these differences and sum up both α and δ:

$$\Delta i=|\Delta(\alpha)i\times\cos(\delta)|+|\Delta(\delta)i|. \quad (4)$$

At the same time, reject those differences whose absolute values are less than a given value, for example $\Delta=0.2''$. This value is selected taking into account the positions accuracy of stars in the ground catalogues. Consider only those differences which are more than Δ. Let the number of such differences equal *n*. Divide n by N and then centuple, i.e., to evaluate the number of differences of large Δ in percent. Denote this as $L(j)$. Perform such calculations for a number of periods (usually more than ten). Further, take $L(j)$ as a function from $T(j)$. Construct the plot of dependence of $L(j)$ on $T(j)$. Curves $L(j)$ are well described by the Lorential function. This is used in astrophysics for the approximation of photometric structures of spectral lines. It has the analytical expression

$$L(x)=\Delta(x)/[(x-x_0)2+\{\Delta(x/2)\}2], \quad (5)$$

where x_0 is the abscissa of the minimum of a Lorential, and $\Delta(x)$ is an average width of a structure Lorential, similar to the equivalent width of the photometric profile of a spectral line.

3. Analysis of modern catalogues

In figure 1, the schedules of dependence of $L(j)$ on $T(j)$ for the PPM catalogue are indicated at $\Delta=0.2''$ separately for the northern zone on the declination PPM(*n*) and the southern zone PPM(*s*). It is known that the observations of the northern and southern stars PPM were made in different years.

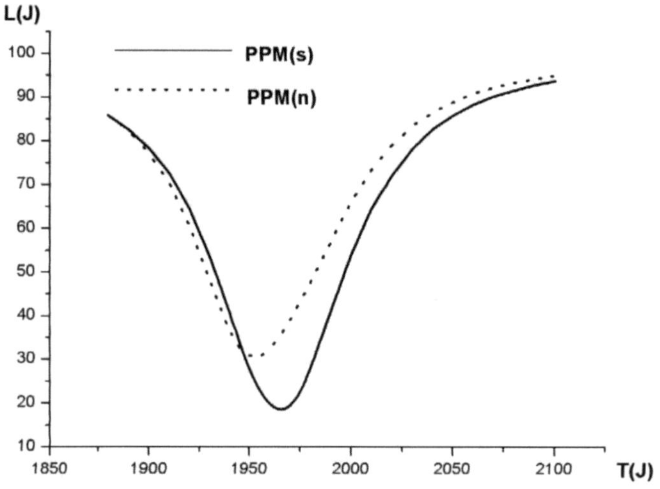

Figure 1. The dependencies of $L(j)$ on $T(j)$ for the PPM catalogue separately for the northern zone on the declination PPM(*n*) and the southern zone PPM(*s*).

This is clearly illustrated in figure 1. Thus, the abscissa of a minimum curve $L(j)$ gives information about the epoch of the observations of the researched catalogues. It is obvious that the ordinate of minimum $L(j)$ gives information about the accuracy of the positions of stars in the researched catalogue. The lower the ordinate at a given Δ, the more precise the stars' positions are in the catalogue. PPM(n) has a minimum at 30.6%, and PPM(s) at 18.6%. So the stars' positions in PPM(s) are more exact than in PPM(n).

Figure 2 presents curves $L(j)$ for the Tycho-2, ACT, TRC, FON and Tycho catalogues at $\Delta=0.2''$. It is evident from a comparison of curves for the Tycho and Tycho-2 catalogues that Tycho-2 (0.4%) has more exact star positions than Tycho (48.8%). The accuracy of the ACT and TRC catalogues is almost identical at 1.48% and 1.87%. At 31.7%, the accuracy of the FON catalogue is worse than ACT and TRC, but greater than the Tycho catalogues. The steepness of curve $L(j)$ gives information about the accuracy of the stars' proper motions in the catalogues researched. The less steep the curve, the more exact the proper motions in the given catalogue. This is for simple reasons. In figure 1 it is visible that the stars of the Tycho-2 catalogue have the most exact proper motions. The ACT and TRC catalogues have approximately identical proper motions. The FON catalogue gives proper motions which are close to the proper motions of the stars in the Tycho-2 catalogue. The Tycho catalogue has the least accurate proper motions.

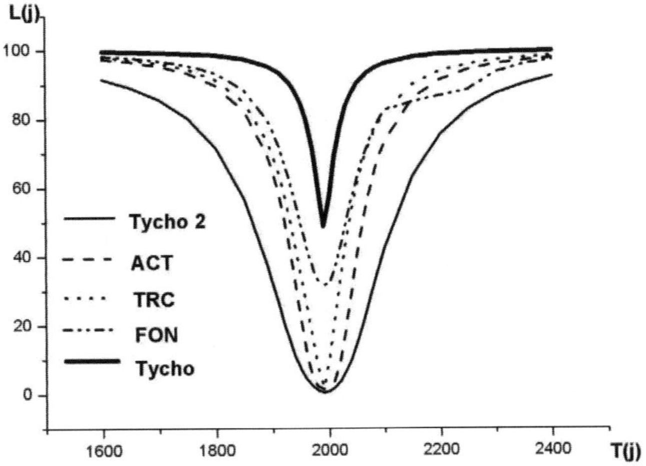

Figure 2. The curves $L(j)$ for the Tycho-2, ACT, TRC, FON and Tycho catalogues.

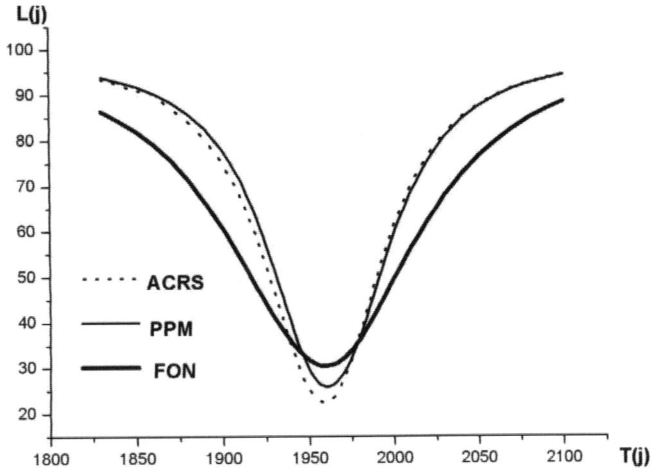

Figure 3. The curves $L(j)$ for PPM, ACRS and FON catalogues.

In figure 3, the PPM, ACRS and FON catalogues are compared. The positions of stars in the ACRS catalogue are a little more exact than in PPM, and are the least exact in the FON catalogue: 22.4%, 26.1% and 30.5% respectively. The stars of the FON catalogue have the most exact proper motions. ACRS and PPM have proper motions which are close on accuracy. Evaluations of the accuracy of the ACRS and PPM catalogues correspond to the results of research on their accuracy conducted at the Pulkovo Observatory [10].

In figure 4, the curves $L(j)$ are indicated for the UCAC2 and Tycho-2 catalogues at the value Δ=0.2". The accuracy of the positions and proper motions of the stars in these catalogues are approximately equal.

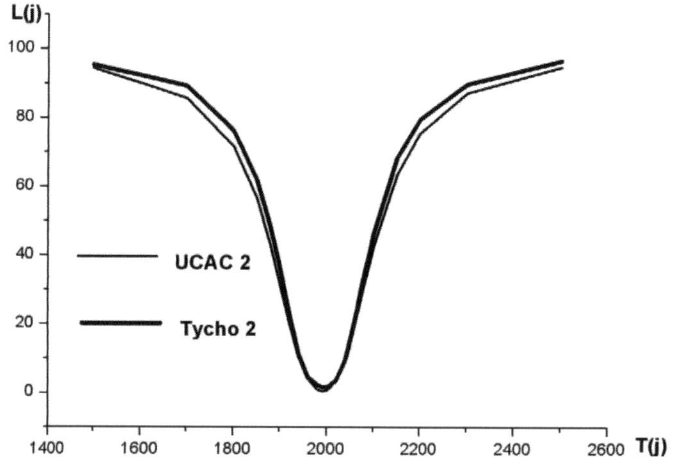

Figure 4. The curves $L(j)$ for the UCAC2 and Tycho-2 catalogues.

4. Summary and conclusions

This method of evaluating the accuracy of the catalogues permits us to determine external global errors by a graphic method, including random and regional errors. The term "external" means those errors received by way of comparison with the Hipparcos catalogue. That is, these errors are received accurate within the Hipparcos catalogue accuracy. Common external errors of positions in the UCAC2 catalogue are on α and δ: 20 mas and 19 mas, while errors of proper motions on α and δ make 2 mas/yr and 2 mas/yr.

As evaluated by the authors, errors of observations for positions in the UCAC2 catalogue for stars with brightness 10÷14 m are about 20 mas. For proper motions, errors of observation for stars which are brighter than 12.5 m are 4÷6 mas/yr, while for stars which are weaker than 12.5 m these errors are about 1÷3 mas/yr.

This problem is discussed in more detail in the following work [9, 11].

Acknowledgments

This work was funded by the subsidy allocated to Kazan Federal University for the state assignment in the sphere of scientific activities and within the of a research projects RFBR N 12-02-97000 -a and RFBR N 13-02-00792-a.

References

[1] European Space Agency 1997 *The Hipparcos and TYCO Catalogues (ESA SP-1200)* (Noordwijk: ESA Publications Division)

[2] Bastian U, Röser S, Yagudin L and Nesterov V 1993 *PPM Star Catalogue: Positions and Proper Motions of 197 179 Stars South of -2.5 Degrees Declination* (Heidelberg; Berlin; Oxford: Spektrum Acad. Verl.)

[3] Röser S and Bastian U 1991 *PPM Star Catalogue, Positions and Proper Motion of 181 731 Stars North of -2.5 Degrees Declination* (Heidelberg; Berlin; Oxford: Spektrum Acad. Verl.)

[4] Corbin T E and Urban S E 1991 *Astrographic Catalogue Reference Stars* (Washington: U.S. Naval Observatory)

[5] Høg E, Fabricius C, Makarov V V, Urban S, Corbin T, Wycoff G, Bastian U, Schwekendiek P and Wicenec A 2000 *A&A* **355** L27

[6] Urban S E, Corbin T E and Wycoff G L 1998 *Astron. J.* **115** 2161

[7] Høg E, Kuzmin A, Bastian U, Fabricius C, Kuimov K, Lindegren L, Makarov V V and S. Röser 1998 *A&A* **335** L65

[8] Kislyuk V, Yatsenko A, Ivanov G, Pakuliak L and Sergeeva T 2000 The FON astrographic catalogue (FONAC): version 1.0 *Motion of Celestial Bodies, Astrometry and Astronomical Reference Frames: Proc. of Journees 1999 & IX Lohrman-Kolloqium* (Drezden: Lohrman Observatory) pp 13–15

[9] Zacharias N, Urban S E, Zacharias M I, Wycoff G L, Hall D M, Monet D G and Rafferty T J 2004 *Astron. J* **127** 3043

[10] Evdokimov A E, Pickin Y D and Potter H I 1995 *Comparison of union ACRS, PPM, HIP catalogues* (Saint Petersburg: GAO RAS)

[11] Zacharias N, Zacharias M I, Urban S E and Høg E 2000 *Astron. J.* **120** 1148

Journal of Physics: Conference Series **661** (2015) 012016 doi:10.1088/1742-6596/661/1/012016

The variation of the baryon-to-photon ratio during different cosmological epochs due to decay and annihilation of dark matter

E O Zavarygin[1,2] and A V Ivanchik[1,2]

[1] Ioffe Institute, Polytekhnicheskaya 26, St.Petersburg, 194021, Russia

[2] St.Petersburg State Polytechnical University, Polytekhnicheskaya 29, St.Petersburg, Russia

E-mail: e.zavarygin@gmail.com, iav@astro.ioffe.ru

Abstract. An influence of annihilation and decay of the dark matter particles on the baryon-to-photon ratio has been studied for different cosmological epochs. We consider the different parameter values of the dark matter particles such as mass, annihilation cross section, lifetime and so on. The obtained results are compared with the data which come from the Big Bang nucleosynthesis calculation and from the analysis of the anisotropy of the cosmic microwave background radiation. It has been shown that the modern value of the dark matter density $\Omega_{\rm CDM} = 0.26$ is enough to provide the variation of the baryon-to-photon ratio up to $\Delta\eta/\eta \sim 0.01 \div 1$ for decay of the dark matter particles, but it also leads to an excess of the diffuse gamma ray background. We use the observational data on the diffuse gamma ray background in order to determine our constraints on the model of the dark matter particle decay and on the corresponding variation of the baryon-to-photon ratio: $\Delta\eta/\eta \lesssim 10^{-5}$. It has been shown that the variation of the baryon-to-photon ratio caused by the annihilation of the dark matter particles is negligible during the cosmological epochs from Big Bang nucleosynthesis to the present epoch.

1. Introduction

Cosmology is an accurate science today. There are a lot of cosmological parameters which are determined with a high precision. One of them is a baryon-to-photon ratio $\eta \equiv n_{\rm b}/n_\gamma$, where $n_{\rm b}$ and n_γ are the numbers of the baryons and the photons in the Universe respectively. In the standard cosmological model it is considered that the modern value of η was formed a few second after the Big Bang and has not been changed till the present epoch.

Today there are several independent experiments to determine value of η for four cosmological epochs (z is the cosmological redshift of the corresponding epoch):

(i) Big Bang nucleosynthesis (BBN, $z \sim 10^9$, e.g., [1]),

(ii) primordial recombination (CMB, $z \simeq 1100$, e.g., [2]),

(iii) the period of time associated with the Lyα forest ($z \sim 2 \div 3$, e.g., [3]),

(iv) the present epoch ($z = 0$, e.g., [4]).

The analysis of the observation data for two first epochs gives the high accuracy of η parameter determination.

For the processes taking place at the BBN and the CMB epochs η is one of the most important parameters which determines physics of these processes. For these epochs the methods for determination of η are:

(i) a comparison of the observational data on the abundance of primordial chemical elements (such as D, ^{4}He, ^{7}Li) with predictions of the BBN theory,

(ii) an analysis of the anisotropy of the cosmic microwave background radiation.

These methods give us the most accurate values of η which coincide within the observational errors: $\eta_{\text{BBN}} = (6.0 \pm 0.4) \cdot 10^{-10}$ [1] and $\eta_{\text{CMB}} = (6.05 \pm 0.07) \cdot 10^{-10}$ [2]. It is an evidence of validity of the standard cosmological model. Determination of η for (iii) and (iv) epochs has lower accuracy and it is also strongly model-dependent. One can hope that future experiments and observational data will let us determine η with higher accuracy for different cosmological epochs. It, in its turn, can become a powerful tool for study of physics beyond the Standard Model where η can change.

In this work we study an opportunity of η variation due to decay and annihilation of dark matter (DM) particles during different cosmological epochs. There are many candidates for the DM particles including supersymmetric particles (see, e.g., [5, 6] and references therein). Such particles can annihilate and decay with emission of the Standard Model particles (such as baryons, leptons, photons and so on). It can lead to η variation. Since the number of baryons in the Universe is a billion times less than the number of photons this variation is more sensitive to variation of the baryon number density. An influence of the DM decay and annihilation processes on the photon component of our Universe has been studied in a large number of works (see, e.g, [7, 8, 9, 10, 11] and references therein). In this work we study an influence of the DM decay and annihilation processes: $\text{X} \to \chi \text{p}\bar{\text{p}}$ and $\chi\bar{\chi} \to \text{p}\bar{\text{p}}$, where χ and X are the lightest supersymmetric particle (LSP) and the next-to-lightest supersymmetric particle (NLSP), on the baryonic component of the Universe.

We consider different parameter values of the DM particles:

- WIMP's annihilation cross section $\langle \sigma v \rangle^{\text{ann}}_{\chi\bar{\chi}} = 3 \cdot 10^{-26} \text{sm}^3/\text{s}$ (see, e.g., [5]),
- lifetime $t_{\text{BBN}} \ll \tau \lesssim t_0$, where $t_{\text{BBN}} \simeq 3$ minutes and $t_0 \simeq 13.8$ billion years are the ages of the Universe at the end of the BBN epoch and today respectively,
- mass of the DM particles within the range $10\,\text{GeV} \div 1\,\text{TeV}$.

Instead of the standard cosmological model which assumes no relic antimatter, we define η as follows:

$$\eta(z) = \frac{n_{\text{b}}(z) + n_{\bar{\text{b}}}(z)}{n_\gamma(z)} = \frac{n_{\text{b}}^{\text{BBN}}(z) + \Delta n_{\text{p}}(z) + \Delta n_{\bar{\text{p}}}(z)}{n_\gamma^{\text{BBN}}(z)} = \eta_{\text{BBN}} + \Delta\eta(z), \tag{1}$$

where $n_{\bar{\text{b}}}$ is the number of antibaryons in the Universe; $n_{\text{b}}^{\text{BBN}}$ and n_γ^{BBN} are the numbers of baryons and photons, respectively, which are related with $\eta_{\text{BBN}} = n_{\text{b}}^{\text{BBN}}/n_\gamma^{\text{BBN}}$; Δn_{p} and $\Delta n_{\bar{\text{p}}}$ are the numbers of the baryons and the antibaryons, respectively, which are the products of the decay and the annihilation of DM.

2. The variation of the baryon-to-photon ratio due to decay and annihilation of dark matter

The time evolution of the number densities $n_{\text{X}}(t)$ of the NLSP, $n_\chi(t)$ the LSP, $n_{\text{p}}(t)$ the protons and $n_{\bar{\text{p}}}(t)$ the antiprotons is described by the system of Boltzmann equations:

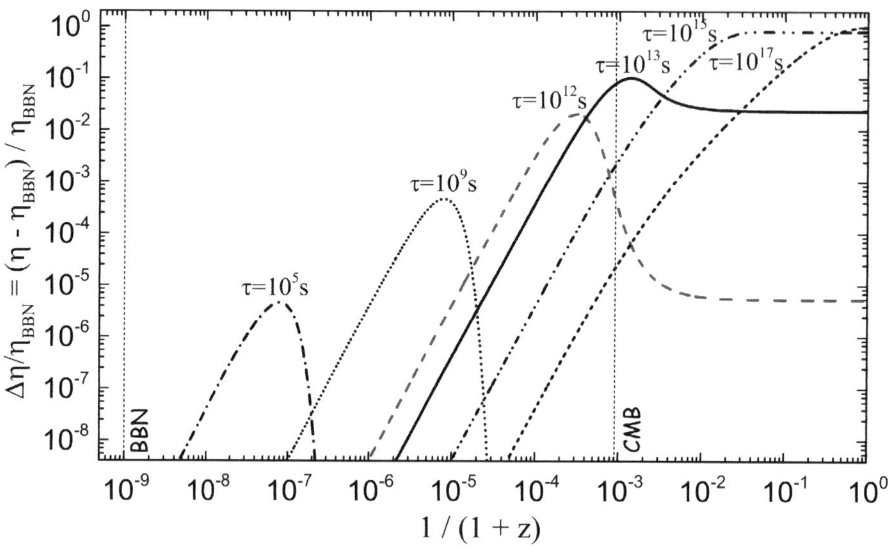

Figure 1. The variation of η due to decay of the dark matter particles with lifetimes $10^5\,\mathrm{s} \le \tau \le 10^{17}\,\mathrm{s}$ ($\alpha = 1$, $m_\chi = 10\,\mathrm{GeV}$). Vertical lines indicate the BBN and the CMB epochs.

$$
\begin{aligned}
\frac{dn_{\mathrm{X}}}{dt} + 3Hn_{\mathrm{X}} &= -\frac{1}{\tau}n_{\mathrm{X}}, \\
\frac{dn_{\chi}}{dt} + 3Hn_{\chi} &= \frac{1}{\tau}n_{\mathrm{X}}, \\
\frac{dn_{\mathrm{p},\bar{\mathrm{p}}}}{dt} + 3Hn_{\mathrm{p},\bar{\mathrm{p}}} &= -\langle\sigma v\rangle^{\mathrm{ann}}_{\mathrm{p}\bar{\mathrm{p}}} n_{\mathrm{p}} n_{\bar{\mathrm{p}}} + \frac{1}{\tau}n_{\mathrm{X}} + \frac{1}{2}\langle\sigma v\rangle^{\mathrm{ann}}_{\chi\bar{\chi}} n_{\chi} n_{\bar{\chi}},
\end{aligned}
\tag{2}
$$

where $H(t) = \dot{a}/a$ is the Hubble expansion rate, $a(t)$ is the scale factor of the Universe, $\langle\sigma v\rangle^{\mathrm{ann}}_{\chi\bar{\chi}} = 3\cdot 10^{-26}\mathrm{cm}^3/\mathrm{s}$ [5] and $\langle\sigma v\rangle^{\mathrm{ann}}_{\mathrm{p}\bar{\mathrm{p}}} = 10^{-15}\mathrm{cm}^3/\mathrm{s}$ [12] are the thermally averaged total cross section for annihilation of $\chi\bar{\chi}$ and $\mathrm{p}\bar{\mathrm{p}}$, respectively, times relative velocity which are almost the constants in the wide energy range, 1/2 in the right side of the third equation is because of the annihilation of the Majorana particles ($\chi \equiv \bar{\chi}$). We use the values of cosmological parameters from [2].

We use the following initial conditions for the system (2):

$$
z^0 = z_{\mathrm{BBN}} = 10^9, \quad t^0 = \frac{1}{2H(z_{\mathrm{BBN}})}, \quad n^0_{\mathrm{p}} = \eta_{\mathrm{BBN}} n_\gamma(z_{\mathrm{BBN}}),
$$

$$
n^0_{\bar{\mathrm{p}}} = 0, \quad n^0_\chi = (1-\alpha)\frac{\Omega_{\mathrm{CDM}}\rho_{\mathrm{c}}}{m_\chi c^2}, \quad n^0_{\mathrm{X}} = \alpha\frac{\Omega_{\mathrm{CDM}}\rho_{\mathrm{c}}}{m_\chi c^2},
$$

where $0 \le \alpha \le 1$ is the fraction of NLSP to all DM number density at the BBN epoch, $\Omega_{\mathrm{CDM}} = 0.26$ is the cold DM density parameter, ρ_{cr} is the critical density of the Universe, m_χ is the mass of the LSP, c is the speed of light.

We found out that annihilation of the DM particles does not lead to significant change of the baryon-to-photon ratio during the cosmological epochs from the BBN to the present epoch. The variation of η due to the only annihilation of DM ($\chi\bar{\chi} \to \mathrm{p}\bar{\mathrm{p}}$) with masses

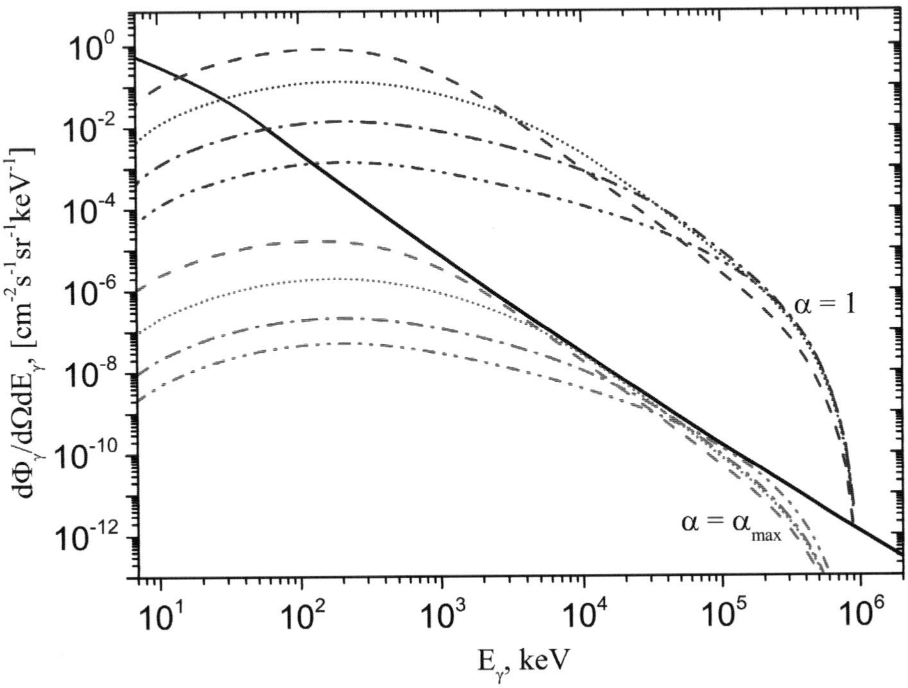

Figure 2. The diffuse gamma ray background due to annihilation of the protons and the antiprotons, the products of the decays of the DM particles with lifetimes $\tau = 10^{14}$ s (dashed curves), 10^{15} s (short-dotted curves), 10^{16} s (dash-dotted curves) and 10^{17} s (dash-dot-dotted curves), blue curves correspond to $\alpha = 1$, red curves correspond to $\alpha = \alpha_{\rm max}$, black solid curve corresponds to fitting the observational data from [13].

$m_\chi = 10\,\mathrm{GeV}\div 1\,\mathrm{TeV}$ even at the BBN epoch (where the rate of the annihilation is the highest) is negligible: $|\Delta\eta(z)/\eta_{\rm BBN}| < 10^{-11} \div 10^{-13}$.

The variation of η due to decay of the NLSP, in its turn, is sizeable. In Fig. 1 we present the results of calculations of η variation due to decay of the NLSP ($\mathrm{X} \to \chi \mathrm{p}\bar{\mathrm{p}}$) for the case $m_\chi = 10\,\mathrm{GeV}$ and $\alpha = 1$. One can see that the late decay of the DM ($\tau \gtrsim 10^{12}$) can lead to η variation up to $\Delta\eta(z)/\eta_{\rm BBN} \sim 0.01 \div 1$ (a particular value depends on τ) what is potentially observable.

The obtained results must not contradict observational data. We use the data [13] on the diffuse gamma ray background (DGRB) in order to make constraints on the DM decay model and on the maximum possible variation of η due to these NLSP decays.

3. The diffuse gamma ray background

Presence of antiprotons in the Universe would lead to emergence of the complementary DGRB due to annihilation of the protons and the antiprotons which can be observed today for the decays of the NLSP taking place in $z \lesssim 1000$ [8, 14] ($\tau > t_{\rm CMB} \simeq 400000$). The DGRB (the number of photons per unit area, time, solid angle and energy range) due to the annihilation of the protons and the antiprotons is given by

$$\frac{d\Phi_\gamma}{d\Omega dE_\gamma} = \frac{c}{4\pi}\langle\sigma v\rangle^{\rm ann}_{\rm p\bar{p}} \int\limits_0^{1000} dz \frac{n_{\rm p}(z) n_{\bar{\rm p}}(z)}{H(z)(1+z)^3}\frac{dN_\gamma}{dE_\gamma}\left[E_\gamma(1+z)\right], \quad (3)$$

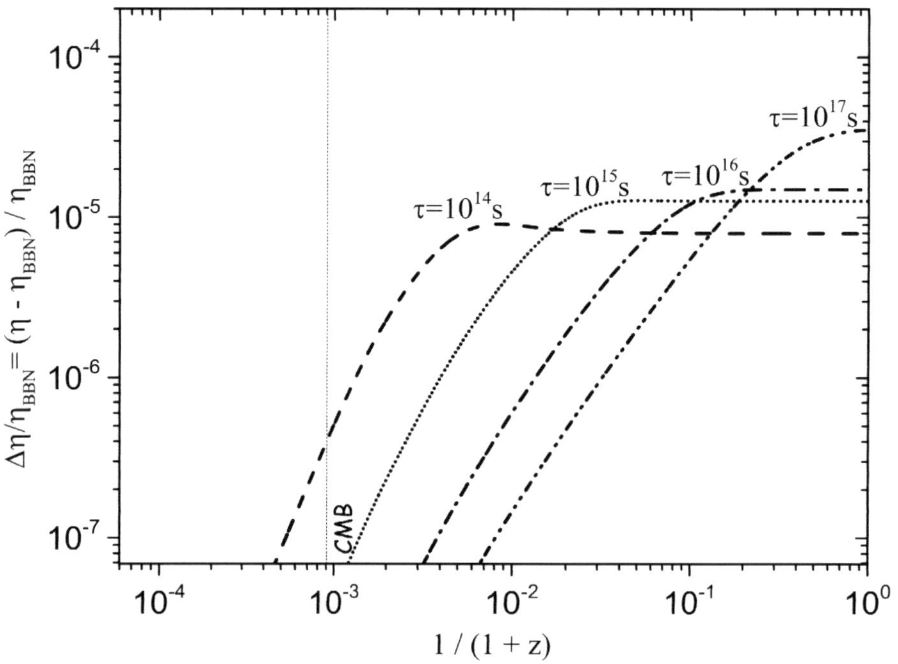

Figure 3. The variation of η due to decay of the dark matter particles with lifetimes $10^{14}\,\mathrm{s} \leq \tau \leq 10^{17}\,\mathrm{s}$ which are consistent with the observational data on the diffuse gamma ray background ($\alpha = \alpha_{\max}$, $m_\chi = 10\mathrm{GeV}$). Vertical line indicates the CMB epoch.

where dN_γ/dE_γ is the gamma spectrum (the average number of photons per energy range and per annihilation) from proton-antiproton annihilation [15]. In Fig. 2 we present this DGRB for the cases $m_\chi = 10\,\mathrm{GeV}$, $\alpha = 1$ and $\alpha = \alpha_{\max}$. The cases with $\alpha \lesssim \alpha_{\max}$ are allowed by the observational data [13]. For the DM mass $m_\chi = 10\,\mathrm{GeV}$ the values of $\alpha_{\max}$ are $2 \cdot 10^{-5}$ for $\tau = 10^{14}\,\mathrm{s}$, $1.5 \cdot 10^{-5}$ for $\tau = 10^{15}\,\mathrm{s}$, $1.5 \cdot 10^{-5}$ for $\tau = 10^{16}\,\mathrm{s}$ and $3.5 \cdot 10^{-5}$ for $\tau = 10^{17}\,\mathrm{s}$.

In Fig. 3 we present our final results on the variation of η for $m_\chi = 10\,\mathrm{GeV}$, $\alpha = \alpha_{\max}$ and $t_{\mathrm{CMB}} < \tau \lesssim t_0$ which are consistent with the observational data on the DGRB. One can see that the variation of baryon-to-photon ratio can be $\Delta\eta/\eta_{\mathrm{BBN}} \lesssim 10^{-5}$. Note that α and m_χ are included into the initial conditions of (2) as a ratio. Hence, the obtained results are the same for the cases of the heavier LSP with keeping the value of α/m_χ.

It should be noted that in this work an approximation of uniform density of the Universe has been used, while density inhomogeneities $\Delta\rho/\rho$ become significant at low redshifts ($z \lesssim 30$). However, given that in the present epoch most of baryons is contained in extragalactic gas where $\Delta\rho/\rho \lesssim 1 \div 10$ and that proton-antiproton annihilation virtually stops at the redshifts $z \sim 100$ (see Fig. 1), the results revealed in Fig. 1 (at $z \lesssim 30$) will not be affected noticeably by taking into account growth of inhomogeneities. At the same time, constraints stemming from the DGRB can be enhanced, especially in case of late X particle decays ($\tau \gtrsim 10^{16}\,\mathrm{s}$).

4. Conclusion

An influence of the annihilation and the decay processes of the dark matter particles:

$$\chi\bar{\chi} \longrightarrow \mathrm{p}\,\bar{\mathrm{p}}, \qquad \mathrm{X} \longrightarrow \chi\,\mathrm{p}\,\bar{\mathrm{p}},$$

on the baryon-to-photon ratio has been studied for different cosmological epochs.

We have shown that the annihilation of the dark matter particles with masses $m_\chi = 10\,\mathrm{GeV}\div 1\,\mathrm{TeV}$ and WIMP's annihilation cross section $\langle\sigma v\rangle^{\rm ann}_{\chi\bar\chi} = 3\cdot 10^{-26}\mathrm{sm}^3/\mathrm{s}$ does not lead to significant variation of the baryon-to-photon ratio during the cosmological epochs from Big Bang nucleosynthesis to the present epoch: $|\Delta\eta(z)/\eta_{\rm BBN}| < 10^{-11} \div 10^{-13}$.

It has been shown that the decay of the dark matter particles (in case of LSP masses $m_\chi = 10\,\mathrm{GeV} \div 1\,\mathrm{TeV}$) can lead to the variation of the baryon-to-photon ratio up to $\Delta\eta(z)/\eta \sim 0.01\div 1$, but it also leads to an excess of the gamma ray flux due to annihilation of the protons and the antiprotons.

We determined our constraints on the decay of the dark matter particles with lifetime $t_{\rm CMB} < \tau \lesssim t_0$: the fraction of such NLSP to all DM number density at the BBN epoch $\alpha \lesssim 10^{-5} \div 10^{-3}$ for the LSP masses $m_\chi = 10\,\mathrm{GeV} \div 1\,\mathrm{TeV}$, and on the corresponding variation of the baryon-to-photon ratio: $\Delta\eta(z)/\eta_{\rm BBN} \lesssim 10^{-5}$.

Acknowledgments

The work was supported by the Russian science foundation grant 14-12-00955.

References

[1] Steigman G 2007 *Annual Review of Nuclear and Particle Science* **57** 463
[2] Ade P A R et al 2014 *Astron. Astrophys.* **571** 66
[3] Kirkman D, Tytler D, Suzuki N, O'Meara J M and Lubin D 2003 *ApJS* **149** 1
[4] Fukugita M and Peebles P J E 2004 *ApJ* **616** 643
[5] Jungman G, Kamionkowski M and Griest K 1996 *Phys. Rep.* **267** 195
[6] Bertone G, Hooper D and Silk J 2004 *Phys. Rep.* **405** 279
[7] Ullio P, Bergstrom L, Edsjo J and Lacey C 2002 *Phys. Rev.* D **66** 123502
[8] Chen X and Kamionkowski M 2004 *Phys. Rev.* D **70** 043502
[9] Ando S and Komatsu E 2006 *Phys. Rev.* D **73** 023521
[10] Bertone G, Buchmuller W, Cove L and Ibarra A 2007 *JCAP* **11** 003
[11] Cirelli M, Moulin E, Panci P, Serpico P D and Viana A 2012 *Phys. Rev.* D **86** 083506
[12] Weniger C, Serpico P D, Iocco F and Bertone G 2013 *Phys. Rev.* D **87** 123008
[13] Gruber D E, Matteson J L, Peterson L E and Jung G V 1999 *ApJ* **520** 124
[14] Zdziarski A A and Svensson R 1989 *ApJ* **344** 551
[15] Backenstoss G et al 1983 *Nucl. Phys.* B **228** 424

Journal of Physics: Conference Series **661** (2015) 012017 doi:10.1088/1742-6596/661/1/012017

The influence of strong connections on dynamics of neuron population with long-tailed synaptic weight distribution

E A Chizhkova[1], A V Chizhov[1,2]
[1]Ioffe Physical-Technical Institute of the Russian Academy of Sciences, Saint-Petersburg, Russia
[2]Sechenov Institute of Evolutionary Physiology and Biochemistry of the Russian Academy of Sciences, Saint-Petersburg, Russia

Abstract. This work is devoted to exploring the influence of long-tailed synaptic weights distribution on neural population dynamics. We show numerically that both strong sparse and very weak connections pay crucial role for spontaneous activity whereas they almost do not influence on driven population activity. Moreover, according to obtained results, we suggest that neurons with strong connections form multiple chains, which are the core of spontaneous activity, whereas neurons with weak connections pay little role of partially enhancing of population activity.

1. Introduction

Recently, cortical synaptic weights have been shown experimentally to obey a long-tailed unimodal distribution peaking near the lowest values which can be fitted with lognormal law [1]. These results gives birth to series of theoretical and experimental works dedicated to finding a shape of synaptic weights distribution in different neural brain structures [2,3,4], genesis of such distribution [5,6], features of networks dynamics with such connection strengths distribution [7,8,9] and numerical description of feed-forward network with different, including long-tailed, synaptic weights distribution [10].

In work of J. Teramae, et al. [7] – which was an inspiration for this study – some features of network with lognormal weights distribution were theoretically obtained, such as producing of spontaneous activity and conditions of the fidelity of spike transmission along the strongest synapses. The results were obtained using numerical simulations of system constituted from integrate-and-fire neurons with lognormal coupling strengths distribution, which form were determined according to experimental results. All connections were divided into weak-dense and strong-sparse and it was concluded that strong-sparse synapses form multiple unidirectional synaptic pathways while weak-dense connections redistribute excitatory activity routed reliably on strong connections over the network as optimal noise sources to sustain spontaneous firing of recurrent networks.

In this work we follow the trend of studying the features of such network, more exactly the influence of neurons with strong connections and with only weak once and the influence of strong and weak connections on population activity. In the next section we will demonstrate the influence of presence of sparse strong connections and weak ones on current-driven activity and on spontaneous activity via eliminating them from the system, then we will do something similar to see role of neurons having strong connections and not in network.

2. The influence of strong and weak connections on neuron population activity.

In order to exploring the influence of sparse strong and weak dense connections on population activity we firstly reproduced the network system and repeated numerical simulation presented in work Teramae et al, specifically we obtained steady low-frequency irregular spontaneous activity (fig.1A). Simulations were performed for population with lognormal coupling strengths distribution constructed from 10000 excitatory and 2000 inhibitory integrate-and-fire randomly connected neurons, to trigger a spontaneous firing Poisson spike trains were applied to all neurons during 100 ms (see Appendices for more details).

After that we wanted to see behavior of the system driven with step of current after spontaneous activity. Then we eliminated strong connections (> 10 mv) just before driven state to see how strong connections influence on this activity. In another case strong connections were eliminated from the start. Figure 1 shows that current-driven activity of population after spontaneous activity differ from such activity after silent state with less amplitude followed by faster setting of stationary state. The last two figures are almost equivalent to first ones and show only little difference in firing rate of stationary phase that leads to conclusion that the strong connections almost do not influence on driven activity, but have crucial role for spontaneous activity (silent state in second figure is result of deleting strong connections, not of turning off initial Poisson spike train).

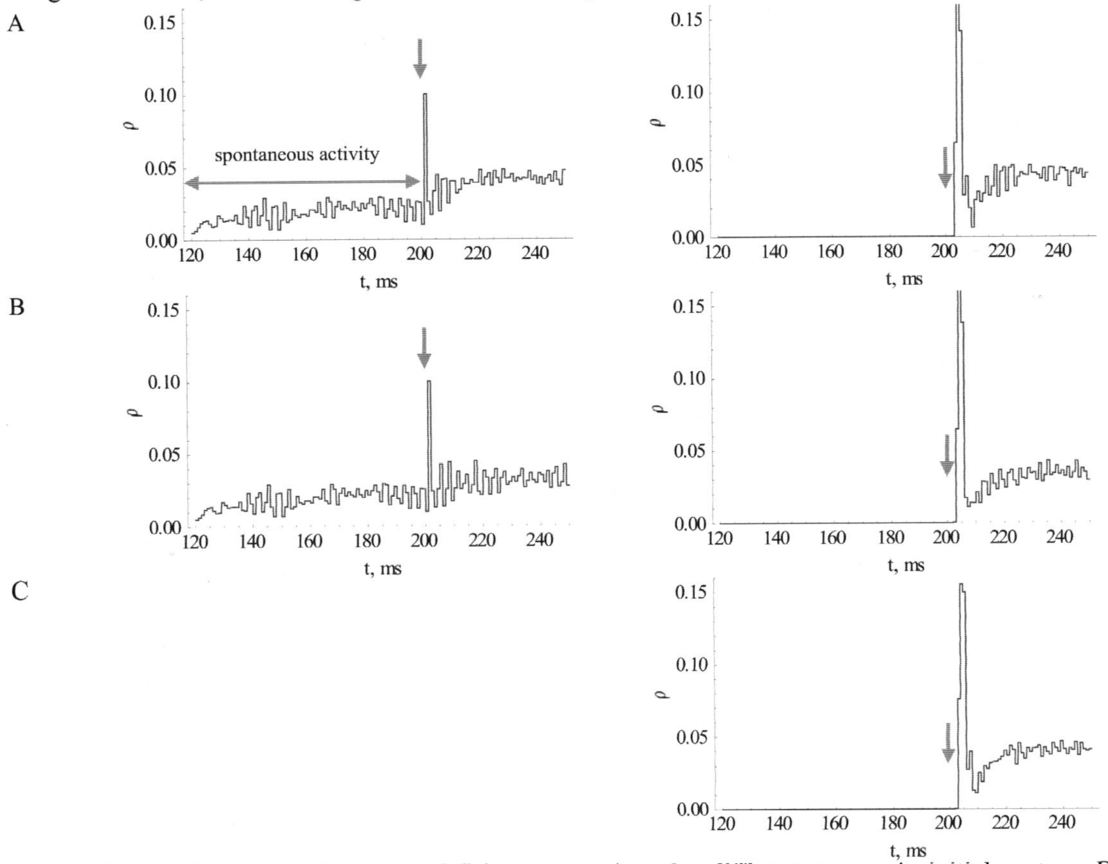

Figure 1. Population activity (rate of firing neurons) under different cases. A: initial system, B system with eliminated strong connections (>10 mV) just before driven activity (left graph) and from the start (right graph), C system with eliminated weak connections (<0.3 mV) from the start. Arrows point the time of current supply.

Similarly, we eliminated weak connections (<0.3 mV) to see their role in spontaneous activity. Spontaneous activity having disappeared in this case either (fig. 1C), weak connections pay crucial role for such activity as well.

3. The influence of neurons with strong connections and without on neurons population activity.

Similarly, to explore the role of neurons having strong connections and those having only weak ones we made simulations with eliminating such neurons. Although we set connection probability to only 14%, it appears that near 97% of neurons have at least 1 strong connection with EPSP more than 10 mV due to huge number of neurons and therefore huge number of connections – 1000 connections per neuron. There is no surprise that system continued to produce spontaneous activity without 3% of neurons with only week connections.

Then we eliminated neurons without as input thus as output strong connections that represented near 50% of all neurons. After that, system continued to produce spontaneous activity and even without dramatic casualties in frequency (fig. 2A). More surprising result was keeping spontaneous generation by system constituted from neurons having strong connections only in one directions or not having ones, but with less frequency (fig.2B).

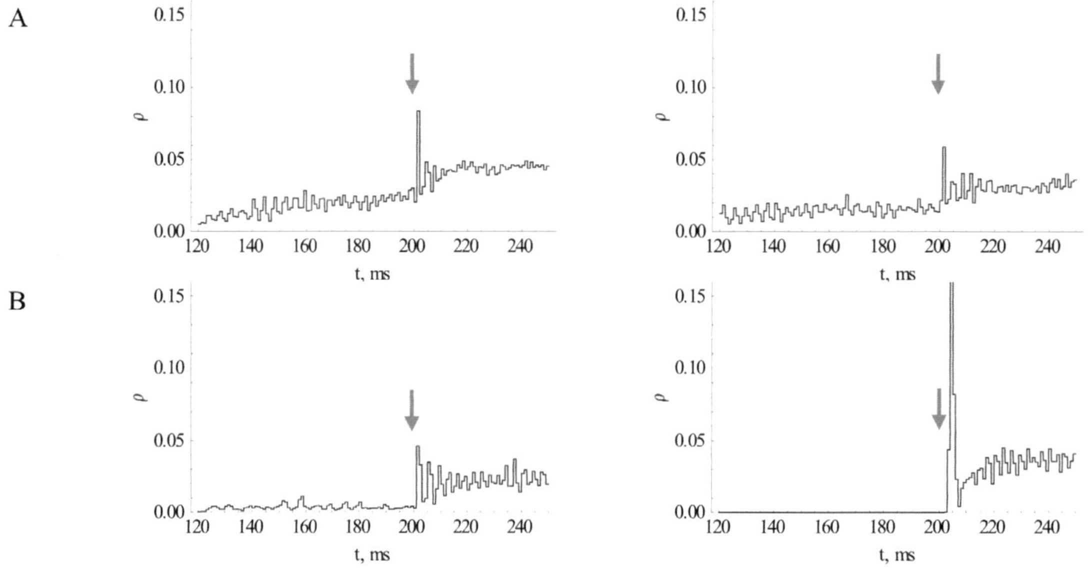

Figure 2. Population activity (spiking neurons rate) under different cases. A: system with eliminated neurons without as input thus as output strong connections >10 mV (left graph) and without strong connections >16 mV (right graph), B system with neurons having strong connections only in one directions or not having ones (left graph) and without strong connections >16 mV (right graph). Arrows point the time of current supply.

It seems that the main role in forming spontaneous activity belongs to chains consisted of neurons with strong connections, serving as sources of input stimulus via weak connections for each other and for the rest of neurons. This looks more likely for real systems that were shown experimentally and theoretically to have correlated strengths of synaptic connections, implying that strong connections are even more clustered [1,8]. Nevertheless, spontaneous activity can be formed without chains made of strong connections, as in figure 2B (left), or by only weak connections (<10 mV). Actually, we have even met parameters with which system is keeping to generate low-frequency spontaneous activity without strong connections (fig. 3) Such activity has less frequency than original system and thus has secondary role in forming spontaneous activity and should have rather undirected propagation of excitation.

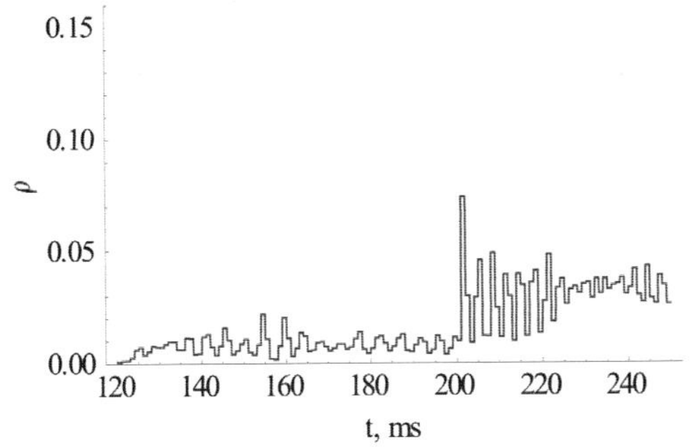

Figure 3. System with eliminated strong connections >10 mV.

After eliminating neurons without connections with EPSP more than 16 mV (connections are suggested to be formed by more than one synapse), the rest part (30%) kept to produce activity. It was the least found part of system capable to produce spontaneous activity without dramatic casualties in frequency (fig. 2A). Without this part system kept silence before driven activity (fig. 2B).

4. Discussion

In this study, we have explored role of neurons having strong connections and those having only weak ones and role of strong and weak connections themselves in spontaneous and driven population activity. Results are briefly summarized in table 1, presenting changes in driven and spontaneous activity under different cases. We found that strong connections almost don't have influence on driven activity, which complements the result obtained in [10] for feed-forward networks with lognormal and Gaussian synaptic current distributions, and described some features of driven activity following after spontaneous one. On the other hand, strong and weak connections are crucial for producing of spontaneous activity, what was illustrated in [7] as well. Then we showed neurons with strong connections to pay main role in forming spontaneous activity and found the least part that could be derived with our methods reproducing spontaneous activity with frequency close to original system frequency.

Table 1. Influence of strong and weak connections and neurons with and without strong connections on population activity. Here, sc, wc are strong and weak connections, sn, wn are neurons with and without strong connections, respectively.

	spontaneous activity	driven activity
sc (>10 mV)	crucial	~ no influence
wc (<0.3 mV)	crucial	~ no influence
sn (>16 mV)	crucial	~ no influence
wn (<16 mV)	little enhancing of frequency	

These results help to understand which model we can use for different cases for mathematical description of population activity. For example, it seems appropriate using statistical models for current-driven state of system whereas spontaneous activity should be described by direct numerical simulation at least that part that forms such activity or by more sophisticated approach.

It's notable that presented system is rather instable in respect of system parameters. This implies that there is rather little range of parameters values when system produces spontaneous activity and does not present epileptic-like activity. Instability can be a result of using simple conductance-base model for neuron without adaptive currents which could prevent too high activity and expand range parameters presenting stable spontaneous activity.

5. Appendices

We use the following integrate-and-fire model for describing activity of one neuron:

$$\frac{dV}{dt} = -\frac{1}{\tau_m}(V - V_L) - g_E(V - V_E) - g_I(V - V_I),$$

where V is the membrane potential, τ_m is the membrane time constant, $V_{L,E,I}$ are reversal potentials for leak, excitatory and inhibitory postsynaptic currents, $g_{E,I}$ are the excitatory and inhibitory synaptic conductances normalized by the membrane capacitance ([ms^{-1}]) and described by

$$\frac{dg_X}{dt} = -\frac{g_X}{\tau_s} + \sum_j G_{X,j} \sum_{s_j} \delta(t - s_j - d_j), \qquad X = E, I$$

where $\delta(t)$ is the delta function, G_j, d_j, s_j are the weight, delay and spike timing of synaptic input from the j-th neuron, respectively, τ_s is the decay constant. The values of G_j for excitatory-to-excitatory connections are distributed such that the amplitude of EPSPs x measured from the resting potential obey a lognormal distribution

$$p(x) = \frac{\exp\left[-\frac{(logx - \mu)^2}{2\sigma^2}\right]}{\sqrt{2\pi}\sigma x},$$

where σ=1.0 and μ-σ^2=log(0.2), the latter corresponds the mode=0.2. Excitatory-to-inhibitory, inhibitory-to-excitatory and inhibitory-to-inhibitory synapses' weights have constant values of G_j= 0.018, 0.002 and 0.0025, respectively. The probabilities of inhibitory and excitatory connections are 0.5 and 0.14, respectively. Excitatory-to-excitatory synaptic transmissions fail with probability p_E= a/(a+EPSP), where a=0.1 [mV]

Picked the other parameters values are the following: V_L, V_E, V_I are -70[mV], 0[mV] and -80 [mV], respectively, spike threshold is -50 mV and V is -60 mV after spiking, the refractory period is 1 ms, τ_m is 20 ms for excitatory and 10 ms for inhibitory neurons, τ_s is 4.7 for excitatory and 3.9 for inhibitory neurons and these values are picked in order to system didn't oscillate, synaptic delays are chosen randomly between d_0-1 to d_0+1 ms, where d_0=2 for excitatory-to-excitatory connections and d_0=1 for other connection types. For initiation driven activity we use step of current different for different neurons with mean value 4[nA/nF] and variance 2[nA/nF] (current is normalized by the membrane capacitance).

6. References

[1] Song S, Sjostrom P, Reihl M, Nelson S, Chklovskii D 2005 *PLoS Biology* **3** e68

[2] Hromadka T, Deweese MR, Zador AM 2008 *PLoS Biology* **6** e16
[3] Lefort S, Tomm C, Floyd, et al 2009 *Neuron* **61** 301
[4] Ikegaya Y, Sasaki T, Ishikawa D, Honma N, Tao K, et al 2012 *Cereb Cortex* **23** 293
[5] Zheng P, Dimitrakakis C, Triesch J 2013 *PLoS Comput Biol* **9** e1002848
[6] Gilson M, Fukai T 2011 *PLoS One* **6** e25339
[7] Teramae J, Tsubo Y, Fukai T 2012 *Sci Rep* **2** 485
[8] Koulakov A, Hromádka T, Zador A 2009 *J Neurosci* **29**(12) 3685
[9] Klinshov V, Teramae J, Nekorkin V, Fukai T 2014 *PLoS One* **9** e94292
[10] Iyer R, Menon V, Buice M, Koch C, Mihalas S 2013 *PLoS Comput Biol* **9** e1003248

Journal of Physics: Conference Series **661** (2015) 012018 doi:10.1088/1742-6596/661/1/012018

A micromechanical device that monitors arterial pressure during general anesthesia and in intensive care units

A V Andreeva[1], V V Luchinin[1], K A Kuzmina[1], A S Klyavinek[2] and A E Karelov[2]

[1] SPbSETU "LETI", Saint–Petersburg 197376, Russia

[2] I. I. Mechnikov NWSMU, Saint-Petersburg 191015, Russia

E-mail: twist5487@mail.ru

Abstract: A vibroacoustic fiber optic system that consists of micromechanical components designated for use in medicine and biology is reviewed. A theoretical analysis of a fiber optic microphone is done and its optimal construction parameters are determined. Указана The possibility of using the developed system with magnetic resonance tomography to noninvasively measure man's arterial pressure is specified.

1. Introduction

Circulation is foremost among all vital signs in importance. Arterial pressure is one of the basic parameters of the human body's circulatory system. Arterial pressure level is a key determinant of patient treatment and affects an infection's duration and outcome. Precisely measuring systolic and diastolic arterial pressure presupposes the use of hemodynamic monitoring during general anesthesia and of magnetic resonance imaging (MRI) in intensive care units. Creating and using new biomedical devices to ensure more precise diagnostics and determine arterial pressure therefore constitute a pressing issue. Presented in this plan is a compelling design for fiber optic sensors (FOS) and systems meant for various medical applications. An optical fiber's very small diameter (smaller than 250 μm), high flexibility and low loss upon exposure to light render possible the production of subminiature biomedical FOS and their direct installation at the location of measurement. A principal advantage of FOS is their insusceptibility to electromagnetic and radio frequency signals. This property makes FOS ideal for uninterruptedly diagnosing patients undergoing MRI examinations or computerized tomography, situations in which applying other types of sensors is problematic. This article considers the creation of acoustic micromechanical FOS, which will help ensure two-way acoustic connections with patients and the remote monitoring of arterial pressure through the use of auscultation during MRI examinations.

2. A micromechanical fiber optic instrumentation system

Micromechanical fiber optic instrumentation systems based on monitoring the micro-displacements of sensors under the influence of measurable physical phenomena have made prevalent the production of FOS [1]. This monitoring can be done through the use of interferometers with external primary transducers.

The optical fiber end-face interferometer (OFEI) is regarded as one of the simplest such devices [2]. Highly sensitive, this interferometer is remarkable in its simple construction, compactness, tolerance to external influences and low cost. An OFEI construction schematic is shown in figure 1. The operating principle of an OFEI lies in the introduction of an emission from laser diode 1 to optical fiber 2 and its consequent transmission through bifurcation 3 to optical fiber 4. An emission fragment reflects from the end of optical fiber 4 while another emission fragment flashes into the air, reflects from mirror 5 and returns back to optical fiber 4. The emission fragment reflected from the end of optical fiber 4 interferes with the emission fragment reflected from mirror 5 and photoelectric receiver

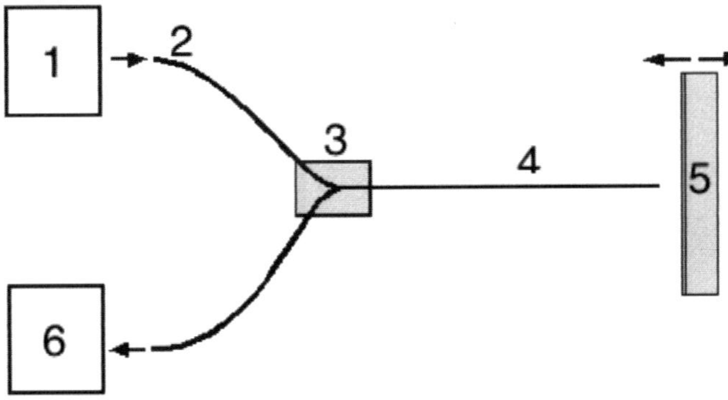

Figure 1. OFEI Construction Schematic

6 registers the emission power. Interference generally results in the superposition of not two, but several waves that emerge due to the repeated transformations of the light beam between the end of the optical fiber and the mirror.

3. A micromechanical fiber optic microphone

Using a micromechanical fiber optic microphone as a pressure measurement sensor necessitates a highly-sensitive membranous component to ensure the membrane's linear displacement even upon small changes in external pressure.

Membranous components are made from silicon wafers through the application of integrated-group methods and micromechanical technologies [3]. First, grooves of predetermined width and depth are etched on the front of the silicon water in the form of several concentric rings. A layer of Si_3N_4 0.2-0.3 μm thick then forms on the grooved side of the silicon wafer, after which a layer of metal settles in the central part of each component. Silicon finally peels off the back of the wafer, as a result of which is obtained a thin corrugated Si_3N_4 membrane with a reflecting center region fixed on a silicon base. Figure 2 is a photograph of a membranous component made specifically for fiber optic microphones at Saint Petersburg State Electrotechnical University's Microtechnology and Diagnostics Center.

Figure 2. Membranous Component

Figure 3 is a construction schematic of the fiber optic microphone's primary transducer. A commercially-produced ceramic capillary can function as the base (1). A cement bonding connects the capillary to a standard single-mode optical fiber 9/125 μm in length (2), after which the end of the optical fiber is glazed so as to create a reflecting mirror surface for optical emission. The membranous component (4) is then connected to a commercially-produced ceramic centralizer (3). The membranous component and the centralizer on which it is anchored are fixed on the ceramic capillary, after which the primary transducer is finely tuned.

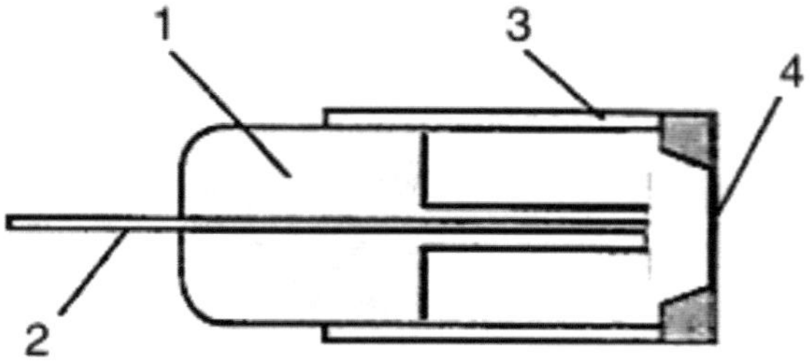

Figure 3. Construction schematic of a fiber optic microphone's primary transducer

Figure 4 is a photograph of a primary acoustic transducer with a membranous component 4x4 mm in size.

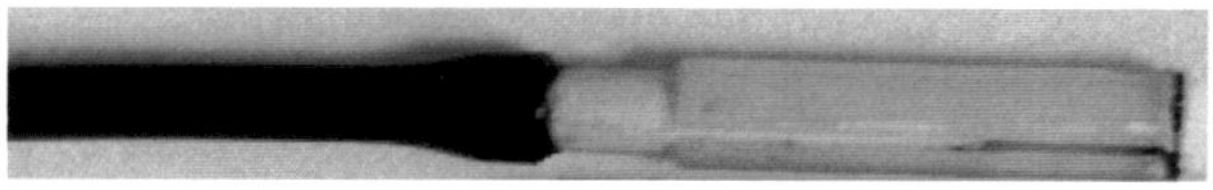

Figure 4. Primary acoustic transducer with membranous component 4x4 mm in size

Values calculated for the integral level of equivalent sound pressure determined by microphones' internal noises at the working frequency range evidence the sensor's high sensitivity (about 20-24 dB). The microphones' measured dynamic range amounts to 60-65 dB.

4. Measuring arterial pressure in different situations

Most noninvasive methods currently used to measure arterial pressure in clinical practice are classified according to the way in which this parameter is determined. Most commonly-employed measurement methods are auscultatory, as proposed Russian internist N. S. Korotkov, and oscillometric. Every method has its own advantages and limitations. Using Korotkov's method, for example, requires that the microphone be properly situated and encounter no disturbances during the course of observation. Considering subjective factors such as choosing locations for the seal and microphone is of particular concern. Despite the difficulty of standardizing its measurement conditions, Korotkov's method is routinely recommended (WHO, 1935). The oscillometric method does not require microphone placement, but it is not standard due to its low precision of measurement.

The standard device for measuring arterial pressure using Korotkov's method consists of an occlusal seal and something to blast air inside of the seal and measure its inside pressure [4]. Using this device requires skill and is inconvenient when it is necessary to obtain a precise result quickly. Strong electromagnetic emissions that cause measurement errors render impossible the use of electronic tonometers in life-saving situations, such as during an MRI examination done under general anesthesia.

Keeping this in mind, an idea for creating an automatic device that noninvasively measured arterial pressure by using the aforementioned micromechanical fiber optic microphone was put forth. This would ensure the complete and safe monitoring of arterial pressure for patients in need of an MRI examination or of surgery done during an MRI examination.

Figure 5 is a typical oscillogram of Korotkov tones taken from a healthy patient exposed to a strong electromagnetic force (up to 3 T).

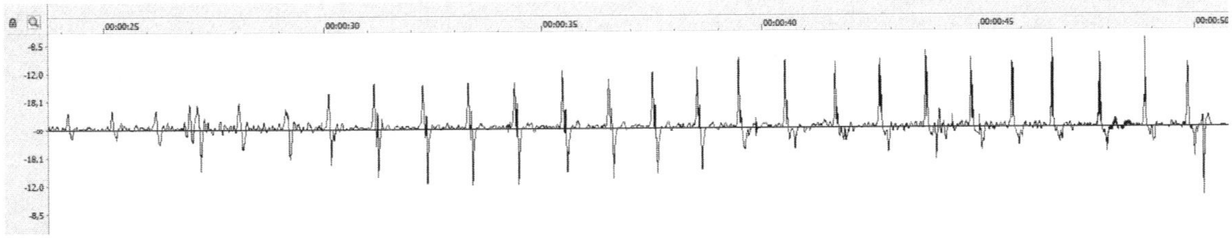

Figure 5. Signal taken by a fiber optic microphone from a human exposed to a strong electromagnetic force

A fiber optic microphone rather than a conventional microphone was used for the auscultatory diagnostics. This result shows the potential for using fiber optic microphones to gauge arterial pressure through a specially-designated acoustic method.

5. Conclusion

It is at present possible to name the following characteristics as those by which the scientific practicality of future measuring instruments will be ensured:

- no acoustic or electrical disturbances upon exposure to strong electromagnetic forces;
- the possibility of being used as a distributed topographic system to validate results;
- the use of a wireless system for transmitting information through a communication channel;
- the possibility of obtaining additional information (a diagnostic of various anomalies in the heart and blood vessels);
- the application of optical sampling to decrease both the affect of the human factor on pressure monitoring and the number of measurement errors;

This work's results can be used for the further technical realization of vibroacoustic fiber optic systems (microphones) that measure arterial pressure in special situations, as well as for the development of a complete medical system that answers all contemporary concerns.

This work was done by the V. L. Vanevsky Department of Anesthesiology and Intensive Care Medicine at the I. I. Mechnikov Northwest State Medical University.

References

[1] Vetrov A A, Sergushichev A N and Shirshov A A 2010 *Fiber-optic vibroacoustic sensors* (Saint - Petersburg: SPBETU «LETI»)

[2] Vetrov A A, Komissarov S S and Sergushichev A N 2008 Fiber-optic end-face interferometer - a universal element of building displacement sensors *Optical Journal* **1** 3-6

[3] Vetrov A A, Komissarov S S, Luchinin V V and Sergushichev A N *Fiber - optic sensing system* RUS Patent 2 279 112, 27. 06. 2006

[4] Tikhonenko V M 2005 Advantages of the method of Korotkov with blood pressure monitoring *J. Vestnik arrhythmology* **40** 36-38

Solvent type influence on thymidine UV-sensitivity

A I Nikolaev, S V Paston
Laboratory of Molecular Biophysics, Department of Molecular Biophysics,
St. Petersburg State University, Petergof, Ulyanovskaya st. 3, St.Petersburg, Russia

E-mail: teterev3000@mail.ru

Abstract. Thymine is the most sensitive DNA nucleobase to UV-irradiation. In the thymidine solution the photoreactions probability is in dependence on the solvent properties which determine stacking-mediated thymidine association and lifetime of excited states. In this work we investigated the degree of UV-irradiation induced thymidine damages in water, salt (NaCl) and ethanol solvents using UV absorption and circular dichroism (CD) spectrometry, and MS ESI method. In the explored systems thymidine association degree rose in the following solvent order: 1 M NaCl, water, ethanol. UV-absorbance and CD intensity fell in the greater extent in the ethanol and water-salt solutions than in water. So the experiment showed that the association degree of thymidine in the solution does not play a main role in its photosensitivity.

1. Introduction

Thymidine (dT) is the most vulnerable to the direct influence of ultraviolet (UV) irradiation nucleoside in DNA. Namely thymine photoproducts cause mutagenic and cancerogenic effects of UV light on a cell [1]. Thymidine can undergo various types of photochemical modifications, i.e. dimerization (with formation of cyclobutane dimers, 4-6-photoproducts, etc.) and hydratation in a DNA cites without secondary structure [1]. These changes in the chemical structure lead to the spectral alternations in the chromophore absorption region which are more manifested in the example of poly-T [2] and frozen thymine solutions [3]. Many researchers indicated the role of monomer association and their mutual orientation in the dimerization process [3-6]. Interaction with solvent plays a significant role in direct and indirect action of radiation on biological macromolecules and their components [7-9]. In the case of UV-irradiation of dT solutions the solvent influences on thymidine photodimerization process via effecting on the monomer association degree, structure of associates and lifetime of excited states [4, 10].

The aim of the present work was to investigate the solvent effects on thymidine self-association and UV-induced dimerization. We studied the UV-absorption and circular dichroism (CD) spectra, and also the mass-spectra with electrospray ionization (MS ESI) of dT in water, ethanol and water-salt solutions before and after UV-irradiation exposure. ESI is relatively soft ionization procedure which does not destroy the analyzed molecules and allows registering complexes formed by ion-dipole and dipole-dipole interactions in a solution [11]. MS ESI is widely used for investigation of nucleoside self-assembling [12, 13] and for identification the DNA photoproducts [14].

2. Experimental section

2.1. Materials

We have used thymidine (Sigma, USA), dissolved in double-distilled water. Thymidine concentration was determined using the known value of its molar extinction coefficient in water ε (275 nm)= 8500

M^{-1} cm^{-1} [15]. We employed sodium chloride NaCl of reagent grade to prepare the saturated solution and ethanol for clinical usage (95% v/v). All solvents were tested by MS ESI and no significant admixtures were revealed. Preparations were carried out with weighting.

2.2. *UV-exposure conditions*

A low-pressure Hg lamp DRB8 (λ=254nm) was used as a source of UV-radiation. Dose rate I=4.14 mJ/(s cm^2). UV-exposure of thymidine solutions ($C=10^{-3}$–10^{-4}M) was carried out under the aerobic conditions in a quartz cells with optical path length of 1 cm and surface area S=4 cm^2. The adsorbed dose D in J/mol (dT) was determined according to the Lambert–Beer law:

$$D = \frac{I(1-e^{-A_{254}})St}{\nu_{dT}},$$

where A_{254} is the optical density of the irradiated solution at λ=254 nm, t – time of irradiation (sec), ν_{dT} is the amount of substance of thymidine (dT) in the irradiated sample (mol). Hence, from the expression one can see that the calculated doses adsorbed by the solutions were corrected. This could be achieved through permanent measurement of solutions absorption spectra during the irradiation what allows considering the changes in the solutions absorption.

2.3. *Methods*

UV absorption spectra of solutions were obtained by spectrophotometer SF-56 (Russia), CD spectra were measured by autodichrograph Mark V (Jobin Ivon, France), the examined parameter is the difference in molar extinction coefficients of nucleoside solution $\Delta\varepsilon=f(\lambda)$ for right and left polarized light. MS study was performed by «MaXis» (Bruker Daltonik GmbH, Germany) using ESI-QTOF analysis, in the positive ions mode. To promote the positive ions formation the solution for MS were prepared at pH(3.0±0.1) by adding HCl.

3. Results and discussions.

Solvent sufficiently influenced spectral properties of thymidine. Its molar extinction coefficient obtained in the different medium under investigation is given in the table 1. It should be noted that the shape of thymidine UV absorption spectrum and its maximum position (around 267 nm) did not alter in water-salt or ethanol solutions as well as in water solution (figure 1).

Table 1. Thymidine molar extinction coefficient in the different solvents.

Solvent	ε (λ_{max}=268 nm)±50, $M^{-1}cm^{-1}$
water	9560
ethanol 84% v/v	7860
1 M NaCl	9740
3.6 M NaCl	10200
5M NaCl	10700

CD spectra of unirradiated solutions are shown in figure 2a. Summarizing the observed spectral alterations of dT due to the solvent change one can suggest the following. It is known that in water solution nucleosides have the tendency to stacking associates formation [16, 17]. The electrolyte concentration increase leads to the demolition of these associates or changes in their mutual orientation. It was revealed from the growing hyperchromic effect in dT UV absorption spectra with the rising in ionic strength of solutions (see table 1, figure 1). On the contrary, in ethanol dT solutions we could see the hypochromic effect in comparison with water medium (table 1, figure 1) and also the considerable monotonous the CD bands intensity upward growth (especially the positive band, which rocketed from $\Delta\varepsilon(274)=1.4$ in water to $\Delta\varepsilon(274)=4.6$ in ethanol 84% v/v) attending to ethanol concentration increase. It can be caused by stronger thymidine molecules stacking association in ethanol than in water (this conclusion is corroborated by the data of nucleosides solubility in water and ethanol [18]), or by changes in the dT orientation in the associate.

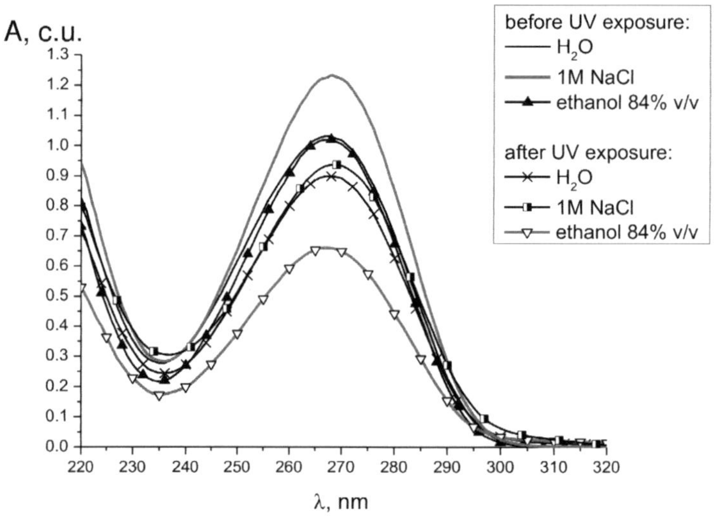

Figure 1. UV absorption spectra of thymidine solutions, conditions and type of solvent are indicated on the graph. Thymidine concentrations in water $1.08*10^{-4}$M, in 1M NaCl $1.06*10^{-4}$M, in ethanol $1.3*10^{-4}$M. The absorbed dose for these solutions irradiated for approximately 5h: D(water) = $6.0*10^{8}$ J/mol (dT), D(1M NaCl) = $4.2*10^{8}$ J/mol (dT), D(1M NaCl) = $3.6*10^{8}$ J/mol (dT).

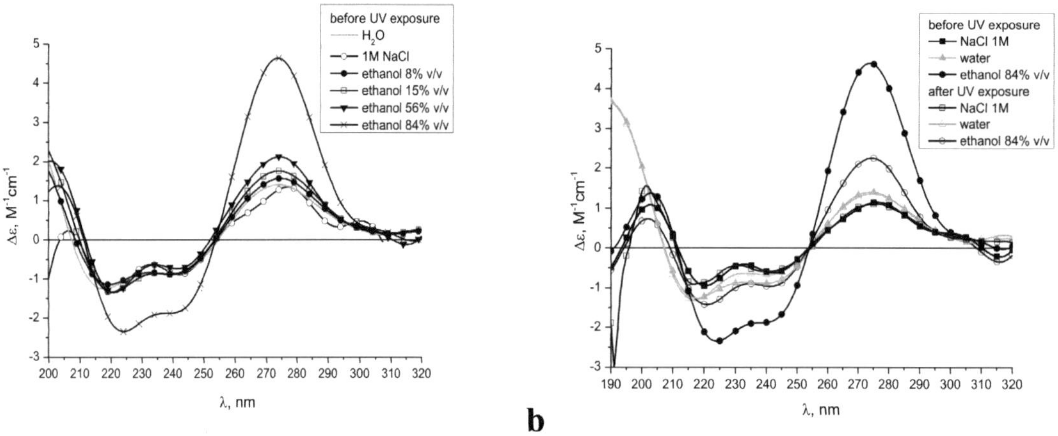

a **b**

Figure 2. CD spectra of thymidine solutions, conditions and type of solvent are indicated on the graph. Thymidine concentrations and absorbed doses are as in Figure 1. (a) CD spectra of solutions before UV-irradiation. All range of ethanol concentrations (points are 8, 15, 56, 84% v/v) is represented. (b) CD spectra of solutions before and after UV-irradiation. There are demonstrated 1M NaCl, water, and 84% v/v ethanol solutions only.

The MS results confirmed the supposition about the rise in dT association (or aggregation) with the increase in ethanol concentration. In Figure 3 the examples of MS ESI data are presented, all the obtained MS results are given in table 2. The intensity of MS peaks assigned to the various dT associates mainly larger in ethanol than in water: eight, and two and a half as much in unradiated, and irradiated solutions, respectively. However, one should be careful in comparing MS data acquired in different solvents: in the medium with smaller dielectric constant (ethanol) the molecular ionization is more preferable. At high electrolyte concentration the mass spectrometrically detected analyte ion intensity decrease sharply [19], so we did not obtain the reliable MS data for dT solutions in 1M NaCl.

a

b

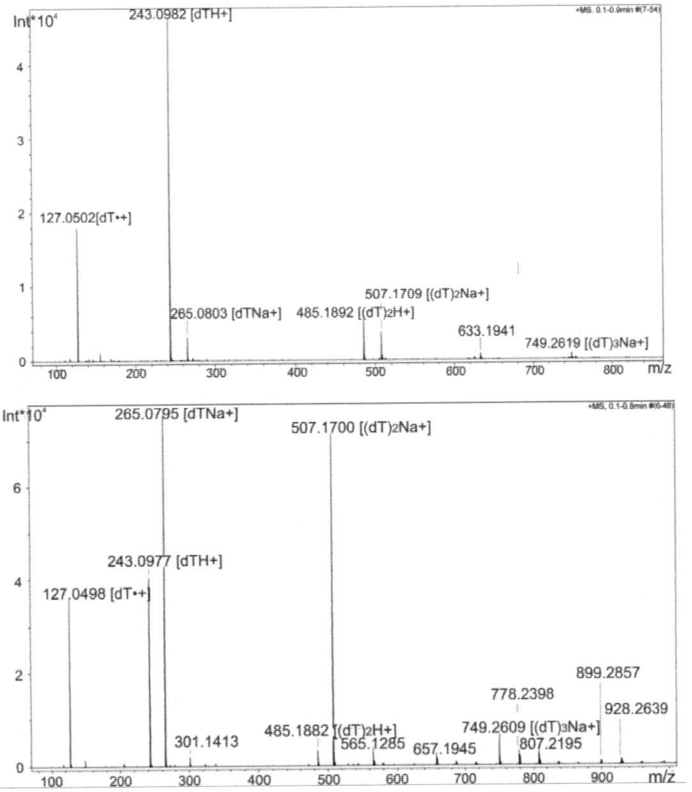

Figure 3. Mass-spectra of dT in water (a) and ethanol 84 % v/v (b) solution before UV-irradiation.

Table 2. The main identified peaks in mass-spectra of thymidine solutions

m/z	ion	water				ethanol 84% v/v			
		before UV-exposure		after UV-exposure		before UV-exposure		after UV-exposure	
		I	I/I(243),%	I	I/I(243),%	I	I/I(243),%	I	I/I(243),%
265.08	dTNa+	3160	7	1000	1	75700	185	28000	34
243.09	dTH+	45800	100	87000	100	40900	100	82500	100
485.19	$(dT)_2H+$	**5900**	**13**	**15000**	**17**	**3040**	**7**	**10000**	**12**
507.17	$(dT)_2Na+$	**3700**	**8**	**1000**	**1**	**71300**	**174**	**29000**	**35**
749.26	$(dT)_3Na+$	**500**	**1**	-		**6100**	**15**	**3000**	**4**
127.05	T•+	17900	39	40000	46	36500	89	50000	61
625.21	$(dT)_5$ K+H+	-		-		-		**730**	**1**
149.02	TNa+			-		-		2000	2
sum of associates peaks intensities		10100		16000		80440		42730	

The absorbance and CD intensity of dT solutions decreased monotonously with the UV-irradiation dose rise (see figure 1, 2b, 4), at that UV-irradiation affected dT spectral properties in larger extent in ethanol and water-salt solutions than in water. Photodimers formation should lead to the decrease in UV absorbance at 268 nm and rise in absorbance at shorter wavelengths [2]. Note, that the number of dT associates, according to the MS data, after the UV-exposure increased in water (by about 58%) but declined in ethanol (by about 47%) (table 2). This result is unexpected because the probability of photodimerization rises with the increase in association degree of dT in a solution. It is possible that the decrease in solvent polarity leads to a fall in quantum yield of thymidine photodimerization, maybe through decrease in lifetime of dT excited states, or by rearrangement of dT in an aggregate to the less proper for the photodimerization process orientation. It should be mentioned also that there is the information about chromophore destruction under the action of high doses (about 1W/m^2) of UV-irradiation [20], so at the conditions of our experiment we cannot exclude this process.

Consequently our results indicated that the association degree of thymidine in a solution did not play a main role in its photosensitivity. We can assume that photoprocesses in dT solutions depend on the thymidine-solvent interaction causes the structure of dT associates and the lifetime of dT excited states.

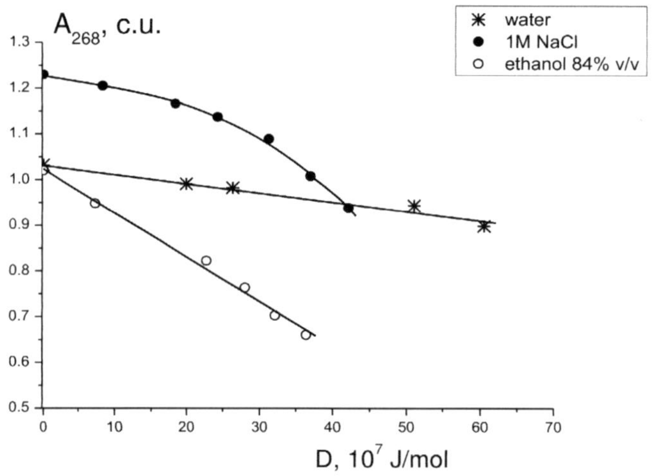

Figure 4. UV-radiation dose dependence of thymidine solutions absorbance intensity in maximum of the spectrum

Acknowledgments

A part of scientific research was performed at the Center for Chemical Analysis and Materials Research of Research park of St. Petersburg State University. Authors are thankful to Dr. Chikhirzhina E. V. for the help in CD measurement and Dr. Moroshkina E.B. for useful discussion. This work is supported by the Russian Foundation for Basic Research (project No 13-03-01192) and SPSU (11.38.644.2013).

References

[1] Smith K C, Hanawalt P C 1972 *Molecular Photobiology* (NY, London: Academic Press)
[2] Mu W, Zhang D, Xu L, Luo Zh, Wang Yu J 2005 Activity assay of His-tagged E. coli DNA photolyase by RPHPLC and SE-HPLC *Biochem. Biophys. Methods, 63,* pp 111–124
[3] Rubin A B 2004 *Biophysics, Biophysics of Cel Processes* (Moscow: MSU Publishers)
[4] Fisher G J, Johns H E 1970 Ultraviolet photochemistry of thymine in aqueous solution *Photochemistry and Photobiology, 11,* pp 429-444
[5] Durbeej B, Eriksson L A 2002 Reaction mechanism of thymine dimer formation in DNA induced by UV light *J. Photochem. Photobiol A, 152,* pp 95-101
[6] Johnson A T, Wiest O 2007 Structure and Dynamics of Poly(T) Single-Strand DNA: Implications

toward CPD Formation *J. Phys. Chem. B, 111,* pp 14398-14404

[7] Paston S V, Sushko M L, Mel'nik B S, Rumyantceva I M, Zyryanova I M, Zarubina O P, Frisman E V 2002 A comparison of the influence of substances stabilizing and destroying the water structure on the conformational changes of DNA after gamma-irradiation of its solutions *Biofizika, 47,* pp 453-458

[8] Dommes O A, Paston S V 2011 Investigation of radiation-induced damages in DNA bases and secondary structure *Eur. Biophysics J. with Biophysics Lett. 40 (Suppl.1),* p 97

[9] Ravanat J-L, Douki T, Cadet J 2001 Direct and indirect effects of UV radiation on DNA and its components *J. Photochem. Photobiol. B,* 63, pp 88-192

[10] Lakowicz J R 1983 *Principles of Fluorescence Spectroscopy* (NY, London: Plenum Press)

[11] Snyder A 1996 *Biochemical and Biotechnological Applications of Electrospray Ionization Mass Spectrometry; ACS Symposium Series; American Chemical Society* (Washington, DC)

[12] Armentano D, De Munno G, Di Donna L, Sindona G, Giorgi G, Salvini L 2004 Self-Assembling of Cytosine Nucleoside Into Triply-Bound Dimers in Acid Media. A Comprehensive Evaluation of Proton-Bound Pyrimidine Nucleosides by Electrospray Tandem Mass Spectrometry, X-rays Diffractometry, and Theoretical Calculations *J. Am. Soc. Mass Spectrom., 15,* pp 268-279

[13] Ding J, Barlow T, Dipple A, Vouros P 1998 Separation and Identification of PositivelyCharged and Neutral Nucleoside Adducts by Capillary Electrochromatography-Microelectrospray Mass Spectrometry *J. Am. Soc. Mass Spectrom., 9,* pp 823-829

[14] Douki T, Court M, Cadet J 2000 Electrospray–mass spectrometry characterization and measurement of far-UV-induced thymine photoproducts *J. Photochem. Photobiol. B: 54,* pp 145–154

[15] Beaven G H, Holiday E R, Johnson E A 1955 *Optical properties of nucleic acids and their components. The Nucleic Acids* (NY: Academic Press)

[16] Sigel H 1998 Metal ion-assisted stacking interactions and the facilitated hydrolysis of nucleoside 5'-triphosphates *Pure & Appl. Chem., 70,* pp 969-976

[17] Danilov V I, Van Mourik T 2008 A PM6 quantum chemical study of the H-bonded and stacked associates of the adenine and thymine DNA bases: The nature of base stacking *Molecular Physics, 106,* pp 1487-1494

[18] Concise Encyclopedia Chemistry 1994 (Berlin: De Gruyter)

[19] Constantopoulos T L, Jackson G L, Enke C G 1999 Effects of Salt Concentration on Analyte Response Using Electrospray Ionization Mass Spectrometry *J Am Soc Mass Spectrom, 10*, pp 625–634

[20] Rubin L B, Menshonkova T N, Simukova N A, Budowsky E I 1981 The effects of high-intensity UV-radiation on nucleic acids and their components – I. Thymine *Photochemistry and Photobiology, 34,* pp 339-344

Journal of Physics: Conference Series **661** (2015) 012020 doi:10.1088/1742-6596/661/1/012020

Thermodynamics of interaction and structure of DNA complexes with phenacylimidazo[5,1-a]isoquinoline derivatives

D N Osinnikova, E B Moroshkina and D M Glushkina

St. Petersburg State University, 1 Ul'ianovskaia str., Peterghof, St. Petersburg 198504, Russia

E-mail: osinnikovadasha@yandex.ru

Abstract. Interaction of native calf thymus DNA (ctDNA) with phenacylimidazo[5,1-a]isoquinoline derivatives was studied by the methods of spectrophotometry, viscometry, isothermal titration calorimetry (ITC) and dynamic birefringence. It was found that both of investigated compounds form complexes with the DNA molecule, the structure of compounds affects the mode of binding these ligands to DNA. The primary binding mode can not be described by the classical models of groove binding or intercalation. It has been suggested that the primary mode of binding is "partial intercalation".

1. Introduction

Nowadays, the problem of the synthesis of new drugs, including drugs based on known compounds, whose effectiveness has been proved, is relevant. Alkaloids are a broad class of biological active compounds. Their chemical structure and mechanisms of action are various. One of well-known antispasmodic drug, isoquinoline derivative, is papaverine. The mechanism of its biological activity is to inhibit the phosphodiesterase [1].

The DNA molecule is a target for many biological active compounds possessing a flat heterocyclic chromophore such as anthracycline antibiotics, isoquinoline alkaloids and many others [2]. The interaction of these compounds with DNA may occur in various ways. The most common ones are groove binding and intercalation.

In the first case the molecule of the compound is located in the minor groove of the DNA double helix [3]. In the case of intercalation, flat hydrophobic chromophore of the molecule integrates into the DNA double helix between adjacent pairs of nitrogenous bases, causing local changes in the structure of the double helix [4].

Investigated in present work compounds - phenacylimidazo[5,1-a]isoquinoline derivatives, were synthesized as analogues of papaverine. They possess a flat heterocyclic chromophore that allows both intercalation and DNA groove binding [5].

Thermodynamic parameters of the interaction of these compounds with DNA, as well as mode of their binding with DNA double helix and structure of the complexes formed, were determined with calorimetric, spectroscopic and hydrodynamic methods.

2. Materials and methods

2.1. Materials

The interaction of DNA with phenacylimidazo[5,1-a]isoquinoline derivatives (figure 1) was studied by the methods of spectrophotometry, viscometry, ITC and dynamic birefringence.

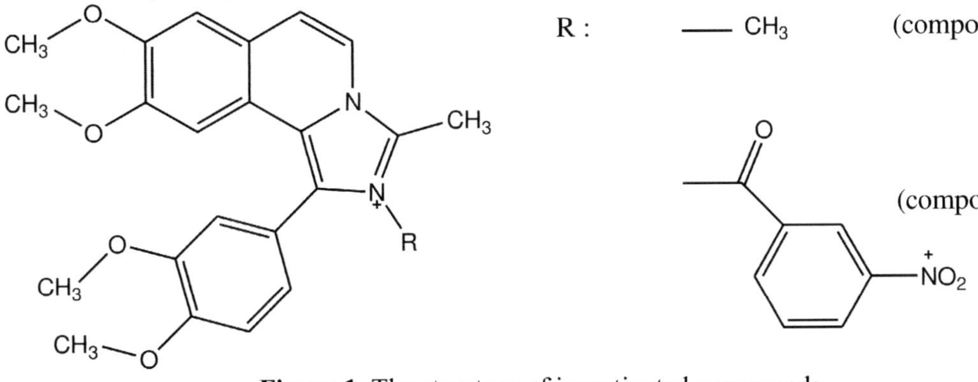

(compound 1)

(compound 2)

Figure 1. The structure of investigated compounds

The compounds were synthesized in the Research Institute of Hygiene, Occupational Pathology and Human Ecology FMBA of Russia by Krivorotov D. V. [6].

High-molecular-weight calf thymus DNA of firm "Sigma" (USA) has been used. Extinction coefficient ε_{260} = 6400–6700 $M^{-1}cm^{-1}$. The complexes were prepared by mixing the DNA and ligand solutions of the necessary concentrations. The ionic strength of the solution (μ) was 0.001 M and subsequently remained unchanged.

2.2. Isothermal titration calorimetry (ITC)

ITC experiments were performed using a TA Instruments Nano ITC 2G Microcalorimeter at Thermogravimetric and Calorimetric Research Centre St.Petersburg State University. Reaction cell volume was 1.4 mL. Aliquots of investigated compounds (7 µL) were injected from a 250-µL rotating syringe (300 rpm) into the isothermal sample cell containing 1.4 mL of DNA solution. All the experiments were carried out at 21°C. Delay time between each injection was 500 s. Each injection generated a heat burst curve (µJ per second), the area under which was determined by integration using Origin software version 8.6 that gave the measure of the heat of reaction associated with the injection. The heat associated with dilution of each ligand injection was subtracted from the corresponding heat associated with the ligand-DNA injection to give the heat of ligand binding for that injection. To determine the thermodynamic parameters of binding - binding constant, enthalpy of complex formation and the number of binding sites, the equation 1 was used. It corresponds to the model - 1 ligand molecule for 1 binding site [7].

$$Q = \frac{nM\Delta HV_0}{2}\left[1 + \frac{X}{nM} + \frac{1}{nKM} - \sqrt{\left(1 + \frac{X}{nM} + \frac{1}{nKM}\right)^2 - \frac{4X}{nM}}\right], \tag{1}$$

where Q – the heat released or absorbed during the titration, ΔH – binding enthalpy, V_0 – cell volume, X – ligand concentration, M – DNA concentration in mol base pares (bp), K – intrinsic binding constant, n – number of binding sites.

2.3. Spectrophotometric titration (SPT)

Absorption spectra were measured by Shimadzu UV-1800. The titration was carried out in a wide range of ratios of concentrations ligand/DNA when concentration of the ligand was constant. The obtained data were analyzed in two ways: the neighbor exclusion model of McGhee and von Hippel (equation 2) [8] and Scatchard model with 1 ligand molecule for 1 binding site (equation 3) [9]:

$$\frac{r}{C_f} = k(1 - lr)\left[\frac{(1 - lr)}{[1 - (l - 1)r]}\right]^{l-1}, \tag{2}$$

where r – molar ratio of binding ligand to DNA (bp), C_f – concentration of free ligand, k – binding constant, l – the binding site size in bp;

$$\frac{r}{C_f} = K(n - r), \tag{3}$$

where r – molar ratio of binding ligand to DNA (bp), C_f – concentration of free ligand, K – binding constant, n – the number of binding sites.

2.4. Viscometry

The viscosity of solutions of pure DNA and DNA-ligand complexes has been measured with the magnetic rotational viscometer [10]. Measurements were carried out at the experimental setup to determine the viscosity of non-Newtonian fluids. Velocity gradient of viscometer $g = (0.1 - 0.5)\ s^{-1}$.

According to Flory formula [11], the intrinsic viscosity of the macromolecule is related to its parameters by the ratio:

$$[\eta] = F\frac{\left(\overline{h_0^2}\right)^{3/2}}{M_w}\alpha^3, \tag{4}$$

where $[\eta]$ - the intrinsic viscosity, F - Flory constant for a given polymer-solvent system, $\overline{h_0^2}$ – the mean-square distance between the ends of the chain, M_w – molecular weight, α – the coefficient of linear swelling.

For freely jointed chain:

$$h_0^2 = LA, \tag{5}$$

where L – contour length, A - length of statistical segment.

2.5. Dynamic birefringence

In parallel with the viscometry studies the dynamic birefringence of the same solutions was studied. The value of dynamic birefringence Δn was measured in the titanium dynamo-optimeter with internal rotor using the optical device with half-shadow elliptical compensator [12]. The Peterlin ratio [13] was calculated to determine the changes in thermodynamic rigidity of macromolecule upon complex formation (A_r/A_0). The Peterlin ratio $\Delta n/g(\eta - \eta_0)$ proportional $(a_1-a_2)S$, where Δn – value of the dynamic birefringence, g – gradient flow rate, η и η_0 – viscosity of the solution and solvent respectively, (a_1-a_2) – polarizability difference of monomer residues along and perpendicular to the axis of the statistical segment of the macromolecule, S – the number of monomers in the segment. Analysis of the results was carried out according to the formula:

$$\frac{A_r}{A_0} = \frac{S_r}{S_0} = \left(\frac{\Delta n}{g(\eta - \eta_0)}\right)_r \Big/ \left(\frac{\Delta n}{g(\eta - \eta_0)}\right)_0 \bullet \frac{(a_1 - a_2)_0}{(a_1 - a_2)_r}, \tag{6}$$

3. Results and discussion

The thermodynamic parameters of the interaction and the stoichiometry of the complex were determined by ITC and SPT.

The results of calorimetric titration of DNA by compound 1 are shown in Figure 2. For compound 2 the results are similar.

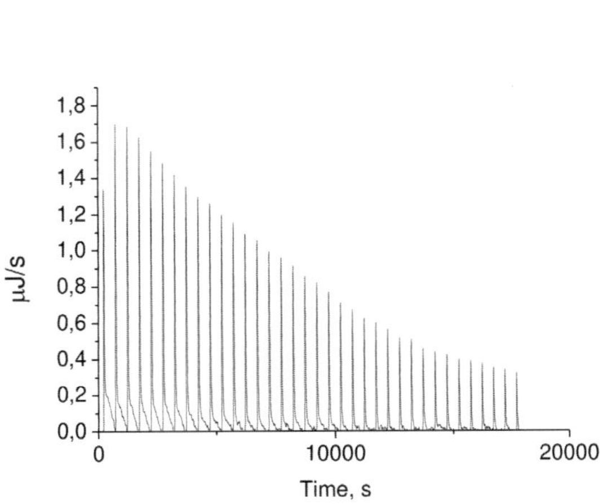

Figure 2. The raw ITC data of compound 1 (640 μM) binding to ctDNA (200μM)

The resulting thermograms are shown in Figure 3.

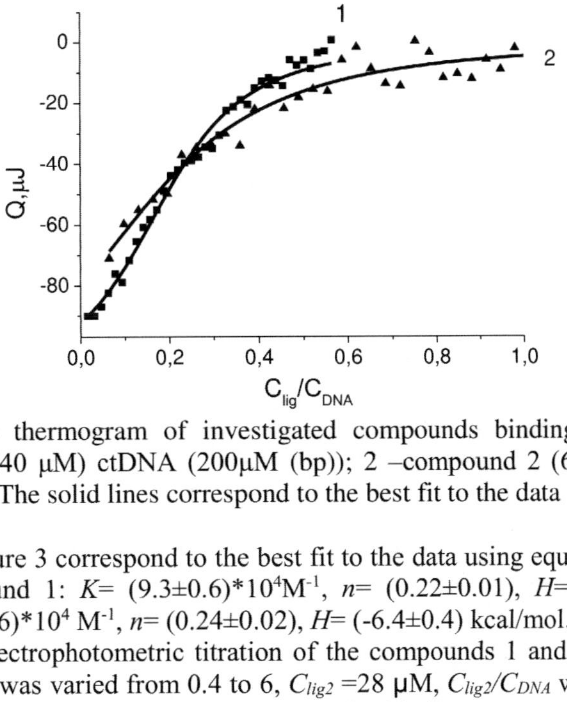

Figure 3. The thermogram of investigated compounds binding to ctDNA (1-compound 1 (640 μM) ctDNA (200μM (bp)); 2 –compound 2 (600 μM), ctDNA (100 μM (bp)). The solid lines correspond to the best fit to the data

The solid lines in figure 3 correspond to the best fit to the data using equation 1 with the following parameters: for compound 1: K= (9.3±0.6)*10^{4}M^{-1}, n= (0.22±0.01), H= (-6.2±0.2) kcal/mol; for compound 2: K= (7.6±0.6)*10^4 M^{-1}, n= (0.24±0.02), H= (-6.4±0.4) kcal/mol.

The results of the spectrophotometric titration of the compounds 1 and 2 are shown in Figure 4. C_{lig1} = 24μM, C_{lig1}/C_{DNA} was varied from 0.4 to 6, C_{lig2} =28 μM, C_{lig2}/C_{DNA} was varied from 0.2 to 4.7. The presence of an isobestic point allows us to plot the binding curves and determine the binding parameters. Binding isotherms for both compounds are shown in figure 5. Thermodynamic parameters of binding for compound 1 were calculated using the McGhee and von Hippel model (equation 2) k= (1.88±0.07)*10^{5}M^{-1}, l= (2.24±0.06), and using Scatchard model: K= (4.7±0.3)*10^{5}M^{-1}, n= (0.36±0.03). For compound 2 the character of binding is more complex, when r> 0.5 there is a second way of binding, therefore determination of binding parameters from the absorption spectra is not possible.

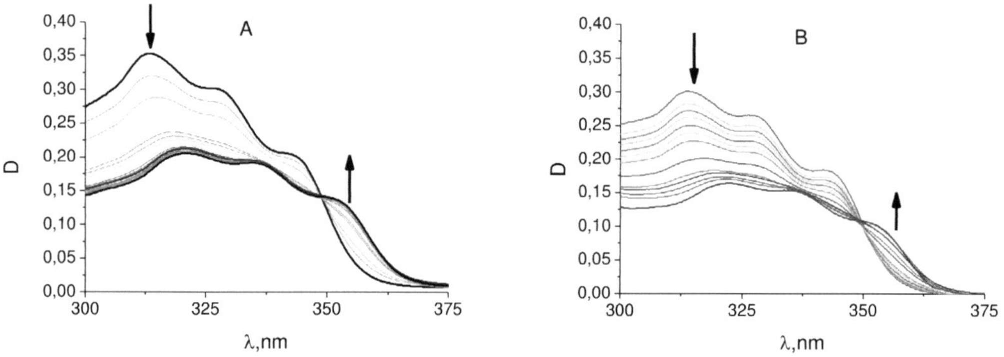

Figure 4. Absorption spectra of complexes DNA with compound 1(C_{lig}=24 μM) (A), with compound 2(C_{lig}=28 μM) (B)

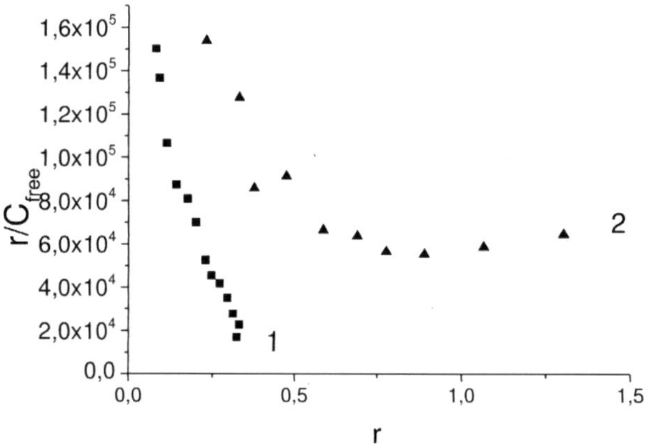

Figure 5. Binding isotherms 1- complexes DNA-compound 1; 2- complexes DNA-compound 2

Comparison of the results obtained from both methods shows that the results obtained from the spectrophotometric titration give significantly higher binding constant value regardless of the model used for the calculations. It can be assumed that the reason for the difference lies in the different titration procedure applied at ITC and SPT.

A mode of binding of investigated compounds with DNA was determined by using viscometry method and method of dynamic birefringence [14]. For this purpose, parallel measurement of the intrinsic viscosity and optical anisotropy (the Peterlin ratio [9]) of the DNA-ligand complexes was carried out (r = 0.2). The results of measurement are given in the table 1. The absence of significant changes of Peterlin ratio during the binding of ligands with DNA indicates constancy of the thermodynamic rigidity of macromolecule. Therefore, the relative change of the contour length of the macromolecule (L_r/L_0), calculated from the values of intrinsic viscosity for complex DNA-compound 1 is $(L_r/L_0)_1$= (1.13±0.05) and for complex DNA-compound 2 is $(L_r/L_0)_2$= (1.09±0.05). In the case of classical intercalation binding when r= 0.2, the change of the contour length (L_r/L_0)=1.2. We can assume that in this case there is a partial or incomplete intercalation [5, 15] of chromophores of investigated compounds into the DNA double helix.

Table 1. The results of measurements of viscosity and dynamic birefringence

	$[\eta]$ (dL/g)	$\Delta n/g(\eta - \eta_0) \cdot 10^8$ ($cm \cdot s^2/g$)
DNA	150±10	25±1
DNA-compound 1	175±10	25±1
DNA-compound 2	174±10	26±1

4. Conclusions

Both of the investigated compounds form complexes with the DNA molecule. The presence of bulky substitute in the isoquinoline chromophore affects the nature of the binding to macromolecule. Thus, compound 2 besides the primary binding mode has secondary one. It was shown that both compounds have similar values of enthalpy, binding constants, and the number of primary binding sites determined by ITC. The binding constant and the number of primary binding sites determined by spectrophotometry have higher values regardless of the model used for the calculation. Changes of macromolecular parameters upon binding of both compounds with DNA are not consistent with the expected in the case of classical intercalation or groove binding. It can be assumed that the compounds partially intercalated into the DNA double helix.

The work was supported by the Russian Foundation of Basic Research (grant 13-03-01192A) and by the grant of St. Petersburg State University (11.38.644.2013).

References

[1] Belousov Y B 2002 *Rus. Med. Zh. (Rus)* **10** 669
[2] Li H H, Aubrecht J, Fernace A J 2007 *Mutat. Res.* **623** 98-108
[3] Dervan P B 2001 *Bioorg. Med. Chem.* **9** 2215-1135
[4] Ihmels H and Otto D 2005 *Top Curr Chem* **258** 161–204
[5] Moroshkina E B, Krivtsova M A and Glibin E N 2002 *Biofizika (Rus)* **47** 444–448
[6] Krivorotov D V, Vorob'ev M V, Polukeev V A and Glibin E N 2006 *Zhurnal Organicheskoi Khimii Vypusk 4(Rus)* **42** 594-596
[7] Breslauer K J, Freire E and Straume M 1992 *Methods Enzymol.* **211** 533–567
[8] McGhee J D and von Hippel P H 1974 *J. Mol. Biology* **86** 469-482
[9] Scatchard G 1949 *Ann. N. Y. Acad. Sci.* **51** 660
[10] Frisman E V, Shchagina L V and Vorob'ev V I 1965 *Kolloidn. J.* **27** 130
[11] Fox T and Flory P 1950 *J. Polymer Sci.* **6** 74
[12] Frisman E V and Cvetkov V N 1952 *Zhurnal Eksperimentalnoi I Teoreticheskoi Fiziki (Rus)* **23** 690-702
[13] Peterlin A J. 1950 *Polymer Sci.* **12** 45-52
[14] Veselkov A N, Moroshkina E B, Soboleva O I and Frisman E V 1984 *Mol. Biol. (Rus)* **18** 481-87
[15] Moroshkina E B 2011 *Vestnik SPbGU Seria 4* **4** 114-123

Using of the interictal EEGs for epilepsy diagnosing

O Yu Panischev, S A Demin, E M Zinatullin

Kazan Federal University, Institute of Physics, Kazan, 420008 Russia

E-mail: opanischev@gmail.com

Abstract. In this work we apply a new method to determine the differences in characteristics of the cortical electroencephalographic (EEG) activity, measured during interictal stage (i.e., period between seizures), between healthy subjects and patients with epilepsy. To analyze the dynamical and spectral properties of bioelectric activity we use power spectra and phase portraits which are introduced on the basis of the Memory Function Formalism (MFF). We discover the significant differences in the types of power spectra of the EEG for healthy subjects and patients. We reveal the cerebral cortex areas for which the EEG activity of considered groups of subjects has a different structure of the phase portraits. The proposed approach can be used as an additional method for diagnosis of epilepsy during interictal stage.

1. Introduction

Epilepsy is one the most common chronic neurological diseases, which is characterized by recurrent seizures, due to different abnormalities of the brain activity [1]. Inasmuch as the seizures can occur in different abnormalities of the brain and central nervous system, it is necessary to note that the epilepsy diagnosing during the interictal period is a complicated problem. To solve it, the EEG before, during and after seizure are used [2]. EEG monitoring, as a noninvasive method to define epileptogenic cortex, effectively reveals the characteristics of brain activity in several epilepsy-related syndromes. The epileptic cortical brain activities have specific spike-wave patterns of EEG discharge. Another important problem in treatment of epilepsy is the long-range prediction of recurrent seizure in patients. Epileptogenesis episodes can be separated by different extended periods, when disease is not manifested. To detect epileptic seizures the long-term EEG monitoring during ictal and interictal stages is used [3].

A reduce in synchronization of interictal EEG activity have been found in epilepsy patients contrast to healthy people [4, 5].

To develop an additional method of the EEG-based diagnostic of the epilepsy during the interictal stage, in this paper we use the memory function formalism [6, 7, 8].

2. Background of the memory function formalism method

Within the framework of MFF [6, 7, 8], we consider stochastic dynamics of the EEG signal as the sequence $\{x_j\}$ of a random value X:

$$X = \{x(T), x(T+\tau), x(T+2\tau), \dots, x(T+(N-1)\tau)\} = \{x_0, x_1, x_2, \dots, x_{N-1}\}, \tag{1}$$

$$x_j = x(T+j\tau), \qquad \langle X\rangle = \frac{1}{N}\sum_{j=0}^{N-1} x_j, \qquad \delta x_j = x_j - \langle X\rangle, \qquad \sigma^2 = \frac{1}{N}\sum_{j=0}^{N-1}\delta x_j^2, \tag{2}$$

where T is the initial time point, τ is the time interval of signal discretization, $\langle X\rangle$ is the mean value of X, δx_j is fluctuation.

To describe the probabilistic relations in the sequence of X we use the normalized time-dependent correlation function

$$a(t = m\tau) = M_0(m\tau) = \frac{1}{(N-m)\sigma^2}\sum_{j=0}^{N-m-1}\delta x_j \delta x_{j+m}, \tag{3}$$

which is rewritten further as a scalar product of vectors of the initial state

$$\mathbf{A}_k^0 = \mathbf{A}_k^0(0) = \{\delta x_0, \delta x_1, \dots, \delta x_{k-1}\} \tag{4}$$

and of system state in time moment t

$$\mathbf{A}_{m+k}^m = \mathbf{A}_{m+k}^m(t) = \{\delta x_m, \delta x_{m+1}, \dots, \delta x_{m+k-1}\} \tag{5}$$

as follows:

$$a(t) = \frac{\langle \mathbf{A}_k^0(0)\mathbf{A}_{m+k}^m(t)\rangle}{\langle |\mathbf{A}_k^0(0)|^2\rangle}. \tag{6}$$

The technique of projection operators was used to calculate the kinetic and relaxation parameters ($\hat{L}$ is the Liouville's quasioperator, see [6, 7, 8] for details):

$$\lambda_n = \frac{\langle \mathbf{W}_{n-1}\hat{L}\mathbf{W}_{n-1}\rangle}{\langle |\mathbf{W}_{n-1}|^2\rangle}, \qquad \Lambda_n = i\frac{\langle |\mathbf{W}_n|^2\rangle}{\langle |\mathbf{W}_{n-1}|^2\rangle}, \tag{7}$$

the dynamic orthogonal variables:

$$\mathbf{W}_0 = \mathbf{A}_k^0(0), \qquad \mathbf{W}_1 = (i\hat{L} - \lambda_1)\mathbf{W}_0, \qquad \mathbf{W}_2 = (i\hat{L} - \lambda_2)\mathbf{W}_1 - \Lambda_1\mathbf{W}_0 - \dots, \tag{8}$$

the memory functions of the n order $(a(t) = M_0(t))$:

$$M_n(t = m\tau) = \frac{\langle \mathbf{W}_n\{1 + i\tau\hat{L}_{22}\}^m \mathbf{W}_n\rangle}{\langle|\mathbf{W}_n|^2\rangle}, \tag{9}$$

and their power spectra:

$$\mu_n(\nu) = \left|\tau \sum_{j=0}^{N-1} M_n(j\tau)\cos 2\pi\nu j\tau\right|^2. \tag{10}$$

3. Experimental Data

Data of EEG activity were analyzed in 8 healthy people (with no reported psychiatric or neurological disorders, mean age of 32.5 years) and 8 patients with epilepsy (mean age of 28.4 years), the subjects were within the age range 18-65 years. All experimental procedures were performed with the consent of the subjects. Multichannel EEG activity was recorded using 16 electrodes: F1, F2, F7, F3, F4, F8, T3, C3, C4, T4, T5, P3, P4, T6, O1, O2 on the scalp according to the standard 10-20 International electrode placement system. The electrodes are numbered (Fig. 1) from 1 to 16. During the EEG recording the eyes of subject were closed. The sampling rate was 200 Hz and the signal was filtered between 0.1-70 Hz. A notch filter of 50 Hz was also used. An epoch length of $\sim$ 10 s of uninterrupted EEG data, which were free from any visual complexes, (e.g., the spike wave complexes for seizure patients) were chosen for analysis. Baseline drift was subtracted by a polynomial of 2nd order [4].

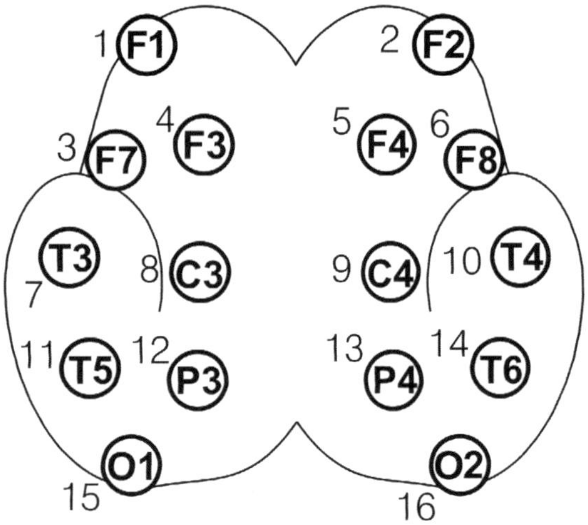

Figure 1. Positions of the 16 electrodes including their number and their designations. The schemata are based on the internationally established 10-20 system. Midline electrodes are exempted from this study.

4. Dynamical and spectral analysis of the interictal EEG

Analysis of EEG activity showed that in all subjects the power spectra has 4 types relative to the frequency bands. I-type spectra ranged within 1–4 Hz frequency band without peak

(Fig. 2a). The II-type and III-type had the peak within 5–16 Hz and 49–52 Hz frequency bands, respectively (Fig. 2b, c); while IV-type spectra had both peaks of frequency bands as II-type and III-type (Fig. 2d).

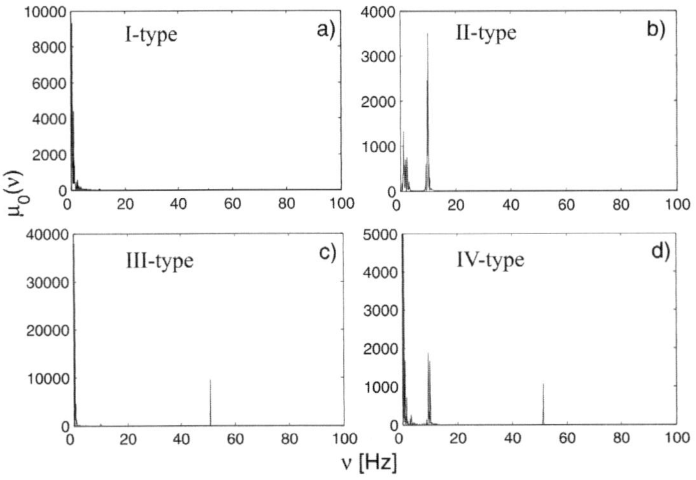

Figure 2. Power spectra of EEG activity for four frequency bands from all the subjects (a typical form of the spectra).

Fig. 3 illustrates four types of power spectra presented for different areas of EEG recording in healthy subjects and patients with epilepsy. In control group of subjects the I-type spectra of EEG activity was found in the fronto-polar (F1, F2), temporal, parietal, occipital cortical fields, while in patient with epilepsy I-type spectra was not observed in similar fields. EEG electrodes which differ in type of spectra between the studied groups: F1, F2, T5, P3, P4, T6, O1, O2.

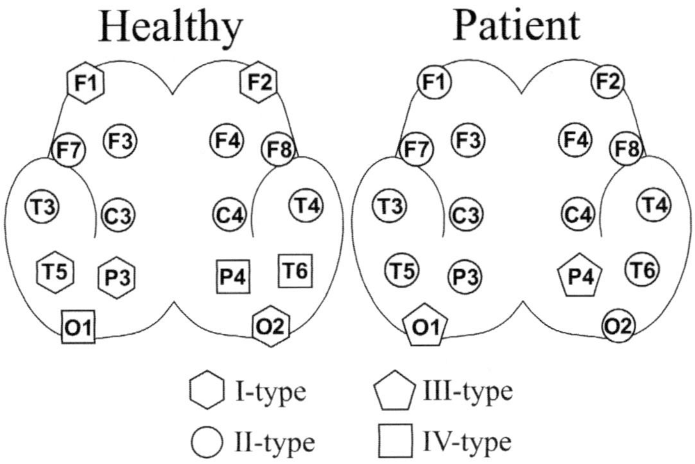

Figure 3. Types of spectra for the different areas of EEG activity in healthy subjects and patients (group average).

The epilepsy existence leads to alterations in size, form and location relative the origin of coordinates of the phase portraits (Fig. 4). The phase portraits are the simple graphical method for qualitative analysis of the studied dynamics [6, 7]. EEG activity variation is wider and has asymmetric shape in patients in contract to healthy subjects. Phase portraits of

EEGs from healthy subjects are characterized by compact symmetrical structure with separate circumferential points. In case of epilepsy for several electrodes (F1, T5, O2) we see the specific structure with high number of circumferential points. These electrodes F1 (Frontal lobe), T5 (Parietal lobe), O2 (Occipital lobe) have a high diagnostic significance for epilepsy identification during the interictal stage.

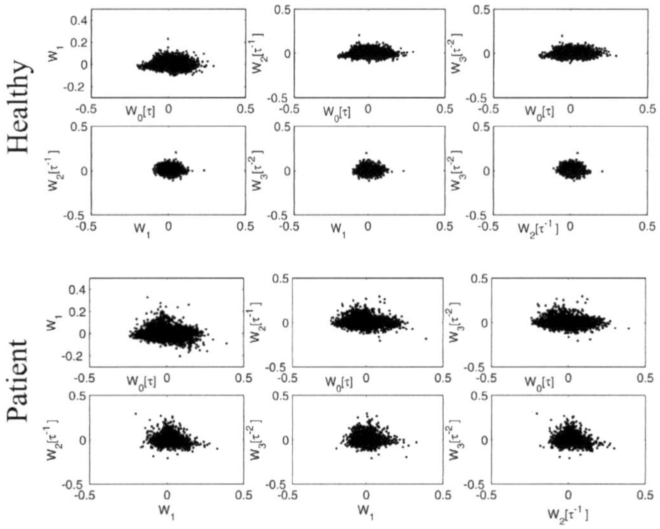

Figure 4. Phase portraits of the dynamic orthogonal variables of EEG signal from F1 electrode for one of healthy subjects and epilepsy patients.

5. Conclusions

Studying the interictal EEGs is one of the important problem for early detection of the epileptogenesis. Using the modern methods of time series analysis gives a deeper understanding of the processes at the pathology [9]. The MFF-based approach allows to reveal the dynamic and spectral properties of the EEG activity at the epilepsy during the interictal stage. Particularly, we have determined the types of the power spectra for EEG from the various cerebral cortex areas of healthy subjects and epilepsy patients. To determine qualitatively the pathology we use the phase portraits of the certain cerebral cortex areas. The proposed method could be applied (as an additional method) for the epilepsy diagnosis during period between seizures.

6. Acknowledgments

The work is performed according to the Russian Government Program of Competitive Growth of Kazan Federal University. This work was supported by the Russian Foundation for Basic Research, grant No 14-02-31385_mol_a.

References

[1] Chang B S, Lowenstein D H 2003 *N. Engl. J. Med.* **349** 1257

[2] Sander J W, O'Donoghue M F 1997 *BMJ* **314** 158

[3] Xie S, Krishnan S 2014 *TSWJ* **2014** Article ID 419308

[4] Bhattacharya J 2001 *Acta Neurobiol. Exp.* **61** 309

[5] Mormann F, Kreuz Th, Andrzejak R G, David P, Lehnertz K, Elger Ch E 2003 *Epilepsy Res.* **53** 173

[6] Demin S A, Yulmetyev R M, Panischev O Yu, Hänggi P 2008 *Phys. A* **387** 2100

[7] Panischev O Yu, Demin S A, Bhattacharya J 2010 *Phys. A* **389** 4958

[8] Panischev O Yu, Demin S A, Kaplan A Ya, Varaksina N Yu 2013 *Biomed. Engin.* **47** 153

[9] Panischev O Yu, Demin S A, Panischeva S N, Bhattacharya J 2015 *NPCS* **18** 230

Synchronization of EEG activity in patients with bipolar disorder

O Yu Panischev, S A Demin, I G Muhametshin, N Yu Demina

Kazan Federal University, Institute of Physics, Kazan, 420008 Russia

E-mail: opanischev@gmail.com

Abstract. In paper we apply the method based on the Flicker-Noise Spectroscopy (FNS) to determine the differences in frequency-phase synchronization of the cortical electroencephalographic (EEG) activities in patients with bipolar disorder (BD). We found that for healthy subjects the frequency-phase synchronization of EEGs from long-range electrodes was significantly better for BD patients. In BD patients a high synchronization of EEGs was observed only for short-range electrodes. Thus, the FNS is a simple graphical method for qualitative analysis can be applied to identify the synchronization effects in EEG activity and, probably, may be used for the diagnosis of this syndrome.

1. Introduction

The instrumental methods, such as the electroencephalography for diagnosis of various psychiatric disorders (BD, schizophrenia) are used restrictedly. However, at the present time few studies point out directly to a high diagnostic capability of these methods. For example, Timashev and coauthors on the basis of the FNS have formulated the criteria of degree of the susceptibility to schizophrenia in children/adolescents [1]. Panischev et al. have demonstrated that the method based on the memory function formalism [2] can discover the susceptibility to schizophrenia by analyzing the collective phenomena in the cerebral cortex bioelectric activity [3]. El-Badri and coauthors found alterations in spectral power of electroencephalograms (EEG) from BD patients [4]. Bhattacharya et al. discovered that for EEG of BD patients the long-range alpha coherence between frontal-temporal regions is reduced [5]. Kam and colleagues found the increase of high frequency power and disruptions in neural synchronization at BD compared with healthy subjects [6]. Kim et al. revealed in EEG of BD patients the decrease of mean synchronization in the alpha band, and the significant decreases of fronto-central and centro-parietal connections [7]. Thus the BD existence is reflected in frequency characteristics and synchronization properties of EEG activity.

Bipolar affective disorder (earlier called manic-depressive illness) is a severe anticipate mental disorder characterized by alternations of episodes of elevated mood, depression and periods of normality. The diagnosis of BD is performed by psychiatrist based on the anamnesis and clinical features of subject according to criteria assumed by World Health Organization (ICD-10) and American Psychiatric Association (DSM-5). The BD occurs as either separate disease or syndrome of other mental diseases (for instance, schizophrenia).

In this paper within the framework of the FNS we study the properties of the frequency-phase synchronization EEG activity at bipolar disorder by using two-point cross-correlator. We will show that the BD existence leads to reducing the level of synchronization of EEG signals from the long-range electrodes (if at least one electrode was situated in between or the two electrodes were from opposite hemispheres see [5], Fig. 2). The FNS-method, based on phenomenological equations for separating the studied signal into components depending of time scale, contains the parameters describing the frequency-phase synchronization [1, 8, 9].

2. Background of the Flicker-Noise Spectroscopy analysis

The FNS-analysis [1, 8, 9] assumes two types of irregularities in responses relative to a time series:

1) spike-like irregularity (i.e. chaotic component) occurs at small or short time scale,

2) jump-like irregularity (i.e. resonant component) occurs at more longer time scale.

To describe these components of irregularity at different frequency ranges of EEG activity the power spectrum $S(f)$ and structure function $\Phi^{(2)}(\tau)$ were calculated:

$$S(f) = 2\int_0^{\tau_M} \langle V(t)V(t+t_1)\rangle_{T-\tau} \cos 2\pi f t_1 \mathrm{d}t_1, \tag{1}$$

$$\Phi^{(2)}(\tau) = \langle [V(t) - V(t+\tau)]^2 \rangle_{T-\tau}, \tag{2}$$

which are derived on the basis of time correlation function

$$\psi(\tau) = \langle V(t)V(t+\tau)\rangle_{T-\tau}, \tag{3}$$

$$\langle(\ldots)\rangle_{T-\tau} = \frac{1}{T-\tau}\int_0^{T-\tau} (\ldots)\mathrm{d}t, \tag{4}$$

where τ is the lag-time ($0 < \tau < \tau_M, \tau_M \leqslant T/2$). The stochastic component of structure function $\Phi^{(2)}(\tau)$ includes only jump-like irregularities, while the stochastic component of power spectrum $S(f)$ includes spike-like and jump-like irregularities.

FNS-method describes the dynamic changes of correlation of $V_i(t)$ and $V_j(t)$ variables for points i and j by means of the "two-point" correlation expression (cross-correlator, CCF):

$$q_{ij}(\tau;\theta_{ij}) = \left\langle \left[\frac{V_i(t) - V_i(t+\tau)}{\sqrt{\Phi_i^{(2)}(\tau)}}\right] \left[\frac{V_j(t+\theta_{ij}) - V_j(t+\theta_{ij}+\tau)}{\sqrt{\Phi_j^{(2)}(\tau)}}\right] \right\rangle_{T-\tau-|\theta_{ij}|}, \tag{5}$$

where θ_{ij} is the "time shift" parameter. The value of θ_{ij} corresponding to maximum values of cross-correlation $q_{ij}(\tau;\theta_{ij})$ characterizes the cause-and-effect relation ("flow direction") between signals $V_i(t)$ and $V_j(t)$. When $\theta_{ij} > 0$, the flow moves from point i to point j, when $\theta_{ij} < 0$, from j to i. When the distance between points i and j is fixed, the value of θ_{ij} can be used to estimate the rate of information transfer between these two points. Analysis of cross-correlators allow to study the collective phenomena in distributed system.

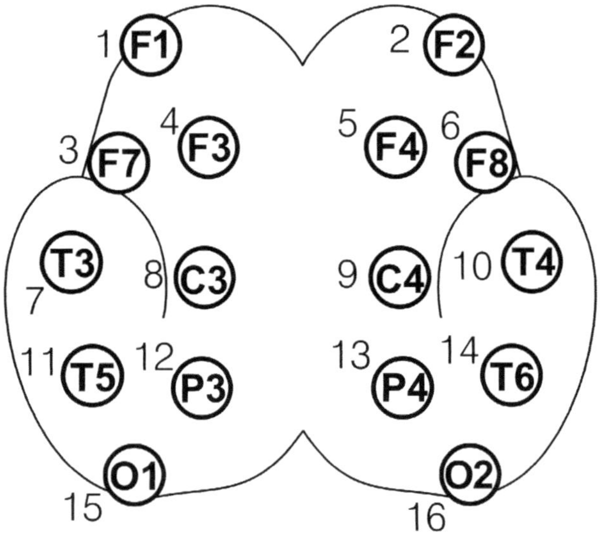

Figure 1. Positions of the 16 electrodes including their number and their designations.

3. E perimental Data

Data of EEG activity were analyzed in 8 healthy people (with no reported psychiatric or neurological disorders) and 8 patients with BD. Mean age was 32.5 in control group and 30.4 in group of BD patients. Multichannel EEG activity was recorded using 16 electrodes placed at standard positions (see Fig. 1). An average reference was used. The subjects were placed in a sound proof, light attenuated air-conditioned (20^0C) room and instructed to relax and close their eyes for some time during the data acquisition period. The sampling rate was 200 Hz and the signal was filtered between 0.1-70 Hz. A notch filter of 50 Hz was also used. Baseline drift was subtracted by a polynomial of 2nd order [5, 10]. All experimental procedures were performed with the consent of the subjects.

4. Reducing the level of the EEG long-range synchronization at bipolar disorder

The three-dimensional plots of cross-correlation function $q_{ij}(\tau, \theta_{ij})$ allows to differentiate the degree of synchronization effects in EEG signals from the long-range and short-range electrodes of healthy subjects and BD patients. Fig. 2 demonstrates the combinations of electrodes with high level of the signal synchronization with a strongly marked oscillating CCFs structure. The EEGs of healthy subjects are characterized by higher synchronization for long-range and short-range electrodes. For BD patients this synchronization level is observed only for short-range electrodes.

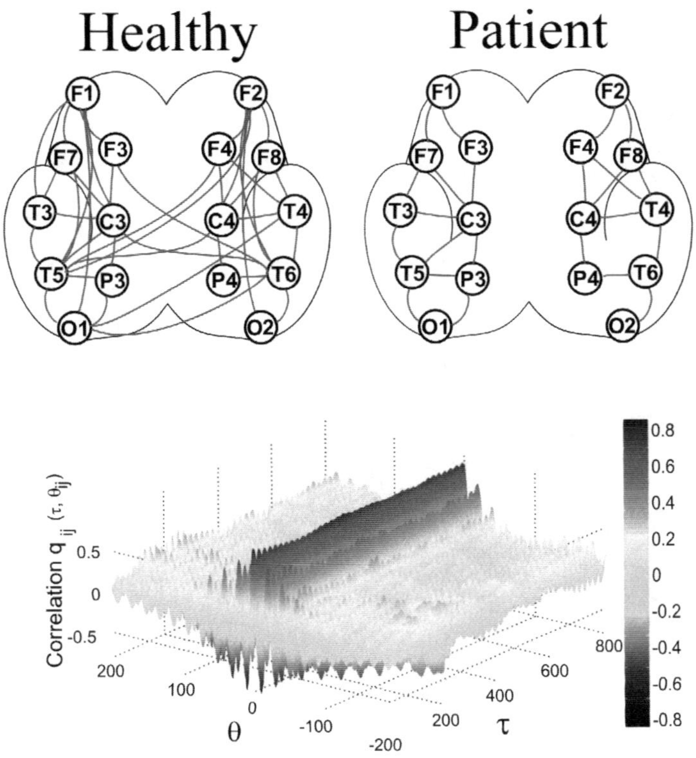

Figure 2. EEG signals with high synchronization in healthy subjects and BD patients (group average).

The another picture is observed in EEGs of BD patients for long-range electrodes (Fig. 3): the synchronization degree is lower than for the short-range electrodes. It is reflected in diffuse CCFs structure with a feeble oscillations.

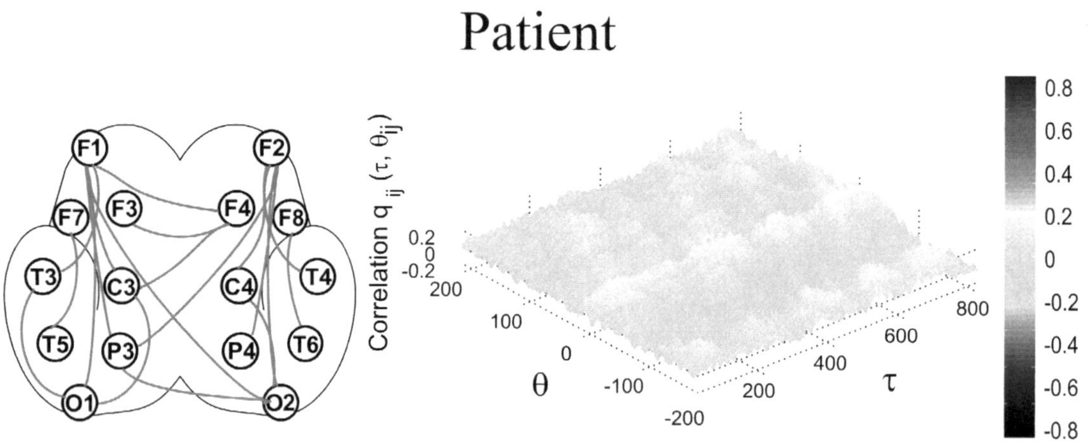

Figure 3. Lower long-range synchronization for BD patients (group average).

5. Conclusions

Developing the instrumental methods is the important problem for diagnosing the psychiatric disorders. EEG analyzing is one of the perspective approaches to solve ones. In this paper we study the synchronization effects in EEG of healthy subjects and BD patients by means of FNS cross-correlation analysis. Using the three-dimensional two-point cross-correlator is a simple graphical method for qualitative analysis of level of the frequency-phase synchronization. We reveal that the EEGs from healthy subjects demonstrate the high synchrony in short-range and long-range electrodes whilst the EEG signals of BD patients this synchronization level only for short-range electrodes. The signals from long-range electrodes for BD patients are characterized by lower synchrony. Thus the presented method is perspective for developing the new approaches for revealing the psychiatric disorders.

6. Acknowledgments

The authors are grateful to Prof. Timashev S.F. (Karpov Institute of Physical Chemistry, Moscow) for help in discussion of results. The work is performed according to the Russian Government Program of Competitive Growth of Kazan Federal University. This work was supported by the Russian Foundation for Basic Research, grant No 14-02-31385_mol_a.

References

[1] Timashev S F, Panischev O Yu, Polyakov Yu S, Demin S A and Kaplan A Ya 2012 *Phys. A*, **391** 1179

[2] Panischev O Yu, Demin S A and Bhattacharya J 2010 *Phys. A* **389** 4958

[3] Panischev O Yu, Demin S A, Kaplan A Ya and Varaksina N Yu 2013 *Biomed. Engin.*, **47** 153

[4] El-Badri S M, Ashton C H, Moore P B, Marsh V R and Ferrier I N 2001 *Bipolar Disord.*, **3** 79

[5] Bhattacharya J 2001 *Acta Neurobiol. Exp.*, **61** 309

[6] Kam J W Y, Bolbecker A R, O Donnell B F, Hetrick W P and Brenner C A 2013 *J. Psych. Res.*, **47** 1893

[7] Kim D J, Bolbecker A R, Howell J, Rass O, Sporns O, Hetrick W P, Breier A and O Donnell B F 2013 *Neur. Clin.*, **2** 414

[8] Timashev S F 2007 *Flicker-Noise Spectroscopy: Information in Chaotic Signals* (Moscow: Fizmatlit) [in Russian]

[9] Timashev S F, Polyakov Yu S, Yulmetyev R M, Demin S A, Panischev O Yu, Shimojo S and Bhattachatya J 2010 *Laser Phys.* **20** 604

[10] Bhattacharya J, Kanjilal P P and Nizamie S H 2000 *IEEE Trans. Biomed. Engin.*, **46** 738

Determination of the Diamagnetic and Paramagnetic Impurity Magnetic Susceptibility in Ge:As near the Metal Insulator phase Transition

P.V.Semenikhin[1], A.I.Veinger[1], A.G.Zabrodskii[1], T.L.Makarova[2], T.V.Tisnek[1] and S.I.Goloshchapov[1]

[1]Ioffe Institute, 194021 St .Petersburg, Russia
[2]Lappeenranta University of Technology, 53850, Lappeenranta, Finland

E-mail: petr3295@gmail.com

Abstract. Low-temperature magnetic susceptibility of heavily doped Ge: As samples has been investigated by methods SQUID magnetometry and ESR spectroscopy near the metal-insulator phase transition. Paramagnetic component of the impurity magnetic susceptibility was investigated by ESR previously. Using both techniques make possible to determined the diamagnetic component of impurity susceptibility. The value of the impurity diamagnetic susceptibility equals to $5x10^{-8}$ cm^3/g and corresponds to the localization radius of the As donor-electron near the metal-insulator phase transition.

1. Introduction

The magnetic susceptibility is a dimensionless proportionality constant that indicates the degree of magnetization of a material in response to an applied magnetic field. The full magnetic susceptibility χ of a doped semiconductor is comprised of diamagnetic magnetic susceptibility of the lattice χ_{LD} and magnetic susceptibility defined by impurities χ_I (impurity paramagnetic susceptibility χ_{IP} and impurity diamagnetic susceptibility χ_{ID}):

$$\chi = \chi_{LD} + \chi_I = \chi_{LD} + \chi_{IP} + \chi_{ID}. \quad (1)$$

Small impurities with one unpaired electron or hole result in temperature-dependent paramagnetic contribution to magnetic susceptibility of semiconductor χ_{IP}. These electrons, moving on their orbits in impurities at low temperatures, also make their diamagnetic contribution χ_{ID} to the full magnetic susceptibility.

SQUID method (Superconducting Quantum Interference Device) is usually used to measure the magnetic susceptibility. SQUID magnetometer measures the full magnetic susceptibility where the biggest contribution is made by lattice susceptibility χ_{LD}.

The method of electron spin resonance (ESR) was used earlier for the similar measurements [1]. It enables to define only paramagnetic component of impurity susceptibility χ_{IP}. A strong influence of spin the interaction near the metal-insulator phase transition (MI transition) in Ge:As [2-4] on paramagnetic susceptibility has been established. Electron location radius is separated closed to MI transition, what should result in increased value χ_{ID}.

Using both SQUID and ESR methods we would like to define behavior of the impurity diamagnetic susceptibility χ_{ID} and influence of spin interaction on it.

Lattice diamagnetic susceptibility of the substance is defined as follows:

$$\chi_{LD} \approx -(Ne^2/6mc^2)\sum_{k}^{Z} <r_k> = -2.83x10^{10}\sum_{k}^{Z} <r_k> , \quad (2)$$

where N = $6x10^{23}$ – Avogadro constant, e –electron charge, m – electron weight, c – speed of light, $<r_k> \approx 10^{-16}$ cm^2 – averaged square of electron orbit, rotating around one of the lattice atoms. For Ge $\chi_{LD} = -8x10^{-7}$ cm^3/g.

Based on experimental data [5] an empiric formula was composed for temperature dependence within range from helium to room temperatures:

$$\chi_{LD} = (-5.9 \text{ x } 10^{-7} + 8.9 \text{ x } 10^{-11}T) \text{ cm}^3/\text{g}. \quad (3)$$

The component, which is provide the temperature correction, is much less than constant part of the magnetic susceptibility. Lattice magnetic susceptibility practically does not depend on temperature.

Paramagnetic susceptibility χ_{IP} in case of no spin interaction (temperature area is above 1 K) follows Curie's law:

$$\chi_{IP} = n_s \mu_B^2 p_{ef}^2/3kT, \quad (4)$$

where n_s – spin concentration, μ_B – Bohr magneton, p_{ef} – effective magnetic moment on atom, k – Boltzmann constant. For samples being studied (close to metal-insulator phase transition) concentration of impurities $n_s \sim 10^{17}$ cm^{-3}, at T = 2 K, k = $1.38x10^{-16}$ erg/K and $\mu_B = 9.27x10^{-21}$ erg/Gs, p_{ef} = 1 we shall have $\chi_{IP} \sim 10^{-8}$ cm^3/g. Adduced estimate is extreme from the top, and in case of antiferromagnetic alignment the value of paramagnetic susceptibility is reduced.

Diamagnetic susceptibility of electrons localized on impurities:

$$\chi_{ID} = -(n_s e^2/6m^* c^2) <r_B^2>, \quad (5)$$

where m* - effective electron weight of state density, r_B – Bohr radius of electron orbit for impurities. At ns = 10^{17} cm^3 and $r_B = 5x10^{-7}$ cm we shall have $\chi_{ID} \approx 7x10^{-9}$ cm^3/g.

2. Samples and method used to measure magnetic susceptibility

2.1. Samples

Magnetic susceptibility was measured on a series of samples Ge:As, which had been used before in papers [2-4]. Phased reduction of electron concentration compared to original value for a series of samples ~3.6 x 10^{17} cm^{-3} was performed on account of introducing compensation impurity Ga during transmutation neutron doping. Sample parameters are given in Table 1.

Table 1. Samples

Sample No.	Electron concentration n_H, 10^{17} cm^{-3}	Potassium concentration N_{As}, 10^{17} cm^{-3}	Compensation $K = N_{Ga}/ N_{As}$
1	3.49	3.63	0.04
2	3.25	3.76	0.14
3	2.35	4.24	0.44
4	1.91	4.48	0.57

2.2. ESR method

The impurity paramagnetic susceptibility χ_{IP} was measured by ESR method. ESR spectrometer E-112 "VARIAN" with cryostat ESR-910 "OXFORD INSTRUMENTS" and digital registration system was used. This spectrometer registered derivative of resonance absorption line. Magnetic susceptibility is proportional to first integral of absorption line [3]. To define absolute value of impurity paramagnetic susceptibility χ_{IP} double integrated signal from the sample I_S was compared with the second integral from reference sample I_{st}. The spin concentration of the reference sample by "VARIAN" is equal 2.58 x 10^{15} cm^{-2}, χ_{st} = 1.8 x 10^{-10} emu at T = 300 K. Impurity paramagnetic susceptibility of the sample was defined by formula:

$$\chi_{IP} = 1.8 \times 10^{-10} I_S / I_{st} \tag{6}$$

2.3. SQUID method

MPMS-XL-1 device operating within temperature range 1.7 – 100 K and magnetic fields up to 10 kOe was used to measure magnetic susceptibility with the help of SQUID method.

3. Experimental results

Temperature dependences of the impurity paramagnetic susceptibility χ_{IP} obtained for these samples from ESR signal analysis are given in figure 1. It also shows $\chi_{IP} \sim 1/T$ dependence, meeting Curie's law, for comparison.

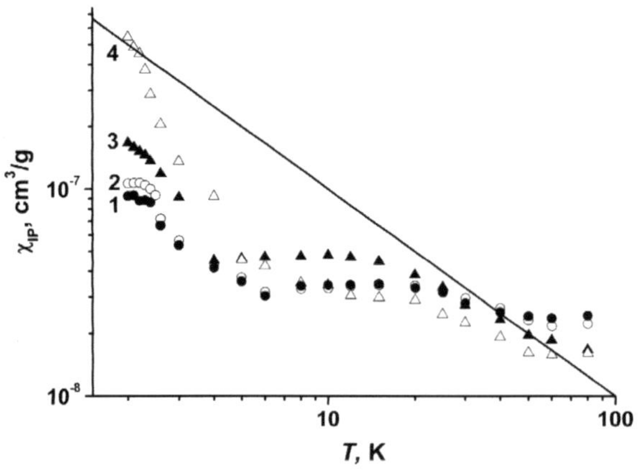

Figure 1. ESR measurements of the paramagnetic susceptibility for Ge:As samples at different temperatures; 1 – 4 – samples, right line – Curie's law.

Journal of Physics: Conference Series **661** (2015) 012023 doi:10.1088/1742-6596/661/1/012023

The figure shows that all curves have similar characteristic features. At temperatures 50 ≥ T ≥ 20 K $\chi_{IP}(T)$ dependence close to Curie's law is seen. At temperatures 20 ≥ T ≥ 6 K $\chi_{IP}(T)$ dependence saturated. Such unusual behavior could be explained by occurrence of antiferromagnetic spin interaction.

At low temperatures T ≤ T*, where T* ≈ 3 ÷ 6 K, impurity paramagnetic susceptibility increases sharply (stronger than the one stipulated by Curie's law).

We explain such behavior by transition from antiferromagnetic to ferromagnetic spin coupling at low temperatures [2,6].

Major contribution to SQUID is made by the lattice diamagnetic magnetic susceptibility, and to separate the impurity magnetic susceptibility χ_I from values χ measured by SQUID method, lattice susceptibility should be deducted.

$$\chi_I = \chi - \chi_{LD}. \quad (7)$$

Obtained values of the impurity magnetic susceptibility χ_I for the samples are given in figure.2.

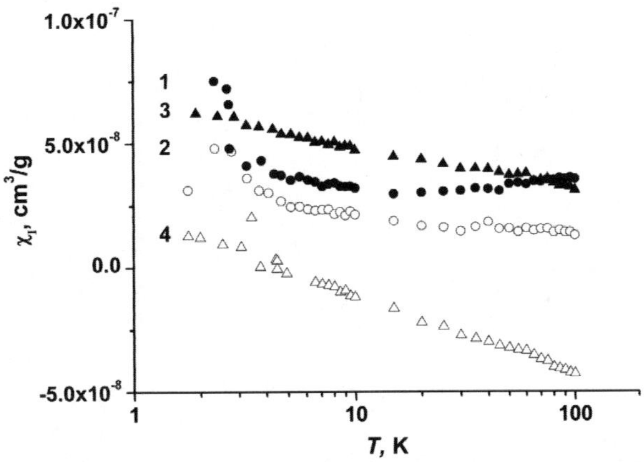

Figure 2. Temperature dependences of impurities magnetic susceptibility obtained with SQUID method.

Temperature dependences χ_{IP} and χ_I for four samples are shown in figure 3. The major difference is that χ_{IP} increases at low temperatures stronger than χ_I.

Let's define χ_{ID} as follows:

$$\chi_{ID} = \chi_I - \chi_{IP}. \quad (8)$$

Figure 4 shows obtained results for the samples. It can be seen that at T ≥ T*, the temperature dependence of difference χ_I - χ_{IP}, is weak. But the difference χ_I - χ_{IP} is sharply reduced at lower temperatures T ≤ (3 – 5) K.

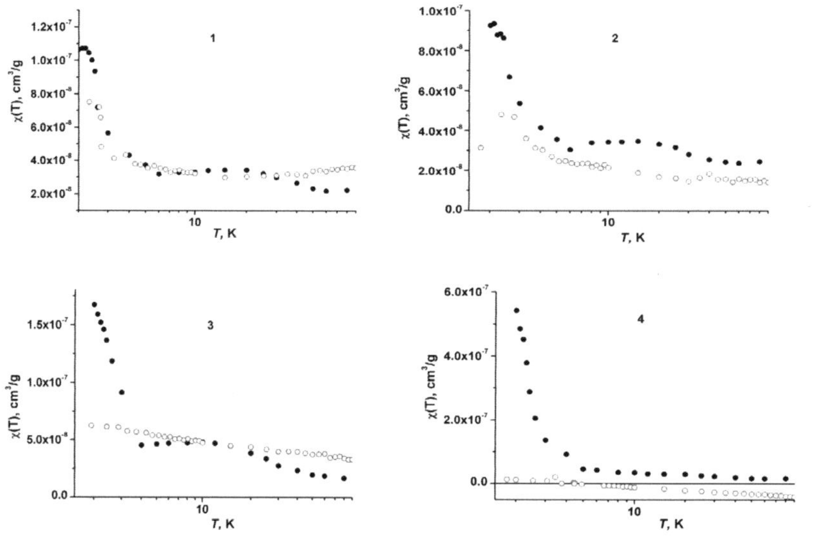

Figure 3. Comparison of temperature dependences χ_{IP} and χ_I; points – χ_{IP}, circles – χ_I; numbers on the graphs correspond to different samples.

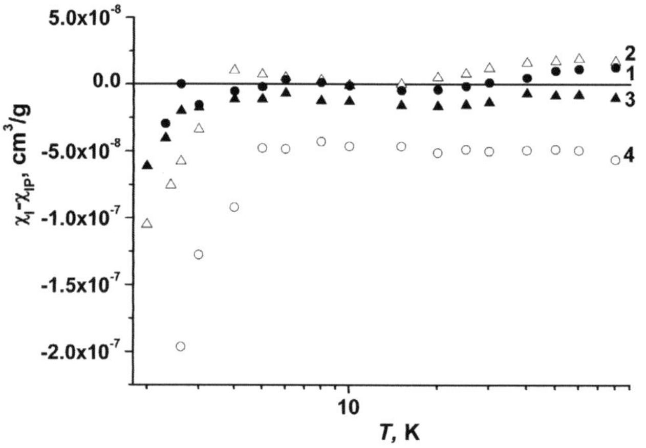

Figure 4. Temperature dependences of difference χ_I - χ_{IP}

4. Discussion

The value of the impurity diamagnetic susceptibility χ_{ID} should be negative at temperatures higher than T*. It could be seen from figure 4 that for the samples 1, 2 value χ_I - χ_{IP} is positive at $T \geq T^*$. It could be explained by random ingress of "dirt" to SQUID measuring device.

Expected low values of difference χ_I - χ_{IP} were registered for samples 1, 2. Average value χ_I - χ_{IP} = 5 x 10^{-8} cm^3/g is the sought value of the impurity diamagnetic susceptibility χ_{ID}.

Based on it and using formula for χ_{ID} and average sample value $n_s = 2x10^{17}$ cm^{-3}, let's estimate radius of electron location in impurity As.

$$<r> = (6|\chi_{ID}|\, m^* c^2 / n_s e^2)^{1/2} \approx 10^{-6}\ \text{cm}. \quad (9)$$

This value exceeds Bohr radius of small impurity As in low-alloy Ge almost twice. Obtained result corresponds to approximation of Ge:As system to metal-insulator phase transition, which is accompanied by increased radius of location.

The measurements of the temperature dependences obtained with two methods are different at low temperature. In higher compensated samples ESR measurements show a sharp increase of $\chi_{IP}(T)$ at $T \leq T^*$. In SQUID measurements this effect is not witnessed.

We tried to explain such difference by increased radius of electron orbits at their parallel orientation, similar to the structure of hydrogen molecule. It is formed only in case of anti-parallel oriented spins. At parallel spin orientation molecule is not formed, in our case at transition from anti-parallel coupling to parallel one, electrons start moving on orbits with a great radius as a result of repulsion.

Thus, changes in the exchange interaction leads not only to a transition from the antiferromagnetic to ferromagnetic ordering in impurity spins, but also to increase the value of the diamagnetic susceptibility due to the increase of the electron orbits.

5. Conclusions

Combined application of ESR and SQUID methods enables dividing contributions of the spin (paramagnetic) susceptibility and orbit (diamagnetic) susceptibility on full magnetic susceptibility of impurities.

In the area of low temperatures transition from anti-parallel to parallel coupling of localized spins results in increased radius of electron orbit, which leads to increased impurity diamagnetic susceptibility.

References

[1] Ch. P. Poole, Jr., Electron Spin Resonance: A Comprehensive Treatise on Experimental Techniques (John Wiley and Sons, New York, 1967; Mir, Moscow, 1970).

[2] A. I. Veinger, A. G. Zabrodskii, T. V. Tisnek, S. I. Goloshchapov, P. V. Semenikhin, AIP: Conference Proceedings. Vol. 1610, 119, (2013)

[3] A. I. Veinger, A. G. Zabrodskii, T. L. Makarova, T. V. Tisnek, S. I. Goloshchapov, P. V. Semenikhin, AIP: Conference Proceedings. Vol. 1610, 129, (2013)

[4] A.I. Veinger, A.G. Zabrodskii, T.L.Makarova, T.V. Tisnek, S.I. Goloshchapov and P.V.Semenikhin. JhETP, 116, 796, (2013)

[5] D.K. Stevens, J.W. Cleland, J.H. Crawford, H.C. Schweinler. Phys.Rev. 100, (1955), 1084.

[6] H.Kamimura. In book "Crystalline Semiconducting Materials and Devices". Eds by P.N.Butcher, N.H.March M.Tosi, (Plenum, 1986).

Discrete spectrum for quantum graph with local disturbance of the periodicity

Igor Yu. Popov[1], **Irina V. Blinova**[2], **Anton I. Popov**[3]

Department of Higher Mathematics, ITMO University, Kroverkskiy pr. 49, St. Petersburg, 197101, Russian Federation

E-mail: [1]popov1955@gmail.com, [2]irin-a@yandex.ru, [3]popov239@gmail.com

Abstract. The problem of discrete spectrum for quantum graph with local disturbance of the periodicity is described. The Hamiltonian is determined as 1D Schrödinger operator on each edge and some boundary conditions at each vertex. The spectral analysis of the quantum graph having the form of branching strips with hexagonal (honeycomb) structure is considered in more details.

1. Introduction

Theoretical investigation of nanostructures, especially, of carbon nanostructures, attracts great attention due to their unusual properties [1, 2, 3] and the existing or expected applications. The studying of complex nanostructures on the background of the mathematical theory of differential operators on graphs — a rapidly developing field of modern mathematical physics, whose significance is related with possible applications to solid state physics and nanoelectronics as a whole [4].

A start of intensive studying of differential operators on metric graphs was in 80th (see works of N.I. Gerasimenko and B.S. Pavlov [5], P. Exner and P. Šeba [6]. One can mention that the quantum graph problem is related with the corresponding discrete operator. In recent years the model of quantum graphs was widely used and allowed one to get many interesting results. An overview of new results and open problems can be found in the book [2].

Quantum graph is determined by a second order self-adjoint differential operator defined at the edges of the metric graph, viewed as intervals, and certain boundary conditions at the vertices. The differential operators describe the movement of a quantum particle along the edges. The choice of the operator is related with the physical motivation. It can be 1D free Schrödinger (for non-relativistic particle) or Dirac (e.g., for an electron in graphene) operator or the Hamiltonian with external electric or magnetic field. The boundary conditions specify the transition probabilities across the vertices. Quantum graphs are used as simplified models in many situations involving wave propagation. One-dimensional type of the structure allows one to obtain many results in an explicit form. On the other hand, the graph structure can be sufficiently complicated, and it allows one to describe quite complex and useful models. The description of the operator spectrum is a natural problem for quantum graphs, because it is related with interesting physical phenomena.

2. Local perturbations and discrete spectrum

Purely periodic quantum graphs have no discrete spectrum (it can have only continuous spectrum or eigenvalues of infinite multiplicity) but any local perturbation create a possibility of its appearance [7, 8, 9]. The information about the bound states is very essential for description of physical properties of nanosystems. There is a natural question: "Under what condition does the disturbance lead to a bound state appearance?". At present, one has no general answer. However, a collection of results in this direction has been obtained.

At first, let us consider the Schrödinger operator on a line with $\delta-$potentials at the points $nT_1, n \geq 0; nT_2, n < 0$. It can be considered as two semi-infinite quantum graphs with different edges lengths coupled at 0. In principle, an eigenvalue can appear, but Albeverio with co-authors [10] showed that the discrete spectrum is empty. The analogous problems for the Schrödinger operator on a line with locally perturbed smooth periodic potential were studied by many authors (see, e.g., [11, 12, 13, 14, 15, 16]). There are also a number of works dealing with the corresponding 2D or 3D nanosystems. Considerable attention of physicists is attracted to the construction of nonodevices based on nanotubes [17]. An interesting example of such nunostructure was recently created experimentally [18, 19]. It is a junction of two nanotubes of zigzag and armchair types (see Fig. 1).

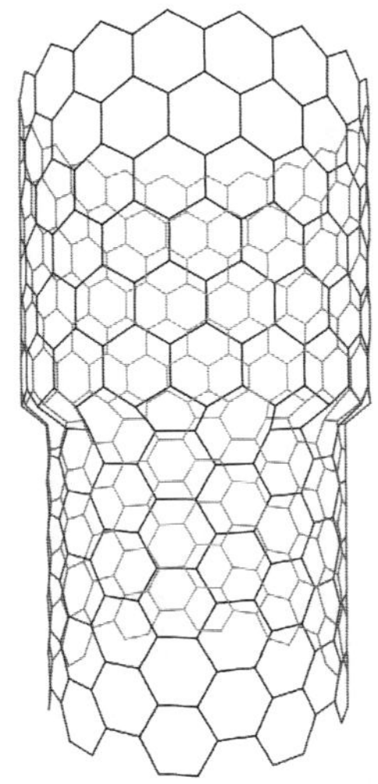

Figure 1. Armchair-zigzag junction

In this system the crystal structure is broken at the interface. This is a reason for appearance of localized states [20]. Different mathematical models were suggested to describe the effect [21, 22, 23, 24]. Particularly, the authors of [23, 24] deal with quantum graph model. They obtained necessary and sufficient conditions ensuring the existence of bound states localized near the interface.

Another interesting locally perturbed periodic physical nanosystems are nanostructures of chain type. There are works describing these systems in the framework of the quantum graph model. The first mathematical result in this direction is obtained in the work [7]. The authors solve the one-dimensional spectral problem on an infinite quantum graph consisting of identical rings connected at the touching points by δ-couplings with a parameter $\alpha \in R$. The "bending" deformation of the chain (i.e. variation of the position of one touching point at one ring) is considered. Using the method of the transfer-matrices, the authors show that this local perturbation leads to an appearance of eigenvalues in the open spectral gaps. The eigenvalues positions depend on the bending angle. The continuous spectrum does not depend, naturally, on the bending angle. It is determined by the transfer-matrix only, i.e. it coincides with the continuous spectrum for the straight chain. The analogous problem for Y-branched chain was considered in [9]. It is interesting to note that in the both cases the discrete spectrum is empty if there are no δ-potentials at the touching points, i.e. if one has the Kirchhoff condition at the graph vertices. At present, it is not known how general is this statement, e.g., does it depend on the graph structure?

To evaluate this hypothesis, we consider a quantum graph Γ, consisting of three symmetrically coupled semi-infinite strips, shown in Figure 2. It can be a model of a nanosystem with honeycomb lattice, e.g., carbon nanoribbon. Although each strip is invariant in respect to a semi-group of translations, the system is not periodic, and the discrete spectrum of the Hamiltonian can exist. The spectral analysis of the system is made in [25]. Let us describe it in more details. Let $\mathcal{E}$, $\mathcal{V}$ be, correspondingly, the sets of the graph edges and vertices.

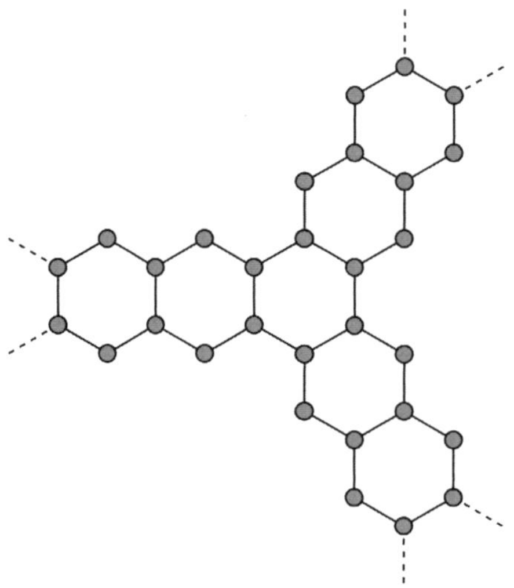

Figure 2. A quantum graph consisting of three symmetrically coupled strips

We consider free Schrödinger operator on each edge of the graph Γ

$$H = -\frac{d^2}{dx^2} \tag{1}$$

and δ-type conditions at the vertices (i.e. the continuity of functions and a "jump" of its derivatives). Correspondingly, the domain of the operator 1 is as follows

$$dom\ H = \left\{ f \in C(\Gamma) \bigcap H^2(\Gamma - \mathcal{V}) \sum_{e\in\mathcal{E}_v} \frac{df}{dx_e}(v) = \alpha f(v) . \right\} \tag{2}$$

Here H^2 is the Sobolev space W_2^2, α is some fixed number, $\mathcal{E}_v := \{e \in \mathcal{E} \,|v \in e\}$ is the set of edges adjacent to the vertex v. The sum is taken over all edges e incident to the vertex v and the derivatives are taken along e in the directions away from the vertex v (outgoing direction). One can see that the operator is self-adjoint. This type of the operator is often used in the quantum graph theory (see, e.g., papers [7, 26]). The structure of the spectrum is described in the following theorem (it is schematically shown in Fig. 3, 4).

Main theorem. *The structure of the spectrum of the Hamiltonian H for the graph is as follows.*

a) *Positive half-axis.*

The continuous spectrum has band structure. A point $E = k^2$ *belongs to the continuous spectrum if*

$$\left[\begin{array}{l} |b^+| \leq 2, \\ |b^-| \leq 2, \end{array}\right.$$

$$b^{\pm} = 1 \mp 2\cos kT + 3\cos 2kT + \frac{\alpha}{k}\sin 2kT.$$

Here T *is the edge length.*

There are points of the discrete spectrum in the gaps if

$$|\alpha|\, T > \alpha_m T = 2\sqrt{2\sqrt{3}-3} \quad arccot\sqrt{2\sqrt{3}-3}.$$

b) *Negative half-axis.*

There is no spectrum on the negative half-axis for $\alpha \geq 0$. *For* $\alpha < 0$ *the lower bound of the continuous spectrum is* $E_{thr} = -\chi_0^2$, *where* χ_0, $\chi_0 > 0$, *is the root of the equation*

$$\cosh \chi T(-1 + 3\cosh \chi T + \frac{\alpha}{\chi}\sinh \chi T) = 2.$$

There is no discrete spectrum below the threshold E_{thr} *if* $\alpha_c \leq \alpha < 0$.

There is at least one eigenvalue below the threshold E_{thr} *if* $\alpha < \alpha_c$. *Here*

$$\alpha_c = -T^{-1}\sqrt{3+2\sqrt{3}} \quad \ln(1+\sqrt{3}+\sqrt{3+2\sqrt{3}}).$$

One can see that in this case one has a situation analogous to that for more simple graphs of chain type: empty discrete spectrum if there are no $\delta-$potentials at the graph vertices.

3. Discussion

Chain type nanostructures are widely used for constructing of elements of nano devices. The availability of the quantum graph model in our case is related with the fact that only electrons having energies, close to the Fermi energy, are involved in the electron transport. It allows one to reduce 3D Schrödinger equation with chain of short-range potentials to 1D one, i.e. to the quantum graph (see, e.g., [27]). Transport properties of the system play a crucial role for nano device creation. In its turn, it strongly depends on the Hamiltonian spectrum. Bound states has an essential effect on electron transport, especially in chain-type nanostructures.

It can be shown by the following qualitative explanation. The density of electrons in a solid (semiconductor) is about $10^{19} cm^{-3}$, thus the distance between them is about $10^{-6} cm^{-1}$ which is typically at least two orders of magnitude larger than the minimal period in the solid. It shows that one-electron approximation is appropriate in many cases. Nevertheless, the interaction between electrons cannot be neglected since various collective effects for the whole system of electrons in the solid are caused by the long-range Coulomb interaction. A practical understanding of the multi-particle aspect on a quantum network might follow the ideology of coupled clusters theory, see for instance [28] and references therein. In our case of branched strips, the effect of the bound state is essentially greater. This state is localized in a neighborhood of the junction. If this energy level is occupied by an electron, then due to Coulomb repulsion, it has great influence on electrons travelling through the junction. The main result of the present paper describes the condition for the bound state appearance which is important for nanoelectronic devices (to ensure a proper transport properties).

This work was partially financially supported by the Government of the Russian Federation (grant 074-U01), by the Ministry of Education and Science of the Russian Federation (GOSZADANIE 2014/190, Projects No 14.Z50.31.0031 and No. 1.754.2014/K)), by grant MK-5001.2015.1 of the President of the Russian Federation.

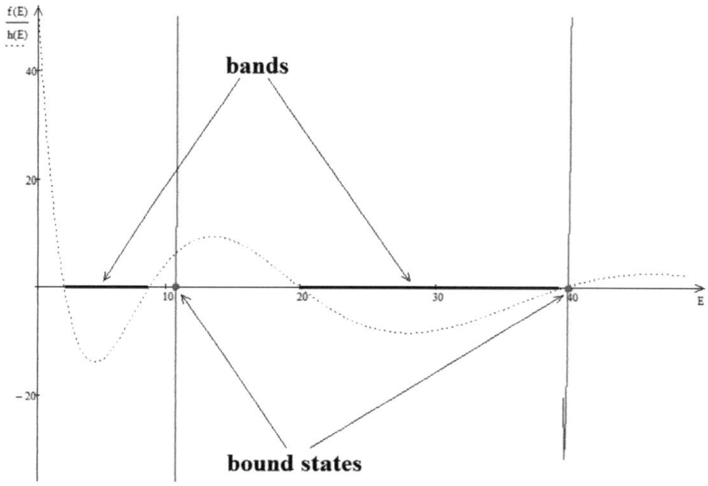

Figure 3. Spectrum for negative alpha

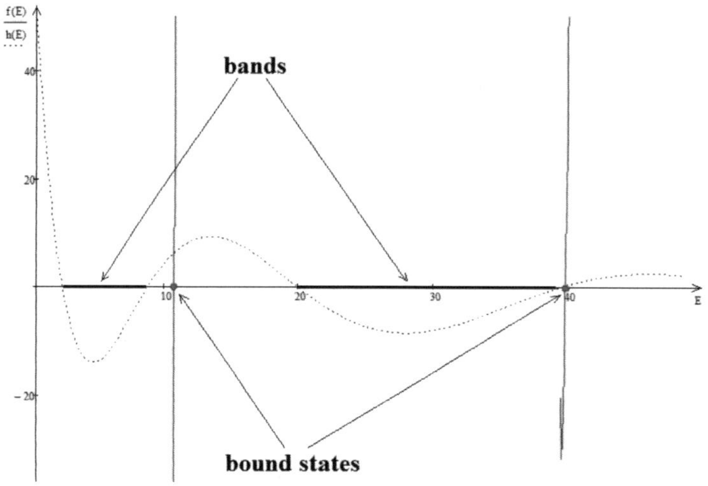

Figure 4. Spectrum for positive alpha

References

[1] Z.Xu (Ed.) 2012 *Graphene: Properties, Sinthesis and Applications* (Nova Science Publ. Inc.)
[2] P.Exner, et.al. (Eds.) 2008 Analysis on Graphs and Its Applications. *Proc. Symp. Pure Math.* (AMS, Providence, Rhode Island)
[3] O.Post 2010 *Spectral Analysis on Graph-like Spaces* (Springer, Berlin, 2010)
[4] P.Kuchment, R.Zeng 2001 Convergence of spectra of mesoscopic systems collapsing onto a graph *J. Math. Anal. Appl.* **258** 671
[5] N. I. Gerasimenko, B. S. Pavlov 1988 Scattering problems on noncompact graphs. *Theoret. Math. Phys.* **74**, pp 230–240
[6] P. Exner, P. Seba 1989 Free quantum motion on a branching graph. *Rep. Math. Phys.* **28**, pp 7–26
[7] P.Duclos, P.Exner, O.Turek 2008 On the spectrum of a bent chain graph. *J. Phys. A: Math. Theor.* **41**, 415206
[8] I.Yu. Popov, M.I. Gavrilov, S.I. Popov 2012 Two interacting particles in deformed nanolayer: discrete spectrum and particle storage. *Physica Scripta*, **86**, 035003/1-5
[9] I.Yu. Popov, P.I. Smirnov 2013 Spectral problem for branching chain quantum graph. *Phys. Letters* **A 377**,

pp 439–442

[10] S.Albeverio, F.Gesztesy, R.Hoegh-Krohn, H.Holden with an appendix by P.Exner 2005 *Solvable Models in Quantum Mechanics: Second Edition.* (Providence, R.I.: AMS Chelsea Publishing)

[11] E.L.Korotyaev, K.M.Schmidt 2012 On the resonances and eigenvalues for a 1D half-crystal with localised impurity. *J.reine angew. Math.* **670**, pp 217–248

[12] F.Rofe-Beketov 1964 A finiteness test for the number of discrete levels which can be introduced into the gaps of the continuous spectrum by perturbation of a periodic potential. *Dokl. Akad. Nauk SSSR.* **156**, pp 515–518

[13] F.Gesztesy, B.Simon 1993 A short proof of Zheludev s theorem. *Trans. Amer. Math. Soc.* **335** (1), pp 329–340

[14] V.A.Zheludev 1967 The eigenvalues of a perturbed Schrodinger operator with periodic potential (in Russian). *Probl. Mat. Fiz.* **2**, pp 108–123

[15] N.Firsova 1986 The Levinson formula for a perturbed Hill operator. *Theor. Math. Phys.* **62** (2), pp 196–209

[16] E. Korotyaev 2000 Lattice dislocations in a 1-Dimensional model. *Commun. Math. Phys.* **213**, pp 471–489

[17] D.Wei, Y.Liu 2008 The Intramolecular Junctions of Carbon Nanotubes. *Advanced Materials* **20**, pp 2815–2841

[18] C. Jin, K.Suenaga, S.Iijima 2008 Plumbing carbon nanotubes. *Nature Nanotechnology* **3**, pp 17–21

[19] Y.Yao, Q.Li, J.Zhang, R.Liu, L.Jiao, Y.T. Zhu, Z.Liu 2007 Temperature-mediated growth of single-walled carbon-nanotube intramolecular junctions. *Nature Materials.* **6**, pp 283–286

[20] A. Rochefort, P. Avouris 2002 Quantum Size Effects in Carbon Nanotube Intramolecular Junctions. *Nano Letters.* **2** (3) pp 253–256

[21] H. Santos, A. Ayuela, W. Jaskolski, M. Pelc, L. Chico 2009 Interface States in Carbon Nanotube Junctions: Rolling up graphene. *Phys. Rev. B.* **80** 035436

[22] R.N.Wang, X.H.Zheng, L.L.Song, Z.Zeng 2011 Ab initio study on the electronic transport properties of carbon nanotube intramolecular junctions. *Physica status solidi (a).* **208**, pp 2803–2808

[23] E. Korotyaev, I. Lobanov 2007 Schrödinger Operators on Zigzag Nanotubes. *Annales Henri Poincare.* **8** (6), pp 1151–1176

[24] I.S.Lobanov, I.Yu.Popov 2012 Scattering by zigzag-armchair nanotube junction. *Nanosystems: Physics, Chemistry, Mathematics.* **3** (2), pp 6–28

[25] I.Yu.Popov, A.N.Skorynina, I.V.Blinova 2014 On the existence of point spectrum for branching strips quantum graph. *J. Math. Phys.* **55**, 033504/1-20

[26] V. Kostrykin, R. Schrader 1999 Kirchhoff s rule for quantum wires. *J. Physics A: Math. Gen.* **32**, pp 595–630

[27] P.Kuchment 2002 Graph models for waves in thin structures. *Waves in Random Media*, **12** (4) R1-R24

[28] R.J. Bartlett, M. Musial 2007 Coupled -cluster theory in quantum chemistry, *Reviews of Modern Physics*, **79**, 291-351.

Journal of Physics: Conference Series **661** (2015) 012025 doi:10.1088/1742-6596/661/1/012025

Layered system with metamaterials

K V Pravdin[1] and I Yu Popov[1]

[1] ITMO University, 49 Kronverkskiy Ave., St. Petersburg, 197101, Russia

E-mail: construeman@gmail.com

Abstract. The layered system composed of metamaterial and vacuum layers is examined. We assume the metamaterial is the isotropic, homogeneous, dispersive and non-absorptive media. The permittivity and permeability of the metamaterial are equal and described by using a single Lorentz contribution. The electric Green's function is obtained for the case, when the permittivity and permeability of the metamaterial are equal –1. For this case we observe the no reflection effect. Also, the photonic band gap structure of the similar infinite layered system is described. For changing layers widths, the absence of the no reflection effect is shown.

1. Introduction

Metamaterials are known as negative index materials (NIMs) since for certain frequencies they have the negative refractive index [1]. By using NIMs, new covered surfaces and cloaking materials can be created [2], as well as superlenses with the resolving power, exceeding the diffraction limit [3]. Among systems with NIMs (NIM systems), layered NIM systems are widely known. The simplest two-layer NIM system is considered in [4]. The three-layer NIM system, which is the model for the superlens, is studied in [3, 5, 6]. The studies of the multilayer NIM systems (i.e., systems with the layers count greater than three) are presented in [7, 8]. A number of NIM studies are dedicated to obtaining the Green's function which describes an electromagnetic field of a point source [4, 6, 9]. An electromagnetic field value at any point can be determined with the Green's function. The layered systems are considered also as one-dimensional photonic crystals (1DPCs) [10, 11]. Numerous investigations of 1DPC were carried out in recent years [12, 13]. But most of these ones consider systems without the frequency dispersion. The first goal of our work is to obtain expressions for the electric Green's function for a multilayer NIM system, composed of arbitrary finite number of layers filled with the metamaterial and vacuum. The magnetic Green's function can as well be obtained by the same way. The second goal of our work is to describe the photonic band gap (PBG) structure for the similar but infinite NIM system (photonic crystal). We assume the metamaterial is the isotropic, homogeneous, dispersive and non-absorptive media, the permittivity and permeability stand equal and are described by using a single Lorentz contribution [4, 14]. We are interesting in the electric field in the case (NIM situation), when the permittivity and permeability of the metamaterial are equal –1 [4].

2. Green's function

2.1. Finite NIM system

We study the NIM system composed of $(N+M+1)$ parallel layers, where $N, M \geq 3$ are natural odd integers (figure 1). The $\mathbf{e}_1$, $\mathbf{e}_2$ unit vectors of the Cartesian basis $\{\mathbf{e}_1, \mathbf{e}_2, \mathbf{e}_3\}$ set a plane of layer's surfaces. The $\mathbf{e}_3$ unit vector set the x_3 axis. Every layer has an index. From the right side of the zero layer, N layers are located, from the left side there are M layers. Thus, the n layer index is $n=-M,\ldots 0,\ldots N$. All even layers (as well as the zero layer) are Δ_1 in width, and filled with a metamaterial. All odd layers are Δ_2 in width, and filled with a vacuum. The last-to-left-side (with $-M$ index) and last-to-right-side (with N index) layers are the half spaces unbounded along the direction of the x_3 axis and are empty (vacuum). The point source is located in the zero layer and defined by coordinates of the vector $\mathbf{y}=\{y_1,y_2,y_3\}$, which is located in the Cartesian basis $\{\mathbf{e}_1,\mathbf{e}_2,\mathbf{e}_3\}$, i.e., $0<y_3<\Delta_1$. We assume the translation invariance along the plane of layer's surfaces.

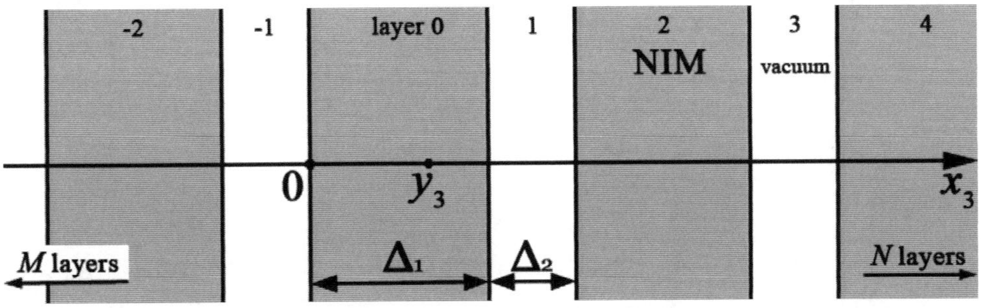

Figure 1. The NIM system composed of (N+M+1) parallel layers filled with the metamaterial (grey) and the vacuum (white). The point source is located in the zero layer at y_3 coordinate of the x_3 axis.

We consider the Maxwell's equations in a differential form

$$\frac{d\mathbf{D}}{dt}(\mathbf{x},t)=\nabla\times\mathbf{H}(\mathbf{x},t),\quad \frac{d\mathbf{B}}{dt}(\mathbf{x},t)=-\nabla\times\mathbf{E}(\mathbf{x},t),\quad \nabla\cdot\mathbf{D}(\mathbf{x},t)=0,\quad \nabla\cdot\mathbf{B}(\mathbf{x},t)=0, \tag{1}$$

where $\mathbf{x}=\{x_1,x_2,x_3\}$ is the vector located in the Cartesian basis $\{\mathbf{e}_1,\mathbf{e}_2,\mathbf{e}_3\}$, ∇ is the Hamilton operator, $\times$ is a cross product symbol as well as a numerical product symbol, $\cdot$ is an inner product symbol as well as a matrix product symbol. Also, we consider the auxiliary field equations

$$\mathbf{D}(\mathbf{x},t)=\varepsilon_0\mathbf{E}(\mathbf{x},t)+\mathbf{P}(\mathbf{x},t),\quad \mathbf{B}(\mathbf{x},t)=\mu_0\left[\mathbf{H}(\mathbf{x},t)+\mathbf{M}(\mathbf{x},t)\right], \tag{2}$$

$$\mathbf{P}(\mathbf{x},t)=\varepsilon_0\int_{t_0}^{t}\chi_e(\mathbf{x},t-s)\cdot\mathbf{E}(\mathbf{x},s)ds,\quad \mathbf{M}(\mathbf{x},t)=\int_{t_0}^{t}\chi_m(\mathbf{x},t-s)\cdot\mathbf{H}(\mathbf{x},s)ds, \tag{3}$$

where ε_0 and μ_0 are the electric and magnetic constants, which we set $\varepsilon_0=\mu_0=1$ for brevity [4], $\chi_e(\mathbf{x},t)$ and $\chi_m(\mathbf{x},t)$ are the electric and magnetic susceptibility tensors. We use the causality condition $\chi_e(\mathbf{x},t)=\chi_m(\mathbf{x},t)=0$ for $t<t_0$ (we assume $t_0=0$) and the passivity condition [4].

The considered system composed of layers divided by plane unbounded surfaces. The boundary conditions for each one are presented in the general form as follows:

$$(\mathbf{E}_1-\mathbf{E}_2)\times\mathbf{e}_3=\mathbf{0},\quad (\mathbf{H}_1-\mathbf{H}_2)\times\mathbf{e}_3=\mathbf{0},\quad (\mathbf{D}_1-\mathbf{D}_2)\cdot\mathbf{e}_3=0,\quad (\mathbf{B}_1-\mathbf{B}_2)\cdot\mathbf{e}_3=0. \tag{4}$$

$\mathbf{E}_j=\mathbf{E}_j(\tilde{\mathbf{x}},t)$, $\mathbf{H}_j=\mathbf{H}_j(\tilde{\mathbf{x}},t)$, $\mathbf{D}_j=\mathbf{D}_j(\tilde{\mathbf{x}},t)$, and $\mathbf{B}_j=\mathbf{B}_j(\tilde{\mathbf{x}},t)$ stand for the limits with $\mathbf{x}\to\tilde{\mathbf{x}}$, where $\mathbf{x}=\mathbf{x}^{\perp}+x_3\mathbf{e}_3$, $\tilde{\mathbf{x}}=\mathbf{x}^{\perp}+\tilde{x}_3\mathbf{e}_3$, $\mathbf{x}^{\perp}=\{x_1,x_2,0\}$ and $x_3\to\tilde{x}_3$, j is the index distinguishing limits calculated on different surface sides ($j=1,2$). We use the Laplace transform with t time

$$\hat{f}(z)=\int_0^{\infty}\exp[izt]f(t)dt,\quad f(t)=\frac{1}{2\pi}\int_{\hat{A}}\exp[-izt]\hat{f}(z)dz, \tag{5}$$

and the Fourier transform with x_1 and x_2 coordinates

$$f_{\mathbf{k}^{\perp}}(x_3)=\int_{-\infty}^{+\infty}\int_{-\infty}^{+\infty}\exp[i(k_1x_1+k_2x_2)]f(\mathbf{x})dx_1dx_2=\int_{R^2}\exp[i\mathbf{k}^{\perp}\cdot\mathbf{x}^{\perp}]f(\mathbf{x})d\mathbf{x}^{\perp}, \tag{6}$$

$$f(\mathbf{x})=\frac{1}{(2\pi)^2}\int_{-\infty}^{+\infty}\int_{-\infty}^{+\infty}\exp[-i(k_1x_1+k_2x_2)]f_{\mathbf{k}^\perp}(x_3)\mathrm{d}k_1\mathrm{d}k_2=\frac{1}{(2\pi)^2}\int_{\mathrm{R}^2}\exp[-i\mathbf{k}^\perp\cdot\mathbf{x}^\perp]f_{\mathbf{k}^\perp}(x_3)\mathrm{d}\mathbf{k}^\perp,$$

where k_1 and k_2 are the coordinates of the two-dimensional wave vector $\mathbf{k}^\perp=\{k_1,k_2,0\}$ with the length $\kappa=|\mathbf{k}^\perp|=\sqrt{k_1^2+k_2^2}$ and the unit vector $\mathbf{e}_{\mathbf{k}^\perp}$, i.e., $\mathbf{k}^\perp=\kappa\mathbf{e}_{\mathbf{k}^\perp}$, $\mathbf{k}=\{k_1,k_2,k_3\}$ is the full wave vector. Therefore, $\mathbf{e}_3\times\mathbf{e}_{\mathbf{k}^\perp}$ is a parallel to the plane of layer's surfaces unit vector. The set of $\mathbf{e}_{\mathbf{k}^\perp}$, $\mathbf{e}_3\times\mathbf{e}_{\mathbf{k}^\perp}$, $\mathbf{e}_3$ unit vectors forms the Cartesian basis.

We consider the isotropic homogeneous metamaterial in layers. Therefore, the electric and magnetic permeabilities do not depend on coordinates of the $\mathbf{x}$ vector, i.e., $\boldsymbol{\varepsilon}(\mathbf{x},z)=\varepsilon(z)\mathbf{U}$, $\boldsymbol{\mu}(\mathbf{x},z)=\mu(z)\mathbf{U}$, where $\mathbf{U}$ is the 3×3 unit matrix. Also, we assume that the metamaterial is the dispersive non-absorptive media. In that case, the susceptibilities consist of a sum of Lorentz contributions [14]. We deal with a single dispersive Lorentz contribution [4]. We assume that the permittivity and permeability of the metamaterial stand equal and

$$\varepsilon(z)=\mu(z)=1-\frac{\Omega^2}{z^2-\omega_0^2}, \tag{7}$$

where Ω, ω_0 are constants, $z=\omega+\mathrm{i}\alpha$, $\alpha\to0$, $\alpha>0$, and $\varepsilon(z)=\mu(z)=1$ in vacuum layers. Note that the $\hat{\omega}=\sqrt{\omega_0^2+\Omega^2/2}$ frequency (NIM frequency [4]) leads to the NIM situation, i.e., $\varepsilon(\hat{\omega})=\mu(\hat{\omega})=1$.

With the Laplace transform (5) we obtain the following Helmholtz equation for the electric field function $\mathbf{E}(\mathbf{x},t)$:

$$\mathbf{L}^e(\mathbf{x},z)\cdot\hat{\mathbf{E}}(\mathbf{x},z)=\mathbf{g}^e(\mathbf{x},z),$$

$$\mathbf{L}^e(\mathbf{x},z)\cdot\hat{\mathbf{E}}(\mathbf{x},z)=z^2\boldsymbol{\varepsilon}(\mathbf{x},z)\hat{\mathbf{E}}(\mathbf{x},z)-\nabla\times\left[\boldsymbol{\mu}^{-1}(\mathbf{x},z)\cdot\nabla\times\hat{\mathbf{E}}(\mathbf{x},z)\right],$$

$$\mathbf{g}^e(\mathbf{x},z)=\mathrm{i}z\mathbf{E}(\mathbf{x},0)-\nabla\times\left[\boldsymbol{\mu}^{-1}(\mathbf{x},z)\cdot\mathbf{H}(\mathbf{x},0)\right],$$

where $\mathbf{L}^e(\mathbf{x},z)$ is the electric Helmholtz operator, $\mathbf{g}^e(\mathbf{x},z)$ is the function of the initial electric field configuration. Let us introduce the $\mathbf{G}^e(\mathbf{x},\mathbf{y},z)$ electric Green's function [4] that satisfies

$$\mathbf{L}^e(\mathbf{x},z)\cdot\mathbf{G}^e(\mathbf{x},\mathbf{y},z)=\delta(\mathbf{x}-\mathbf{y})\mathbf{U}, \tag{8}$$

where $\delta(\mathbf{x}-\mathbf{y})$ is the Dirac delta function. Then, the $\mathbf{E}(\mathbf{x},t)$ electric field function is given by the inverse Laplace transform (5) of

$$\hat{\mathbf{E}}(\mathbf{x},z)=\int\mathbf{G}^e(\mathbf{x},\mathbf{y},z)\cdot\mathbf{g}^e(\mathbf{y},z)\mathrm{d}\mathbf{y}.$$

Note that the magnetic Green's function $\mathbf{G}^m(\mathbf{x},\mathbf{y},z)$ and the magnetic field function $\mathbf{H}(\mathbf{x},t)$ can be obtained in a similar way. Therefore, below we consider only electric Green's function.

2.2. *Electric Green's function*

By using the Laplace (5) and Fourier (6) transforms for the Helmholtz equation (8) and the boundary conditions (4) we obtain the system of equations for the electric Green's function in every layer with transverse electric (TE) and transverse magnetic (TM) modes. We solve this system by using the recurrence relation approach, and obtain results in the NIM situation with $z\to\pm\hat{\omega}$. The electric Green's function in the NIM situation is expressed in the following way: for $\hat{\omega}>\kappa$ in the layers filled with the metamaterial, i.e., with $n=-(M-1),\ldots-2,0,2,\ldots(N-1)$,

$$\mathbf{G}^e_{\mathbf{k}^\perp}(x_3,y_3,\pm\hat{\omega})=\pm\frac{1}{2\mathrm{i}\rho(\hat{\omega})}\exp\left[\pm\mathrm{i}\rho(\hat{\omega})\left(|n|\Delta_2-|x_3-y_3|\right)\right]\left\{(\mathbf{e}_3\times\mathbf{e}_{\mathbf{k}^\perp})^{\mathrm{T}}\otimes(\mathbf{e}_3\times\mathbf{e}_{\mathbf{k}^\perp})\right.$$

$$+\frac{\rho^2(\hat{\omega})}{\hat{\omega}^2}\left[\mathbf{e}^{\mathrm{T}}_{\mathbf{k}^\perp}\otimes\mathbf{e}_{\mathbf{k}^\perp}\pm\mathrm{sign}(x_3-y_3)\frac{\kappa}{\rho(\hat{\omega})}\left(\mathbf{e}^{\mathrm{T}}_{\mathbf{k}^\perp}\otimes\mathbf{e}_3+\mathbf{e}_3^{\mathrm{T}}\otimes\mathbf{e}_{\mathbf{k}^\perp}\right)+\frac{\kappa^2}{\rho^2(\hat{\omega})}\mathbf{e}_3^{\mathrm{T}}\otimes\mathbf{e}_3\right]\Bigg\},$$

in the layers filled with a vacuum, i.e., with $n=-M,\ldots-3,-1,1,3,\ldots N$,

$$\mathbf{G}^e_{\mathbf{k}^\perp}(x_3,y_3,\pm\hat{\omega})=\pm\frac{1}{2\mathrm{i}\rho(\hat{\omega})}\exp\left[\pm\mathrm{i}\rho(\hat{\omega})\left(-|n|\Delta_1+|x_3-\Delta_1+y_3|\right)\right]\Bigg\{(\mathbf{e}_3\times\mathbf{e}_{\mathbf{k}^\perp})^{\mathrm{T}}\otimes(\mathbf{e}_3\times\mathbf{e}_{\mathbf{k}^\perp})+$$

$$+\frac{\rho^2(\hat{\omega})}{\hat{\omega}^2}\left[\mathbf{e}^{\mathrm{T}}_{\mathbf{k}^\perp}\otimes\mathbf{e}_{\mathbf{k}^\perp}\pm\mathrm{sign}(x_3-\Delta_1+y_3)\frac{\kappa}{\rho(\hat{\omega})}\left(\mathbf{e}^{\mathrm{T}}_{\mathbf{k}^\perp}\otimes\mathbf{e}_3+\mathbf{e}_3^{\mathrm{T}}\otimes\mathbf{e}_{\mathbf{k}^\perp}\right)-\frac{\kappa^2}{\rho^2(\hat{\omega})}\mathbf{e}_3^{\mathrm{T}}\otimes\mathbf{e}_3\right]\Bigg\},$$

for $\hat{\omega}<\kappa$ in the layers filled with the metamaterial, i.e., with $n=-(M-1),\ldots-2,0,2,\ldots(N-1)$,

$$\mathbf{G}^e_{\mathbf{k}^\perp}(x_3,y_3,\pm\hat{\omega})=\frac{1}{2\rho(\hat{\omega})}\exp\left[-\rho(\hat{\omega})\left(|n|\Delta_2-|x_3-y_3|\right)\right]\Bigg\{(\mathbf{e}_3\times\mathbf{e}_{\mathbf{k}^\perp})^{\mathrm{T}}\otimes(\mathbf{e}_3\times\mathbf{e}_{\mathbf{k}^\perp})+$$

$$+\frac{\rho^2(\hat{\omega})}{\hat{\omega}^2}\left[\mathbf{e}^{\mathrm{T}}_{\mathbf{k}^\perp}\otimes\mathbf{e}_{\mathbf{k}^\perp}-\mathrm{sign}(x_3-y_3)\frac{\mathrm{i}\kappa}{\rho(\hat{\omega})}\left(\mathbf{e}^{\mathrm{T}}_{\mathbf{k}^\perp}\otimes\mathbf{e}_3+\mathbf{e}_3^{\mathrm{T}}\otimes\mathbf{e}_{\mathbf{k}^\perp}\right)-\frac{\kappa^2}{\rho^2(\hat{\omega})}\mathbf{e}_3^{\mathrm{T}}\otimes\mathbf{e}_3\right]\Bigg\},$$

in the layers filled with a vacuum, i.e., with $n=-M,\ldots-3,-1,1,3,\ldots N$,

$$\mathbf{G}^e_{\mathbf{k}^\perp}(x_3,y_3,\pm\hat{\omega})=\frac{1}{2\rho(\hat{\omega})}\exp\left[-\rho(\hat{\omega})\left(-|n|\Delta_1+|x_3-\Delta_1+y_3|\right)\right]\Bigg\{(\mathbf{e}_3\times\mathbf{e}_{\mathbf{k}^\perp})^{\mathrm{T}}\otimes(\mathbf{e}_3\times\mathbf{e}_{\mathbf{k}^\perp})+$$

$$+\frac{\rho^2(\hat{\omega})}{\hat{\omega}^2}\left[\mathbf{e}^{\mathrm{T}}_{\mathbf{k}^\perp}\otimes\mathbf{e}_{\mathbf{k}^\perp}-\mathrm{sign}(x_3-\Delta_1+y_3)\frac{\mathrm{i}\kappa}{\rho(\hat{\omega})}\left(\mathbf{e}^{\mathrm{T}}_{\mathbf{k}^\perp}\otimes\mathbf{e}_3+\mathbf{e}_3^{\mathrm{T}}\otimes\mathbf{e}_{\mathbf{k}^\perp}\right)+\frac{\kappa^2}{\rho^2(\hat{\omega})}\mathbf{e}_3^{\mathrm{T}}\otimes\mathbf{e}_3\right]\Bigg\},$$

where the $\mathrm{sign}(x)$ function equals $+1$ for $x\geq 0$ and -1 for $x<0$, and the $\rho(\omega)$ function is defined as

$$\rho(\omega)=\sqrt{|k_3|^2}=\sqrt{\left||\mathbf{k}|^2-|\mathbf{k}^\perp|^2\right|}=\sqrt{\left|\omega^2\left(1-\frac{\Omega^2}{\omega^2-\omega_0^2}\right)^2-\kappa^2\right|},\ \text{where } \rho(\hat{\omega})=\sqrt{\left|\hat{\omega}^2-\kappa^2\right|}. \quad (9)$$

Note that expressions for Green's function have only one term responsible for propagation and no terms responsible for reflection, i.e., there is no reflection for the frequencies $\pm\hat{\omega}$, which correspond to $\varepsilon(z)=\mu(z)=-1$. The magnetic Green's function can as well be obtained with reasoning similar to the presented above.

One notices that the Green's function as a resolvent kernel is defined for regular values of the spectral parameter only, i.e., outside the operator spectrum (see, e.g., [15]). It is, of course, valid for any non-real value. As for the real axis, one has the Green's function only at points which do not belong to the spectrum. Correspondingly, the obtained Green's function formula is valid only if the NIM frequencies $\pm\hat{\omega}$ do not belong to the spectrum of the electric Helmholtz operator $\mathbf{L}^e(\mathbf{x},z)$. In a limiting case of periodic 1DPC (when N and M tend to infinity) the Green's function at the real axis exists only in the spectral gaps.

3. Photonic band gap structure

3.1. *Infinite NIM system*

Now we examine the system similar to the studied above, but composed of infinite count of parallel layers. $\Delta_1+\Delta_2$ is the period of the system and the system is a periodical 1DPC. We consider the Maxwell's equations (1) with the auxiliary field equations (2)-(3). Now we set $\varepsilon_0\mu_0=1/c^2$, where c is the speed of light in vacuum. We use the similar to (5) Fourier transform with t time

$$\hat{f}(\omega)=\int_{-\infty}^{+\infty}exp[-\mathrm{i}\omega t]f(t)\mathrm{d}t,\quad f(t)=\frac{1}{2\pi}\int_{-\infty}^{+\infty}\exp[\mathrm{i}\omega t]\hat{f}(\omega)\mathrm{d}\omega, \quad (10)$$

and consider the permittivity and permeability of the metamaterial as in (7) with $z=\omega$. They are negative and the metamaterial behaves like the NIM when the ω frequency is inside the (ω_0,ω_2) interval (NIM interval), where $\omega_2=\sqrt{\omega_0^2+\Omega^2}$, $\varepsilon(\omega_2)=\mu(\omega_2)=0$, and $\varepsilon(\omega_0+0)=\mu(\omega_0+0)=-\infty$.

With the Fourier transform (10) we obtain the following Helmholtz equation for the electric field function $\mathbf{E}(\mathbf{x},t)$:

$$\nabla\times\nabla\times\hat{\mathbf{E}}(\mathbf{x},\omega)=(\omega/c)^2\varepsilon(\omega)\mu(\omega)\hat{\mathbf{E}}(\mathbf{x},\omega). \tag{11}$$

By using the Fourier transforms (10) and (6) for the Helmholtz equation (11), the boundary conditions (4) and the Floquet-Bloch theorem [15, 16] we obtain the following dispersion equation:

$$\left(e^{-i\theta(\Delta_1+\Delta_2)}\right)^2-\left[\frac{\sigma^+_{1,2}\sigma^+_{2,1}}{4}\left(1+e^{i2k_3^{(1)}\Delta_1}e^{i2k_3^{(2)}\Delta_2}\right)+\right.$$
$$\left.+\frac{\sigma^-_{1,2}\sigma^-_{2,1}}{4}\left(e^{i2k_3^{(1)}\Delta_1}+e^{i2k_3^{(2)}\Delta_2}\right)\right]e^{-ik_3^{(1)}\Delta_1}e^{-ik_3^{(2)}\Delta_2}e^{-i\theta(\Delta_1+\Delta_2)}+1=0, \tag{12}$$

where θ is a yet undefined wave vector, called the Bloch wave vector, $\sigma^{\pm}_{m,p}=\left(\varepsilon_m k_3^{(p)}\pm\varepsilon_p k_3^{(m)}\right)/\varepsilon_m k_3^{(p)}$ with indices $m=1$ and $p=2$, or $m=2$ and $p=1$, and ε_m and $k_3^{(m)}$ are calculated for the metamaterial if $m=1$, and vacuum if $m=2$. The dispersion equation is the same for the TE and TM modes. It follows from the equality of the permittivity and permeability (7). Therefore, we have the identical PBG structure for TE and TM modes.

3.2. Solutions

The Helmholtz equation has non-trivial solutions when the dispersion equation (12) holds true. By using the Fourier transforms (10) and (6), the boundary conditions (4) and the Floquet-Bloch theorem [16, 17], we obtain the following solutions in metamaterial layers (subscript $_1$) and vacuum layers (subscript $_2$) for TM mode:

$$E_1(x_3,\omega,\kappa)=C(\omega,\kappa)\left\{\sigma^-_{1,2}\sigma^+_{1,2}\left(e^{ik_3^{(2)}\Delta_2}-e^{-ik_3^{(2)}\Delta_2}\right)e^{-ik_3^{(1)}\Delta_1}e^{i\theta(\Delta_1+\Delta_2)}e^{ik_3^{(1)}x_3}-\right.$$
$$\left.-\left[\left(\sigma^+_{1,2}\right)^2\left(e^{-ik_3^{(1)}\Delta_1}-e^{ik_3^{(2)}\Delta_2}e^{i\theta(\Delta_1+\Delta_2)}\right)-\left(\sigma^-_{1,2}\right)^2\left(e^{-ik_3^{(1)}\Delta_1}-e^{-ik_3^{(2)}\Delta_2}e^{i\theta(\Delta_1+\Delta_2)}\right)\right]e^{ik_3^{(1)}\Delta_1}e^{-ik_3^{(1)}x_3}\right\}, \tag{13}$$

$$E_2(x_3,\omega,\kappa)=2C(\omega,\kappa)\left\{\sigma^-_{1,2}\left(e^{-ik_3^{(1)}\Delta_1}-e^{-ik_3^{(2)}\Delta_2}e^{i\theta(\Delta_1+\Delta_2)}\right)e^{-ik_3^{(2)}\Delta_1}e^{ik_3^{(2)}x_3}-\right.$$
$$\left.-\sigma^+_{1,2}\left(e^{-ik_3^{(1)}\Delta_1}-e^{ik_3^{(2)}\Delta_2}e^{i\theta(\Delta_1+\Delta_2)}\right)e^{ik_3^{(2)}\Delta_1}e^{-ik_3^{(2)}x_3}\right\}, \tag{14}$$

where $C(\omega,\kappa)$ is an unknown function, which can be obtained from some initial conditions. For the TE mode we obtain the same expressions (13)-(14) but with swapped $\sigma^{\pm}_{1,2}$ for $\sigma^{\pm}_{2,1}$. The solutions (13)-(14) in the NIM situation for TE and TM modes are expressed for $\hat{\omega}>\kappa$ as follows:

$$E_1(x_3,\hat{\omega},\kappa)=-4C(\hat{\omega},\kappa)\left(e^{i\rho\Delta_1}-e^{i\rho\Delta_2}e^{i\theta(\Delta_1+\Delta_2)}\right)e^{-i\rho\Delta_1}e^{i\rho x_3}, \tag{15}$$

$$E_2(x_3,\hat{\omega},\kappa)=-4C(\hat{\omega},\kappa)\left(e^{i\rho\Delta_1}-e^{i\rho\Delta_2}e^{i\theta(\Delta_1+\Delta_2)}\right)e^{i\rho\Delta_1}e^{i\rho x_3}, \tag{16}$$

and for $\hat{\omega}<\kappa$ as follows:

$$E_1(x_3,\hat{\omega})=4C(\hat{\omega},\kappa)\left(e^{\rho\Delta_1}-e^{\rho\Delta_2}e^{i\theta(\Delta_1+\Delta_2)}\right)e^{-\rho\Delta_1}e^{\rho x_3}, \tag{17}$$

$$E_2(x_3,\hat{\omega})=4C(\hat{\omega},\kappa)\left(e^{\rho\Delta_1}-e^{\rho\Delta_2}e^{i\theta(\Delta_1+\Delta_2)}\right)e^{\rho\Delta_1}e^{-\rho x_3}, \tag{18}$$

where $\rho=\rho(\hat{\omega})$ is given by equation (9). Note that the solution (15), as well as the solutions (16)-(18), has only one term responsible for propagation and no terms responsible for reflection, i.e., there is no reflection at the NIM frequency $\hat{\omega}$. But this effect is observed only when the dispersion equation (12) holds true, i.e., only for permitted bands (at the spectrum).

3.3. Numerical results

To study the PBG structure of the considered 1DPC we use a numerical approach. We consider the equality (12) and seek for which ω and κ values it holds true with any θ value, which belongs to the $(0, 2\pi/\Delta)$ interval. We fix the constants $\omega_0 = 30$ THz, $\Omega = 90$ THz. Then, the NIM frequency is $\hat{\omega} = 70.36$ THz and the NIM interval is (30 THz, 94.87 THz). We consider the intervals for ω values from 42 till 96 THz (then the ω / c normalized frequency has values from 1.4×10^6 till 3.2×10^5 m^{-1}) and for κ values from 0 till 0.8 ηm^{-1} (i.e., from 0 till 8×10^5 m^{-1}). We use four following cases of the Δ_1 and Δ_2 parameters: 1. $\Delta_1 = \Delta_2 = 10$ ηm (see (a) in figure 2); 2. $\Delta_1 = 20$ ηm and $\Delta_2 = 10$ ηm (see (b) in figure 2); 3. $\Delta_1 = 10$ ηm and $\Delta_2 = 20$ ηm (see (c) in figure 2); 4. $\Delta_1 = 10$ ηm and $\Delta_2 = 100$ ηm (see (d) in figure 2).

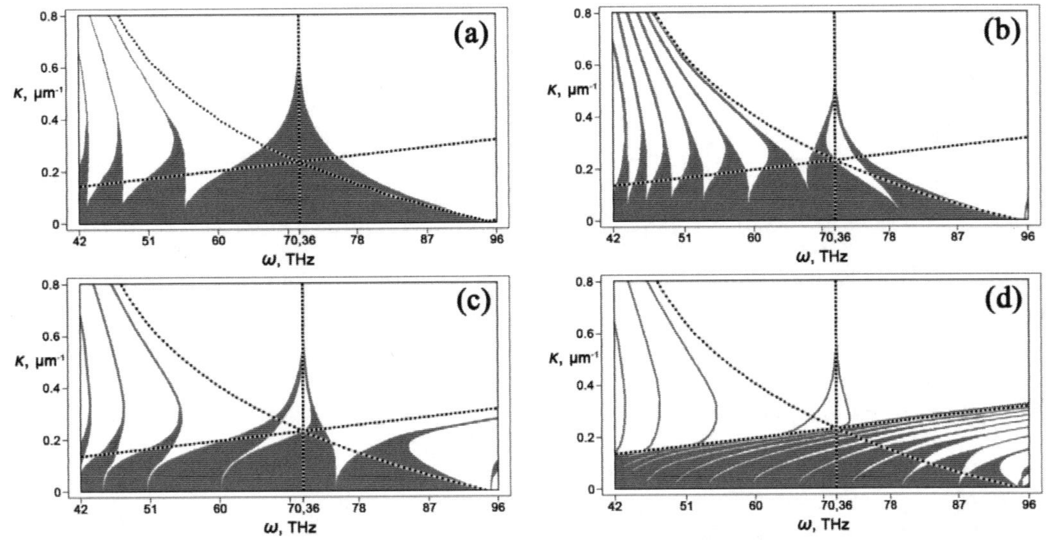

Figure 2. Dependences of PBG structure on the ω frequency and κ values for TE and TM modes simultaneously. Permitted bands are grey, forbidden bands are white. The vertical straight dot line shows the NIM frequency 70.36 THz. The dot curves divide the space on four areas. The bottom and top areas correspond to the radiative and evanescent regime, respectively, in the metamaterial and vacuum simultaneously. The right area corresponds to the evanescent regime in the metamaterial and the radiative regime in the vacuum. The left area corresponds to the radiative regime in the metamaterial and the evanescent regime in the vacuum. The 1, 2, 3, and 4 cases are presented in (a), (b), (c), and (d), respectively.

In the case 1, we observe the NIM frequency belonging to a permitted band for all κ values. Therefore, there is no reflection for any κ value in the NIM situation. With changing the Δ_1 or Δ_2 parameters, i.e., in the cases 2, 3, and 4, we observe the faster narrowing of permitted bands with the increasing of κ value, than it is in the case 1. In cases 3 and 4 for small κ value permitted bands shift to the zero ω value. The contained the NIM frequency permitted band splits into two bands. As in the case 2, one observes the absence of reflection in the NIM situation only for a part of κ values set.

4. Conclusion

In this paper, we solved the problem of obtaining the electric Green's function for the layered NIM system with the point source was located inside the certain metamaterial layer. We assumed that the permittivity and permeability of the metamaterial stand equal and are defined with the single Lorentz contribution. In the NIM situation, we obtained relations for the Fourier transformed electric Green's function. The magnetic Green's function can as well be obtained with reasoning similar to the presented above. The obtained formulas are symmetric, relative to the position of the point source. This fact shows the correlation with the physical conception of the electromagnetic field propagated into the system composed of isotropic homogeneous layers. The system can be composed of arbitrary finite number of layers. This fact allows us to use the considered system as a model for simulation or engineering of the real objects, such as superlens systems and multilayer NIM coverings. Besides, the reflection in the NIM situation is absent. This fact was discussed earlier [4, 6] and holds true only if the NIM frequency does not belong to the spectrum of the electric Helmholtz operator. For a limiting case, when we deal with infinite periodic NIM system, the obtained formulas correspond to gaps.

Also, we solved the problem of describing the photonic band gap structure of the similar infinite layered NIM system (1DPC). In the permitted bands we obtained expressions for the scalar field functions, which contained only one term responsible for propagation and no terms responsible for reflection, i.e., there was no reflection at the NIM frequency in permitted bands. By the numerical approach, we observed the NIM frequency belonging to a permitted band, i.e., in the NIM situation there was no reflection for any direction. With changing the layers widths the permitted band contained the NIM frequency was splitted into two bands, i.e., for the NIM frequency there was the absence of reflection not for any direction.

5. Acknowledgements

The work was partially financially supported by the Government of the Russian Federation (grant 074-U01), by the Ministry of Education and Science of the Russian Federation (GOSZADANIE 2014/190, Projects No 14.Z50.31.0031 and No. 1.754.2014/K), by grant MK-5001.2015.1 of the President of the Russian Federation.

References

[1] Burgos S, Waele R, Polman A and Atwater H 2010 *Nat. Mater.* **9** 407
[2] Dubinov A E and Mytareva L A 2010 *Phys. Usp.* **53** 475
[3] Iyer A and Eleftheriades G 2009 *IEEE. Trans. Antennas Propag.* **57** 1720
[4] Gralak B and Tip A 2010 *J. Math. Phys.* **51** 052902
[5] Collin R 2010 *PIER* **B19** 233
[6] Pravdin K and Popov I 2013 *Nanosystems: Phys. Chem. Math.* **4** 570
[7] Lai K, Tsang L and Huang C 2005 *Micro. Opt. Tech. Lett.* **44** 86
[8] Maksimovic M, Hammer M and Jaksic Z 2008 *Proceedings of SPIE* **6896** 689605
[9] Potemkin A, Poddubny A, Belov P and Kivshar Yu 2012 *Phys. Rev. A.* **86** 023848
[10] Pendry J *Phys. Rev. Lett.* 2000 **85** 3966
[11] Wu L, He S, Shen L 2003 *Phys. Rev. B* **67** 235103
[12] Pandey G, Thapa K, Srivastava S, Ojha S 2008 *PIER M* **2** 15
[13] Feng X, Li H 2013 *Eur. Phys. J. D* **67** 40157
[14] Tip A 2004 *Phys. Rev. E* **69** 016610
[15] Birman M S, Solomyak M Z 1987 *Spectral Theory of Self-Adjoint Operators in Hilbert Space* (Dordrecht: Reidel)
[16] Floquet G 1883 *Ann. Ecole Norm. Sup.* **12** 47
[17] Bloch F 1929 *Z. Phys.* **52** 555

Journal of Physics: Conference Series **661** (2015) 012026 doi:10.1088/1742-6596/661/1/012026

Modelling of discrete TDS-spectrum of hydrogen desorption

Natalia I Rodchenkova and Yury V Zaika

Institute of Applied Mathematical Research, Karelian Research Centre, Russian Academy of Sciences, Pushkinskaya Street 11, Petrozavodsk, Russia

E-mail: nirodchenkova@krc.karelia.ru, zaika@krc.karelia.ru

Abstract. High concentration of hydrogen in metal leads to hydrogen embrittlement. One of the methods to evaluate the hydrogen content is the method of thermal desorption spectroscopy (TDS). As the sample is heated under vacuumization, atomic hydrogen diffuses inside the bulk and is desorbed from the surface in the molecular form. The extraction curve (measured by a mass-spectrometric analyzer) is recorded. In experiments with monotonous external heating it is observed that background hydrogen fluxes from the extractor walls and fluxes from the sample cannot be reliably distinguished. Thus, the extraction curve is doubtful. Therefore, in this case experimenters use discrete TDS-spectrum: the sample is removed from the analytical part of the device for the specified time interval, and external temperature is then increased stepwise. The paper is devoted to the mathematical modelling and simulation of experimental studies. In the corresponding boundary-value problem with nonlinear dynamic boundary conditions physical-chemical processes in the bulk and on the surface are taken into account: heating of the sample, diffusion in the bulk, hydrogen capture by defects, penetration from the bulk to the surface and desorption. The model aimed to analyze the dynamics of hydrogen concentrations without preliminary artificial sample saturation. Numerical modelling allows to choose the point on the extraction curve that corresponds to the initial quantity of the surface hydrogen, to estimate the values of the activation energies of diffusion, desorption, parameters of reversible capture and hydride phase decomposition.

1. Introduction

The interest in the interaction of hydrogen and its isotopes with metals, alloys and intermetallic compounds is multifarious [1–6]. It is sufficient to mention problems in power production, protection of metals from hydrogen corrosion, chemical reactor design, rocket production.

High concentration of hydrogen in metal leads to hydrogen embrittlement. Common metallurgical concentrations of dissolved hydrogen range from 0.1 to 100 ppm. The authors of [7] have developed the hydrogen analyzer AV-1. The device allows to measure the hydrogen concentration in a solid sample using industrial laboratory facilities. A cylindrical sample is placed into the quartz vacuum extractor. The extractor is exposed to a given external temperature in the furnace. The contact between the sample and the extractor walls is point-like, the thermal conductivity of quartz is negligible, wherefore the heat transfer is determined by thermal radiation. As the sample is heated, atomic hydrogen diffuses inside the bulk and is desorbed from the surface in the molecular form. The extraction curve is measured by a mass-spectrometric hydrogen analyzer and recorded. The dependence of the desorption flux on the sample temperature at uniform heating called the TDS-spectrum usually has several clear-cut peaks. In addition to the diffusion, the surface processes and the capture of hydrogen atoms by different types of defects are limiting factors. Experiments with monotonous external heating show that background hydrogen fluxes from the extractor walls and fluxes from the sample cannot be reliably distinguished. Thus, the extraction curve is not reliable. In this case discrete TDS-spectrum is used. The sample is removed from the analytical part of the device for the specified time interval, and external temperature is then increased stepwise. The problem is formulated in the following way. It is necessary to have the adequate mathematical model of mass spectrometric analyzer measurements and the computational algorithm for plotting of model dependence of hydrogen desorption flux with respect to the time and to the temperature (of the sample surface). Then it is possible to estimate a hydrogen permeability parameters and to determine which physical-chemical processes are dominant within the different TDS-peaks at the spectrum. This can help specialists to understand how a structural material will behave in a hydrogen environment under specific conditions of long-term exploitation. If these conditions are extreme, the possibilities of experimental researches are limited and the role of mathematical modelling is very important. For numerical simulation, we use heat parameters which are suitable for aluminium and some of its alloys when modelling the desorption flux.

2. Mathematical model

Heating equation. The sample has the form of a cylinder with dimensions: radius of the base is $L = 4 \cdot 10^{-3}\,$m, height is $H = 2 \cdot 10^{-2}\,$m. If heating of the sample is uniform (sufficiently slow, $T = T(t)$, $[T] = $ K), the temperature dynamics can be described by the differential equation

$$\dot{T}(t) = \sigma\varepsilon S(\tilde{c}\rho V)^{-1}\big[T_{\mathrm{e}}^4 - T^4(t)\big], \quad T_0 = T(0) = 293\,\mathrm{K}. \tag{1}$$

Here, $T_{\mathrm{e}} = \mathrm{const} \in [300, 900]$ is temperature of the extractor wall, $S = 2\pi LH$, $V = \pi L^2 H$ are the surface area and the cylinder capacity, $\sigma = 5.67 \cdot 10^{-8}\,\mathrm{J \cdot s^{-1} m^{-2}\,K^{-4}}$ is the Stefan-Boltzmann constant. For aluminium: $\rho = 2.71 \cdot 10^3\,\mathrm{kg \cdot m^{-3}}$ is the bulk density, $\tilde{c} = 1.15 \cdot 10^3\,\mathrm{J \cdot kg^{-1} K^{-1}}$ is the specific heat capacity, $\varepsilon = 7 \cdot 10^{-5}(T + 64.3)$ is the heat absorption coefficient.

Since the assumption about uniform heating does not have adequate accuracy for all values of T_{e}, we consider an alternative distributed model. Since the geometry of the extractor is tubular heating of the sample mainly proceeds through the lateral surface. Thus, we speak of a lower estimate of the heating of symmetry center of the sample. The considered equation (1) is majorized from above. The radiosymmetrical model is:

$$\frac{\partial T}{\partial t} = \varkappa\Big(\frac{\partial^2 T}{\partial r^2} + \frac{1}{r} \cdot \frac{\partial T}{\partial r}\Big), \quad \frac{\partial T}{\partial r}\Big|_{r=+0} = 0, \quad \lambda\frac{\partial T}{\partial r}\Big|_{r=L} = \alpha\big[T_{\mathrm{e}}^4 - T^4(t, L)\big],$$

$T(0, r) = T_0 < T_{\mathrm{e}}$, $r \in [0, L]$. Here, λ is the heat conduction coefficient (for aluminium in the range of $T \in [T_0, T_{\mathrm{e}}]$ $\lambda \approx 236\,\mathrm{J \cdot s^{-1} m^{-1} K^{-1}}$), $\varkappa = \lambda(\tilde{c}\rho)^{-1}$ is the thermal diffusivity, $\alpha = \sigma\varepsilon$. The finishing time t^* of computing is determined by the stationary distribution $T(t, 0) \approx T_{\mathrm{e}}$, $t > t^*$. The auxiliary problem of numerical modelling of heating is the following: to estimate quickly how much the distribution $T(t, r)$ differs from the uniform heating $T(t)$ under given T_{e}, L, H, and thermal and physical characteristics of the material. For instance, when we use $T_0 = 293\,$K, $T_{\mathrm{e}} = 773\,$K and the L, H stated above the assumption about uniform heating of aluminium sample is true: the difference $T(t) - T(t, 0)$ is within a hundredth of a degree. We obtain this result in the distributed model where heat absorption by butt ends is disregarded. Peak temperature is reached in 2.2 hours, heating being practically linear within an hour. Hereinafter, suppose that heating of the sample is uniform.

Diffusion model with defects. Consider the boundary-value problem of thermodesorption for the cylinder taking into consideration diffusion in the bulk, hydrogen capture by defects of two types, penetration from the bulk to the surface and desorption (surface desorption model):

$$\frac{\partial c}{\partial t} = D(T)\Big(\frac{\partial^2 c}{\partial r^2} + \frac{1}{r}\cdot\frac{\partial c}{\partial r} + \frac{\partial^2 c}{\partial z^2}\Big) - a_1[1-W]c + a_2 w + a_3\gamma,\ r\in(0,L),\ z\in(0,H), \quad (2)$$

$$\frac{\partial w}{\partial t} = a_1(T)[1-W]c(t,r,z) - a_2(T)w(t,r,z), \quad W \equiv \frac{w}{w_{\max}},\ t\in(0,t_*), \quad (3)$$

$$\dot\gamma(t) = -a_3\gamma(t),\ \gamma(t_{\rm crit}) = \overline{\gamma} \Rightarrow \gamma(t) = \overline{\gamma}\exp\{[t_{\rm crit}-t]a_3\},\ a_3 = 0,\ t < t_{\rm crit},$$

$$c(0,r,z) = \bar c = \text{const},\ w(0,r,z) = \overline{w} = \frac{a_1\bar c}{a_2 + a_1\bar c\, w_{\max}^{-1}},\ r\in[0,L],\ z\in[0,H],$$

$$\partial_r c\big|_{r=+0} = 0,\ \partial_z c\big|_{H/2} = 0,\ D(t)\equiv D(T(t)),\ b(t)\equiv b(T(t)),\ g(t)\equiv g(T(t)),$$

$$\frac{\partial q_1}{\partial t}(t,z) = -b(t)q_1^2(t,z) - D(t)\frac{\partial c}{\partial r}\Big|_{r=L},\quad c(t,L,z) = g(t)q_1(t,z), \quad (4)$$

$$\frac{\partial q_2}{\partial t}(t,r) = -b(t)q_2^2(t,r) - D(t)\frac{\partial c}{\partial z}\Big|_{z=H},\quad c(t,r,H) = g(t)q_2(t,r), \quad (5)$$

$$\frac{\partial q_3}{\partial t}(t,r) = -b(t)q_3^2(t,r) + D(t)\frac{\partial c}{\partial z}\Big|_{z=0},\quad c(t,r,0) = g(t)q_3(t,r),$$

$$\dot T(t) = \sigma\varepsilon S(\tilde c\rho V)^{-1}\big[T_{\rm e}^4 - T^4(t)\big],\quad \varepsilon = 7\cdot 10^{-5}\big(T(t)+64.3\big),\ T(0) = T_0.$$

Here, $c(t,r,z)$ is the concentration of dissolved (diffusion movable) hydrogen atoms in the metal; $w(t,r,z)$ is H concentration in the defects with reversible capture (for instance, microcavities); $w_{\max}$ is the peak concentration of reversible capture; $\gamma(t)$ is H concentration in traps; they begin to release hydrogen only when some critical temperature $T_{\rm crit}$ is reached (it is typical for the inclusion of hydride phases); a_i are the traps dynamic coefficients ($a_3 > 0$ only at $T \geqslant T_{\rm crit}$); $q_1(t,z)$, $q_{2,3}(t,r)$ are the surface concentrations (on the lateral surface of the cylinder and on the butt ends); g is the fit coefficient for concentrations of hydrogen atoms in the bulk and on the surface (quick dissolution parameter); D, b are the diffusion and the desorption coefficients.

Initial concentration $c(0,r,z) = \bar c$ is constant (it forms during the material production). When necessary, it is possible to take account of the decrease in H concentration in the near-surface layer (for instance, as the result of mechanical and thermal pre-treatment) without essential changes in the numerical algorithm. Similar correction takes place for hydrogen traps of the hydride phase type: $\gamma(0,r,z) = \widetilde{\gamma}(r,z)$, $\partial_t\gamma = -a_3(T)\gamma(t,r,z)$. Suppose that $a_i > 0$ are constant within the considered temperature range ($T\in[T_0,\,T_{\rm e}]$). Changes in the time-dependent case of $a_i(t)\equiv a_i(T(t))$ are insignificant.

For practical purposes we consider capture in a simple integrated form. Detailed description of geometry of defects and their distribution in the sample essentially complicates the model. It is difficult to obtain these parameters experimentally as an input model data. Since the sample sizes are macro the defects may be considered as uniformly distributed. To simplify the model description we consider one generalized trap with reversible capture and one with hydride-like decomposition. It is possible to consider several traps with their individual $a_i^{(j)}$ and $T_{\rm crit}^{(k)}$.

We assume that diffusion and desorption coefficients depend on the temperature in an Arrhenius way: $D(T) = D_0\exp\{-E_D/[RT]\}$, $b(T) = b_0\exp\{-E_b/[RT]\}$. Here $E_{D,b}$ are activation energies. We use a shorthand notation $D(t) = D(T(t))$, $b(t) = b(T(t))$. The time interval $[0,t_*]$ is determined by relative degassing of the sample: $J(t)\approx 0$, $t\geqslant t_*$ ($J(t)\ll J_{\max}$). The conditions $\partial_r c(t,+0,z) = 0$, $\partial_z c(t,r,H/2) = 0$ are consequences of symmetry.

A more precise model of dissolution on the surface (considering the lateral surface for clarity):

$$k^+(T)c(t,L,z)\big[1 - q_1(t,z)q_{\max}^{-1}\big] - k^-(T)q_1(t,z)\big[1 - c(t,L,z)c_{\max}^{-1}\big] = -D(T)\partial_r c(t,L,z).$$

Journal of Physics: Conference Series **661** (2015) 012026 doi:10.1088/1742-6596/661/1/012026

Under saturation ($q_1 \to q_{\max}$, $c \to c_{\max}$) the output flux to the surface and dissolution into the bulk become less. However, when diffusion is considerably slower than dissolution and concentrations are low, we obtain the condition of quick dissolution $c \approx gq$, where $g = k^-/k^+$. If the surface is isotropic (in the sense $E_{k^-} \approx E_{k^+}$), then the parameter g weakly depends on T.

For traps which begin to release hydrogen only when critical temperature is reached (traps of the hydride phase type) we took only the trap capacity and decomposition rate into account. Modelling of the dehydriding process is a separate complicated problem, which results in nonlinear boundary-value problems with moving free phase boundaries and with conditions of the Stefan type (see, for example, [8]).

Since the initial data are symmetrical we have $q_3 = q_2$, so we construct the difference approximation only for a half of the cylinder ($z \in [H/2, H]$) with corresponding boundary conditions ($\partial_z c|_{H/2} = 0$, $\dot{q}_2 = \ldots$). For the defect with reversible capture the constant $\bar{w}$ is determined according to $\partial_t = 0$: $a_1(T_0)[1 - \bar{w}/w_{\max}]\bar{c} - a_2(T_0)\bar{w} = 0$. For the trap of the decomposition type the values of $\bar{\gamma} = \mathrm{const}$, T_{crit}, a_3 are set using information about the hydride.

The presence of derivatives $\dot{q}_i$ (accumulation on the surface [9]) corresponds to the possibility of H migration along the surface until atoms form H_2 molecules which are desorbed from the surface. In the case of bulk desorption (for "porous" material atoms combine in H_2 molecules in the near-surface layer and leave the sample) instead of dynamic boundary conditions we use nonlinear conditions: $D(t)\partial_r c(t, L, z) = -\tilde{b}(t)c^2(t, L, z)$, $D(t)\partial_z c\big|_{z=0,H} = \pm\tilde{b}(t)c^2\big|_{z=0,H}$. To estimate the influence of surface accumulation we consider the bulk and surface desorption coefficients satisfying the condition $\tilde{b} = b/g^2$ ($\dot{q} \approx 0$).

Thus, the aim of the model is to analyze the dynamics of hydrogen concentrations without preliminary sample saturation. The problem has a practical context, so we apply minimal math-based environment to describe basic processes. Further detailed elaboration leads to an increase of the set of parameters. The inverse problem of their estimation becomes complicated [10].

Specification of the problem statement. The aim of the research is to develop the difference scheme and computational algorithm for modelling of desorption flux from the cylindrical sample:

$$J(t) = 4\pi b(t)\left[L\int_{H/2}^{H} q_1^2 dz + \int_0^L q_2^2\, r dr\right],\quad \tilde{J}(t) = 4\pi\tilde{b}(t)\left[L\int_{H/2}^{H} c^2(t, L, z)\, dz + \int_0^L c^2(t, r, H)\, r dr\right].$$

Hydrogen molecules is desorbed, but we compute the desorption flux in atoms ($[J] = 1/\mathrm{s}$). The criterion of computation correctness is the material balance (the total amount of hydrogen including the desorbed hydrogen is constant).

Under uniform monotonous heating it is convenient to consider the TDS-spectrum (the curve $J = J(T)$) together with the flux dependence on t. It is believed that first TDS-spectrum peak corresponds to initial surface hydrogen, another one corresponds to bulk hydrogen. However, while the surface hydrogen is desorbed, hydrogen atoms penetrate from the bulk to the surface. The problem of estimating the corresponding correction is important. Let t_s be the time when the initial amount of surface hydrogen is exhausted. At the numerical modelling stage the time t_s is determined from the equation

$$\int_0^{t_s} J(\tau)\, d\tau = 4\pi\left[L\int_{H/2}^{H} q_1(0, z)\, dz + \int_0^L q_2(0, r)\, r dr\right].$$

Numerical modelling allows to choose the point on the extraction curve that corresponds to the initial quantity of the surface hydrogen, to estimate the values of the activation energies of diffusion, desorption, parameters of reversible capture and hydride phase decomposition. Such problem is the inverse ill-posed problem of model parametric identification. Presented article is only devoted to the model description and direct problem of numerical simulation.

3. Results of numerical modelling

The iterative algorithm is developed on the basis of implicit difference schemes and the Thomas algorithm. Input data are L, H, D_0, E_D, b_0, E_b, g_0, E_g, $\bar{c}$, $w_{\max}$, $\bar{\gamma}$, T_0, T_e, T_{crit}, a_1, a_2, a_3, σ, ρ, $\tilde{c}$. Desorption flux curves are represented in the figures. Varied coefficient values correspond to the curves in the following way: the first value is for the curve with the highest maximum, the last one is for that with the lowest maximum. The small circle shows the time when the initial surface hydrogen is exhausted. The developed software (algorithm for numerical thermodesorption modelling) is applicable in a wide range of parameters. In our figures for demonstrative purpose we confine ourselves to values of the parameters presented below. We define $\bar{\gamma} = 10^{22}\,\mathrm{m}^{-3}$, $w_{\max} = 10^{22}\,\mathrm{m}^{-3}$. The default common parameters are the following: $b_0 = 2\cdot10^{-7}$, $E_b = 12\cdot10^4$, $D_0 = 2\cdot10^{-3}$, $E_D = 6\cdot10^4$, $g_0 = 10^3$, $E_g = 0$, $T_0 = 293$, $T_{crit} = 700$, $T_e = 773$, $a_1 = 0.3$, $a_2 = 0.1$, $a_3 = 10^{-4}$, $L = 4\cdot10^{-3}$, $H = 2\cdot10^{-2}$, $\bar{c} = 10^{22}$.

The algorithm of discrete TDS-spectrum construction in accordance with the experiment is the following. Each stage of the numerical experiment is carried out at a given external temperature $T = T_e$ (initial sample temperature is $T_0 < T_e$), maximum stage duration is 3 hours. The total amount of desorbed hydrogen Q is calculated (as integral of $J(t)$) at given external temperature T_e up to the time when the desorption flux becomes comparable with the background flux in the scale of maximum value $J_{\max}$ (but the interval no more than specified 3 hours). A point on the plane (T, Q) is obtained. Then T_e increases stepwise and we reiterate the calculations starting at temperature T_0 with initial concentration distributions from the previous stage. Dependence J with respect to current temperature T ($T(t) \leftrightarrow t$) is shown in Fig. 1. The impact of different kinetic parameters for the surface desorption can be seen in the next six figures. Fig. 2 illustrates the influence of the preexponential factor b_0 on the desorption flux curve. Changes are: $b_0 = 5\cdot10^{-9}$; $5\cdot10^{-8}$; $2\cdot10^{-7}\,\mathrm{m^2s^{-1}}$. The influence of the activation energy E_b can be seen in Fig. 3, changes in order of decrease of the maximum are: $E_b\cdot10^{-4} = 10$; 11; 12 $\mathrm{J{\cdot}mol^{-1}}$. Changes for Fig. 4 are: $D_0 = 2\cdot10^{-3}$; $5\cdot10^{-4}$; $10^{-4}\,\mathrm{m^2s^{-1}}$. For the next four figures the solid line is for the surface desorption model, the dashed line is for the bulk desorption model. Fig. 5 shows discrete TDS-spectra with temperature step size h_{T_e} equal to 50 K. Discrete TDS-spectra modelling with $h_{T_e} = 75$ K is presented in Fig. 6. The small cross shows the time when the initial surface hydrogen is exhausted.

Thus, parameter variations agree with physical ideas, and the model may be used for kinetic parameter identification of degassing of real materials.

Acknowledgements
This work was supported by the Mathematical Division of Russian Academy of Sciences.

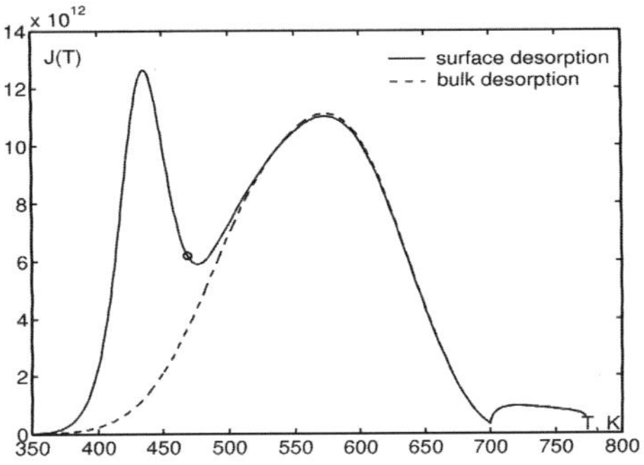

Figure 1. TDS-spectrum $J(T)$.

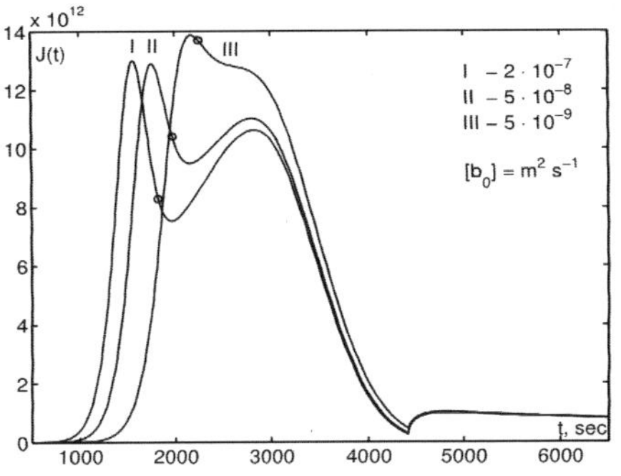

Figure 2. The impact of the parameter b_0.

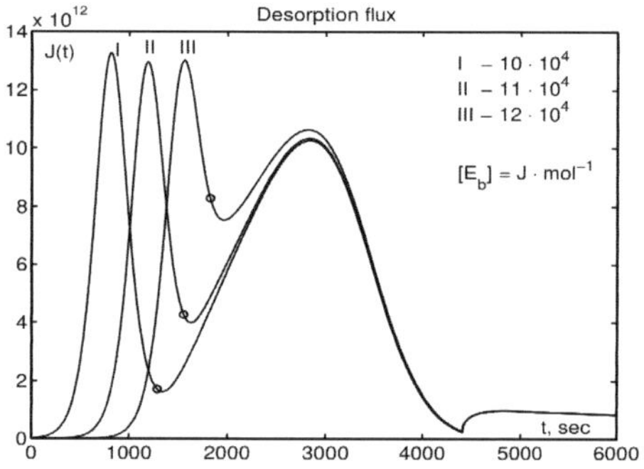

Figure 3. The impact of the parameter E_b.

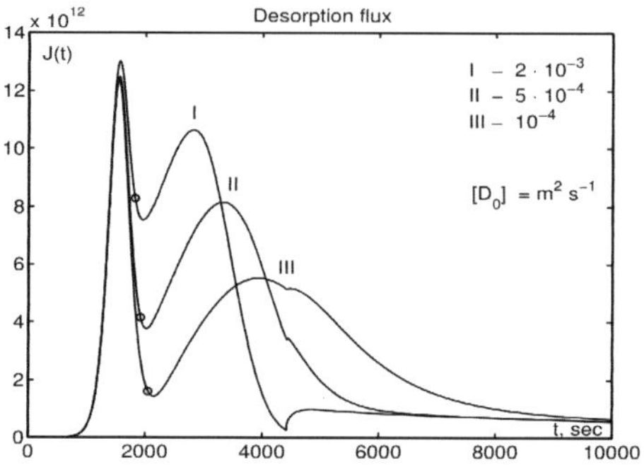

Figure 4. The impact of the parameter D_0.

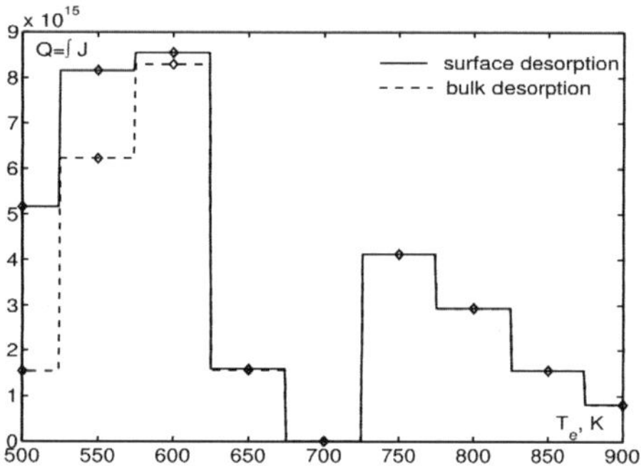

Figure 5. Discrete TDS-spectra, $h_{T_e} = 50\,\mathrm{K}$.

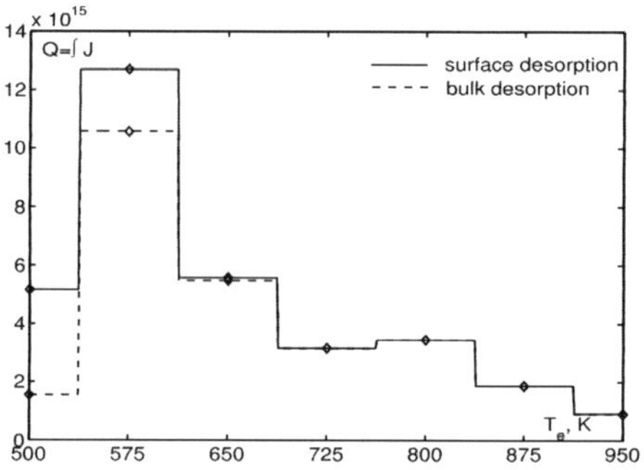

Figure 6. Discrete TDS-spectra, $h_{T_e} = 75\,\mathrm{K}$.

References

[1] Alefeld G and Völkl J (Eds.) 1978 *Hydrogen in Metals* Berlin; New York: Springer vol I Basic Properties p 427, vol II Application-oriented Properties p 387

[2] Pisarev A A, Tsvetkov I V, Marenkov E D and Yarko S S 2008 *Hydrogen Pearmeability through Metals* (Moscow: MEPhI) p 144

[3] Hirscher M (Ed.) 2010 *Handbook of Hydrogen Storage* Wiley–VCH p 353

[4] Gabis I 1999 The method of concentration pulses for studying hydrogen transport *Tech. Phys.* **44** 90–4

[5] Lototskyy M V, Yartys V A, Pollet B G and Bowman R C Jr. 2014 Metal hydride hydrogen compressors: a review *Int. J. Hydrogen Energy* **39** 5818–51

[6] Indeitsev D A and Semenov B N 2008 About a model of structure-phase transformations under hydrogen influence *Acta Mechanica* **195** 295–304

[7] Polyanskiy A M, Polyanskiy V A, Popov-Diumin D B and Kozlov E A 2006 New measuring complex for absolute determination of hydrogen content in hydrogen energetics materials *Alt. Energy Ecology* **38** 29–31

[8] Zaika Yu V and Rodchenkova N I 2009 Boundary-value problem with moving bounds and dynamic boundary conditions: diffusion peak of TDS-spectrum of dehydriding *Appl. Math. Modelling* **33** 3776–91

[9] Zaika Yu V and Rodchenkova N I 2013 Hydrogen-solid boundary-value problems with dynamical conditions on surface *Math. Modelling* (New York: Nova Sci. Publishers) pp 269–302

[10] Zaika Yu V and Bormatova E P 2011 Parametric identification of a hydrogen permeability model by delay times and conjugate equations *Int. J. Hydrogen Energy* **36** 1295–1305

Journal of Physics: Conference Series **661** (2015) 012027 doi:10.1088/1742-6596/661/1/012027

Computer simulation of hydrogen permeability of structural materials through protective coating defect

E K Kostikova and Yu V Zaika

Institute of Applied Mathematical Research, Karelian Research Centre, Russian Academy of Science, 11 Pushkinskaya street, Petrozavodsk, 185910, Russia

E-mail: zaika@krc.karelia.ru

Abstract. In the context of problems of hydrogen and thermonuclear power engineering intensive research of the hydrogen isotopes properties is being conducted. Mathematical models help to specify physical-chemical ideas about the interaction of hydrogen isotopes with structural materials, to estimate the limiting factors and to significantly reduce the expenses of experimental research by means of numerical simulation for different parameters and experimental conditions (including extreme ones). Classical diffusion models are often insufficient. The paper is devoted to the models and numerical solution of the boundary-value problems of hydrogen permeability taking into account nonlinear sorption-desorption dynamics on the surface. Algorithms based on difference approximations. The results of computer simulation of the hydrogen flux from a structural material sample are presented.

1. Introduction

Interaction of hydrogen and its isotopes with solids has many applications [1–8]. It is sufficient to mention power engineering, protection of metals from hydrogen corrosion, design of chemical reactors, rocket and missile engineering. Hydrides help to retain substantial amounts of hydrogen: environmental-friendly energy source. Hence the high expectations are attached to hydrogen batteries and motors that avoid using high pressures or low temperatures. Some special topics are considered in [9–12]. Enthusiasts speak not only of hydrogen energy but also of hydrogen economy [5]. Reduction of hydrogen permeation through protective walls of structural materials is the most important objective when solving complex problems of hydrogen storage, hydrogen transport, and reduction of tritium content in protective systems of future thermonuclear reactors (the ITER project). Barrier of a structural material provides the necessary mechanical resistance of the construction, while superimposed protective coating is supposed to block migration of hydrogen atoms. Some areas of the material might face hydrogen and its isotopes directly due to defects in the protective coating. In [10], the model of hydrogen permeability through a cylindrical membrane with radius of the base L and height H was examined under assumption that diffusion is the only limiting process. The inlet surface $z = 0$ was covered with thin protective coating with a small (radius $r_0 < L$) defect called “pinhole defect”. Hydrogen was able to permeate through it. The remainder of the inlet surface and the lateral surface of the cylindrical membrane were impermeable for hydrogen. Vacuum was created on the output side $z = H$. At the initial time the membrane contained no hydrogen. Pressure p of molecular hydrogen was instantly created at the inlet side. Relatively fast interjacent process was neglected (its speed depends on p, the material, and sample size). Then, in the first approximation, concentration of dissolved hydrogen under defect can be supposed to be constant ($\bar{c}$), equilibrium with respect to the gas phase according to Sievert's law: $\bar{c} \propto \sqrt{p}$). Dissolved atomic hydrogen diffused to the outlet surface, recombinated into molecules, and desorbed. Mass-spectrometer registered the permeation flux. The boundary-value problem neglecting surface processes was analytically studied only for the case of an infinite plate ($L \to +\infty$). The main weakness of such definition of the problem is that the dynamics of surface processes which has attracted a lot of attention recently, is ignored in the model [10].

2. Diffusion model

As a basic model of hydrogen transfer inside the membrane we consider the boundary-value problem of hydrogen permeability with diffusion as the only limiting factor [10]:

$$\frac{1}{D}\frac{\partial c}{\partial t} = \frac{\partial^2 c}{\partial r^2} + \frac{1}{r}\cdot\frac{\partial c}{\partial r} + \frac{\partial^2 c}{\partial z^2}, \quad r \in (0, L), \ z \in (0, H), \ t \in (0, t_*), \tag{1}$$

$$c(0, r, z) = 0, \ r \in [0, L], \ z \in [0, H], \quad c(t, r, 0) = c_0, \ r \in [0, r_0], \tag{2}$$

$$\partial_z c(t, r, 0) = 0, \ r \in (r_0, L], \quad c(t, r, H) = 0, \ r \in [0, L], \ t \geqslant 0, \tag{3}$$

$$\partial_r c(t, L, z) = 0, \ \partial_r c(t, +0, z) = 0, \quad z \in [0, H], \ t \geqslant 0. \tag{4}$$

Here $c(t, r, z)$ is concentration of dissolved atomic hydrogen. The experiment is carried out at constant temperature $T = \overline{T}$. The material is nearly homogeneous, so c_0 and D are constants. For definiteness let $c_0 = \bar{c}$; values of $\bar{c}(T, p)$ can be found in reference books. The defect is small, we assume it round and situated in the center: this is not a fundamental constraint for numerical simulation. At time t_* the permeation flux $J(t)$ reaches the stationary value. The determination is asymptotic: $J(t) \approx \text{const}, \ t \geq t_*$. The boundary condition (2) corresponds to rapid (compared to t_*) establishment of local equilibrium on the inlet side.

The aim is to develop the algorithm of simulation of the flux of hydrogen from the outlet surface. For the model (1)–(4): $J(t) = -D \int_0^L \partial_z c(t, r, H)\, 2\pi r\, dr$. Iterative numerical algorithm is based on implicit difference schemes and the Thomas algorithm. Control of hydrogen balance is used for checking the calculation. We used the Richardson extrapolation to improve accuracy of the calculations. The method of grid condensation was used for $r_0 \ll L$.

Dimensionless form of the boundary-value problem. To present the results we pass to dimensionless variables, using values intrinsic for this problem: $u = c/c_0$, $\rho = r/L$, $\zeta = z/H$, $\tau = (D/L^2)t$. The value of L^2/D is treated as the characteristic time of concentration stabilization due to diffusion. In the areas with length-scale parameters comparable to L (on vertical direction, i.e., z-direction) diffusion is faster for $H \sim L$, due to vacuum from the outlet side. Denoting $\rho_0 = r_0/L < 1$, $R = L/H$, we get the boundary value problem:

$$\frac{\partial u}{\partial \tau} = \frac{1}{\rho}\frac{\partial}{\partial \rho}\Big(\rho\frac{\partial u}{\partial \rho}\Big) + R^2\frac{\partial^2 u}{\partial \zeta^2}, \ \rho \in (0,1), \ \zeta \in (0,1), \quad u(\tau, \rho, 0) = 1, \ \rho \in [0, \rho_0],$$

$$\partial_\zeta u(\tau, \rho, 0) = 0, \ \rho \in (\rho_0, 1], \ \rho_0 < 1, \quad u(\tau, \rho, 1) = 0, \ \rho \in [0, 1], \ \tau \geqslant 0,$$

$$\partial_\rho u(\tau, 1, \zeta) = 0, \quad \partial_\rho u(\tau, +0, \zeta) = 0, \quad u(0, \rho, \zeta) = 0, \ \rho \in [0, 1], \ \zeta \in [0, 1].$$

Varying parameters are $R > 0$ and $\rho_0 \in (0, 1)$. We obtained an anisotropic diffusion equation: $D_\rho = 1$, $D_\zeta = R^2$. The smaller H (i.e., the thinner the membrane), the faster the diffusion in the ζ-direction. Coefficient D_ζ represents the ratio of the geometric characteristics. During the "time" τ the significant "length" in the ρ direction is $\sqrt{\tau}$, and in the ζ direction it is $R\sqrt{\tau}$. Steady-state (i.e., $\partial_\tau u \approx 0$) is determined by the value $\tau_* = (D/L^2)t_*$.

Let us introduce a dimensionless, average (on the base of the π square) outward flux

$$U(\tau) = -\frac{1}{\pi}\int_0^1 R^2 \partial_\zeta u|_{\zeta=1} 2\pi\rho\, d\rho.$$

The function $U(\tau)$ increases monotonically approaching asymptotically a steady-state value $U_* = U(\tau_*)$. If a protective coating is absent (formally consider $\rho_0 = 1$), we have a maximum of $U_* = R^2$: $\partial_\rho u = 0$, $\partial_\zeta u|_{\tau_*} = -1$. Original averaged flux can be expressed as follows:

$$I(t) \equiv \frac{J(t)}{\pi L^2} = \frac{Dc_0}{H} V(\tau) \ (\tau = DL^{-2}t), \quad V(\tau) = R^{-2}U(\tau) = -\frac{1}{\pi}\int_0^1 \frac{\partial u}{\partial \zeta}\Big|_{\zeta=1} 2\pi\rho \, d\rho.$$

The value of Dc_0/H is density of the permeation steady-state flux for the membrane without a protective coating ($\rho_0 = 1$). The number $V(DL^{-2}t) \in (0, 1)$ is a fraction of $I(t)$ in the highest possible density of the output flux Dc_0/H. Letting $\rho_0 = 1$ we have a maximum $V_* = 1$.

The range for the diffusion coefficient is fixed as D: $10^{-7} - 10^{-5}$ cm^2/c. We will consider cylindrical membrane dimension as comparable with a coin (the barrier of a pipeline).

Output parameters of the computing experiment are the values τ_*, τ_0, U_* (V_*), where τ_0 is the delay time calculated by the formula $\tau_0 = \tau_* - S(\tau_*)/\dot{S}(\tau_*) = \tau_* - S(\tau_*)/U(\tau_*)$, $S(\tau) \equiv \int_0^\tau U(\tau)\,d\tau$ ($\dot{S} \equiv dS/d\tau$). In the original variables we obtain $t_0 = t_* - \int_0^{t_*} J(t)dt/J(t_*)$. Geometrically this is the point of intersection of the axis t with asymptotes of the function $Q(t) = \int_0^t J(\tau)d\tau$ (the number of hydrogen atoms permeating through the membrane). The accuracy of calculation of $\tau_0(t_0)$ increases with increasing $\tau_*(t_*)$ taking into account the asymptotic behavior of the stabilized mode permeation flux. The time of stabilization of the permeation flux, the value of the steady-state flux, and the delay time are experimentally measured. These values serve as the input information for the inverse problem of evaluation of the kinetic parameters of hydrogen permeability using measurements; see, e.g. [12] ($\rho_0 = 1$)). Additionally, calculation checking was carried out for $\rho_0 = 1$, this case corresponds to a membrane without a protective coating. Calculations confirmed that hydrogen concentration is $u(\tau, \rho, \zeta) = 1 - \zeta$ ($\tau \geq \tau_*$) and the delay time through the plate of thickness ℓ is equal to $\ell^2/[6D]$, it is $1/[6R^2]$ using our notation. Besides we checked asymptotic behavior of $U_* \to R^2$ for $\rho_0 \to 1$.

Dimensionless characteristics τ_* and τ_0 weakly depend on radius of the defect. We restrict the discussion to the case of $\rho_0 < 0.2$, because this is the adopted limit of a protective coating. Functions $\tau_*(\rho_0)$ and $\tau_0(\rho_0)$ have extrema at $R < 1.5$ (derivative sign changes, in Fig. 1, 2). This is caused by growing influence of the boundary conditions (insulated lateral surface). Functions $\tau_*(\rho_0)$, $\tau_0(\rho_0)$ are monotone for $R \geq 1.5$. The curve for $R = 1.5$ is indicated by a dotted line. "Interfluve value" $R = 1.5$ is rather nominal. It separates relatively small and large values of R at the qualitative level. At the qualitative level we obtain the following results focusing on an "ideal" experimental error under 10% and not distinguishing output model fluxes under 1%. The influence of the boundary conditions with $r = L$ on stationary permeation flux is insignificant for $R = L/H > 3$, $\rho_0 < 0.2$; such membrane can be considered as a plate ($L \to +\infty$). This corresponds to the results by [10]: analytical methods of research are useable for $R > 3$. Practically, it is possible to estimate when permeation flux is proportional to the number of defects (or their density). The developed software allows to analyze spatial distribution of hydrogen in the membrane at any points of time (Fig. 3, 4).

3. Modification of the model taking into account the dynamics on the surface

The boundary conditions (2), (3) are replaced by ratio of the flux balance

$$\mu sp - bc^2(t, r, 0) = -D\frac{\partial c}{\partial z}\Big|_{z=0}, \ r \in [0, r_0], \quad bc^2(t, r, H) = -D\frac{\partial c}{\partial z}\Big|_{z=H}, \ r \in [0, L], \qquad (5)$$

$J(t) = \int_0^L bc^2(t, r, H)\, 2\pi r\, dr$. Here b is the coefficient of bulk desorption (effective recombination [3]). Additionally, the inlet and outlet surfaces may have different properties, this is easily taken into account in the model: $b = b_1$ at $z = 0$ and $b = b_2$ at $z = H$. Formation and decomposition of hydrid phases are more difficult to be taken account (see [11]); we suppose

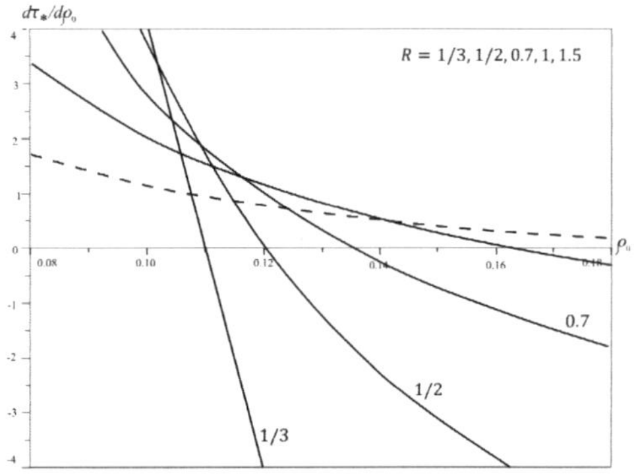

Figure 1. Rate of τ_*, $R \leq 1.5$ change.

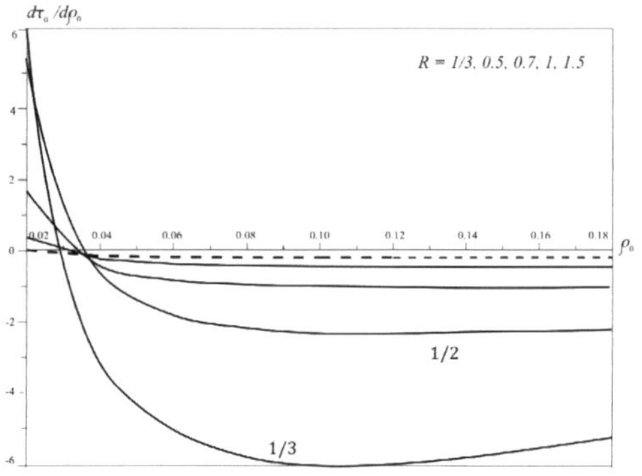

Figure 2. Rate of τ_0, $R \leq 1.5$ change.

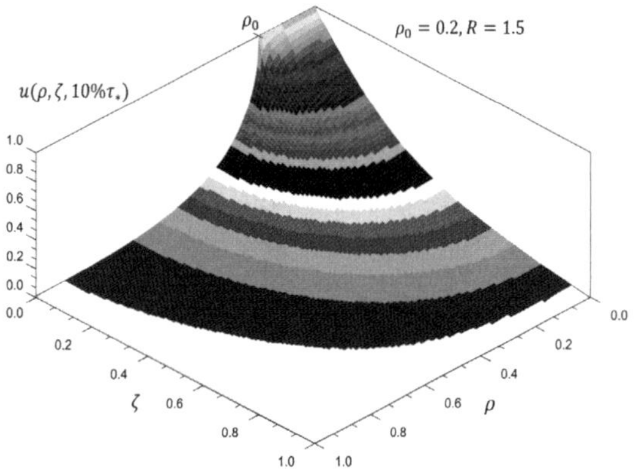

Figure 3. Concentration profile, $\tau = 0.1\tau_*$.

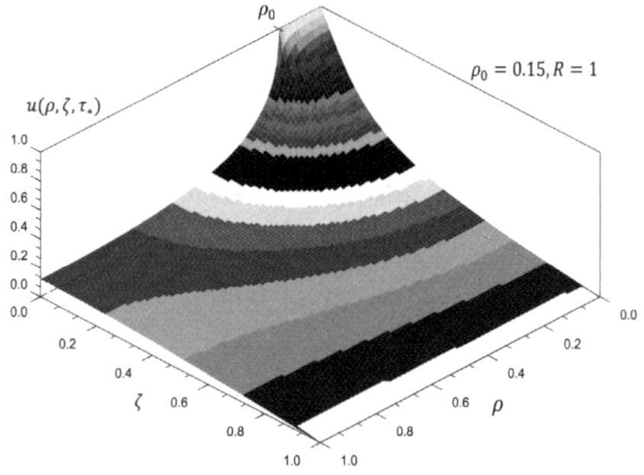

Figure 4. Steady-state concentration level.

they are not important. Vacuum is created at the outlet of the membrane, so resorption is ignored. Output desorption flux is counted in atoms ($[J] = 1/c$). In the equilibrium state the condition $\partial_z c = 0$ is replaced by $\mu s p = b\bar{c}^2$, sort of the Sievert's law $\bar{c} \propto \sqrt{\bar{p}}$.

Numerical simulation makes it possible to estimate how fast the defect concentration under protective coating is stabilized and how concentration $c(t_*, r, 0)$ differs from the equilibrium level $\bar{c}$. In addition, information about sensitivity of $J(t)$ with respect to variations of the diffusion coefficient and surface processes ("derivatives" on D, b, s) is of special interest.

Let us pass to dimensionless variables: $u = c/c_0$, $\rho = r/L$, $\zeta = z/H$, $\tau = tD/L^2$, $R = L/H$, $\rho_0 = r_0/L$. Concentration c_0, $c_0 < \bar{c}$ for $\rho_0 = 1$ can be taken as a normalizing concentration. However, we accept $c_0 = \bar{c}$ ($u \leq 1$), having the extensive "equilibrium" reference information about metals and alloys available. For the purpose of the model we have $\bar{c} = \sqrt{\mu s p / b}$. Conditions (5) can be written in the form

$$\hat{p} - \hat{b}u^2(\tau, \rho, 0) = -\partial_\zeta u\big|_{\zeta=0}, \ \rho \in [0, \rho_0], \quad \hat{b}u^2(\tau, \rho, 1) = -\partial_\zeta u\big|_{\zeta=1}, \ \rho \in [0, 1], \tag{6}$$

$\hat{b} \equiv bc_0H/D$, $\hat{p} \equiv \hat{\mu} s p$, $\hat{\mu} \equiv \mu H / Dc_0$, $c_0 = \bar{c}$. The pressure p of molecular hydrogen on the input surface is constant. The coefficient $\hat{b}$ is equal to the ratio of the local desorption density $b\bar{c}^2$ when concentration is close to the equilibrium concentration $\bar{c}$ to the density $D\bar{c}/H$ of the permeation flux in the diffusion regime, when $c|_H = 0$, $\rho_0 = 1$, $t \geqslant t_*$. Besides, the outlet flux μsp is normalized by value $D\bar{c}/H$. For equilibrium state we have $\mu sp = b\bar{c}^2$, hence $\hat{b} = \hat{p}$. We denote this parametric variable by W. In the present model we obtain $W = HbS\sqrt{p}/D$ $\big(H\sqrt{p}F/D = L\sqrt{p}F/RD\big)$, where $F \equiv \sqrt{b\mu s}$, $S \equiv \sqrt{\mu s/b}$ is dissolution ($\bar{c} = S\sqrt{p}$).

W is called a transporting parameter when protective coating is absent ($\rho_0 = 1$) [3]: if $W \gg 1$ diffusion limits permeation; this is the diffusion limited regime, DLR. The asymptotic condition $W \ll 1$ corresponds to the SLR (surface limited regime). The transitory area is estimated within $W \in (10^{-2}, 10^4)$. For the problem under consideration when hydrogen permeates through a defect of the protective coating ($\rho_0 < 1$) we have three independent parameters R, ρ_0, W. The value of R^2 serves as a "diffusion coefficient" on the ζ direction, and W parametrizes the left parts of the boundary conditions (6): $W[1 - u^2(\tau, \rho, 0)]$, $Wu^2(\tau, \rho, 1)$.

Model with surface desorption. Boundary conditions (2), (3) are replaced by the following:

$$\partial_t q_0 = \mu sp - bq_0^2(t,r) + D\partial_z c\big|_0, \; r \in [0, r_0], \quad \partial_t q_H = -bq_H^2(t,r) - D\partial_z c\big|_{z=H}, \; r \in [0, L], \quad (7)$$

$c(t,r,0) = gq_0(t,r)$, $c(t,r,H) = gq_H(t,r)$, $q_0 = q_H = 0\,(t=0)$, $J(t) = \int_0^L b\,q_H^2(t,r)\;2\pi r\,dr$. Here q_0, q_H is the surface concentration on the inlet and outlet surfaces, $[q] = 1/\text{cm}^2$; g is the coefficient of concordance between the hydrogen atoms concentration in the near-surface bulk and on the surface (a coefficient of fast dissolution, $[g] = 1/\text{cm}$); b is the coefficient of surface desorption, $[b] = \text{cm}^2/\text{c}$ ($b = b_{\text{surf}} = g^2 b_{\text{vol}}$). Dynamic boundary conditions (7) allow to take into account and estimate storage effect of hydrogen atoms on the surface. These conditions are agreed (in the classical context) with zero initial data $c(0,r,z) = 0$ (as $t \to +0$). But keep in mind that, generally speaking, $\partial_t q_0(0,r) = \mu sp \gg 1$.

We may pass to the dimensionless problem, as it was done above. But the introduction of an additional parameter g does not reduce the number of independent numeric hydrogen permeability characteristics. Hence we illustrate the model using physical units of measurement. The following value specifies transport parameter estimation $W = HD^{-1}b_{\text{vol}}\bar{c} = HD^{-1}b\bar{c}/g^2$.

General geometric values of parameters for presented computing experiments (Fig. 5–6) were $L = 1$ cm, $H = 1/3$ cm, $r_0 = 0.1$ cm. Note that the flux decreases for increasing b, because desorption on the inlet surface predetermines hydrogen permeability to a greater extent (decreases the equilibrium level of concentration $\bar{c}$). Interesting effect is visible on Fig. 6. Steady state establishment time t_* decreases monotonically as the diffusion coefficient D (10^{-7}–10^{-5}) increases. In this case stabilized flux level J_* increases at first and then decreases, because hydrogen atoms come up to the lateral isolated surface faster, get "reflexed" and thus increase concentration, simultaneously decreasing gradient ($|\partial_z c|$) and the diffusion flux.

4. Conclusions

The paper presents the hydrogen permeability models of a cylindrical membrane made of a structural material in the presence of protective coating defect on the inlet surface. Iterative numerical algorithm based on implicit difference schemes and the Thomas algorithm is developed. The basic model assumes local equilibrium between gaseous and dissolved (under defect) hydrogen and diffusion as the only limiting factor. Further hierarchy of models consists in sequential adding adsorption-desorption processes described by nonlinear and dynamic boundary conditions. Limit transition is valid: with reduction of hydrogen atoms accumulation on the surface, and with increase of desorption coefficient and inlet pressure, the model with nonlinear dynamic boundary conditions becomes a basic model.

Some qualitative characteristics of permeability are presented. In particular, weak dependence for stabilization and delay times on radius of the defect is detected. It means that the informative value of time characteristics for inverse problem solution is low. The ratio is revealed wherein flux is insensitive to the isolation conditions (when flow is proportional

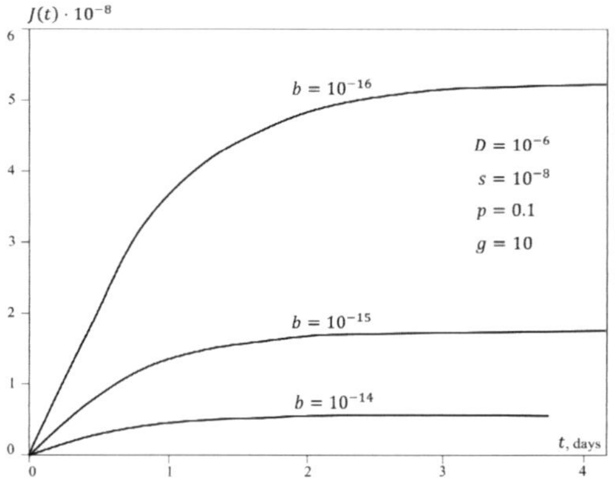

Figure 5. Effect of a desorption coefficient.

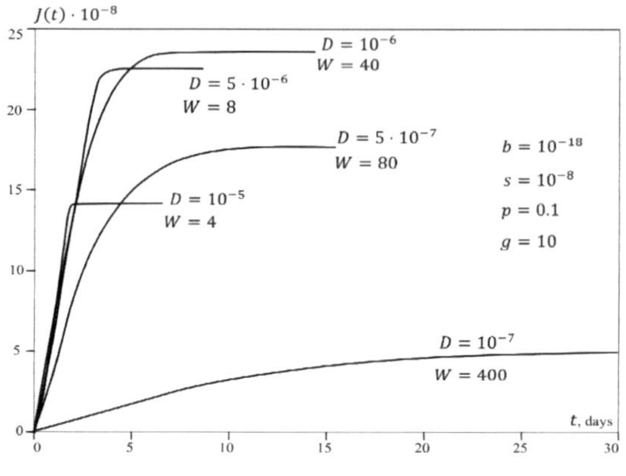

Figure 6. Effect of a didffusion coefficient.

to the number of such defects). It allows to reduce experimental cost. The flux decreases for increasing desorption coefficient, because desorption on the inlet surface predetermines hydrogen permeability to a greater extant (decreases the equilibrium level of concentration) than provides outward flux on the output surface. Steady state establishment time decreases monotonically as the diffusion coefficient increases. In this case, stabilized flux level increases at first and then decreases. Presented nonlinear effects should be taken into account while experimental curves processing to estimate limiting factors of hydrogen permeability.

Acknowledgments
The work is supported by the Basic Research Program (Branch of Mathematics, Russian Academy of Sciences) and the Russian Foundation for Basic Research (project 15-01-00744).

References
[1] Alefeld G and Völkl J (Eds.) 1978 *Hydrogen in Metals* Berlin; New York: Springer vol I Basic Properties p 427, vol II Application-oriented Properties p 387
[2] Zakharov A P (Ed.) 1987 *Interactions of Hydrogen with Metals* (Moscow: Nauka) pp 177–206
[3] Pisarev A A, Tsvetkov I V, Marenkov E D and Yarko S S 2008 *Hydrogen Pearmeability through Metals* (Moscow: MEPhI) p 144
[4] Yukhimchuk A A (Ed.) 2009 *Hydrogen Isotopes* (Sarov: Russian Federal Nuclear Center-VNIIEF) p 697
[5] Ball M and Wietschel M (Eds.) 2009 *The Hydrogen Economy* Cambridge University Press p 646
[6] Hirscher M (Ed.) 2010 *Handbook of Hydrogen Storage* Wiley–VCH p 353
[7] Gabis I 1999 The method of concentration pulses for studying hydrogen transport *Tech. Phys.* **44** 90–4
[8] Lototskyy M V, Yartys V A, Pollet B G and Bowman R C Jr. 2014 Metal hydride hydrogen compressors: a review *Int. J. Hydrogen Energy* **39** 5818–51
[9] Indeitsev D A and Semenov B N 2008 About a model of structure-phase transfomations under hydrogen influence *Acta Mechanica* **195** 295–304
[10] Zajec B 2011 Hydrogen permeation barrier – recognition of defective barrier film from transient permeation rate *Int. J. Hydrogen Energy* **36** 7353–61
[11] Zaika Yu V and Rodchenkova N I 2009 Boundary-value problem with moving bounds and dynamic boundary conditions: diffusion peak of TDS-spectrum of dehydriding *Appl. Math. Modelling*, **33** 3776–91
[12] Zaika Yu V and Bormatova E P 2011 Parametric identification of a hydrogen permeability model by delay times and conjugate equations *Int. J. Hydrogen Energy* **36** 1295–1305

Journal of Physics: Conference Series **661** (2015) 012028 doi:10.1088/1742-6596/661/1/012028

Doping graphene with a monovacancy: bonding and magnetism

M Yu Arsent'ev[1], A V Prikhodko[2], A V Shmigel[1], T L Egorova[1] and M V Kalinina[1]

[1]Institute of Silicate Chemistry of Russian Academy of Sciences, Adm. Makarova emb. 2, St. Petersburg, 199034 Russia
[2]St. Petersburg State Electrotechnical University "LETI", Pr. Popova 5, St. Petersburg, 197376 Russia

E-mail: ars21031960@gmail.com

Abstract. The possibility of creation of nanostructures on graphene and carbon nanotubes was investigated by placing different atoms at a single vacancy of graphene. Using the SIESTA software package and a high-throughput computing approach, we have consistently identified the binding energy and magnetic properties of 49 elements of the periodic table. The results show that all atomic species are stable at room temperature; some elements exhibit interesting magnetic behavior. E.g. Ni is not magnetic, whereas Al, Ga and P have magnetic moment. It is widely accepted that the rarity of metal-doped graphenes could be explained by the tendency of metal atoms to form clusters instead of doping. We analyzed corresponding binding energies and proved the opposite statement.

1. Introduction

Carbon nanotubes (CNTs) and graphene are unique materials for nanotechnology. The unique properties of these materials can be used in numerous applications, from advanced composite materials to artificial muscles and quantum computers [1, 2].

There is an evidence of the ability of the transition metals to dope graphene-like systems (graphene, onion-like carbon); this process can take place during growth or saturation of existing vacancies, like e. g. in carbon nanostructures synthesized using Ni containing catalyst [3]. The presence of substantial amounts of Ni impurities was revealed in the purified carbon nanostructures produced with a Ni - containing catalyst [4]. In particular, doped graphene and CNT systems were investigated by DFT calculations as promising materials for gas sensors [5-7], hydrogen storage devices [8], electrodes of supercapacitors [9, 10]. Unfortunately, the spectrum of the used dopants is narrow [11]. This study deals with a wide range of atoms (49 atoms).

Computer modeling of properties of CNTs using density functional theory (DFT) method requires very large computer resources. Thus, we investigate doping of graphene sheet as a prototype for the CNT. We can expect qualitative agreement between doping of graphene and CNT [12].

There are two types of graphene doping: electrical and chemical [13]. In this study, we investigate the chemical doping of graphene, where doping occurs via chemical routes, namely substitutional doping with heteroatoms.

Thus, in this paper, we investigated the possibilities of chemical doping and magnetic moments for almost 50 atomic species to find a stable for synthesis, nano structure formation and practical applications.

2. Method

We used the spin-polarized density-functional theory (DFT) as implemented in the localized basis-set SIESTA method [14, 15]. The generalized gradient approximation of Perdew, Burke, and Ernzernhof (PBE) was adopted for the exchange correlation potential [16]. The double zeta plus polarization (DZP) basis set was used in the calculations along with the 300 Ryd mesh cut-off for the grid. Brillouin zone sampling was made by a Monkhorst–Pack mesh 24 × 24 × 4 k-points.

A single layer 3×3 supercell with a vacuum width of 12 Å above was constructed. Dopant atoms were placed at the center of the single vacancy. These structures were entered into the SIESTA code. They were allowed to fully relax according to the atomic positions and cell parameters with an accuracy of 0.04 eV/Å for forces and 1.0 GPa for stresses. All atoms were allowed to relax for all calculations.

We calculated the binding energy from the formula:

$$E_b = E_{tot} - (E_{vac} + E_{atom})$$

where E_b is the binding energy of the atom adsorbed on a single vacancy in a graphene sheet. E_{tot} is the energy of the doped graphene, E_{vac} is the energy of the reconstructed naked vacancy and E_{atom} is the total energy as an isolated atom. E_b is a complex quantity including redistribution of the electron density arising from the adsorption of the isolated atom on the single vacancy of the graphene sheet, as well as the relaxation of the stresses being induced by this adsorption.

3. Results

We calculated the binding energy and magnetic moment of 49 atoms shown in figure 1 and figure 2 respectively.

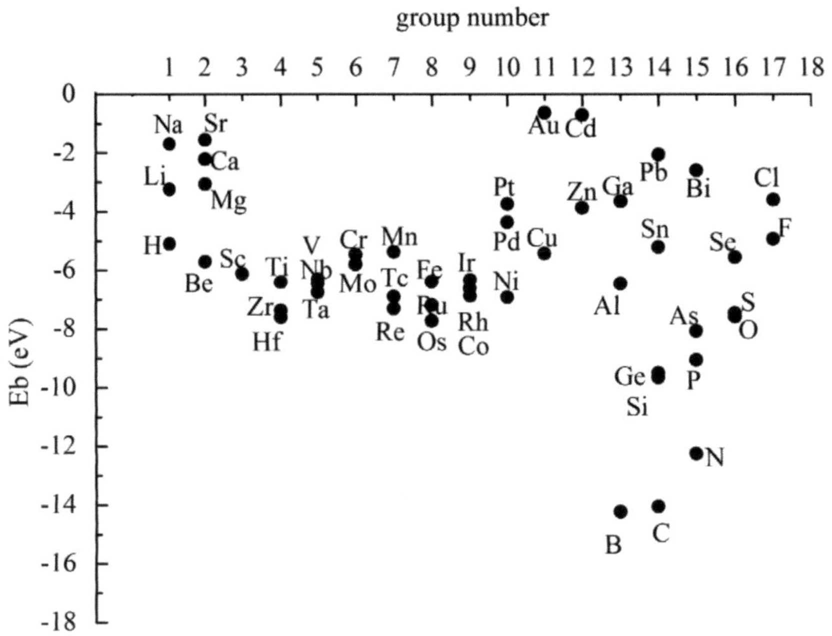

Figure 1. Binding energies Eb of the graphene sheet with atoms adsorbed on single vacancies. The x-axis shows the number of the group in the periodic table.

We found that the types of the atom-single vacancy complexes can be divided into two types: stable and not stable at room temperature. The type of unstable complexes has the binding energy less than 0.4eV. The 0.4eV is roughly the maximum threshold for migration of atoms at room temperature [12]. Thus, all atoms are stable at the room temperature. Moreover, according to first-principles calculation reported by K.Nakada et. al. [17] they can migrate further along the surface of graphene even at the room temperature. The binding energy of a C atom to a graphene sheet with a single vacancy is about -14 eV and C also can migrate along the surface of graphene [17]. As for the stability at high temperatures, the additional simple estimations are required. The additional calculations may be required to check the stability of these species toward oxidation by oxygen of the air.

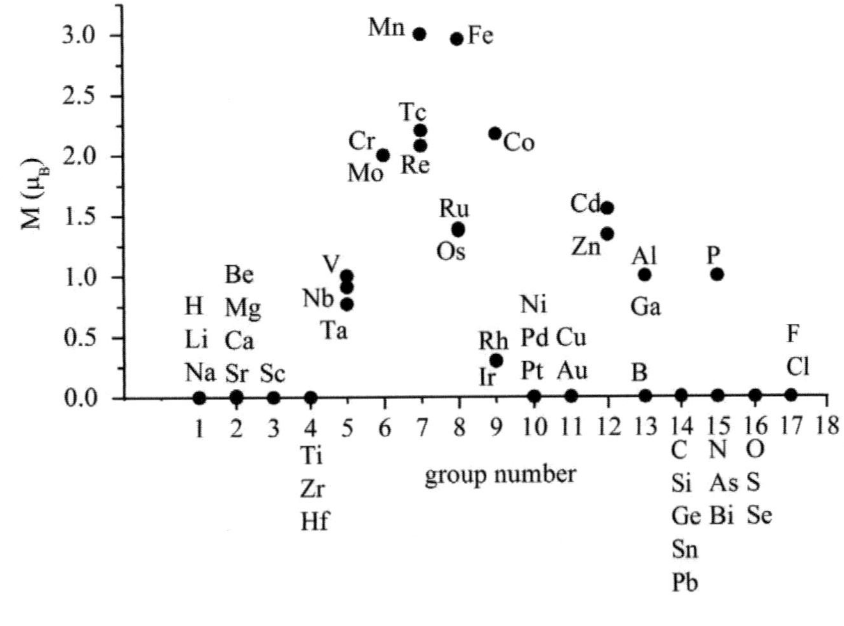

Figure 2. Magnetic moments M of the graphene sheet with atoms adsorbed on single vacancies. The x-axis shows the number of the group in the periodic table.

There is a slight off-layer position of the dopant atoms, because the doping atom–carbon bond is somewhat longer than a carbon–carbon bond. These data are in agreement with other first-principles calculations [7, 11]. But for the most stable N and B atoms the in-plane position is observed.

Magnetic data are in agreement with the calculation reported by A.Krasheninnikov et. al. [11]. Surprisingly, Zn, Cd, and even Al, Ga and P atoms are magnetic. This phenomena can be explained by the model [11].

Figure 3 shows density of states (DOS) for some doped systems. There exist some enhancement in the absolute value of DOS in the rows B-doped-Ga-doped and pristine-Ge-doped graphene, but for the most of doped graphenes the DOS takes complex form like for the Bi-doped one.

4. Discussion

The character of the doping is expected to be similar between CNT, graphene and also graphite, because there is an experimental evidence that CNT have only slightly lower vacancy formation and diffusivity energy [18]. For graphite, the interaction between the layers in graphite is so small that considering only a single graphene sheet gives only negligible differences [19]. Thus, we can expect that our calculated results for graphene can be used as a reference to consider the candidate dopant atoms for CNT and graphite.

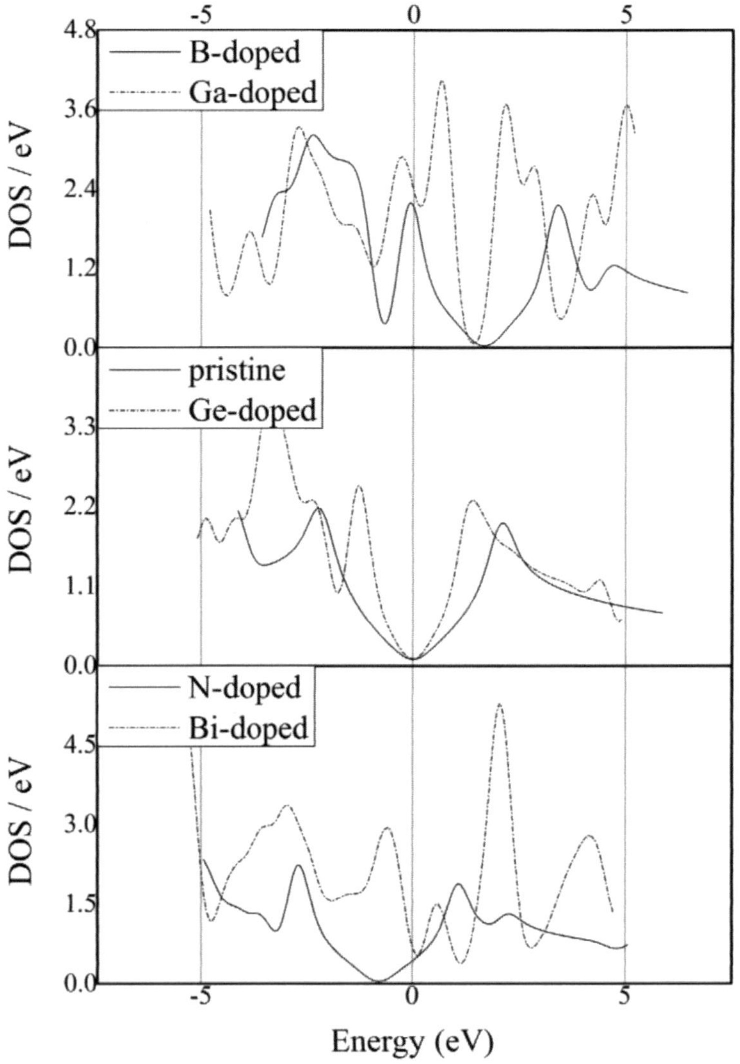

Figure 3. Density of states for some doped graphenes. The Fermi level is at zero energy.

We found that for all atomic species the binding is strong at room temperature. Moreover, most of the atoms are allowed to migrate along the surface, so that doped nanostructures could be obtained by filling of the existing vacancies. It should be taken into account that in their initial state the empty vacancies are saturated by stable O groups, such as ether (C-O-C), carbonyl (C-O), anhydride (O=C-O-C=O) and other groups [19]. Other synthesis routes could use the doping in plasma process [9], chemical vapor deposition (CVD), thermal annealing of graphene oxide, catalytic growth of carbon nanostructures, ball milling, etc [20]. The range of dopants used experimentally is quite wide (B, N, S, F, Cl, I, Br and Se). According to our calculations (figure 1) the binding energy for these dopants is in a wide range from -14.25 to -3.59 eV, so from this energetical point of view there are no obstacles in using of other atomic species as dopant.

Such heteroatom-doped graphene could be also used as electrode material for supercapacitors, providing the binding energy is at least 2.5-4.6 eV (the voltage window of stability for current commercial electrolytes) divided by 2. The division by 2 is necessary because the voltage is distributed across the two electrodes. These values of binding energy will not allow the voltage to pull out the dopant atom from the vacancy of the graphene.

The doping of graphene with metal atoms has rarely been confirmed experimentally. X. Wang et. al. [20] suggest that this is probably because the binding energy of these atoms in graphene vacancy is much lower than their cohesive energy. And because of this these atoms tend to form clusters instead of doping. But if we compare some of our calculation results with the one available from the literature (table 1), we can see that this statement is wrong. As a rule, the binding energy of atoms in cluster ($E_{cluster}$) increases with the cluster size. We took the maximum value of $E_{cluster}$ available from the literature and gave the corresponding cluster size. This allowed us to consider the worst-case scenario, and even in this case, (for the most of metal atoms, except for Au) the doping is energetically more favorable than the forming of clusters. We believe that the only prerequisite for the doping is the presence of sufficient amount of single vacancies/divacancies, which should be free from oxygen or other contaminants. Moreover, as already mentioned above, these metal atoms can migrate along the surface of graphene even at the room temperature, so they can reach these vacancies just after the adsorption and fill them. And in evidence of our assumptions the possible existence of such doped structures was proved experimentally by direct observations of Au – vacancy and Pt – vacancy complexes in carbon nanostructures [18].

Table 1. The calculated binding energies of some metal atoms and Si adsorbed on single vacancy of the graphene sheet (E_{dopant}) in comparison with the corresponding binding energies per atom in clusters ($E_{cluster}$). N is the cluster size.

Element	E_{dopant} (eV/atom)	$E_{cluster}$ (eV/atom)	N	Reference
Al	-6.44	-2.95	34	[21]
Si	-9.64	-4.61	16	[22]
Fe	-6.38	-3.67	100	[23]
Co	-6.88	-3.93	200	[24]
Cu	-5.43	-3.35	13	[25]
Mo	-5.79	-5.96	100	[23]
Pt	-3.73	-3.60	15	[26]
Au	-0.63	-3.14	13	[25]
Pb	-2.06	-1.68	160	[27]

5. Conclusion

We calculated the binding energy and magnetic moment for 49 atoms embedded in single vacancies in a graphene sheet as a prototype for the CNT and graphite. The result shows us that for all of atomic species the binding is strong, even at room temperature, and that they exhibit interesting magnetic behavior. In particular, all transition-metal atoms and Al, Ga, P are magnetic, whereas Ni, Sc, Pd, Pt, Cu, Au are not. We compared our calculated binding energies for metal atoms with the ones for metal clusters available in the literature, and concluded that, in contrast to the suggestions of other authors [20], the doping is more energetically favorable than clusterforming,. We discussed the experimental realization of such systems.

References

[1] De Volder M F L., Tawfick S H, Baughman R H and Hart A J 2013 *Science* **339** 535
[2] Geim A K and Novoselov K S 2007 *Nature Mater.* **6** 183
[3] Santos E J G, Ayuela A, Fagan S B, Mendes Filho J, Azevedo D L, Souza Filho A G and S´anchez-Portal D 2008 *Phys. Rev.* B **78** 195420
[4] Ushiro M, Uno K, Fujikawa T, Sato Y, Tohji K, Watari F, Chun W J, Koike Y and Asakura K 2006 *Phys. Rev.* B **73** 144103
[5] García-Lastra J M, Mowbray D J, Thygesen K S, Rubio A and Jacobsen K W 2010 *Phys. Rev.* B **81** 245429
[6] Dai J, Yuan J and Giannozzi P 2009 *Appl. Phys. Lett.* **95** 232105
[7] Ao Z M, Yang J, Li S and Jiang Q 2008 *Chem. Phys. Lett.* **461** 276
[8] Ao Z M, Jiang Q, Zhang R Q, Tan T T and Li S 2009 *J. Appl. Phys.* **105** 074307
[9] Jeong H M, Lee J W, Shin W H, Choi Y J, Shin H J, Kang J K and Choi J W 2011 *Nano Lett.* **11** 2472
[10] Paek E, Pak A J, Kweon K E and Hwang G S 2013 *J. Phys. Chem.* C **117** 5610
[11] Krasheninnikov A V, Lehtinen P O, Foster A S, Pyykkö P and Nieminen R M 2009 *Phys. Rev. Lett.* **102** 126807
[12] Ishii A, Yamamoto M, Asano H and Fujiwara K 2008 *J. Phys.: Conf. Ser.* **100** 052087
[13] Lv R and Terrones M 2012 *Mater. Lett.* **78** 209
[14] Soler J M, Artacho E, Gale J D, Garcia A, Junquera J, Ordejon P and Sanchez-Portal D 2002 *J. Phys.: Condens. Matter* **14** 2745
[15] Junquera J, Paz O, Sanchez-Portal D and Artacho E 2001 *Phys. Rev.* B **64** 235111
[16] Perdew J P, Burke K and Ernzerhof M 1996 *Phys. Rev. Lett.* **77** 3865
[17] Nakada K and Ishii A 2011 *Graphene Simulation* vol 1, ed J R Gong (Rijeka: InTech) p 376
[18] Gan Y, Sun L and Banhart F 2008 *Small* **4** 587
[19] Carlsson J M, Hanke F, Linic S and Scheffler M 2009 *Phys Rev Lett.* **102** 166104
[20] Wang X, Sun G Z, Routh P, Kim D H, Huang W and Chen P 2014 *Chem. Soc. Rev.* **43** 7067
[21] Aguado A and López J M 2009 *J. Chem. Phys.* **130** 064704
[22] Mahtout S and Belkhir M A 2006 *Acta Physica Polonica* A **109** 685
[23] Elliott J A, Shibuta Y and Wales D J 2009 *Philosophical Magazine* **89** 3311
[24] Zhan L, Chen J Z Y, Liu W K and Lai S K 2005 *J. Chem. Phys.* **122** 244707
[25] Fernández E M, Soler J M, Garzón I L and Balbás L C 2004 *Phys. Rev.* B **70** 165403
[26] Winczewski Sz and Rybicki J 2011 *Comput. Meth. Sci. Technol.* **17** 75
[27] Doye J P K and Hendy S C 2003 *Eur. Phys. J.* D **22** 99

Behavior of charges in the active zone of composite OFET under the source-drain electric field

P A Dement'ev[1], P A Alekseev[1], M S Dunaevskii[1,2] and A N Aleshin[1,3]

[1] Ioffe Physical-Technical Institute, Russian Academy of Sciences, Politekhnicheskaya ul. 26, St. Petersburg, 194021 Russia

[2] National Research University of Information Technologies, Mechanics and Optics (ITMO), St. Petersburg 197101, Russia

[3] St. Petersburg Electrotechnical University "LETI", ul. Professora Popova 5, St. Petersburg, 197376 Russia

e-mail: demenp@yandex.ru

Abstract. Charge accumulation and relaxation processes in the OFET active zone are studied. It is found that positive and negative charges injection from metal electrodes under drift field occurs. Characteristic time of these processes is found to be about 30 sec. Behaviour of charges under the combination of source-drain and gate potentials is also investigated.

1. Introduction

One of the main challenges of diagnostics of organic electronics devices is the visualization of electrophysical processes in their active zone under working conditions. It could provide more detailed understanding in physics of memory, "bias-stress" and other known effects in organic field-effect transistors (OFET). The most useful method for solving this problem is the scanning Kelvin-probe microscopy (KPM). It has been already used for characterization of semiconductor laser diodes and other structures under working conditions [1]. The KPM key difference from the traditional testing techniques is the possibility of electric charge and potential measurements with high spatial resolution.

In our work KPM was used to investigate the charge accumulation and dissipation processes in OFET with active zone composed of poly(9-epoxypropylecarbazole) (PEPK) polymer and Au nanoparticles with the diameter 50-70 nm. We studied the influence of the drift and gate electric fields on these processes.

2. Objects and methods

The composite for active zone was made from PEPK and "SigmaAldrich" Au nanoparticles. Both PEPK and Au nanoparticles were dissolved in chloroform. The obtained solutions were mixed and, after the ultrasound treatment during 8–10 min (f ~ 22 kHz), were applied onto silicon substrates (Si with a ~200nm SiO_2 layer) with prepared Au (source) and Al (drain) electrodes (Figure 1).

KPM measurements were performed on NTEGRA-Aura microscope produced by NT-MDT company (Moscow, Russia) in ambient conditions. Standard DCP11 conducting probes (NT-MDT) were used.

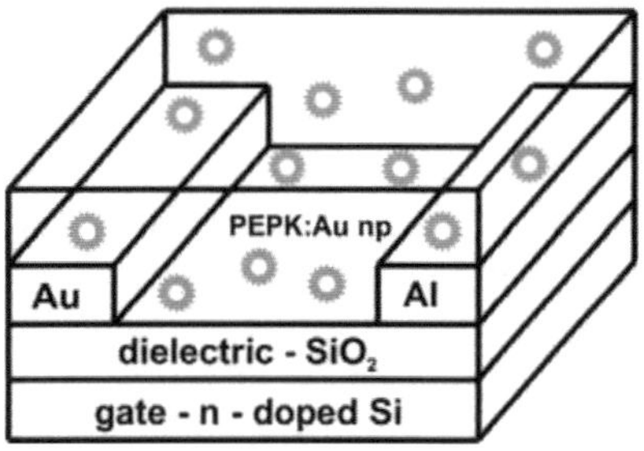

Figure 1. Schematic structure of OFET.

3. Experimental

3.1. Charge injection by source-drain electric field.

Typical experimental procedure was the following. For the observation of negative and positive charges accumulation in OFET active zone by KPM a +2V bias voltage was applied for 2 minutes to Au-electrode (Al electrode was grounded). After the stabilization of electrical potential redistribution process the voltage was turned off that caused the dissipation of accumulated charges. Figure 2 illustrates these two stages of experiment. To eliminate the influence of topography and composite inhomogeneities to the resulting potential distribution the initial unperturbed potential distribution was mathematically subtracted from the whole picture. The initial potential distribution can be seen in region I of figure 2.

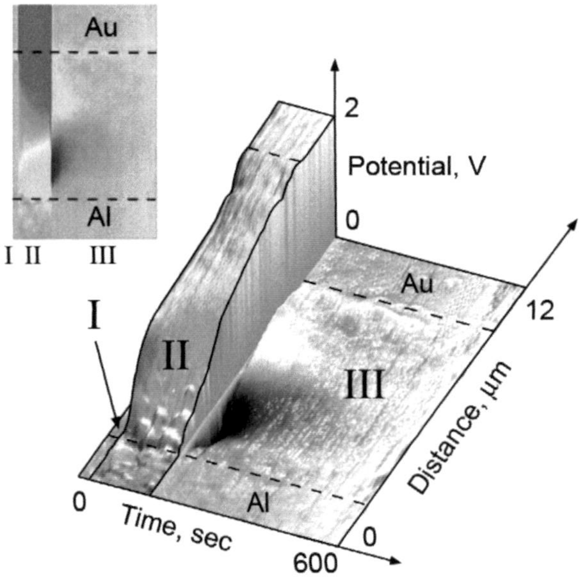

Figure 2. Illustration of unperturbed (Region I) potential, accumulation (Region II) and relaxation (Region III) processes in OFET active zone. Inset – top view of the same time-distance region.

Region II in figure 2 illustrates the change of the potential distribution in the OFET active zone under an external bias of +2V. It can be seen, that at the beginning the main part of the applied voltage falls on the barriers near the borders composite - metal electrodes. Then the potential barriers are reduced with time due to the charge accumulation process. Indeed, after grounding of both electrodes one can see the accumulated positive and negative charges near the former barriers (Region III). Then the accumulated charges dissipate with characteristic time of several minutes. Detailed potential distribution profiles for accumulation and dissipation processes are shown in figure 3 and figure 4 respectively. Potential profiles were measured every 2.5 sec, but only several of them are shown on figures.

The most significant change in the potential drop profile is observed in areas near point A and C on figure 3(a). Its value reaches 0.5 V for negative charges and 0.3 V for positive. It should be noted that the accumulation process is due to the charge injection from the metal electrode instead of the polarization of the composite. This is evidenced by the fact that the negative charges accumulation was observed near the ground electrode and positive charges near the positive electrode. The potential in the areas of strongest accumulation is changed with time by exponential law $\sim 1-e^{-t/\tau}$ with a characteristic saturation time of about 30 seconds.

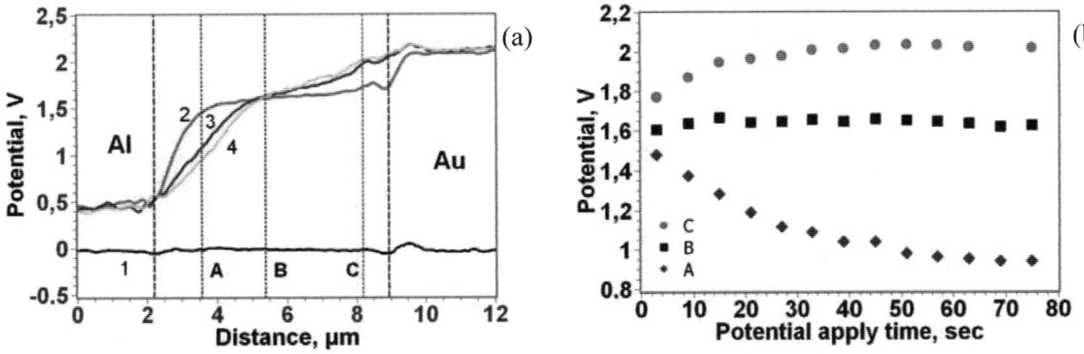

Figure 3. Accumulation of charges under drift field: (a) change in potential distribution: 1 – initial potential; 2-4 – potential distribution in every 25 seconds after applying drift field; (b) time dependence of the potential in several points: A – 1.2 µm from Au electrode, B – in the middle of active zone, C – 1.5 µm from Al electrode.

The relaxation (dissipation) of accumulated charges begins when the external field is switched off. This process is illustrated in region III of figure 2. One can see that not only a reduction of charge-related potentials takes place but also a spatial redistribution (see figure 4).

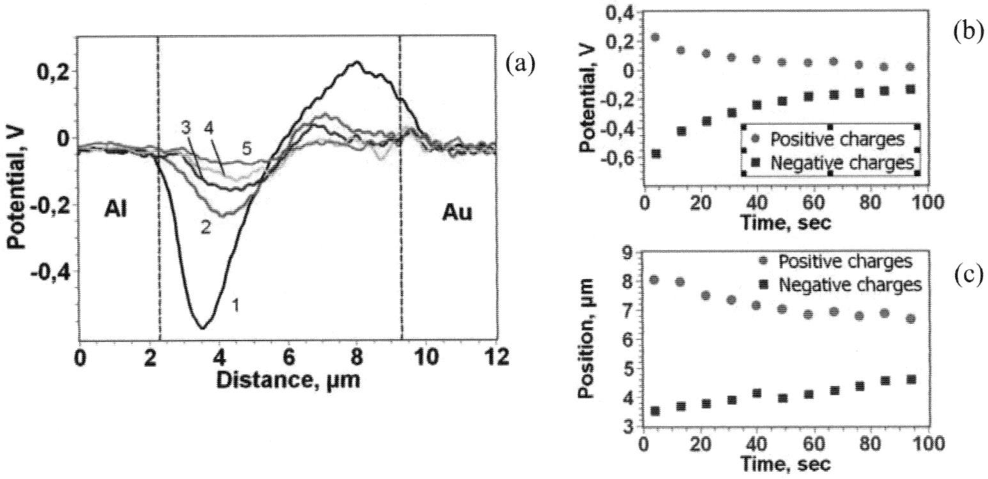

Figure 4. Dissipation of accumulated charges: (a) change in potential distribution, time between curves is about 40 seconds; (b) time dependence of the potential in maxima of charge spots; (c) time dependence of the position of charge spot maxima.

The potential distribution and the time dependence of the size and position of the charge spots maximum are shown in figure 4. It is seen that while the value of the potential at the center of the charge spots decreases exponentially, the spatial position of maximum changes linearly (with velocity $v \sim$ 1mkm/min). The spatial movement of charge spots cannot be explained by the drift mechanism. The estimated mobility of charges is abnormally low $\mu \approx 10^{-9}$ cm^2/V·s, which is several orders of magnitude lower than the expected value of mobility $\mu \approx 4{\cdot}10^{-2}$ cm^2/V·s obtained for the same samples by conventional methods [2]. This suggests that the change in position of the charge spots maxima is related with the different charge dissipation rates at the near electrodes area and at the center of the active zone.

3.2. Charge injection by source-drain and gate electric fields.

Since the OFET working conditions are controlled by gate electrode voltage, the effect of charge accumulation/dissipation under the simultaneous applying of gate and source-drain voltages [3] was studied. The combined measurements were carried out as follows: at first the source-drain voltage was applied for a time sufficient for full charge storage near the contact region, and then the gate voltage of -10 V was applied for 5 minutes as in ref [3]. The potential drop profiles during the accumulation are shown in figure 5a.

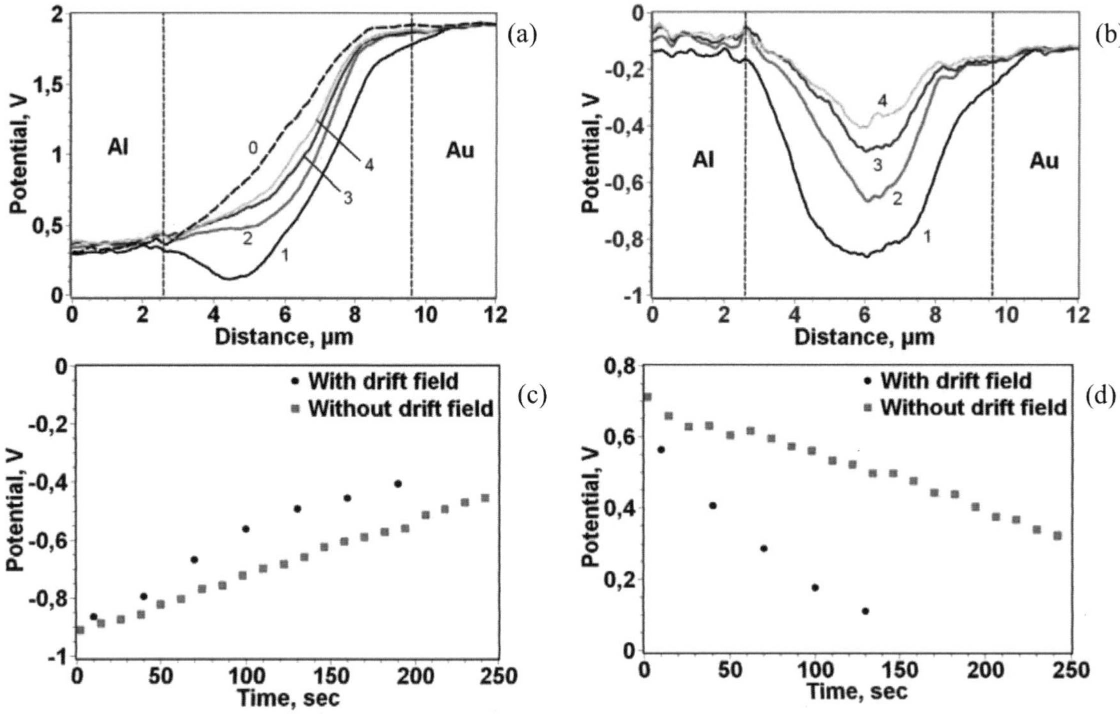

Figure 5. Charge accumulation under simultaneous applying of gate and source-drain voltages: (a) change in potential distribution during accumulation process. Time between profiles is about 60 sec. Dotted line shows the equilibrium profile with grounded gate electrode; (b) difference profiles after subtraction; (c) comparison of time dependence of the potential in the minimum of potential distribution with and without drift field during accumulation process; (d) comparison of time dependence of the potential in the maximum of potential distribution with and without drift field during relaxation process.

Journal of Physics: Conference Series **661** (2015) 012029 doi:10.1088/1742-6596/661/1/012029

Mathematical subtraction of the equilibrium profile of the potential at the gate grounded structure (shown by dotted line) was made to simplify the interpretation of the data. This resulted in the difference profiles shown in figure 5b. The figures show that initially the gate voltage leads to a shift in the voltage distribution profile by -0.8 V, but with time, this shift is reduced due to the screening of gate potential by charges, accumulated in the polymer. When turning off the gate voltage, the charges are dissipating (not shown here).

Amplitudes of potential drop (figure 5b) due to the accumulation/dissipation processes versus time are plotted on figure 5c and figure 5d respectively. For comparison the corresponding data for the only gate voltage case (without applying source-drain voltage) also plotted on figure 5 by red dots. The detected surface potential of the OFET active zone at first after applying the gate voltage coincides for the both cases. On the other hand, in the presence of the applied source-drain voltage (see figure 5c) the accumulation of charges and, respectively, screening of the gate voltage is faster by about 20%. The situation is similar at the beginning of the accumulated charges dissipation – immediately after turning off the gate voltage (see figure 5d), value of the charge in the center of the active zone coincides with that measured in the absence of the applied source-drain field. The relaxation process goes 3–4 times faster with the presence of the applied source-drain drift field. The acceleration of these processes could be associated with an increase in the electric field acting on the charge carriers in the composite layer.

4. Conclusion

Thus, the effect of positive and negative charges injection from metal electrodes into the active zone of OFET was found. By Kelvin probe microscopy technique it was performed a detailed study of the time evolution of the injected charges under complex configuration of electrically biased contacts. It was shown that characteristic time of accumulation and dissipation processes is tens of seconds. Nature of such long characteristic time is still unclear. Nevertheless, in other works on composite OFET [4] characteristic times are even longer.

Injection and accumulation of charges in the active region is an important process that has a significant impact on the current-voltage characteristics of OFET and the account of this effect is required in the design of device structures of organic electronics.

References

[1] Ankudinov A V, Evtikhiev V P, Ladutenko K S, Titkov A N, Laiho R 2006 *Semiconductors* **40** (8) 982–989

[2] Aleshin A N, Fedichkin F S, and Gusakov P E 2011 *Phys. Solid State* **53** (11) 2370–2374

[3] Dement'ev P A, Dunaevskii M S, Aleshin A N, Titkov A N and Makarenko I V 2014 *Phys. Solid State* **56** (5) 1054–1057

[4] Lee K, Weis M, Taguchi D, Manaka T, Iwamoto M 2012 *Chem. Phys. Lett.* **551** 105-110

Obtaining and investigation of C_{60} <A_2B_6> semiconductor compounds with a view to create effective solar cells

M A Elistratova[1,2], I B Zakharova[1] and N M Romanov[1,2]

[1] St. Petersburg State Polytechnical University, 29 Polytechnicheskaya Street, St. Petersburg 195251, Russia
[2]Lappeenranta University of Technology, 34 Skinnarilankatu, Lappeenranta, 53850, Finland

E-mail: elistratovamari@yandex.ru

Abstract. Reproducible vacuum method of thin fullerene films production with Cd<Te> impurity on Si, glass and mica surfaces are developed. Thin composite films were obtained by Knudsen cell method. Initial studies of condensation and surface morphology of the films are investigated SEM methods. Optical spectroscopy was used to confirm the obtained results. Results showed the presence of an additional peak associated with the formation of C_{60}-CdTe molecular complexes. SEM results confirm absence of phase separation.

Introduction

Over the last couple decades carbon becomes one of the most studied chemical element. Discoveries and developments related with carbon allows to create new nanomaterials which will move forward science and industry. The practical scope of this research is wide - machinery, electronics, chemical industry, biology, medicine. This enables the carbon to be found as a strategic material of the future, which is confirmed by Nobel Prize in chemistry and physics, awarded for the discovery of new carbon forms, including fullerenes. Fullerenes and their chemical forms are promising materials for use in semiconductor technology and nanoelectronics [1]. The effect of fullerenes nonlinear transparency opens up the possibility of their using as of the laser radiation optical limiters intensity in the visible and near infrared ranges. The high light absorption efficiency and charge separation in conducting composites polymers and fullerene allows to predict the creation of photodiodes, transistors, solar cells and other opto-electrical devices on their basis [1, 2]. Composite materials based on fullerene with <A_2B_6> donor group find applications in the field of telecommunications and computer technology and optoelectronic devices. They form a new class of materials, which have easier manufacturing and low cost. Structures based on C_{60}<A_2B_6> also are of interest as a material for the efficient solar cells creation [3]. Studies show the presence of the photoexcited carriers transport from impurity molecules to the C_{60} molecule [4, 5]. This effect leads to photovoltaic effect appearance in these structures. CdTe was selected as impurity because of its low sublimation temperature, thus providing a much simpler samples obtaining method. This paper presents the results obtained from photoluminescence investigations and from composition analysis results of such samples.

Obtaining method

The essence of a deposition thin films process implies that materials are heated by the special temperature. At this temperature atomic and molecular kinetic energy become enough to separate atoms and molecules from the bulk material surface to surround space. It becomes at such heating temperature when own material vapor pressure exceed, by several order, residual gas pressure. The atomic stream propagates linearly in this case. Evaporated atoms and molecules have a collisions with the surface and then condense on it. Fullerene and A_2B_6 materials transferred into vapor without liquid phase. Thus evaporation is carried out in short way: the solid phase – the liquid phase. It is called sublimation. Follow conditions was provided at a residual preasure of 10^{-6} Pa. This vacuum was easily achieved by mechanical foreline and high vacuum diffusion pumps. The substrate temperature has a great impact on films structure and also on its electrophysical and optical propirties. In this work substrates were kept at room temperature during the process. Sputtering on a cold substrate lead to amorphous and polycrystalline structure formation and possibly alignment of randomly distributed semiconductor clusters or molecular. The main advantage of this method is sterility of process. It allows to propagate high films quality without any contamination and what is more important without oxygen subject.

Thin fullerene films doped with CdTe were prepared by co-deposition technique from Knudsen cell. Open source evaporating allows to obtain composite films on various substrate. Powder of clear C_{60} and C_{60} with adding of 50 wt.% CdTe impurities was evaporated at 500-650 °C on Si, glass with ITO, KBr and mica substrate. Silicone substrates were cleaned by isopropyl alcohol. The cell was made from wolfram and surrounded by the wolfram capsule. The out part is connected to 0-1.5 V. To provide uniformal mixture distribution evaporator was slowly heated during 10 min. It promotes better oxygen adhesion which penetrate into fullerene. Films thickness was 200-500 nm. In comparison with the initial charge it can lead to a significant changing of a film composition. Obtaining a large area film possibility is an advantage of the Knudsen cell.

Experimental research of the samples

A scanning electron microscope Jeol JSM-6390 was used to study the surfaces' morphology. It worked with resolution of 3 nm. The films' composition in the selected area was measured by energy dispersive micro-analysis console "Oxford INCA Energy" with the utmost sensitivity 0.1wt%. Studies have shown a high homogeneity of the impurity distribution and the absence of the mosaic structure of the composite films. Photoluminescence spectra were measured by using automated installation based on Horiba Jobin Yvon monochromator. It is composed of FHR 640 monochromator with a grating of 1200 mm-1 and the Symphony II (1024 * 256) Cryogenic Open - Electrode CCD detector. . Excitation of photoluminescence were produced by continuous semiconductor laser with 408 nm wavelength. Filter was installed before the slit of the monochromator to avoid scattering lazer radiation falling into the monochromator. CCD chamber was maintained at the temperature of 77 K [7].

Photoluminescence spectra investigation

Obtained at 300 K photoluminescence (PL) spectra are shown at fig. 1. Pristine fullerene films has wide PL peak with 725 nm maximum. The large peak wide explained by two fullerene peaks contribution - at 720 nm and its vibration recurrence at 830 nm. CdTe adding leads to extra peaks appearing at the wavelength of 630 nm. The extra peak intensive increases with amount of impurity. This wavelength according to the energy of singlet-singlet transition in C_{60} energy structure. Such transition is forbidden for an isolated fullerene molecule for the reason of symmetry, but if CdTe forms a molecular complex with C_{60} symmetry of the cluster is reduced (from T_h to C_s). According to the results of quantum chemical calculations, in case of optimization geometry CdTe molecular is located on 6-6 bond of fullerene. However C_{60} molecules and C_{60} complexes interaction is limited. Limitation is going from standing in crystal lattice. C_{60} toluene solution PL spectra presented a big additional peak at 650 nm and a small degree at 600 nm. These values correspond to S_1-S_0 and S_2-S_0 fullerene transitions. It is possible because molecules in liquid phase are solvated by toluene and

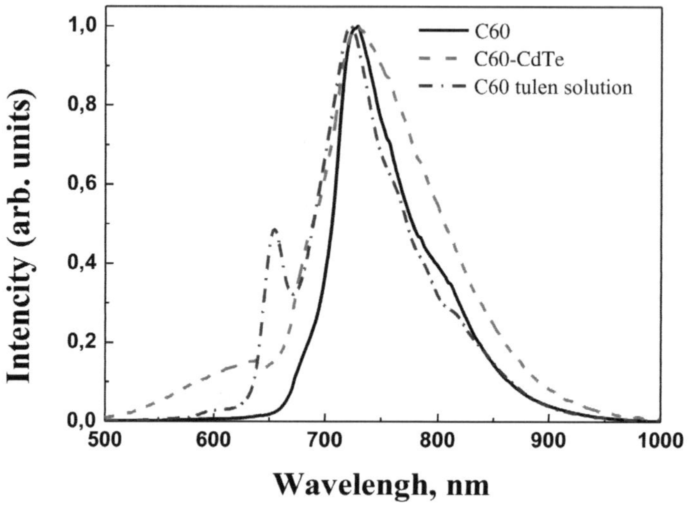

Figure 1. Photoluminescence spectra obtained at 300 K. Pure C_{60} spectrum is showed by solid line. C_{60} with 50% CdTe is showed by dash line. PL spectra of C_{60} toluene solution is showed by dash-dot line.

interact with each other, the degree of symmetry distortion increases and S_1-S_0 and S_2-S_0 transitions became permitted.

Scanning electron microscopy analysis

Figure 2 presents surface topography of C_{60}-CdTe sample contain 50% of impurity obtained by SEM. At most surface is clear without inhomogeneities (spectra 2 and 4). It may contains microcrystalline structure (spectra 1 and 3). X-ray analysis display that these microcrystals are from C_{60}-CdTe mixture and do not contain CdTe nanoparticle.

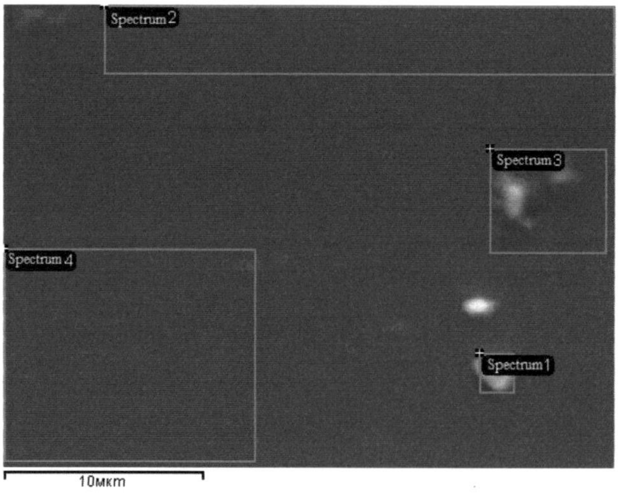

Figure 2. SEM image of C_{60}-CdTe (with 50% of CdTe) sample.

Chemical compositions spectra were measured by energy dispersive consoles (table 1). All spectra contain as C_{60} as CdTe. Points (2, 4) selected on a smooth homogeneous surface show sufficiently good uniformity of composition with an average content value of Cd – 2.37, Te – 5.53%. At opposite to atomic masses of Cd and Te, it mean that CdTe included in film in molecular form. Quantum-chemical calculations verified this fact [8].

Silicon appears in the spectra because the absorption of electrons and output of x-ray radiation is possible from the substrate when the film has thin thickness (about 500 nm). Oxygen also presents in spectra, because fullerene is susceptible of oxygen and water vapor. In comparison with the initial charge, which was 50% to 50%, it is noticed that amount of impurities became lower. Open source method production leads to significant depletion of the impurity in films.

Table 1. X-ray analysis result for sample contain 50% of CdTe impurities.

Spectrum	C	Si	Cd	Te	O	Total
Spectrum 1	14.32	17.84	2.35	5.34	60.15	100.0
Spectrum 2	15.47	15.69	2.33	5.68	60.83	100.0
Spectrum 3	16.39	14.06	2.33	5.53	61.57	100.0
Spectrum 4	15.43	15.73	2.48	5.59	60.77	100.0

Conclusions

It is shown that open source evaporating methods lead to a molecular dispersion of impurities in the amorphous films. X-ray composition analysis results displayed significantly depletion of impurity in film in comparison with the initial charge. It determined impurities intercalation way in fullerene matrix. Films compound analysis also showed that CdTe is in molecular form and generate molecular complexes with C_{60}. Photoluminescence spectra were measured for obtained samples. Additional peaks appeared. They related with singlet-singlet (S_1-S_0, S_2-S_0) transition of fullerene, which is forbidden for isolated molecule. C_{60}-CdTe complexes formation remove the symmetry prohibition. This removing also appears in solution for the reason of toluene solvating and interaction between molecules. These results well agree with SEM analysis and quantum-chemical calculations.

References

[1] Christoph J, Brabec N, Sariciftci S 2001 *Adv. Funct. Mater*, **1**, 15
[2] Park S, Srivastava D, Cho K 2001 *Nanotechnology*, **12**, 245
[3] Bang J H, Kamat P V 2011 *ACS Nano*, **5**, 9421
[4] Chaban V V, Prezhdo V V and Prezhdo O V 2013 *J. Phys. Chem.* Lett. **4** 1
[5] Guldi D M, Zilbermann I, Anderson G, Kotov N A, Tagmatarchis N, Prato M 2004 *J. Am. Chem. Soc.* **126** 14340
[6] Drechsel J, Fröb H 2008 *Vacuum's Best*, **20**, 15
[7] Elistratova M, Romanov N, Zakharova I, Kvyatkovskii O, 2014 *J. Phys.: Conf. Ser.* **541**, 012021
[8] Kvyatkovskii O E, Zakharova I B, and Ziminov V M 2014 *Physics of the Solid State*, **56**, 1289.

Observation of linear I-V curves on vertical GaAs nanowires with atomic force microscope

P Geydt[1*], P A Alekseev[2], M Dunaevskiy[2], E Lähderanta[1], T Haggrén[3], J-P Kakko[3], H Lipsanen[3]

[1]Department of Mathematics and Physics, Lappeenranta University of Technology, P.O. Box 20, 53851 Lappeenranta, Finland
[2]Ioffe Physico-Technical Institute, RAS, 194021 Saint-Petersburg, Russia
[3]Department of Micro- and Nanosciences, Micronova, Aalto University, P.O. Box 13500, FI-00076 Aalto, Finland

*E-mail: pavel.geydt@lut.fi

Abstract. In this work we demonstrate the possibility of studying the current-voltage characteristics for single vertically standing semiconductor nanowires on standard AFM equipped by current measuring module in PeakForce Tapping mode. On the basis of research of eight different samples of p-doped GaAs nanowires grown on different GaAs substrates, peculiar electrical effects were revealed. It was found how covering of substrate surface by SiO_x layer increases the current, as well as phosphorous passivation of the grown nanowires. Elimination of the Schottky barrier between golden cap and the top parts of nanowires was observed. It was additionally studied that charge accumulation on the shell of single nanowires affects its resistivity and causes the hysteresis loops on I-V curves.

1. Introduction

Semiconductor nanowires (NWs) attracts significant attention as structural elements of electronic and optoelectronic devices [1]. Remarkable progress was achieved in application of NWs for solar cells, e.g. efficiency of such structures has grown already on one order during the last decade [2]. Typical solar cell based on NWs is a regular array of vertically oriented NWs grown on a semiconductor wafer. After the growth the array is covered by a transparent dielectric matrix, which is then covered with a transparent electrode. Despite the achieved progress, efficiency of NW-based solar cells is relatively low. Exist several factors limiting the efficiency of solar cells, e.g. high surface recombination velocity [3], difficulty of creating the ohmic contact with nanowires etc. Thus, study of the influence of surface modifying on electrical and photoelectrical properties of nanowires and electrical contacts to them is still topical.

Electrical contacts are required to study photoelectrical and electrical properties of NWs. Two most developed paths for creating such contacts are: contacting the horizontally lying nanowire with electron beam lithography (EBL) [4] and using a nanoprobe to form the contact with a vertically standing NW [5]. Creating contacts with EBL has few drawbacks, connected with laboriousness of this procedure. Besides that, substrate and resistance of the barriers, which appears between the contact and NW, are affecting the electrical properties of the NW [6]. Application of nanoprobes seems to be the most preferable method to investigate characteristics of individual nanowires. Geometry of such experiment corresponds well with working conditions of solar cells [7]. Moreover, using such method doesn't require laborious processing and allows measuring large quantities of NWs.

Precise positioning of the nanoprobe is required for measuring the current-voltage characteristics (I-V curves). This is usually achieved by using a system, where the nanoprobe is accurately positioned on top of a NW in a scanning electron microscope (SEM) chamber. I-V curves of Ge [8], InAs [9], GaAs [10] etc. were measured with such method. Newly appeared results of measurements of I-V curves by scanning probe microscopy: scanning tunneling microscopy (STM) [11], conductive atomic force microscopy (C-AFM) [12] and photocurrent (PC)-AFM [7] demonstrate possibility to obtain conductivity and photoconductivity maps of NW arrays. Worthily note that obtaining the conductivity maps has been carried only for covered arrays, because uncoated NWs are flexible and become bended during scanning.

Recently, I-V curves were measured with nanometer scale lateral resolution on free standing NWs without SEM controlling procedure by the STM technique [11]. Despite the advantages of this method such as high lateral resolution and low resistance of probe-NW contact, construction of the microscope impedes the additional illumination and requires vacuum conditions. Instead of STM, the AFM provides wider opportunities for studying the photoelectrical properties of NWs. However, it is rather difficult to control the force of probe-NW interaction. In standard contact and semicontact AFM modes it is not possible to obtain stable electrical contact as NW can bend and break during scanning. In 2009 a technique called PeakForce Tapping® was introduced, which allows to control the maximum force, i.e. "peak force" of probe-surface interaction. This method was extensively used for nondestructive scanning of conductive arrays of carbon nanotubes in polymer matrix [13].

Present study is concerned with modification of surface and its influence on electrical and photoelectrical properties of GaAs NWs, including NWs with p-n junction. Two types of modification were examined: 1. passivation of the structure's surface by epitaxial GaP layer; 2. covering with SiO_x, spin-on-glass (SOG) layer. Measurements were carried by C-AFM method in PeakForce mode. It is worth noting that individual current-voltage characteristics were obtained for single nanowires.

2. Materials and Methods

NWs were grown by VLS technique with Au-catalyst in MOVPE chamber. Eight different samples were investigated in this work and they were varied by three parameters: (1) two types of substrates with p- and n-type doping, with carrier concentration $1.3\text{-}1.6\cdot10^{19}$ cm^{-3} and $1.6\text{-}3.9\cdot10^{18}$ cm^{-3}, respectively; (2) two alternatives of GaP-passivation [14], phosphorous passivation parameters are indicated in Table 1; (3) two variants of coating, i.e. with SOG and without SOG layer of thickness ~1 um (See Fig. 1 (b), (d)).

Table 1. Growth parameters for GaAs NWs, used for further classification in this article.

Passivation	**no**	**GaP-passivated**
Growth time	120 sec	45 sec
Height	2.3 μm	1.1 μm
Diameter	~ 100 - 200 nm	
DEZn flow	1 sccm	5 sccm

With such combination of samples it was expected to distinguish the influence of pn-junction, GaP-passivation and SOG coating separately on the IV-curves of individual nanowires. Additionally, it was expected to study the carrier excitement due to light illumination and localized NW surface charges.

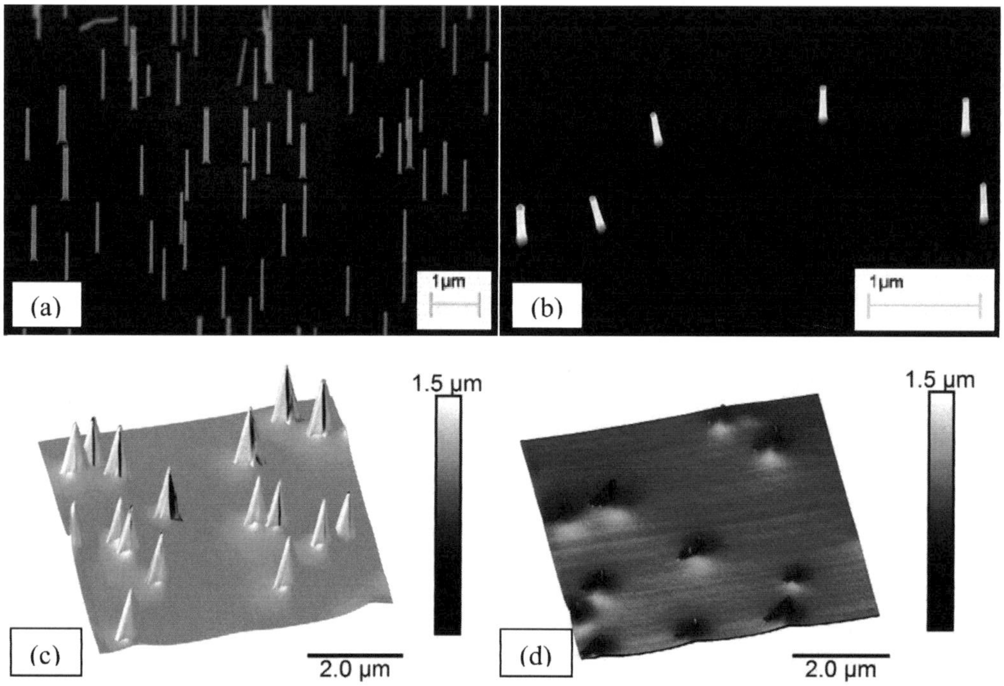

Figure 1. SEM (a), (b) and AFM 3D (c), (d) images of GaAs NWs array. a), c) uncoated; b), d) coated by 1 μm SiO_X layer. The conductivity skin (black conductive spots) is imposed on AFM topography image. The conductivity was measured with -2V applied to the substrate.

AFM studies were performed using the PeakForce Tapping® technique, based on the measurement of force-distance curves at each point of the scan, with periodic modulation of the tip-surface distance. During the topography imaging, it is also possible to measure the I-V curves of each point at the contact between the probe and surface. Studies were performed on Bruker Multimode 8 AFM under ambient conditions. The conductive PFTuna probes (Bruker) with Pt/Ir coating were used. In this device, the cantilever deflection is controlled by using an optical system consisting of a red laser and a 4-quadrant photodiode. Since the diameter of the laser beam exceeds the width of the cantilever, a portion of the radiation reaches the scanned sample, so all measurements were performed in the overilluminated conditions.

SiO_X deposition does not lead to breaking of NWs away from the substrate and all NWs remain conducting, which is seen on Figure 1 (c), (d) as colored conductivity spots near top parts of NWs. Due to the convolution of the pyramidal shape of the AFM probe and the actual cylindrical shape of NWs, uncoated NWs have the pyramidal shape in the 3D image. Topography image of NWs have almost no noise, moreover changing of scanning direction did not alter the image which indicates absence of NWs bending during the scanning procedure. Interestingly, the areas of the highest conductivity, are located not in the top of the pyramid, but on their side faces. This effect is caused by practical observation of fact that conductive coating thickness at the end of the probe's tip is small. Moreover, contact area of the tip and the NW Au cap is smaller than the contact area of pyramid faces of the probe and the Au cap, therefore face of the pyramid provides better conductivity. Furthermore, conductive coating can be ripped off the probe's tip because of the scanning effort, which is a feature of the probes with Pt/Ir coating. Wear-resistant probes with conductive diamond-like coating could be used, but their resistance is in the scale of tens of MΩ, which drastically limits their applicability in NW I-V measurements.

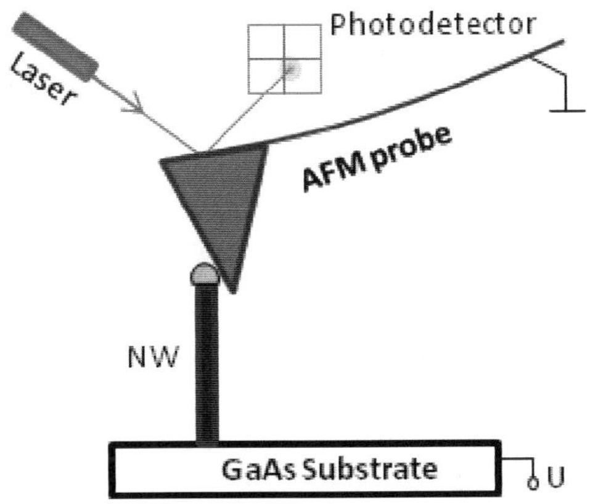

Figure 2. Experimental scheme for AFM measurement of NW I-V curves.

Measurement of the I-V curves was performed as following: firstly, the conductivity map of NWs array was taken, and then the I-V curves were measured in the points with the highest conductivity. The experimental scheme is shown in Figure 2. Worth noting that in this configuration, the probe was grounded and voltage was applied to the substrate. The bottom contact was created by bonding of substrate to a metal wafer using special conductive silver paste. Since the area of such Ag contact and its conductivity are high, then such contact may be considered as ohmic. Moreover, substrates were highly doped and their resistance were only several Ohms.

3. Experimental results

3.1. p-NWs on p-substrate

Figure 3 shows the current-voltage characteristics for uncoated and coated by SiO_X layer NWs of unpassivated (a) and GaP-passivated (b) samples, respectively, grown on the p-substrate. I-V curves for the passivated NWs are almost linear (coated) and linear ohmic (uncoated), while for the unpassivated NWs they are non-linear.

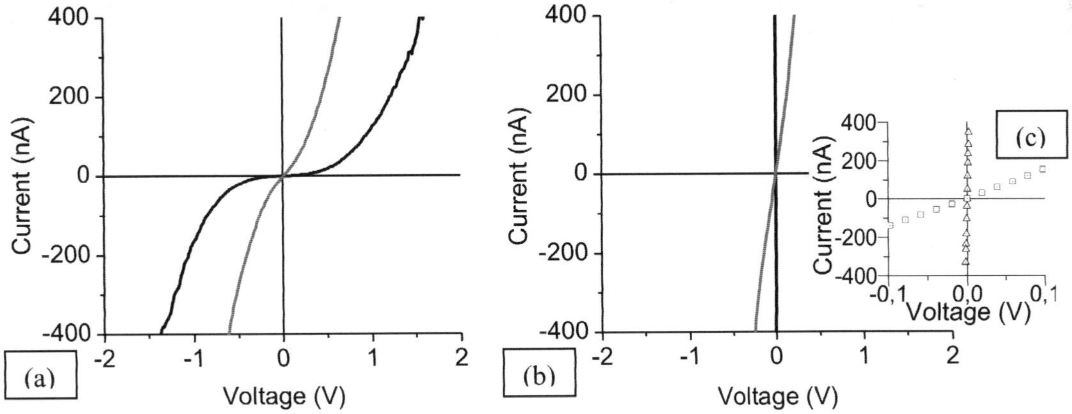

Figure 3. I-V curves of GaAs NWs grown on p-substrate: uncoated (black) and coated (red) by SiO_X layer for (a) unpassivated and (b) passivated samples with detailed inset (c) indicating the voltage step of the measurements 0,39 mV (black) and 19,53 mV (red).

The linear shape of the I-V curves for uncoated passivated NWs indicate the absence of barriers between NW and substrate, as well as between Au cap and NW (Fig. 3 (c)). Thus, in the case of passivated samples Au cap has formed ohmic contact with NW that is not consistent with the previously known data where Schottky barrier between Au cap and GaAs NWs was observed [15]. A possible explanation of this discrepancy, is the high doping level of studied NWs, which can lead to the formation of an ohmic contact [16], that is confirmed by the fact that in unpassivated NWs, which were less doped, the rectifying contact was formed. In addition, possible that GaP-passivation also contributed to the creation of an ohmic contact.

It is needed to be emphasized that for uncoated passivated NWs, the slope of I-V curve corresponds to resistance of 7 kΩ, which corresponds to the resistance of heavily doped GaAs NWs. Further, it seems possible to estimate the range of resistance R of p-NW assuming that level of doping was $N_A = 10^{18}$-10^{19} cm^{-3}, mobility $\mu = 400$ cm^2 / V·s, length $l = 1$ μm and radius $r = 80$ nm. Geometrical dimensions were taken from the AFM topography data. Then, assuming that for highly doped and passivated NWs depletion near surface region is negligible, and according to the equation $R = l/en\mu\pi r^2$, where e-electron charge, we find that R = 0.7 – 7 kΩ. When calculating, we took the value of mobility corresponding to bulk GaAs. However, due to scattering by surface defects, as well as on the bulk structural defects, the charge carrier mobility in NWs is typically several orders of magnitude smaller [17]. In this case, assuming that the doping level is 10^{19} cm^{-3}, the mobility will be only one order of magnitude less than in bulk GaAs. Such high values of mobility are likely caused by the reduction of the surface influence due to the GaP-passivation.

The abovementioned arguments do not take into account the influence of the contact resistance between probe and NW, which can be higher than nanowire self-resistance and be responsible for the slope of I-V curves. However, taking into account the contact resistance will not lead to a change in the order of the estimated values. Additionally, it can be concluded that contact resistance between probe and uncoated NW is even less than 7 kΩ. Increased resistance of coated passivated NWs we associate with a significant increase in contact resistance, because after applying SiO_X coating the height of the NWs is decreased up to 100 – 200 nm and in this case only the probe's tip with relatively high resistance is contacting the NW. For higher NWs (unpassivated) such problem was not observed. Besides that, SiO_X coating led to increase in the current through the NW, which can be explained by the passivation effect of SiO_X coating.

3.2. p-NWs on n-substrate

Figure 4 shows the I-V curves for unpassivated (a) and GaP-passivated (b) NWs on n-substrate. Presence of p-n junction between the NW and the substrate affects the shape of I-V curve from linear to the diode shape. Fig. 4 also demonstrates that SiO_X coating does not noticeably change the I-V curves of passivated sample, while it significantly changes its character for the unpassivated sample.

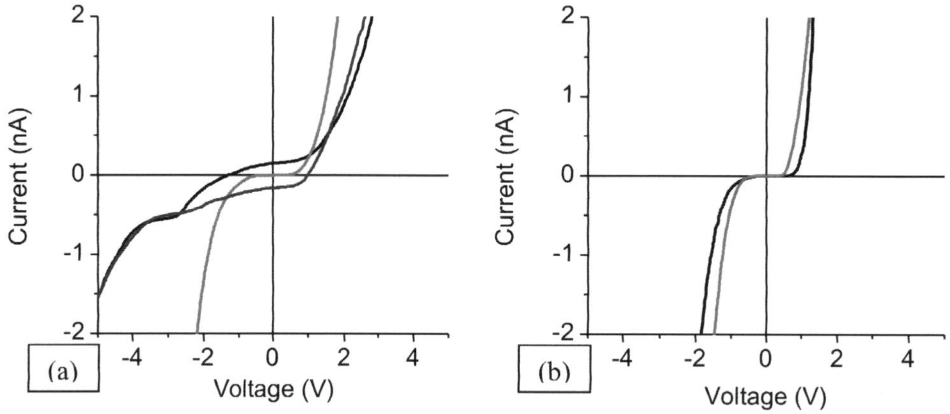

Figure 4. I-V curves of the GaAs NWs grown on n-substrate: uncoated (black/blue) and coated (red) by SiO_X layer for (a) unpassivated and (b) passivated samples. Black/blue curves indicate current during increasing/decreasing of the input voltage for uncoated NWs.

I-V curves of unpassivated NWs have a pronounced hysteresis. Moreover, these I-V curves have saturation in the -2 – -3V range, whose presence can be explained by influence of the Schottky barrier between NW and Au cap. Indeed, in this case the current is determined by serial resistance of two diodes connected towards to each other, i.e. p-n junction and Schottky barrier. Saturation in -2 – -3V range corresponds to the case with open Schottky barrier, and still closed p-n junction. Sufficiently high reverse current in this range is caused by effects of illumination and photoconductivity.

It is worth noting that such shape of I-V curves occurs in approximately 80% of uncoated unpassivated NWs. In rest 20% cases, the I-V curves had shape similar to the passivated samples. Differences in the shape of the I-V curves can be explained by various height of the Schottky barrier in the structure of unpassivated NWs. Schottky barrier height can be changed under the influence of charge accumulated by the NW surface states [18]. It was previously shown that this charge accumulation on the surface is possible for p- GaAs NWs near the Schottky barriers with applying an external electric field [19]. Hysteresis of I-V curves confirms charging and discharging of the NW surface near the barriers area.

Coating of NWs by SiO_X layer leads to elimination of the hysteresis. Thus, SiO_X passivates surface and blocks the capturing of charges on the NWs shell. It is known that presence of the pn-junction between NW and substrate can cause the photo voltage [20]. Since in our case I-V curves were measured under illumination, they were expected to be shifted along the axes. However, this effect was not observed even for passivated samples. This can be explained by the small pn-junction area, and by high doping level, which significantly reduces the lifetime of non-equilibrium charge carriers.

4. Conclusion

The effect of surface modification on the electrical and photoelectric properties of GaAs nanowires was investigated with AFM in PeakForce Tapping® mode. This method was successfully applied for measuring I-V curves of single freestanding vertical NWs. Contact resistance between the probe and NW was found to be less than 7 kΩ that allows investigating heavily doped GaAs NWs. The absence of Schottky barrier between Au catalyst cap and heavily doped NW with a GaP passivation layer was found. Charge accumulation on the surface of unpassivated GaAs NWs is found to change the height of the Schottky barrier between the cap and NW and produce hysteresis observed on I-V curves. The SiO_X coating leads to surface passivation and suppresses the process of charge trapping on the surface.

References

[1] Duan XF et al., 2001 *Nature* **409** 66
[2] LaPierre R et al., 2013 *Phys. Stat. Sol.* **7** 815
[3] Calarco R et al., 2005 *Nano Lett.* **5** 981
[4] Cui Y et al., 2000 *J. Phys. Chem. B* **104** 5213
[5] Talin AA et al., 2008 *Phys. Rev. Lett.* **101** 076802
[6] Leonard F et al., 2011 *Nat. Nanotechnol.* **6** 773
[7] Giacomo M. et al., 2011 *Nano Lett.* **11** 2490
[8] Leonard F et al., 2009 *Phys. Rev. Lett.* **102** 106805
[9] Talin AA et al., 2010 *Semicond. Sci. Technol.* **25** 024015
[10] Katzenmeyer AM et al., 2010 *Nano Lett.* **10** 4935
[11] Rainer T et al., 2013 *Nano Lett.* **13** 5182
[12] Dementyev PA et al., 2010 *Semiconductors* **44** 610
[13] Desbief S et al., 2012 *Nanoscale* **4** 2705
[14] Haggren T et al., 2014 *Appl. Phys. Lett.* **105** 033114
[15] Suyatin DB et al., 2014 *Nat. Comm.* **5** 3221
[16] Brillson LJ et al., 1993 *Contacts to Semiconductors: Fundamentals and Technology* (NJ,Noyes)
[17] Lin A et al., 2012 *Nanotechnology* **23** 105701
[18] Piscator J and Engström O., 2008 *J. Appl. Phys.* **104** 054515
[19] Dunaevskii MS et al., 2013 *Tech. Phys. Lett.* **39** 209
[20] Cirlin GE et al., 2010 *Nanoscale Res. Lett.* **5** 360

Electro-physical characteristics of MIS structures with HgTe-based single quantum wells

S Dzyadukh, S Nesmelov, A Voitsekhovskii, D Gorn
Radiophysics faculty, Nanoelectronics and Nanophotonics Laboratory, Siberian physical-technical institute, Tomsk state university, 36 Lenina Av., 634021 Tomsk

E-mail: gorn_dim@sibmail.com

Abstract. The paper presents brief research results of the admittance of metal-insulator-semiconductor (MIS) structures based on $Hg_{1-x}Cd_xTe$ grown by molecular-beam epitaxy (MBE) method including single HgCdTe/HgTe/HgCdTe quantum wells (QW) in the surface layer. The thickness of a quantum well was 5.6 nm, and the composition of barrier layers with the thickness of 35 nm was close to 0.65. Measurements were conducted in the range of temperatures from 8 to 200 K. It is shown that for structure with quantum well based on HgTe capacitance and conductance oscillations in the strong inversion are observed. Also it is assumed these oscillations are related with the recharging of quantum levels in HgTe.

1. Introduction

HgCdTe MBE method provides new opportunities to optimize the characteristics of highly sensitive infrared detectors by growing heteroepitaxial structures based on HgCdTe with a pre-defined distribution of stochiometric composition in epitaxial films. Today the most common is creation of varyband layers with increased stochiometric composition of CdTe near the borders of the epitaxial layer. It allows reducing the influence of surface recombination on the charge carrier lifetime in the bulk of the epitaxial film, as well as to reduce the series resistance and improve the threshold characteristics of photodiodes based on p-n-junctions. Investigations of characteristics of MIS structures based on HgCdTe with inhomogeneous distribution of stochiometric composition are actual because of the need in passivation of infrared optoelectronic devices based on HgCdTe (laser diodes based on heterostructures or surface-emitting lasers [1], far-infrared range photodetectors [2]).

It should be noted that studies of the characteristics of MIS structures based on HgCdTe with inhomogeneous composition distribution are still extremely rare. The influence of surface graded-gap layers on the electrical characteristics of MIS structures based on MBE HgCdTe heteroepitaxial structures is studied in [3-5]. The properties of a MIS structure based on MBE n-$Hg_{0.75}Cd_{0.25}Te$ comprising CdTe/HgTe superlattice in the surface region of the semiconductor were experimentally investigated [6]. Also the influence of the arrangement of periodic barrier layers with sharply increasing stochiometric composition on capacitive and photoelectric characteristics of HgCdTe-based MIS structures was studied [7]. In studies of p-HgCdTe, obtained by volumetric methods, it is shown that at high hole concentration ($p > 10^{17}$ cm^{-3}) quantization of density of states in the inversion layer is possible, that leads to quantum oscillations in the strong inversion of photo-electromotive force (EMF) [8], capacitance and conductance [9].There are scientific works devoted to the research of quantum-

size structures by electrophysical methods using surface-barrier structures, e.g. Schottky barriers with InGaAs/GaAs quantum-wells [10], [11] or Schottky barriers with GeSi/Si quantum dots [12].
The aim of this scientific work is the investigation of electrophysical characteristics of MIS structures based on heteroepitaxial HgCdTe with a single HgCdTe/HgTe/HgCdTe quantum well by admittance methods in wide frequency (1 kHz till 2 MHz) and temperature (8 K to 300 K) ranges.

2. Experimental methods and samples

Investigated MIS structures were fabricated by molecular-beam epitaxy method on the basis of mercury cadmium telluride on the substrates of GaAs (013) in Institute of Semiconductor physics of Siberian branch of Russian Academy of Sciences (Novosibirsk, Russia). Some of the investigated structures did not include a quantum well in the surface region. A structure with a single quantum well had a CdTe composition in the $Hg_{1-x}Cd_xTe$ barrier layer of $x = 0.65$, and the quantum well thickness was 5.6 nm. Barrier layers had a thickness of 35 nm. A low-temperature double-layer SiO_2/Si_3N_4As dielectric applied over the CdTe 40 nm layer was used as the dielectric coating. Figure 1 shows the basic layout of the layers in the structure of the quantum well, and Figure 2 shows the distribution of CdTe composition in the $Hg_{1-x}Cd_xTe$ epitaxial film for this structure.
Measurements were carried out with the use of an automated admittance spectroscopy installation based on non-optical Janis cryostat and Agilent E4980A immittance meter. The measurements were conducted with the change in voltage from negative to positive values (forward sweep scan direction measurements).

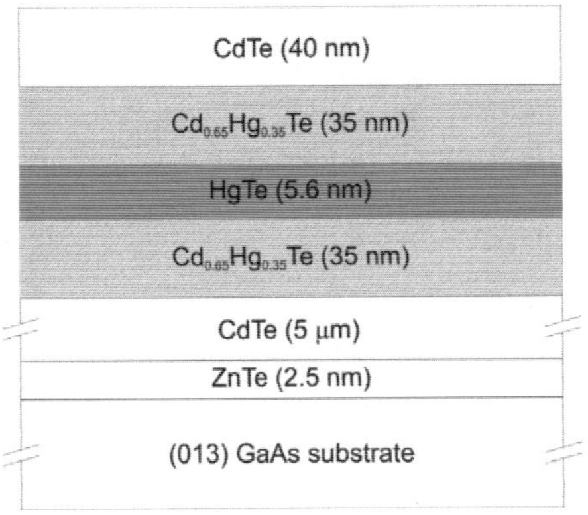

Figure 1. Scheme of the layers position in structure with single quantum well with the thickness of 5.6 nm.

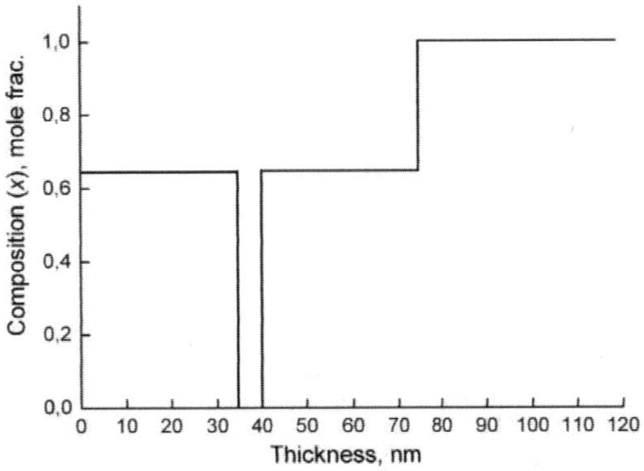

Figure 2. Composition distribution in operating layer of $Hg_{1-x}Cd_xTe$ in the structure with single quantum well with the thickness of 5.6 nm.

3. Results and discussion

Figures 3 and 4 show dependencies of capacitance and differential conductance on bias voltage for MIS structure with 5.6 nm single quantum well measured at a temperature of 7.7 K for various frequencies of alternating test signal. From Fig. 3 and 4 it may be seen that HgCdTe has a p-type conductivity, and in the area of strong inversion at the range of frequencies from 500 Hz to 100 kHz non-monotonic change in capacitance and differential conductance is observed. Capacitance maxima in this range of frequencies are observed at the same bias voltages, but for differential conductance with the change of frequency maxima can be replaced by minima for some bias voltages (Fig. 4). At 500 Hz capacitance-voltage characteristic is of low-frequency type and maxima become less pronounced than at frequencies of 5-20 kHz. At 100 kHz capacitance-voltage characteristic is close to high-frequency type and capacitance maxima also appear less clearly, than at frequencies of 5-20 kHz. The capacitance value at 100 kHz in strong inversion is close to 0.7 pF.

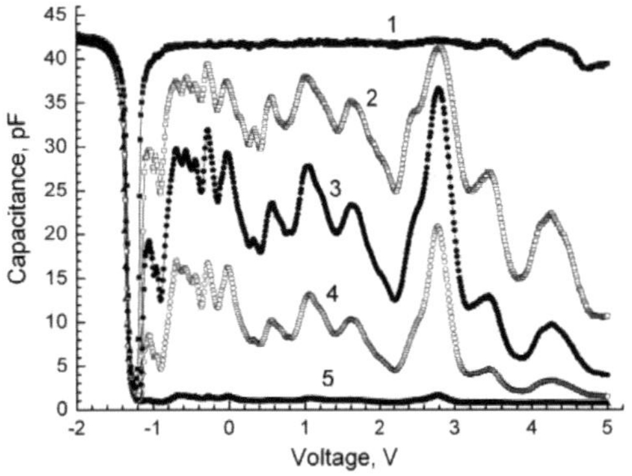

Figure 3. Capacitance-voltage characteristics of MIS structures based on HgCdTe with quantum well measured at 7.7 K under the forward direction voltage sweep at different frequencies: 1 – 500 Hz, 2 – 5 kHz, 3- 10 kHz, 4 – 20 kHz, 5 – 100 kHz.

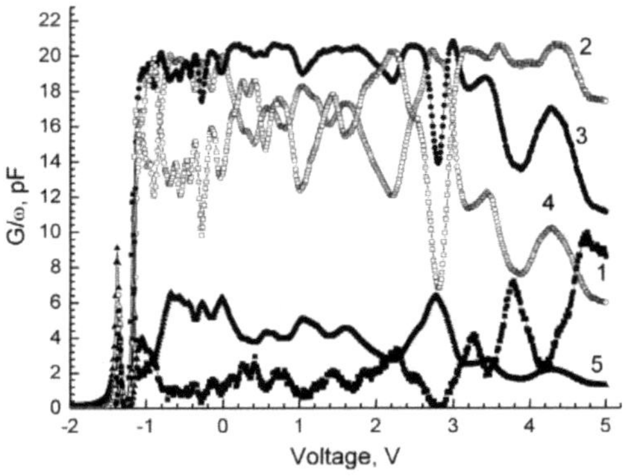

Figure 4. Dependence of differential conductance on voltage for MIS structures based on HgCdTe with quantum well measured at 7.7 K under the forward direction voltage sweep at different frequencies: 1 – 500 Hz, 2 – 5 kHz, 3- 10 kHz, 4 – 20 kHz, 5 – 100 kHz.

Figure 5 shows capacitance-voltage characteristic for a MIS structure with a quantum well measured at 10 kHz under the forward direction voltage sweep at different temperatures in the range of temperatures from 7.8 to 77 K. From fig. 5 it is seen that capacitance maxima in strong inversion are pronounced at low temperatures, and their amplitudes and position on the voltage axis change when increasing the temperature from 7.8 to 10 K. In the range of temperatures from 30 to 77 K maxima become diffused and are pronounced weakly.

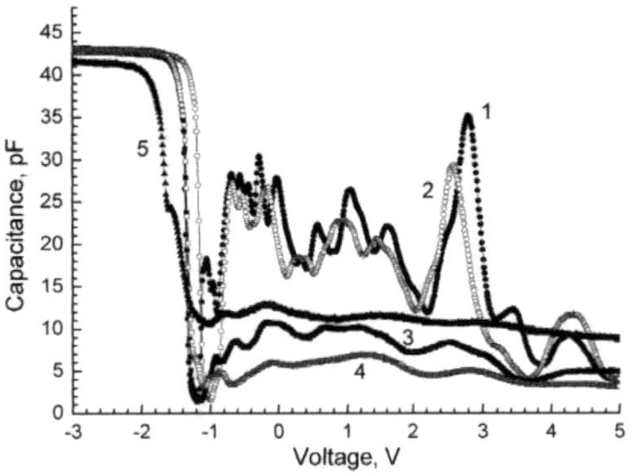

Figure 5. Capacitance-voltage characteristics of MIS structures based on HgCdTe with quantum well measured at 10 kHz under the forward direction voltage sweep at different temperatures: 1 – 7.8 K, 2 – 10 K, 3 – 30 K, 4 – 50 K, 5 – 77 K.

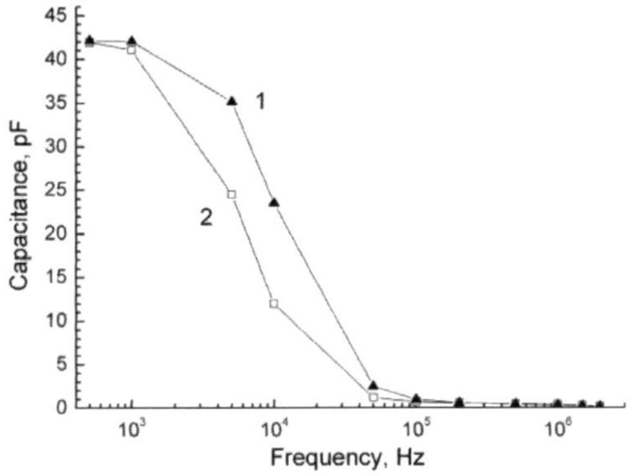

Figure 6. Frequency dependencies of capacitance of MIS structure with quantum well, measured at 8 K and different voltages: 1 – 1 V (maximum), 2 – 2 V (minimum).

Figure 6 shows capacitance dependencies on frequency for a MIS structure with a quantum well at 8 K and voltages corresponding to a maximum (1 B) and a minimum (2 B) on the capacitance-voltage characteristic. It is seen from figure 6 that a transition of capacitance-voltage characteristic from low-frequency to high-frequency type takes place with the change in frequency from 500 Hz to 100 kHz. Capacitance in the maximum at intermediate frequencies is significantly higher than in the minimum.

It must be noticed that other investigated MIS structures based on HgCdTe without quantum wells have capacitance-voltage characteristics that do not have non-monotonous changes of capacitance in strong inversion. Figure 7 shows capacitance-voltage characteristics of MIS structures based on varyband *p*-HgCdTe ($x = 0.22$) without quantum wells measured at 20 K and different frequencies of test signal. Electrophysical measurements at lower temperatures for *p*-HgCdTe ($x = 0.22$-0.36) with conversion of conductivity type after annealing are less accurate because of the sharp increase in epitaxial film's bulk resistance during cooling from 20 K to 8 K. For comparison, figure 8 shows capacitance-voltage characteristics of MIS structure based on HgCdTe with a single quantum well, measured at 30 K and various frequencies of test signal. It is seen from figures 7 and 8 that for MIS structures based on HgCdTe without quantum wells there are no non-monotonic changes of capacity in strong inversion.

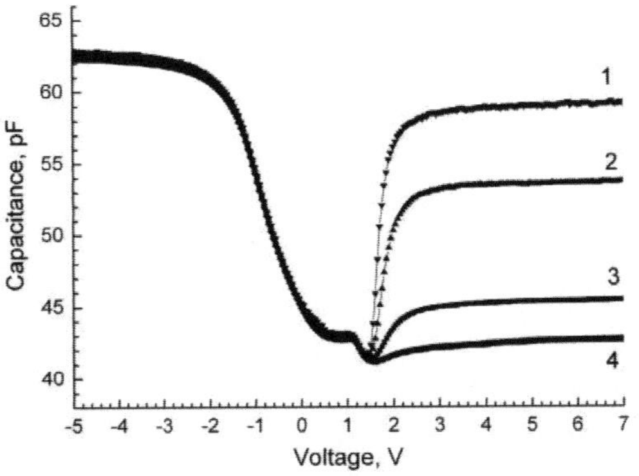

Figure 7. Capacitance-voltage characteristics of MIS structure based on varyband p-HgCdTe ($x = 0.22$) without quantum wells measured at 20 K under the forward direction voltage sweep at different frequencies: 1 – 10 Hz, 2 – 20 kHz, 3- 50 kHz, 4 – 100 kHz.

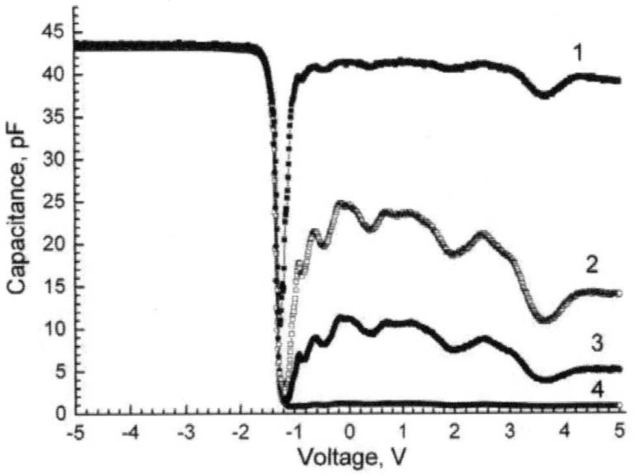

Figure 8. Capacitance-voltage characteristic of MIS structure based on HgCdTe containing quantum well measured at 30 K under the forward direction voltage sweep at different frequencies: 1 – 1 kHz, 2 – 5 kHz, 3 - 10 kHz, 4 – 50 kHz.

Thus, it is shown experimentally that capacitance maxima in strong inversion are related with the presence of a single HgTe quantum well in the structure. From the slope of capacitance-voltage characteristic in the depletion mode it is possible to define charge carrier concentration in HgCdTe, which is equal to 6.5×10^{17} m^{-3} at 7.7 K for the sample with quantum well, that is too small for the material with $x = 0.22$, but is suitable for wide band gap material with $x = 0.65$. Having known the majority charge carrier concentration it is possible to calculate capacitance value of a MIS structure in strong inversion, which was found to be 0.8 pF (experimental value is approximately 0.7 pF). The calculation of surface potential dependence on voltage for MIS structure based on HgCdTe containing quantum well is carried out.

We will briefly describe the most likely mechanism for the CVC maxima appearance in the strong inversion for the MIS structure comprising a single quantum well. Surface potential (as well as the potential in the field of quantum well) for p-HgCdTe depends on the bias voltage in the strong inversion because of the effects of degeneracy and conduction band nonparabolicity. When the Fermi level in the field of quantum well approaches to the level of quantization, quantization level recharges following a change of the AC test bias voltage. Capacity of dimensional quantization level in the quantum well contributes to the full capacity of the structure and appears at intermediate frequencies in the strong inversion. When a capacitance-voltage characteristic has a low-frequency view capacitance of the MIS structure in the strong inversion tends to the dielectric capacitance and maxima associated with dimensional quantization levels do not appear. When a capacitance-voltage characteristic has a high-frequency view the carrier concentration in the inversion layer does not have enough time to follow the test signal and contribution of the quantization levels capacitance at full capacity decreases because the quantum well is located at a distance of 35 nm from the interface. Because the charge carriers thermal energy increases with the temperature increasing dimensional quantization effects are mild.

Energies of dimensional quantization levels in a quantum well can be approximately defined by analogy with [8]. To do this, it is necessary to define the flat-band voltage in MIS structure comprising QW in terms of flat-band capacitance value and to obtain surface potentials corresponding to maxima of capacity in the strong inversion. Then it is necessary to build the potential spatial distribution in the HgCdTe surface layer and to find the potential under the given bias voltage in the quantum well. Knowing the position of the Fermi level in the flat-band mode, position of dimensional quantization levels can be find relative to the allowed energy bands in HgCdTe barriers. Crucial issue when using this method is to use the correct mathematical expressions for the main parameters of HgCdTe at low temperatures (i.e. densities of states, intrinsic concentration etc.), the proper construction of the ideal CV characteristics at low temperatures, as well as the correct numerical calculation of energy-band diagrams in the surface HgCdTe layer with taking into account the effects of degeneracy and nonparabolicity of the conduction band.

Thus the flat-band capacitance value is 33.56 pF in our case. Hence the flat-band voltage is -1.36 V. Band-gap energy of barrier layer is 0.838 eV. Fermi level location is ($E_v + 0.010$ eV) or ($E_c - 0.828$ eV). Then two first maxima of capacitance for quantum well MIS structure under the direct sweep bias voltage correspond to energies ($E_c - 0.520$ eV) and ($E_c - 0.210$ eV). Here E_c is the energy of conduction band edge of the barrier layer near the border with well.

Another way for estimation of dimensional quantization energies levels is the using of theoretical model described by us earlier in [13]. This model based on self-consistent solution of Poisson and Shrödinger equations for heterostructure comprising QW with taking into account real parameters of MBE-grown CdHgTe solid solution such as electron affinity and concentrations of ionized acceptors and donors. Calculation of dimensional quantization levels for structure under consideration at 7.7 K gives values ($E_c - 0.597$ eV) for first level of electrons E_{c1}, ($E_c - 0.347$ eV) for second level of electrons E_{c2}, ($E_c - 0.486$ eV) for first level of heavy holes E_{h1}, ($E_c - 0.517$ eV) for second level of heavy holes E_{h2}. Then two first maxima of capacitance observed on the CVC in the experiment may be associated with the presence of dimensional quantization levels. It was confirmed by estimation. But the accuracy of used calculation methods does not allow give more precise results and does not allow interpret each observed peak.

4. Conclusions

Thus the admittance of MIS structures based on MBE HgCdTe with a single 5.6 nm quantum well located in the surface layer of semiconductor was experimentally investigated. It is shown that the presence of a single quantum well thickness of 5.6 nm can lead to the appearance of capacitance-voltage and conductance-voltage characteristics peaks in the strong-inversion mode.

Capacitance maxima in the strong inversion mode are observed in an intermediate case between the low frequency and high frequency capacitance-voltage characteristics (5-50 kHz for a sample with a single quantum well). The capacitance maxima in strong inversion mode are particularly pronounced at low temperatures (8-10 K), although they are weakly apparent up to 77 K. Maxima in capacitance and conductance in strong inversion mode are not observed for samples without quantum wells. The dependences of the surface potential on voltage at various approximations are built. It is shown that considering the effects of degeneracy and non-parabolicity of the conduction band, the surface potential at strong inversion depends on the bias voltage. It is assumed that the dimensional quantization levels recharge capacitance in the quantum well contributes to the overall capacitance of the structure and appears in the capacitance-voltage characteristics in the strong inversion. An approximate method for determining the energy levels of quantum wells with the use of data obtained from the capacitance-voltage characteristics measurements is proposed.

5. Acknowledgements

This scientific work was carried out in the network of Tomsk State University Competitiveness Improvement Program and State task of Ministry of education and science of Russian Federation (№ 16.1032.2014/K).

6. References

[1] Pautrat J.L., Hadji E., Bleuse J., Magnea N. 1997 *Journal of Electronics Materials* **26** (6) 667.
[2] Grein C.H., Jung H., Singh R., Flatte M.E. 2005 *Journal of Electronics Materials* **34** (6) 905.
[3] Voitsekhovskii A.V., Nesmelov S.N., Dzyadukh S.M., Varavin V.S., Dvoretskii S.A., Mikhailov N.N., Sidorov Yu.G., Yakushev M.V. 2010 *Opto-Electronics Review* **18** (3) 259.
[4] Voitsekhovskii A.V., Nesmelov S.N., Dzyadukh S.M. 2012 *Thin Solid Films* **522C** 261.
[5] Voitsekhovskii A.V., Nesmelov S.N., Dzyadukh S.M. 2014 *Thin Solid Films* **551** 92.
[6] Goodwin M.W., Kinch M.A., Koestner R.J. 1988 *J. Vac. Sci. Technol* **A6** (4) 2685.
[7] Voitsekhovskii A.V., Nesmelov S.N., Dzyadukh S.M., Varavin V.S., Dvoretskii S.A., Mikhailov N.N., Sidorov Y.G. 2011 *Russian Physics Journal* **54** (3) 263.
[8] Voitsekhovskii A.V., Davydov V.N. Photoelectric MIS Structures Based on Narrow-Band Semiconductors, *Radio I Svyaz'*, 1990, Tomsk, 327 p.
[9] Kinch M.A.1981 *Semicond. Semimetals* **18** 313.
[10] Zubkov V.I. Semiconductor nanohetrostructures diagnostics by the admittance spectroscopy method. Saint-Petersburg, "Elmor", 2007, 220 p.
[11] Zubkov V.I. 2007 Semiconductors **41** (3) 320.
[12] Yakimov A,I,, Dvurechenskii A.V., Nikiforov A,I,, Bloshkin A.A., Nenashev A.V., Volodin V.A. 2006 *Phys. Rev. B.* **73** 115333.
[13] A.V. Voitsekhovskii, D.I. Gorn, I.I. Izhnin, A.I. Izhnin, V.D. Goldin, N.N. Mikhailov, S.A. Dvoretskii, Yu.G. Sidorov, M.V. Yakushev, V.S. Varavin. 2013 *Russian Physics Journal* **55** (8) 910.

Preparation of cadmium selenide colloidal quantum dots in non-coordinating solvent octadecene

D S Mazing, A M Brovko, L B Matyushkin, O A Aleksandrova, V A Moshnikov

Saint Petersburg Electrotechnical University "LETI", ul. Prof. Popova, 5, 197376 Russia

E-mail: dmazing@yandex.ru, leva.matyushkin@gmail.com

Abstract. Nearly monodisperse cadmium selenide quantum dots (QDs) were synthesized in non-coordinating solvent octadecene through phosphine-free method using oleic acid as surfactant. Selenium powder suspension in octadecene obtained by ultrasound processing was used as one of precursor solutions. Influence of multiple selenium precursor injections on nanocrystal growth process was investigated. Nanoparticles were characterized by means of absorption and photoluminescence spectroscopies.

1. Introduction

Extensive development of colloidal semiconductor QDs field is caused by their important features (tunable absorption and emission spectra, high luminescence quantum yield, processibility) which give way to wide range of prospective applications in light emitting devices, photovoltaic cells, photodetectors. Facility of nanocrysral surface processing and higher stability against photodegradation in comparison with conventional organic fluorophores stimulates application of colloidal QDs in biological and medical detection [1, 2]. Cadmium selenide (CdSe) is semiconductor with band gap of 1.7 eV which is widely used as a QDs material for visible spectrum.

Classical hot-injection method introduced in the beginning of 1990s [3] allowed preparation of nearly monodisperse colloidal nanocrystals of AIIBVI semiconductors. Using rapid injection of precursors at elevated temperatures it enabled separation of nucleation and growth stages making synthesis procedure more controllable. One of its major drawbacks however was high toxicity and pyrophoricity of the components being used (namely, dimethyl cadmium as cadmium precursor, mixture of trioctylphosphine and trioctylphosphine oxide as coordinating solvent) which necessitates employing of air free techniques. In this respect one of the options of medium for colloidal QDs production is water [4, 5]. However as a rule these methods yield nanoparticles of poorer quality and size distribution.

Later less hazardous compounds were proposed as immediate sources of cadmium and selenium species such as cadmium – fatty acid complexes and elemental selenium, fatty acids and long-chained alkylamines might be used as stabilizing agents [6, 7]. Due to low toxicity and cost non-coordinating solvent octadecene proves to be a good alternative to coordinating phosphine-based solvents. Its boiling point is high enough (317 °C) and its melting point is below room temperature making it convenient to use. Moreover being non-coordinating compound octadecene doesn't participate in nanoparticle stabilization making synthesis procedure more controllable.

2. Materials and methods

Cadmium oxide (CdO), elemental selenium (powder), Octadecene 90% technical grade (ODE), oleic acid (OA), acetone were used without additional purification.

Absorption spectra were acquired on a PE-5400UV UV–vis spectrophotometer (LLC "Ekohim"). Photoluminescence measurements were carried out on equipment based on fast scanning spectrometer Ocean Optics USB-4000 using laser excitation on $\lambda = 445$ nm.

Mean size of nanoparticles was determined using polynomial equation from the work [8]:

$$a = (1.6122 \cdot 10^{-9})\lambda^4 - (2.6575 \cdot 10^{-6})\lambda^3 + (1.6242 \cdot 10^{-3})\lambda^2 - 0.4277\lambda + 41.57, \quad (1)$$

where a is average size of particles, λ is a wavelength of the first excitonic peak. Derived from experimental data acquired by means of TEM studies this equation supposedly is more reliable than effective mass approximation model for smaller nanoparticles.

3. Experiment and results

For the first CdSe QDs synthesis two precursor solutions were prepared. Firstly 1 mmol of CdO, 20 mmol of OA and 40 ml of ODE were loaded into 100 ml three neck flask and heated to 160 °C under nitrogen flow to obtain clear yellowish solution of cadmium oleate in ODE. For preparation of selenium precursor solution 0.5 mmol of selenium powder, 10 mmol of OA and 3.4 ml of ODE were treated in ultrasound bath at room temperature to form homogeneous suspension. Cadmium precursor solution was heated further up to 240 °C and selenium suspension was rapidly injected which resulted in immediate nucleation. To monitor nanoparticle growth aliquots were taken in 10 seconds, 40 seconds, 2, 16 and 120 minutes after injection. Nanocrystals were further precipitated from crude solution by addition of certain amount of acetone with subsequent centrifugation and were redissolved in hexane. Due to hydrophobic surface thus prepared nanocrystals are soluble in other common non-polar solvents of choice such as toluene and chloroform. Figures 1 and 2 show absorption and photoluminescence spectra.

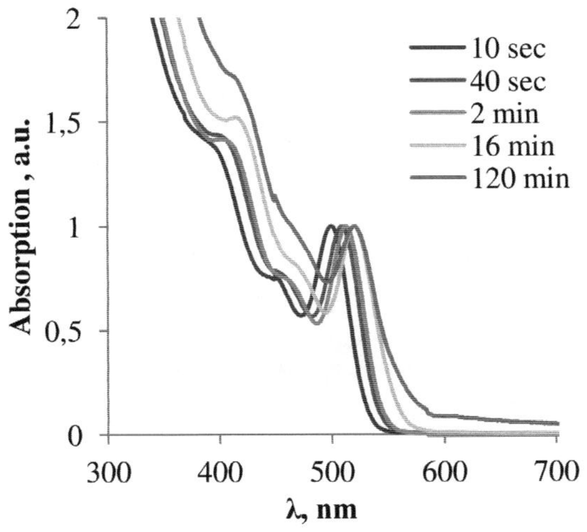

Figure 1. Absorption spectra of CdSe QDs.

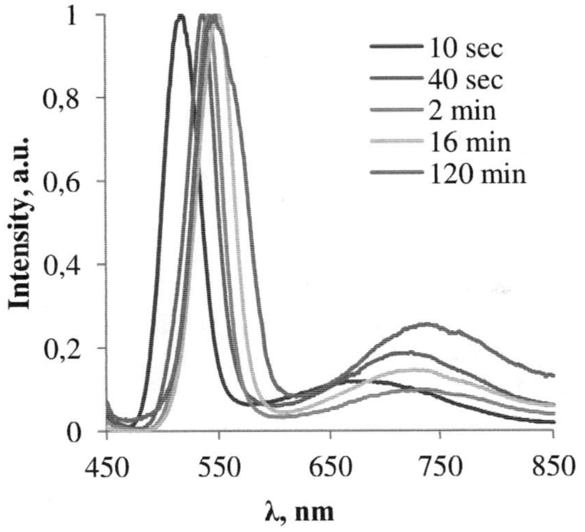

Figure 2. Photoluminescence spectra of CdSe QDs.

As can be seen from absorption spectra nanocrystal growth almost stopped in 16 minutes after injection of selenium precursor solution. Photoluminescence spectra feature two distinctive bands. The left one corresponds to interband radiative recombination and its form reflects nanocrystal size distribution. Presence of wider band on longer wavelengths may be explained by insufficient passivation of nanoparticle surface by stabilizer molecules that causes radiative recombination via trap states inside band gap. Interband photoluminescence peak full width on half maximum parameter (FWHM) decreased at the first stage of the reaction from 38 nm (10 sec) to 31 nm (40 sec and 2 min). Samples obtained in 16 minutes and 120 minutes are characterized by larger FWHM of 39 nm and 53 nm. An explanation to this might be given as follows: depletion of free monomers in reaction volume shifts thermodynamic equilibrium towards growth of larger nanocrystals at the expense of smaller ones which leads to nanoparticles size distribution broadening. This effect is known as Ostwald ripening.

The same synthesis was repeated with smaller amount of stabilizer being used. 1 mmol of CdO, 5 mmol of OA and 40 ml of ODE were used for cadmium precursor solution preparation. Selenium suspension consisted of 0.5 mmol of Se and 1 mmol of OA in 5 ml of ODE.

The first excitonic peak of the sample obtained in 2 minutes (figure 3) after injection is positioned at 490 nm which indicates that in comparison with previous implementation smaller nanoparticles were formed at initial stage of the reaction. As in the previous case growth rate was rather slow but nanoparticle size distribution stayed narrow throughout whole experiment. For 60 minutes of reaction photoluminescence FWHM (figure 4) increased from 25 nm to 28 nm only. It's also should be noted that nanocrystals which were formed right after nucleation likely had surface of low quality.

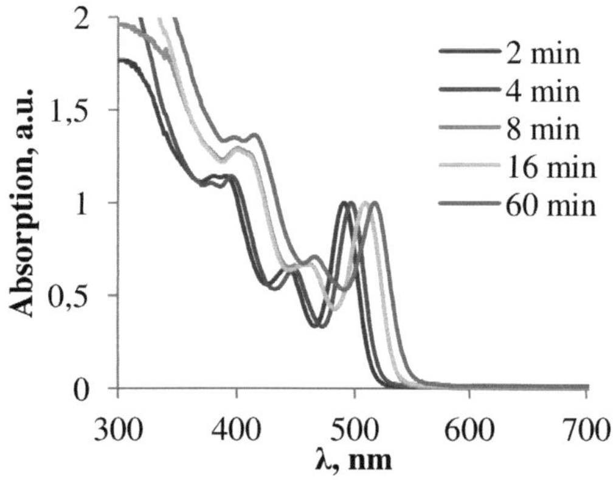

Figure 3. Absorption spectra of CdSe QDs.

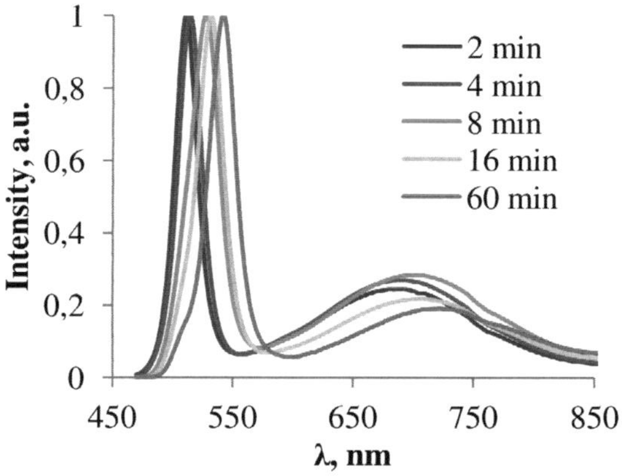

Figure 4. Photoluminescence spectra of CdSe QDs.

The next experiment was based on the first one with only one difference that selenium precursor solution was divided into two equal batches. One half of the solution (0.25 mmol of selenium) was injected as described earlier and the second half was used in the course of the reaction so that initial molar ratio between cadmium and selenium species upon nucleation was [Cd]:[Se] = 4:1. Absorption and photoluminescence spectra are shown in figures 5 and 6.

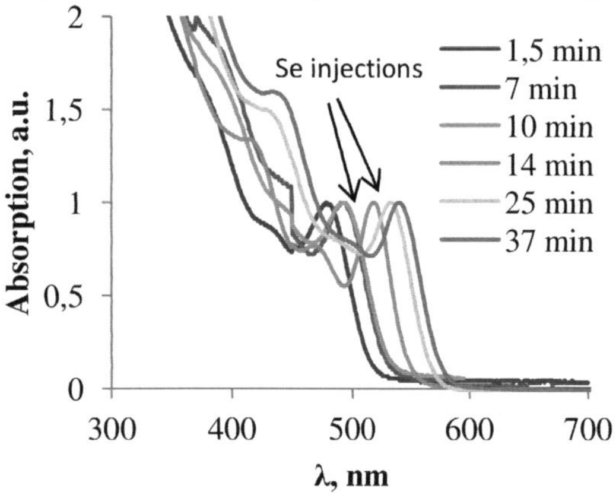

Figure 5. Absorption spectra of CdSe QDs prepared with use of Se precursor multiple injections.

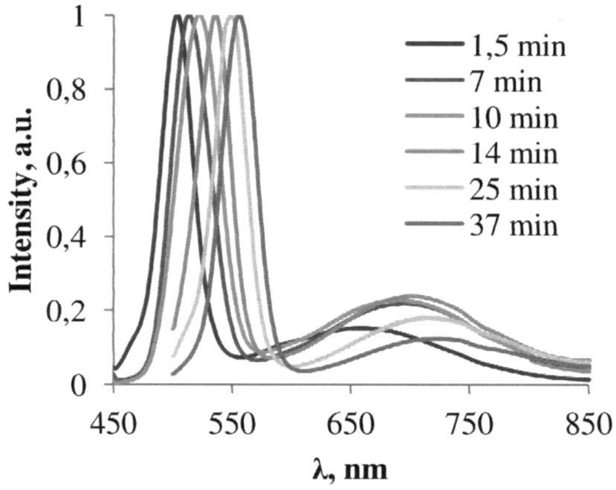

Figure 6. Photoluminescence spectra of CdSe QDs prepared with use of Se precursor multiple injections.

As can be well seen on absorption spectra (figure 5) similar to the previous case growth slows down considerably in 10 minutes after initial selenium injection. Additional injections were performed on 12th and 22nd minutes using two approximately equal parts of the second half of the suspension. It is noticeable that precursor addition resulted in growth resumption as wavelength distance between first excitonic peaks of the third (10 minutes) and forth (14 minutes) samples is about 25 nm. Photoluminescence FWHM (figure 6) gradually decreased from the start of the reaction from 31 to 41 nm for the first three samples. Second injection of selenium suspension resulted in focusing of nanoparticle size distribution so that FWMH after 14 minutes of growth was about 34 nm and it didn't change considerably till the last aliquot was taken on 37th minute. The average size of nanocrystals as calculated from equation (1) ranges from 2.1 to 2.8 nm.

4. Conclusion

As a result colloidal QDs of CdSe were synthesized via phosphine-free hot-injection method using suspension of selenium in ODE as one of precursor solutions. Implementation with relatively high concentration of stabilizing agent is characterized by fast slowing down of growth with subsequent nanoparticle size distribution broadening via Ostwald ripening. Multiple injection of selenium precursor promoted further growth of nanoparticles without substantial deterioration of monodispersity of the system. Synthesis with lower amount of stabilizing oleic acid yielded nanoparticles with better monodispersity and as consequence lower photoluminescence FWHM corresponding to interband transition (less than 30 nm). The average size of prepared CdSe QDs lies between 2 and 3 nm.

Acknowledgments

This work was supported by grant of Russian Science Fund (project №14-15-00324).

References

[1] Musikhin S F, Aleksandrova O A, Luchinin V V, Maksimov A I, Moshnikov V A 2012 *Biotekhnosfera* **5(6)** 40–48
[2] Musikhin S F, Aleksandrova O A, Luchinin V V, Maksimov A I, Matyushkin L B, Moshnikov V A 2013 *Biotekhnosfera* **2(26)** 2–16
[3] Murray C B, Norris D J and Bawendi M G 1993 *J. Am. Chem. Soc.* **115** 8706–15
[4] Rogach A L, Kornowski A, Gao M, Eychmu A and Weller Horst 1999 *J. Phys. Chem. B.* **103** 3065-69
[5] Mazing D S and Aleksandrova O A 2013 *Smart nanocomposites* **4(2)** 17–23
[6] Peng Z A, Peng X A 2001 *J. Am. Chem. Soc.* **123** 183–184
[7] Pu C, Zhou J, Lai R, Niu Y, Nan W and Peng X 2013 *Nano research* **6(9)** 652–670
[8] Yu W W, Qu L H, Guo W Z and Peng X G 2003 *Chem. Mater.* **15** 2854–60

Control of surface plasmon resonance in out-diffused silver nanoislands for surface-enhanced Raman scattering

E S Piliugina[1], F Heisler[2,3], S D Chervinskii[1,4], A K Samusev[2], A A Lipovskii[1,4]
[1] Department of Physics and Technology of Nanostructures, Peter the Great St. Petersburg Polytechnic University, 29 Polytechnicheskaya, St. Petersburg 195251, Russia
[2] The International Research Centre for Nanophotonics and Metamaterials, ITMO University, Kadetskaya line 3b "Red House", St. Petersburg 197101, Russia
[3] Abbe Center of Photonics, Friedrich–Schiller-Universität Jena, Max–Wien-Platz 1, 07743 Jena, Germany
[4] Institute of Photonics, University of Eastern Finland, Yliopistokatu 7, P.O.Box 111, Joensuu 80101, Finland

E-mail: piliugina.ekaterina@gmail.com

Abstract. We present the studies of self-assembled silver nanoislands on the surface of silver ion-exchanged glasses. The nanoislands were formed by out-diffusion of reduced silver atoms from the bulk of the glass to its surface. Control of silver ions distribution in the glass by thermal poling after the ion exchange allowed formation of relatively big, up to 250 nm, isolated silver nanoislands while without the poling an ensemble of silver nanoislands with average size from several to tens of nanometers with random size distribution was formed. The nanoislands were characterized using atomic force microscopy and spectral measurements. We used optical absorption spectroscopy for "random" nanoislands and dark field scattering spectroscopy for isolated ones, corresponding spectra showed peaks in the vicinity of 450 nm and 600 nm, respectively. The "random" nanoislands significantly enhanced Raman scattering from Rhodamine 6G, also the modification of Raman signal from deposited on the surface of the samples bacteriorhodopsin in purple membranes was registered.

1. Introduction

Nowadays, the development of ultrasensitive biosensors is one of the priority scientific and engineering problems, and sensors based on surface-enhanced Raman scattering (SERS) are of particular interest [1]. Recently the ensembles of self-assembled nanoislands that are metal island films (MIF) have attracted attention as SERS-active substrates [2,3]. MIF provide the excitation of surface plasmon resonance (SPR) which increases local electric field of incident light-wave. This E-field enhances Raman scattering from biomolecules or other analytes [4,5]. The spectral position of the SPR can be tuned via varying of MIF formation conditions, which influence size distribution, concentration and shape of the nanoislands [6], and the substrate and cover materials used [7]. These also influence the Raman scattering enhancement which can be as high as ~ $10^{10} - 10^{11}$ [8].

MIF are conventionally fabricated via depositing metals onto dielectric substrates using thermal and e-beam evaporation, sputtering, and chemical or plasma-chemical metal deposition [9]. Here we use the formation of silver nanoparticles on the surface of a silver ion-exchanged soda-lime glass in the course of out-diffusion during annealing the glass in hydrogen [2]. Recently developed technique employing thermal poling treatment of silver-enriched glass with profiled anodic electrode followed by the annealing of the glass in hydrogen [10] has also allowed us to form single nanoparticles. Such 2D-structuring modifies electrodynamic properties of the MIF that expands their applicability as SERS substrates.

2. Sample fabrication

2.1. Self-assembled silver nanoisland MIF fabrication
For SERS experiments we use self-assembled silver nanoisland films. The fabrication process starts from 20 minutes ion exchange step, for which a soda-lime glass slide [11] is placed in $Ag_{0.05}Na_{0.95}NO_3$ melt heated up to 325°C. As the result of thermally-activated silver-sodium ion exchange, the subsurface layer of the glass is enriched with silver ions.

After the ion exchange step the glass substrate is treated for 10 minutes in hydrogen atmosphere at the temperature of 250°C. Hydrogen diffused into the subsurface glass layer reduces silver ions. Due to the low solubility of neutral silver in the glass matrix silver atoms coagulate and form an ensemble of randomly distributed nanoislands on the glass surface [2].

2.2. Single silver nanoisland fabrication
To grow isolated systematically placed silver nanoislands we use modified technique [10]. In the fabrication procedure described in subsection 2.1, after the ion exchange step we include thermal poling of the glass with profiled anodic electrode (square net of 300x300 nm^2 deepenings with the periodicity of 5 μm) to alter silver ions distribution in the glass under the electrode. In the poling, the ion-exchanged glass is heated up to 300°C while applied voltage is 500 V. After the poling for 4 minutes 30 seconds the sample is treated for 30 minutes in hydrogen atmosphere at the temperature of 250°C. As the result silver reduction is prevented in poled regions as schematically shown in Figure 1. To avoid degradation of the sample in the air, the surface with patterned nanoislands is coated with 5 nm titanium dioxide layer using atomic layer deposition (ALD) technique [3,12].

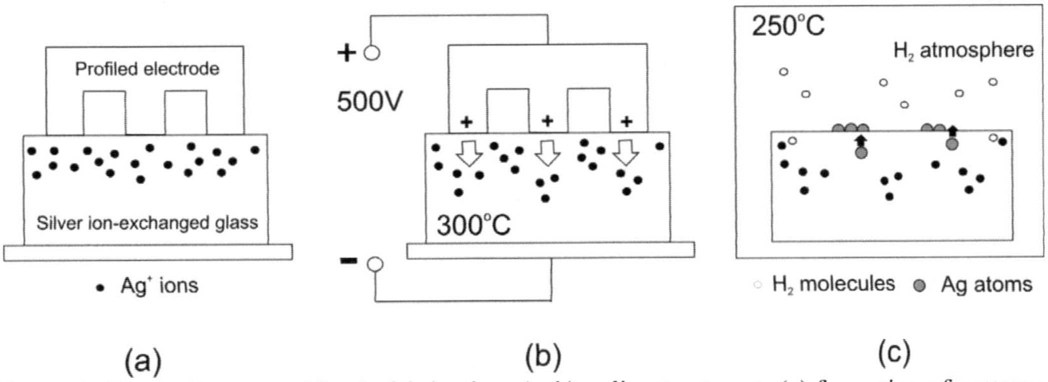

Figure 1. Single silver nanoislands fabrication: (a, b) poling treatment; (c) formation of patterned silver nanoisland film under thermal annealing in hydrogen atmosphere.

3. Studies of the samples
The silver nanoislands were characterized using an "Aist-NT" atomic force microscope (AFM); we also used an UV VIS spectrophotometer Specord 50 to measure optical absorption spectrum of self-

assembled silver nanoisland film and a dark field setup with spectrometer Horiba LabRAM HR UV-VIS-NIR to characterize the spectral properties of single (isolated) nanoislands. In the dark field setup we used s-polarized light from a halogen lamp focused at the sample surface with M Plan Apo 10x/0.26 objective lens. Scattered light was collected with 100x/0.7 objective lens.

AFM images of the samples are shown in Figure 2. It is seen that the glass poling after the ion exchange allowed formation of isolated, relatively big, up to 200-250 nm, silver nanoislands on the surface while without the poling stage randomly distributed silver nanoislands up to tens of nanometers in size were formed.

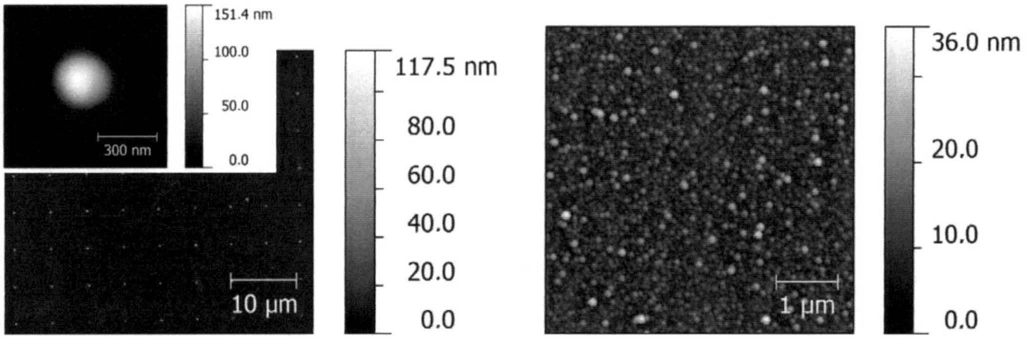

(a) (b)

Figure 2. Typical AFM images of the fabricated samples: (a) patterned silver nanoisland film with isolated nanoislands on the glass surface; (b) self-assembled silver nanoisland film with random distribution of silver nanoislands.

The optical absorption spectrum of "random" nanoisland film is presented in Figure 3a, the absorption peak at 430 nm corresponds to the SPR in the MIF [12]. Being used as a substrate for Raman spectroscopy this sample is capable to provide a signal enhancement for laser excitation wavelength in the vicinity of 430 nm [13]. It is worth to note that one can move the SPR peak within the range of 420-470 nm via varying the mode of the nanoislands growth, as it was shown in Ref. 10 (see Fig. 3b).

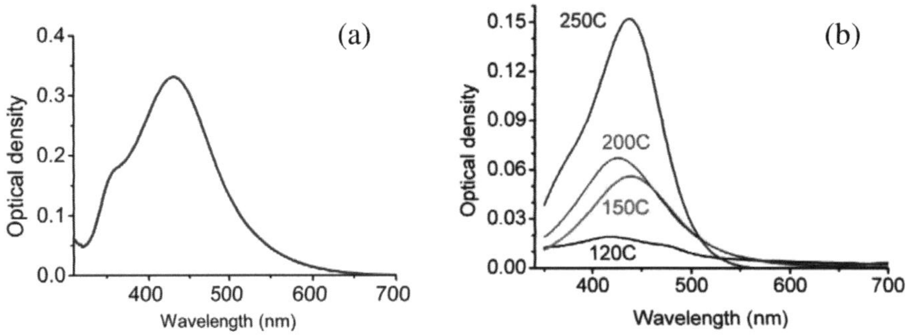

Figure 3. Absorption spectra of the "random" silver nanoisland film under study (a) and of MIF prepared under different annealing temperature (b) [10] (adopted from S. Chervinskii, V. Sevriuk, I. Reduto, A. Lipovskii *Journal of Applied Physics* **114**, 224301 (2013). Copyright 2013, American Institute of Physics).

The dark field spectra and corresponding AFM images of several single nanoislands are shown in Figure 4. All chosen nanoislands have approximately the same round shape in the plane, but vary in height. One can see that the height difference of a few nanometers results in a shift of the SPR peak by about 10 nm. In comparison with "random" silver nanoislands film, the system of single nanoislands is more flexible to control the SPR position and change spectral properties of samples. Detailed studies and modeling of the SPR in single nanoisland structures are presently in progress.

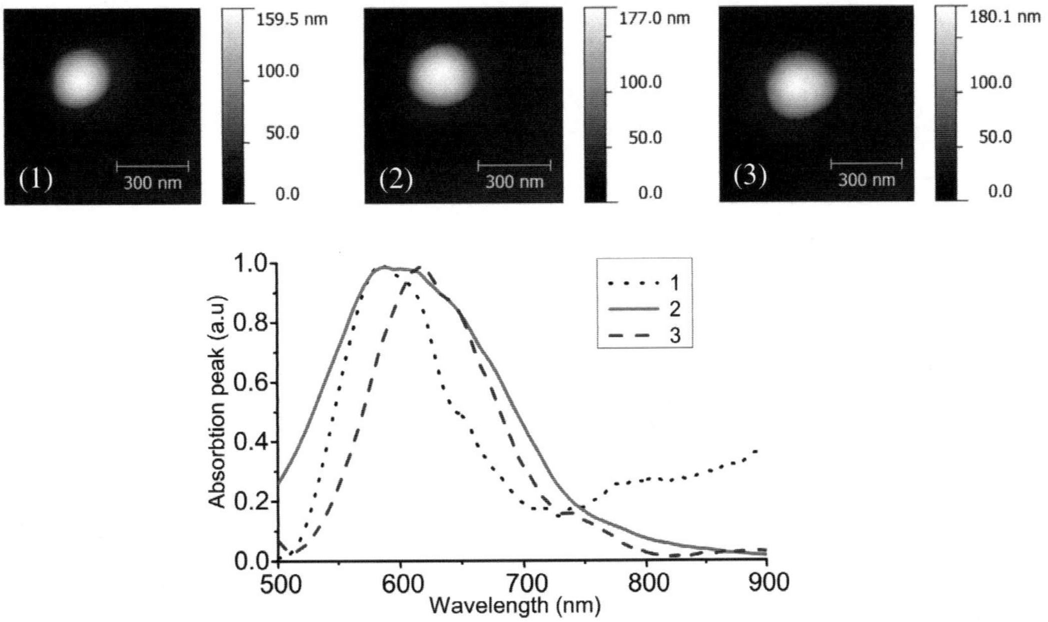

Figure 4. AFM images and corresponding dark field spectra of several single nanoislands.

4. Silver nanoisland films as SERS substrates

We tested the self-assembled silver nanoisland film as a substrate for surface enhanced Raman spectroscopy (SERS) using a dye Rhodamine 6G (R6G) and a protein Bacteriorhodopsin embedded in purple membranes (BR in PM) [14]. For the measurements we used a confocal Raman spectrometer Horiba LabRAM HR UV-VIS-NIR and a frequency-doubled (532 nm) continuous-wave Nd:YAG laser. The laser beam was focused at the sample surface with a 50x/0.55 objective lens. The beam waist was approximately 1.5 μm.

The Raman spectrum shown in corresponds to the monomolecular layer of R6G prepared via drying a drop of 10 μl 10^{-6} M R6G aqueous solution with ~ 7 mm diameter of the spot remained on the substrate. The registered R6G spectrum is in a good agreement with literature [15]. The Raman spectrum from the R6G monolayer deposited on a virgin glass could not be measured because of extra low level. The absence of the reference spectrum did not allow the evaluation of the signal enhancement using standard procedure [16]. Nevertheless, achieved sensitivity is comparable with the reported one for a similar plasmonic structure and measurements conditions [17].

Figure 6 presents BR in PM Raman spectra obtained after drying of 5 μl droplets of water with the membrane concentration of 0.48 and 0.048 mg/ml deposited on both the prepared substrate with "random" nanoislands and a virgin glass. The droplets were dried in the dark at room temperature to form spots of diameter ~ 3 mm. For the concentration of 0.48 mg/ml characteristic Raman spectra corresponding to known purple membranes spectra [18] were registered using both pure glass and silver nanoisland film. One can see more pronounced Raman peaks in the SERS spectrum, for example at ~858 cm^{-1}, ~1375 cm^{-1}, and ~1610 cm^{-1}, in contrast to Raman signal obtained from BR deposited onto the virgin glass. This difference may be because of the influence of silver nanoparticle plasmonic field on retinal photoisomerization process in BR [19, 20] and the dependence of electromagnetic enhancement factor on the distance between chemical bonds of the analyte and the metal surface [21]. The reduction of the BR in PM concentration by an order of magnitude resulted in resolved Raman signal only from the substrate with nanoislands.

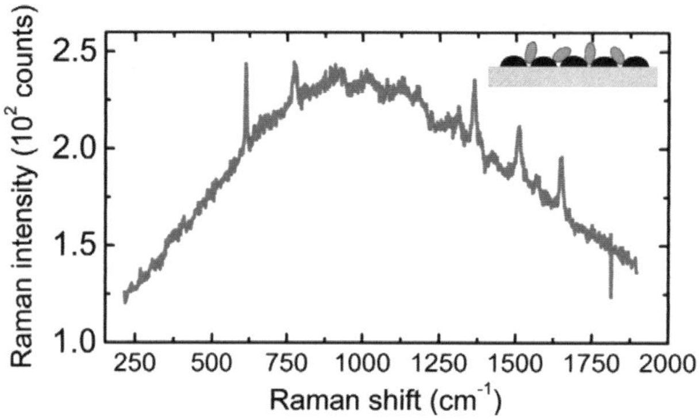

Figure 5 R6G SERS spectrum acquired at laser power of 20 nW, 30 s exposition, 1.5 μm beam waist. Adopted from [14]. Inset: positioning of R6G between the nanoislands.

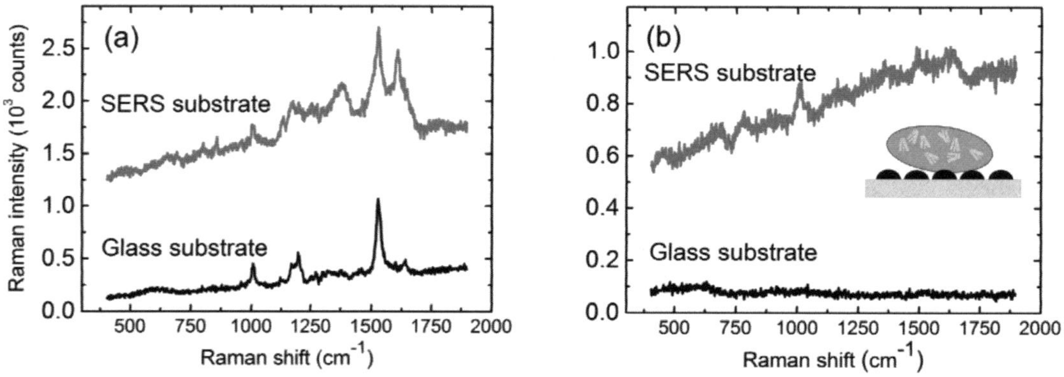

Figure 6.Raman spectra of BR in PM in the concentrations of (a) 0.48 and (b) 0.048 mg/ml. The black and red curves correspond to reference Raman spectra of BR on pure glass substrate and SERS spectra of BR deposited on the nanoisland film respectively. The spectra were acquired at laser power of 200 μW, 50 s exposition, 1.5 μm beam waist. Adopted from [14]. Inset: positioning of BR in PM relatively to nanoislands.

It is also worth to note that a purple membrane thickness is about 5 nm and typical lateral size is of the order of several hundreds of nanometers [22]. Therefore, contrary to R6G, BR in PM is unable to penetrate between the nanoislands where the local enhancement of the electric field of the light wave is maximal as illustrated with the insets in and Figure 6. These results in an essentially weaker Raman signal from BR in PM compared to one obtained from R6G molecules despite the fact that semispherical form of the silver nanoislands provides the highest enchantment in comparison with other design of silver nanostructure [23]. This feature of analytes containing bigger structural units should be taken into account for optimal choice or design of SERS substrates [24].

5. Conclusions

We fabricated the self-assembled silver nanoisland film by means of diffusion during silver-sodium ion-exchange and following hydrogen annealing and isolated single silver nanoislands using thermal poling with profiled electrode as the intermediate step. The samples demonstrated different surface distribution of nanoislands, with size up to 50 nm for nanoislands in MIF and of 250 nm for the isolated nanoislands. This resulted in differing optical characteristics: the SPR peak position and the position dependence on nanoislands size.

Performed SERS measurements of the R6G dye and a protein allowed us to verify the capability of the fabricated silver nanoisland films to enhance Raman signal from different type of analytes. The obtained results properly correspond to known Raman spectra of these analytes.

Acknowledgements

This work was supported by Russian Academy of Science, the Ministry of Education and Science of the Russian Federation project 16.1233.2014/K and Program "5-100-2020" (F. H, A. K. S and S. D. C). E. S. P is also grateful to the Foundation for Assistance to Small Innovative Enterprises.

References

[1] Bantz K C, Meyer A F, Wittenberg N J, Im H, Kurtulus O, Lee S H, Lindquist N C, Oh S-H and Haynes C L 2011 Phys. Chem. Chem. Phys. 13 11551–67
[2] Zhurikhina V V, Brunkov P N, Melehin V G, Kaplas T, Svirko Y, Rutckaia V V and Lipovskii A A 2012 Nanoscale Res. Lett. 7 676
[3] Chervinskii S, Matikainen A, Dergachev A, Lipovskii A and Honkanen S Nanoscale Res. Lett. 9 398
[4] Dieringer J A, McFarland A D, Shah N C, Stuart D A, Whitney A V, Yonzon C R, Young M A, Zhang X and Van Duyne R P 2006 Faraday Discuss 132 9–26
[5] Aroca R and Martin F 1985 J. Raman. Spectrosc. 16 156–62
[6] Horikoshi S, Matsumoto N, Omata Y, Kato T. 2014 J. Appl. Phys. 115, 193506
[7] Royer P, Goudonnet J P, Warmack R J and Ferrell T L 1987 Phys. Rev. B. 35 3753
[8] Camden J P, Dieringer J A, Wang Yingmin, Masiello D J, Marks L D, Schatz G C, Van Duyne R P 2008 J. Am. Chem. Soc. 130 12616
[9] Smith D L 1995 Thin-film Deposition. Principles and Practice (New York: McGraw-Hill)
[10] Chervinskii S, Sevriuk V, Reduto I and Lipovskii A 2013 J. Appl. Phys. 114 224301
[11] Menzel-Glaser: Microscope slides
[12] George M 2010 Chem. Rev. 110 111–31
[13] Kreibig U and Vollmer M 1995 Optical Properties of Metal Clusters (Berlin: Springer)
[14] Heisler F, Piliugina E, Chervinskii S, Samusev A and Lipovskii A A 2014 Sci. Tech. J. Inf. Technol. Mech. Opt. 5 18–22
[15] Zhao J, Jensen L, Sung J, Zou S, Scharz G C, Van Duyne R P 2007 J.Am.Chem.Soc. 129 7647–56.
[16] Le Ru E C, Blackie E, Meyer M and Etchegoin P G 2007 J. Phys. Chem. C 111 13794–803
[17] Chursanova M V, Dzhagan V M, Yukhymchuk V O, Lytvyn O S, Valakh M Ya, Khodasevich I. A, Lehmann D, Zahn D R T, Waurisch C, Hickey S G 2010 Nanoscale Res. Lett. 5 403–9
[18] Terner J, Campion A and El-Sayed M A 1977 Proc. Natl. Acad. Sci. 74 5212–16
[19] Yen C-W, Chu L-K, El-Sayed M A 2010 J. Am. Chem. Soc. 132 7250–1
[20] Biesso A, Qian W, El-Sayed M A 2008 J. Am. Chem. Soc. 130 3258–9
[21] Kennedy B J, Spaeth S, Dickey M, Carron K T 1999 J. Phys. Chem. B 103 3640–6
[22] Lanyi J K 2004 Annu. Rev. Physiol. 66 665–88
[23] Zhu Z, Li Q, Bai B and Fan S 2014 Nanoscale Res. Lett. 9 25–32
[24] Santoro G, Yu S, Schwartzkopf M, Zhang P, Sarathlal Koyiloth Vayalil, Risch J F H, Rübhausen M. A., Hernández M, Domingo C and Roth S V 2014 Appl. Phys. Lett. 104 243107–1

Journal of Physics: Conference Series **661** (2015) 012035 doi:10.1088/1742-6596/661/1/012035

Electron beam modification of glasses containing Ag and Au nanoparticles

M A Prosnikov[1], A I Sidorov[2], O A Podsvirov[3]

[1]Ioffe Physical Technical Institute of RAS, Polytekhnicheskaya 26, St.Petersburg 194021, Russian Federation
[2]ITMO University, Kronverskiy 49, St.Petersburg 197101, Russian Federation
[3]St.Petersburg Polytechnic University, Polytekhnicheskaya 29, St.Petersburg 195251, Russian Federation

E-mail: yotungh@gmail.com

Abstract. The goal of our work was the investigation of the processes of interaction electrons (5-30 keV) with metal nanoparticles in glasses. The objects of our research were the silicate glasses containing Au and Ag nanoparticles. The experiments have shown that electron irradiation with electron energy 5-10 keV of glass containing Au nanoparticles results in the formation of high reflective gold film about 50 nm thick on a glass surface. Electron irradiation (5-30 keV) of glass containing Ag nanoparticles results in the increase of optical density in the irradiated areas and formation of silver clusters (Ag_x, $x<5$) in a form of a ring at periphery. Thus, electron beam can be used to produce photonic and plasmonic devices without etching mask.

1. Introduction

The limitation of modern electronics is huge ohmic losses at high frequencies caused by skin-effect. In 1969, Mayer proposed to use light to transmit signals in planar waveguides [1] at optical frequencies, but that imposes restrictions on the minimum size of devices, due to the diffraction limit. An alternative to integrated optics and dielectric photonics proposes the use of nanoplasmonics, with the collective electron oscillations at optical frequencies exciting at nanoscale metal/dielectric interface [2]. The traditional methods of such devices creation, which are mainly the complex shape thin metal films on dielectric substrate, are the optical or electron beam lithography, are multistage and laborious that restrains the development of nanoplasmonics.

In our previous works we demonstrate the possibility of the creation of Cu [3] and Ag [4] nanoparticles in glasses, V nanorods at glass surface [5], and the dissolving of Ag thin film in silica glass [6], under the effect of electron beam irradiation. Thus, electron beam is the powerful and

[1] To whom any correspondence should be addressed.

precision instrument for glass surface or subsurface volume modification, depending of electron energies.

2. Experimental

In the experiments we used: gold-ruby optical glass containing Au nanoparticles (10-50 nm) based on the SiO_2-B_2O_3-CaO-Na_2O-SnO_3-As_2O_3-K_2O system with 0.025 mol% Au; and the glass containing silver nanoparticles based on Na_2O – ZnO – Al_2O_3 – SiO_2 – NaF – NaCl, where Ag was introduced by ion exchange from $AgNO_3$:$NaNO_3$ (5:95) melt at 360 °C within 30 minutes. After thermal treatment at 500 °C during 1 hour Ag nanoparticles were formed in ~30 um subsurface layer. These samples are intense dark-red. It should be noted, that thermal treatment was not carried out in the reducing atmosphere (e.g. hydrogen), thus not all silver ions were reduced. Therefore the glass contains Ag nanoparticles, ions, atoms, neutral and charged small molecular clusters. Both samples were synthesised and prepared in ITMO University and represent the plane-parallel polished plates 2 mm thick. To remove the surface charge that appeare on glass during electron irradiation, samples were coated with an Al layer 100 nm thick, which was removed after the irradiation by chemical etching in 10% KOH.

Electron irradiation was performed in high-current JEBD-2 scanning electron microscope at electron energy 3-35 keV with the electron current density 40 $\mu A/cm^2$. The electron irradiation dose was varied in a range 20-500 mC/cm^2. The irradiation was carried out at room temperature and according to our calculation, the surface of the samples was heated at less than 250 °C during irradiation at the chosen current and energies.

Heat treatment was performed in computer-assisted muffle furnace (Nabertherm). The optical density spectra were measured by Cary 500 (Varian) spectrophotometer at 300-800 nm spectral region. Luminescence spectra and images were obtained using luminescence microscope MSFU-K (LOMO) with excitation at 400-440 nm. The morphology and thickness of the formed films were studied with ASM Solver-Pro (NT-MDT).

3. Results and Discussion

3.1. Glass containing Ag nanoparticles

After the irradiation of the Ag nanoparticles containing samples by electrons (E = 5-30 keV, Q = 20 and 50 mC/cm^2) one can observe the increase of optical density at irradiated areas. Microphotograph of one of this areas is shown at figure 1. Cross section of absorption through the spot is shown at figure 2. It can be seen the increased absorption at irradiated spot center and, more significantly, on its periphery.

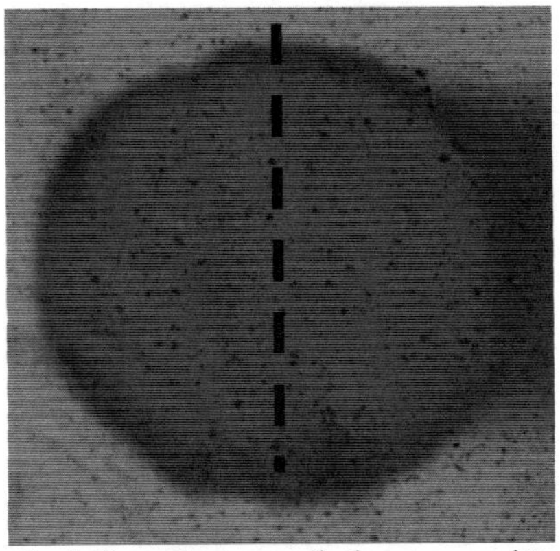

Figure 1. Microphotograph by transmitted light of irradiated zone after electron irradiation with energy 30 keV and dose 50 mC/cm^2. Diameter of spot is 2 mm.

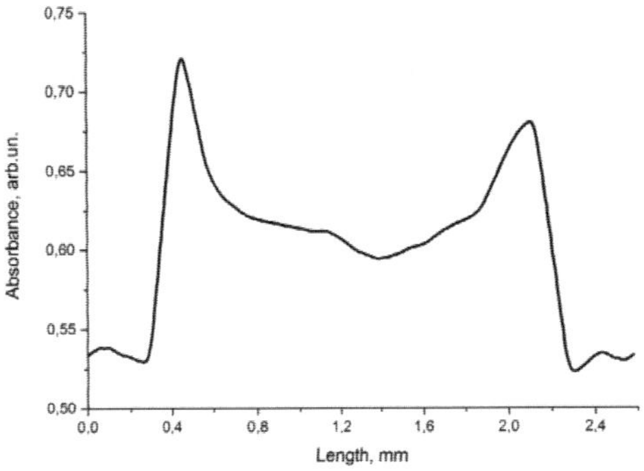

Figure 2. Cross section of absorption through dashed line across the spot at Fig. 1.

The absorption spectra of the irradiated areas for electron energies 5, 10, 20 and 30 keV with the same dose 50 mC/cm^2 are shown at figure 3. The decrease of optical density at 300 nm region can be caused by the reduction of Ag ions and charged molecular clusters [7] by thermalized electrons. The increase of optical density at 350 nm is due to the formation Ag atoms [7], and the deformed band at 440-480 nm results from the formation of Ag nanoparticles. It should be noted, that the high concentration of nanoparticles leads to their dipole interaction, which causes the offset and deformation of plasmon band [8].

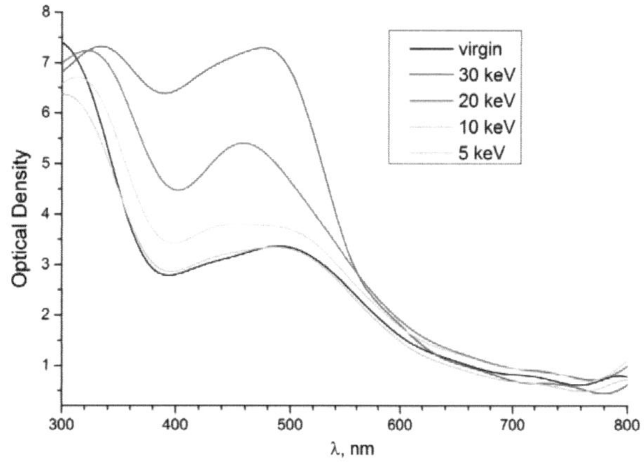

Figure 3. Absorption spectra of the irradiated zones after irradiation with various electron energies (shown at inset) with the same dose - 50 mC/cm^2.

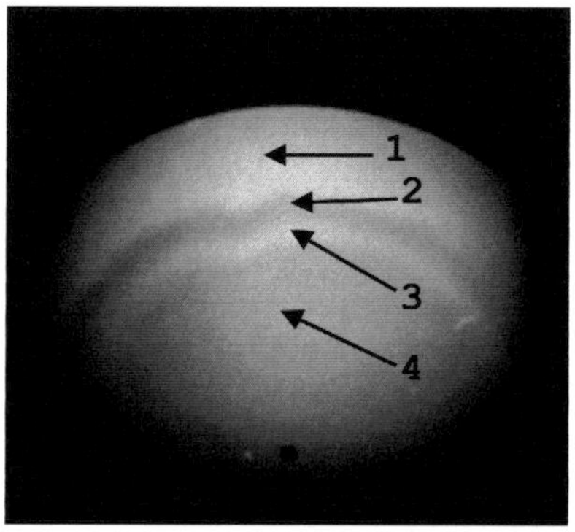

Figure 4. Photo of luminescence of irradiated zone shown at figure 1. Arrows and numbers indicate areas where luminescence spectra at figure 5 were obtained.

Luminescence photograph of one of the spots is shown at figure 4. One can see four zones: 1 – not irradiated zone, 2 – dark ring at periphery, 3 – bright ring, 4 – dark centre of spot. For each of these zones luminescence spectra were obtained (figure 5). For the excitation wavelengths 400-440 nm the contribution to the luminescence makes the neutral silver molecular clusters Ag_2, Ag_3, Ag4 [9,10]. That way, figure 4 gives the information on the concentration of neutral molecular clusters in the irradiated spot.

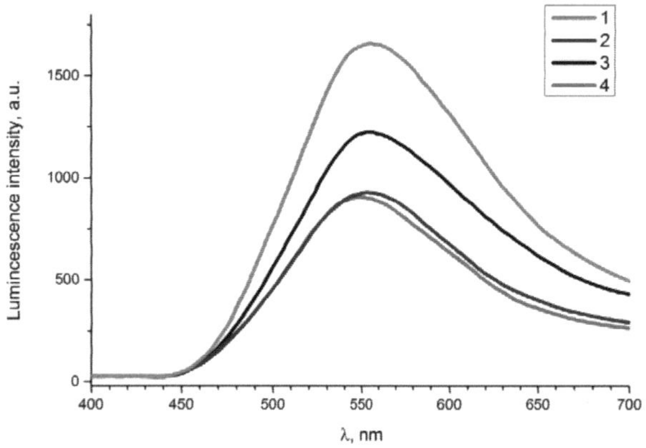

Figure 5. Luminescence spectra of the four zones at figure 4. Excitation wavelength is 400-440 nm.

Electric field strength in glasses due electron irradiation can reach 100 kV/cm [11]. Electric field can cause the decomposition of nanoparticles with metal ion formation, which effect was demonstrated in

[12]. Further, the positive Ag ions can migrate into the region of negative volume charge due to the bell-shaped current distribution in an electron beam cross-section, as is described in [13]. Then the Ag ions are reduced by thermalized electrons and can form nanoparticles of molecular clusters.

3.2. Glass containing Au nanoparticles

After the irradiation of the ruby-glass samples by electrons (E = 5, 10 keV, Q = 500 mC/cm^2), one can observe highly reflective gold films at the irradiated areas, which are shown at figure 6.
AFM showed that the created films are about 50 nm thick.
The effect of solid film appearance at the glass surface due electron irradiation can be described as in previous part, except for periodically volume charge depth, ~200 nm for E = 5 kev vs. ~10 μm for E = 35 keV. The periodically volume charge causes the high electrical field strength [11] wherein gold nanoparticles can decompose [12] becoming ions and which migrate to the negative charge area close to the glass surface, where ions can be reduced by thermalized electrons. Then, Au atoms come out to the surface, creating solid gold film. The higher the energy, the deeper the charge and the lower gold atoms leaves to surface, as is shown at figure 6.

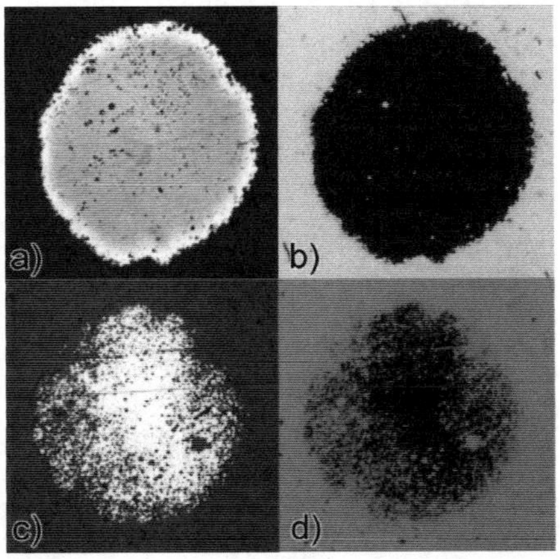

Figure 6. Microphotographs by reflected (a, c) and transmitted (b, d) light of irradiated zones after electron irradiation with a, b – 5 keV, c, d – 10 keV, dose 500 mC/cm^2. Spots diameter are 2 mm.

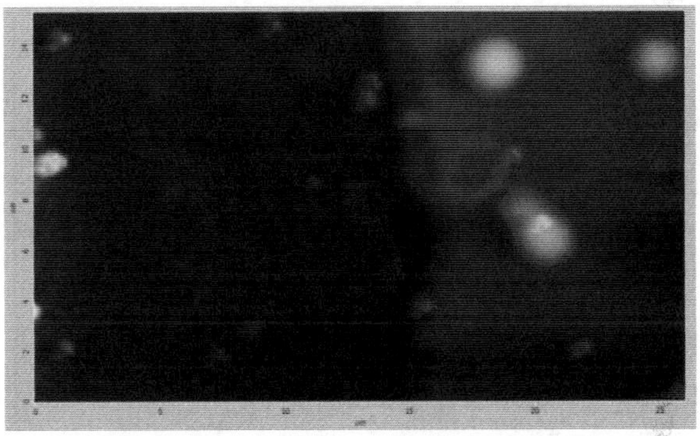

Figure 7. AFM image of periphery of created gold film (right) on ruby-gold glass (left). The film is ~50 nm thick.

4. Conclusion

It is shown experimentally that the action of electron beam irradiation (E>10 keV) on the Ag nanoparticle containing glasses results in the formation of high concentration molecular clusters and nanoparticles in a ring at the periphery of the irradiated area. The higher the energy, the more clearly the evident effect.
The action of electron beam irradiation (E = 5, 10 keV) on the Au nanoparticle containing glasses results in the formation highly reflective ~50 nm (island-like for E = 10 keV, and continuous for E = 5 keV) gold film at the glass surface.

Both effects are caused by the creation of periodically volume charges due the electron irradiation, the decompose of metal nanoparticles in high strength electrical field, by the ion field migration and by their reduction by thermalized electrons.
The observed effects may be used for the creation of photonic and plasmonic devices without the need of etching masks, at any modern electron lithography setup.

Acknowledgments
The authors would like to thank Kretusheva I. and Baryshnikova M. for their assistance with SEM and AFM microscopy analysis.

References

[1] Miller S 1969 *The Bell Systems Technical Journal* **48** 2059
[2] Ozbay E 2006 *Science* **311** 189
[3] Podsvirov O A, Sidorov A I, Tsekhomskii V A, Vostokov A V 2010 *Physics of the Solid State* **52** 1906
[4] Ignat'ev A I, Nashchekin A V, Nevedomskii V M, Podsvirov O A, Sidorov A I, Solov'ev A P, Usov O A 2011 *Technical Physics* **56** 662
[5] Podsvirov O A, Prosnikov M A, Brunov V A, Ignatiev A I, Sidorov A I 2014 *Proc. Vacuum Electron Sources Conference (IVESC)* (St.Petersburg: IEEE) p 1
[6] Brunov V S, Podsvirov O A, Sidorov A I, Prosnikov M A 2014 *Technical Physics* **84** 126 (in russian)
[7] Ershov B G, Janata E, Henglein A, Fojtik A 1993 *The Journal of Physical Chemistry* **97** 4589
[8] Tolga A, Jung-Hoon S, Arto V N 2004 *Nano letters* **4** 1627
[9] Klyukin D A, Sidorov A I, Ignatiev A I, Nikonorov N V 2014 *Optical Materials* **38** 233
[10] Simo A, Polte J, Pfänder N, Vainio U, Emmerling F, Rademann K 2012 *Journal of the American Chemical Society* **134** 18824

[11] Touzin M, Goeuriot D 2006 *Journal of applied physics* **99** 114110
[12] Lipovskii A A, Melehin V G, Petrikov V D 2006 *Technical physics letters* **32** 275
[13] Sidorov A I, Nashekin A V, Nevedomskiy V N, Usov O A, Podsvirov O A 2011 *International Journal of Nanoscience* **10** 1265

Local analysis of porous silicon structure fabricated by non-traditional approach

V Vorobev[1], Y Osin[1], M Ermakov[2,3], V Nuzhdin[2], V Valeev[2], D Tayurskii[1], A Stepanov[1,2]

[1]Kazan Federal University, Kazan, Russia
[2]Kazan Physical-Technical institute of RAS, Kazan, Russia
[3]Pacific National University, Khabarovsk, Russia

E-mail: slava.v679@gmail.com

Abstract. A comparison of experimental electron backscattering diffraction patterns for porous Si formed by ion implantation and thermal annealing is presented. For this purposes Ag-ion implantation into monocrystalline c-Si substrates at energy of 30 keV with dose of 1.5×10^{17} ion/cm^2 was carried out. Surface nanoporous Si structures were studied by scanning electron microscope imaging and electron backscattering diffraction. Amorphization of Si after implantation and recrystallization of porous Si after annealing is observed. Ion implantation is suggested to be effective technique for a formation of nanoporous semiconductor layers, which could be easily combined with the crystalline substrate matrix for various applications.

1. Introduction

Main fields of application with porous silicon (PSi) are microelectronics, optical waveguide, photo sensor and solar cells[1]. The first time PSi was produced and characterized by A. Uhlir in 1956. At that time PSi was formed by electrochemical etching of Si [2].

At the present the electrochemical etching is most applicable technic for fabrication of PSi. Recently, in 2013 it was suggested to use a novel technological way for formation PSi based on low-energy high-dose implantation of noble metal ions [3,4]. Ion implantation widely used in modern industrial process of semiconductor microelectronics. Thus such physical method of creation PSi could easily integrated in industrial semiconductor process.

The precise control of crystalline structure is one of the important tasks in science especially in the field of new material fabrication. For this purpose various analysis methods such as electron microscopy [5-7], Raman scattering and X-ray diffraction are applied. Every of this methods have advantages, but also have some limitation.

Aim of present work is a creating of thin semiconductor layers PSi on surface of a monocrystalline c-Si used Ag^+-ion implantation combined with postimplantation thermal annealing and structural analyze of this materials by electron backscattering diffraction method.

2. Experimental details

A (100)-oriented single crystalline Si wafer as the substrate was used for Ag^+-ion implantation to create PSi structures. The substrates were cleaned in a wet chemical etching process. The semiconductor wafers were implanted with Ag^+ ions at energy of 30 keV and the ion current density of 8 μA/cm^2 with dose of 1.5x10^{17} ion/cm^2 using the ion implanter ILU-3 at residual vacuum of 10^{-5} Torr and room temperature of the irradiated samples. Thermal annealing was realized with VUP-5M during 30 minute at 600°C.

Fabrications of PSi samples were carried out in Kazan Physical-Technical institute. Analyze of local structure and observation of surface sample morphology were performed with scanning electron microscopy (SEM) Merlin (Carl Zeiss), which combined with a detector of electron backscattering diffraction (EBSD) Norlys HKL (Oxford Instruments). Microscopy measurements were carried out in Interdisciplinary Center for Analytical Microscopy of Kazan Federal University.

3. Results and discussion

EBDS used in practice for control various structural parameters of materials [6]. EBDS determines a thickness of sample, which participate in capture diffraction picture. Modeling of profile intensity distribution of backscattering electrons for Si was realized by the Casino 2.48 software for case of accelerating voltage 20 keV and tilt angle 70 to surface sample respective horizontally (Fig. 1). Thus, transmitted analyzing thickness of sample is about 300 nm

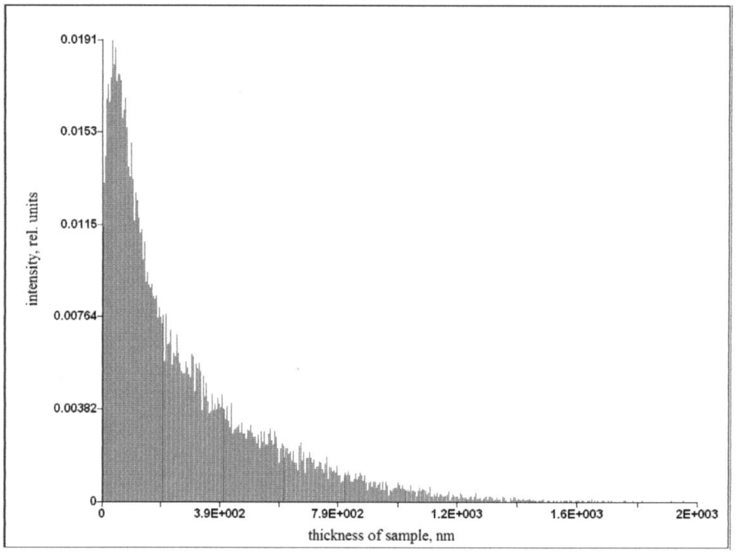

Figure 1. Intensity distribution backscattering electrons in Si

A plane-view SEM image of unimplanted Si wafer looks likes as very smooth without any surface structural inhomogeneity (Fig. 2a). Experimental EBDS pattern of c-Si shows typical monocrystalline Kikuchi structure (Fig. 2b). Identification of Kikuchi band was carried out in program Aztec 2.1 and demonstrated that surface crystal orientation is (100). A plane-view SEM image of PSi formed by Ag-ions with dose 1,5x10^{17} ion/cm^2 is presented in Fig. 3a. In contrast to unimplanted Si (Fig. 2a) the morphology of PSi surface characterized by a structure with the black holes on the implanted Si. On experimental electron backscattering pattern (Fig. 3b) In EBDS image seems asa superposition of low intensity Kikuchi band of monocrystalline substrate and diffusely ring from formed PSi layer.

Thermal annealing was carry out at 600°C in vacuum in order to recrystallize PSi structure. Porous Si layers after annealing procedure is shown Fig. 4a and this image demonstrates some consistence with PSi created by implantation. Diffraction pattern in Fig. 4b shows reconstruction Kikuchi band of annealed PSi to cubic syngony. Identification Kikuchi band indicates about appearance of crystallographic orientation modified PSi to be different from (100).

Journal of Physics: Conference Series **661** (2015) 012036 doi:10.1088/1742-6596/661/1/012036

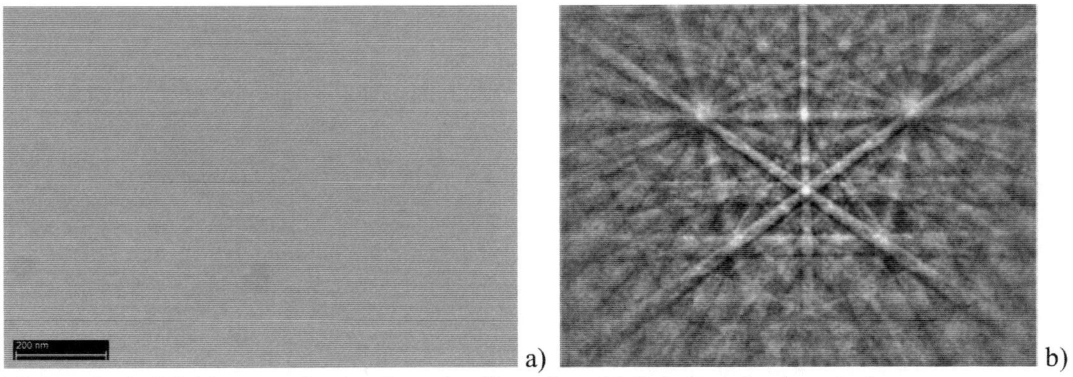

Figure 2. a) SEM image of unimplanted c-Si surface, b) Eelectron backscattering pattern of c-Si.

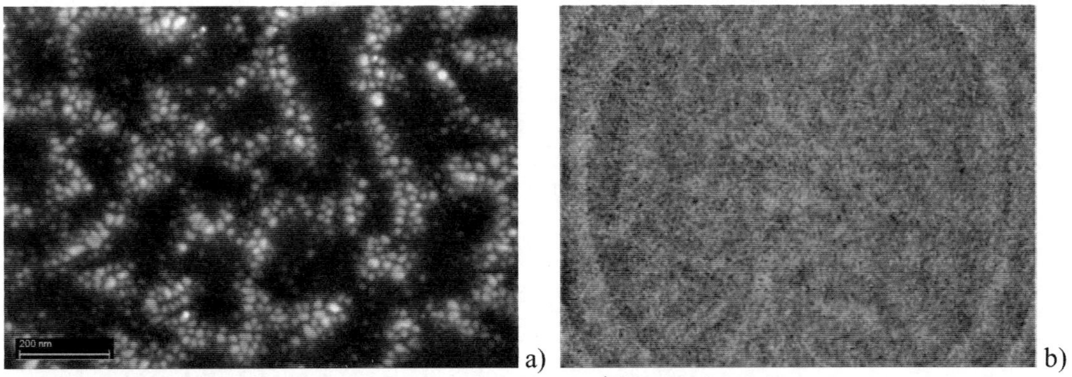

Figure 3. a) SEM image of surface PSi formed by Ag^{+}-ion implantation, (b) Eelectron backscattering pattern of PSi

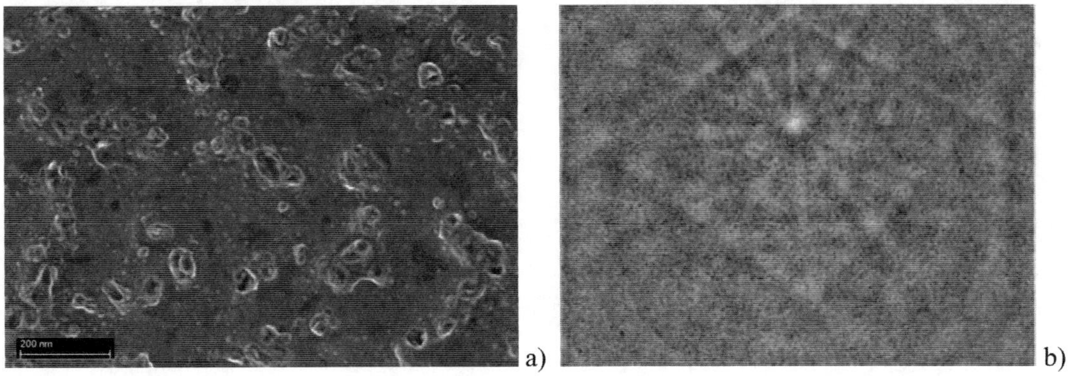

Figure 4. (a) SEM image of annealed PSi (a), and (b) Electron backscattering pattern of annealed PSi

4. Conclusions

In this work PSi layer using high dose low energy Ag-ion implantation of c-Si was fabricated and epyт modified by thermal annealing. For the first time a structure analysis of surface layers PSi by EBDS was carry out. It was concluded that after Ag-ion implantation the surface layer PSi is amorphous which could be partly recrystallized by annealing. Crystallographic structure of annealed PSi is different from unimplanted c-Si (100).

Acknowledgment
This work was partly supported by the Russian Foundation for Basic Research Nos. 13-02-12012 and 14-32-50301. A.L.S. is grateful to the Russian Scientific Foundation No. 14-13-00758.

References
[1] M. J. Sallor, Porous Silicon in Practice: Properties, Characterization and Applications 2011*Welnheim: Willey-VCH*
[2] A. Uglir Electrolytic shaping of germanium and silicon 1956 *Bell Syst. Tech. J.* **35** 333-347
[3] A.L. Stepanov, A.A. Trifonov, Y.N. Osin, V.F. Valeev, V.I. Nuzhdin New way for synthesis of porous silicon using ion implantation 2013 *Optoelectronics and Advanced Materials – Rapid Communications* **7** 9-10 692-697
[4] A.L. Stepanov, A.A. Trifonov, Y.N. Osin, V.F. Valeev, V.I. Nuzhdin Fabrication of Nanoporous Silicon By Ag^{+}-Ion Implantation 2013 *Nanoscience and Nanoengineering* **1** 3 134-138
[5] Angus J. Wilkinson, Peter B. Hirsch Electron diffraction based techniques in scanning electron microscopy of bulk materials 1997 *Micron* **28** 279-308
[6] A. Schwartz, M. Kumar,B. Adams, D. Field Electron Backscatter Diffraction in Materials Science 2009 *Springer* **2** 403
[7] F.J. Humphreys, Characterisation of fine-scale microstructures by electron backscatter diffraction 2004 *Scripta Materialia* **51** 771–776

Explosive crystallization of PZT microstructures by femtosecond infrared radiation

A S Elshin[1], N Yu Firsova[1], V I Emelianov[2], I P Pronin[3], S V Senkevich[3], O M Zhigalina[4] E D Mishina[1] and A S Sigov[1]

[1] Moscow State Technical University of Radio Engineering, Electronics and Automation, Vernadskogo av. 78, 119454, Moscow, Russia.

[2] Lomonosov Moscow State University, Leninskie Gory st., 1, 119991, Moscow, Russia.

[3] Ioffe Physical-Technical Institute of the RAS, Politechnicheskaya st. 26, 194021, Saint-Petersburg, Russia.

[4] Shubnikov Institute of Crystallography of the RAS, Leninsky av. 59, 119333, Moscow, Russia.

Email: elshin_andrew@mail.ru

Abstract. The features of microstructure crystallization into perovskite phase in lead zirconate titanate film by femtosecond laser radiation of near-infrared range were discussed. In-situ crystallization kinetics by method of second harmonic generation (SHG) was studied. The presence of several types of crystallization was shown, including ultra-fast (explosive) crystallization occurring immediately after the start of exposure, and slow (self-sustaining) crystallization, occurring after termination of exposure. The advantage of the second-harmonic generation microscopy for the study of annealed microstructures was shown. The morphology of microstructures was investigated by transmission electron microscopy (TEM).

1. Introduction

Thin-film ferroelectric structures form the basis for a new generation of micro- and nanoelectronic devices [1, 2]. For obtaining the ferroelectric phase from the amorphous film it is annealed. Thermal annealing furnace is used in the vast majority of technologies [3].

Laser annealing is used for local crystallization of amorphous films into the ferroelectric phase in order to minimize surrounding functional area elements heating. Upon annealing by excimer laser with photon energies much higher than the band gap, absorption, and hence the annealing occurs in a thin surface layer [4]. In addition, the mode structure of excimer laser spot does not allow localizing crystallization in the submicron region.

In paper [5] we proposed to use for annealing of ferroelectric film on platinized substrate single-mode femtosecond laser with a wavelength that falls in the transparency of the PZT film and consequently absorbs by platinum layer. This method makes it possible to solve three problems simultaneously. Firstly, the annealing conditions are close to the thermal annealing furnace, as the heating of the film is carried out by the platinum. Second, the heating is carried out locally with a Gaussian distribution of temperature along the radius of the laser spot. Third, the use of femtosecond laser allows diagnosing the formation of the ferroelectric phase during the process of annealing. The basis of this diagnosis is the method of optical second harmonic generation (SHG), which is an effective method for studying phase transitions, including the crystallization of perovskite phase. We also show the advantage of the SHG technique for ex-situ visualization of the annealed regions compared with linear-optical imaging.

2. Experimental setup

Film of a solid solution of lead zirconate titanate (PZT) with thickness 700 nm, with a 10 percent excess lead oxide ($Pb(Zr_{0,54}Ti_{0,46})O_3$+10%PbO) was deposited on platinized silicon substrate by RF magnetron sputtering (without annealing) [6].

Femtosecond Ti: sapphire laser («Avesta-Project») was used as the laser source with a wavelength of 800 nm, pulse duration of 100 fs, a repetition frequency of 100 MHz. Duration of annealing exposure was 1 s - 30 s. Confocal microscope Alpha-300 (Witek) was used for beam focusing during annealing, diagnosing and scanning after all. Focusing lens x40 (numerical aperture NA = 0.65) were used and the size of the laser spot on the surface of the film, measured as the width of the constriction was 1 μm. There was no scanning of the laser spot on the sample during annealing and in-situ monitoring. We used a double-beam scheme in reflection geometry with two mixed coinciding beams – for annealing and diagnosing both focused at the same place. The power density of the annealing beam took the value $W_A = 1{,}0$ MW/cm^2, diagnosing – $W_D = 0.2$ MW/cm^2. Annealing and its registration were performed as follows: on the way of the annealing beam (before mixing with the other one) was set shutter which opens only during annealing (for a time τ_A), diagnosing beam exposed the sample throughout the experiment, thereby second harmonic generation is also registered for the entire duration of the experiment. The moment of opening the shutter was set as zero point of time $t = 0$, that is the beginning of annealing. Thus, when $t < 0$ SHG signal was detected from the non-annealed film at a laser power W_D, when $0 < t < \tau_A$ SHG signal was detected during annealing at the total power of annealing and probing beam ($W_A + W_D$), and when $t > \tau_A$, after closing the shutter, SHG signal was detected during cooling / crystallization at laser power W_D. Since the intensity of the SHG is proportional to the square of the incident power, to compare the results of diagnosing for all areas SHG signal was normalized by the square of the corresponding power in each of the intervals $t < 0$, $0 < t < \tau_A$, $t > \tau_A$.

3. Investigations of laser annealing

3.1. In-situ investigation

Before annealing irradiation is turned on, SHG signal from the amorphous PZT film is absent. After the annealing beam is turned on (moment t = 0 on Fig.1 a), in a short time of the order 0.1 second, PZT film is heated to a temperature of 700 K, due to the optical absorption in Pt layer. These values were estimated in COMSOL Multiphysics software. At these times superfast crystallization of the film material occurs with a transition to noncentrosymmetric-perovskite phase, which is accompanied by a sharp rise in the SHG signal. This ultra-fast crystallization apparently does not occur by the classical mechanism of nucleation and growth of nuclei, which works in a conventional (slow) furnace annealing [7], but by the mechanism of superfast explosive crystallization of PZT or transition of "order-disorder" [8].

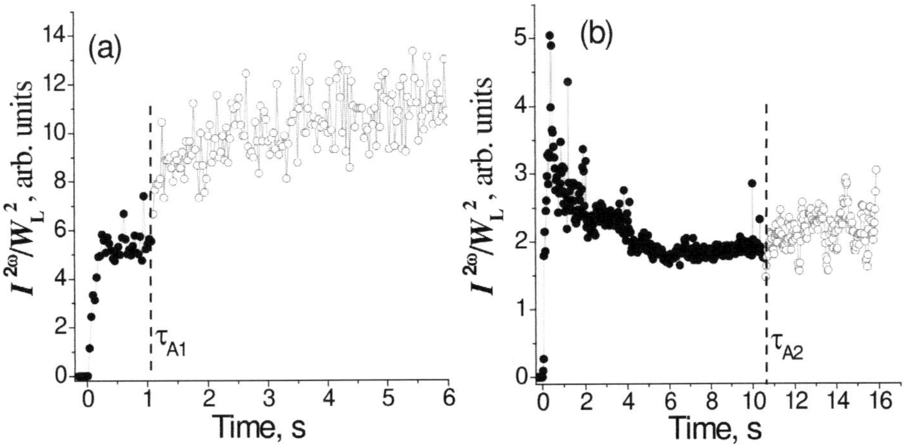

Figure 1. Experimental time dependences of normalized SHG intensities during heating ($0 < t < \tau_A$) and cooling ($t > \tau_A$) at the power density W_A = 1 MW/cm^2, at times τ_{A1} = 1 s, τ_{A2} = 10 s.

Then the growth and the intensity of SHG is saturated and during the entire time of irradiation (τ_{A1} = 1 s), SHG signal remains approximately constant. The inhibition of PZT crystallization due to a sharp increase of the activation energy of crystallization can cause SHG saturation. The large tensile stresses in the amorphous regions surrounding the crystallized areas can be a reason of increasing of activation energy [9]. Such stresses occur due to increased density of the crystallized material [10]. Increasing the annealing time to τ_{A2} (рис. 1 b) lead to a decrease of the SH signal. Reduction of the SHG signal may be connected with the fact that the center of the spot where the temperature is highest may be "overheated" with the formation of defects. After the annealing radiation is turned off sharp (in times of the order or less than 0.1 s) cooling (to room temperature) of the film including the perovskite PZT phase and a Pt layer occurs. Corresponding sharp decrease in their volume take place. Due to the higher thermal expansion coefficient, Pt layer compresses more, so that the PZT film suffers shock compressive stress. We can assume that this shock decrease in activation energy launches slow wave of explosive crystallization, extending from the perovskite phase. Estimates show that the critical temperature of ignition and propagation of a wave may be lower than room temperature [11], due to the huge value of the latent heat of crystallization of PZT, including the elastic interactions energy [12,13], and due to the strain-induced reduction of the activation energy. The described effect is very distinct in the case of short-time annealing τ_{A1} and much weaker with prolonged annealing time. It can be assumed that "overheating" causes defects formation which damps crystallization wave propagation.

3.2. Ex-situ monitoring

To monitor the results of annealing ex-situ confocal optical and nonlinear optical microscopy was used. For imaging the sample was scanned by diagnosing beam with a wavelength of 800 nm with a power density W_D = 0.2 MW/cm^2, wherein a linear microscopy detected emission had wavelength 800 nm, and microscopy SGH - at wavelength 400 nm.

The presence of a defective area in the case of large annealing times is confirmed by SHG microscopy (Fig. 2). For τ_{A1} annealing region is a bright perfect circle, and the cross section is approximated by a Gaussian function with a full width at half maximum FWHM = 1.5 μm, which corresponds to the estimated size of the annealed region. For τ_{A2} annealing region is a circular structure with a dark spot in the center. The decrease in the SH signal in the center of the annealed region indicates a disturbance of the perovskite structure. Line images (Fig. 2 c, d) have the same slightly distorted circular form with a strong blurring of the edge area. Cross sections of line images drawn through the diameter of the spots are approximated by Gaussian functions and differ only by the value FWHM, which is 2 μm and 3 μm for τ_{A1} and τ_{A2}, respectively.

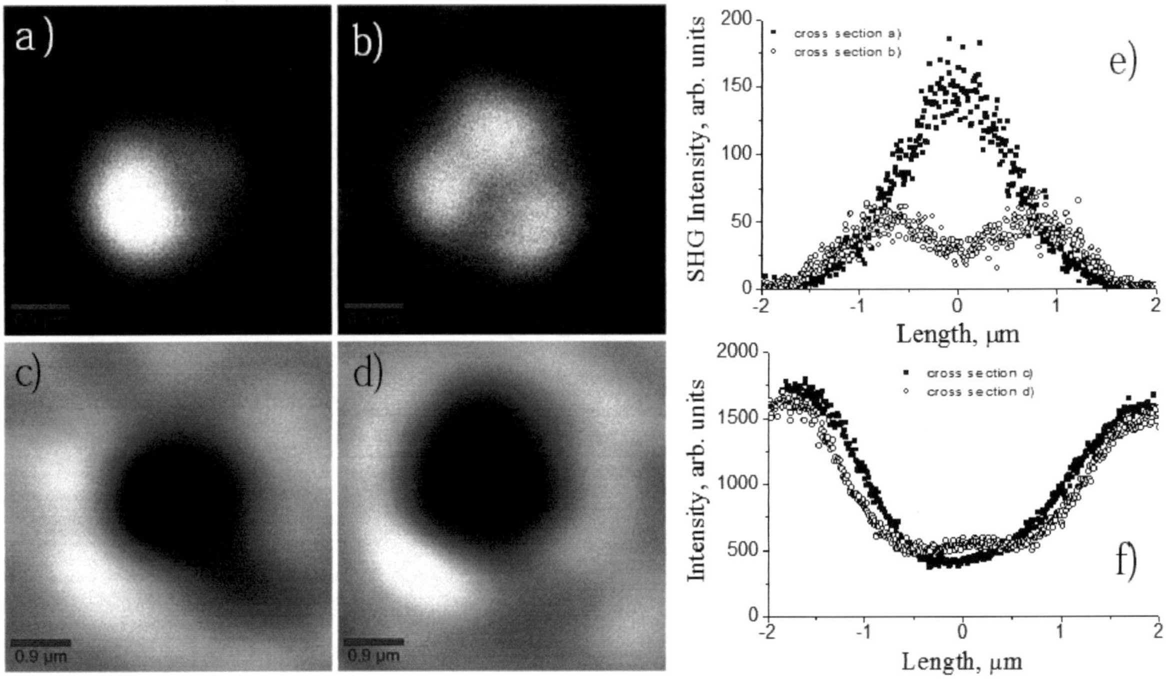

Figure 2. Top images (a, b, e) – nonlinear-optical images of annealed areas (incident wavelength 800 nm and power density WD = 0.2 MW/cm2, detecting wavelength - 400 nm) and their cross-sections. Bottom images (c, d, f) – linear-optical images of annealed areas (incident and detecting wavelength – 800 nm) and their cross-sections; annealing time: a, c – 1 s; b, d – 10 s.

3.3. Transmission electron microscopy

Incisions of obtained microstructures for detailed analysis were made by focused ion beam. These incisions were investigated by scanning electron microscopy (Fig. 3). The film was coated with the layer of platinum to prevent damage to the film by an ion beam during cutting.

The delamination of the film from the lower layer of platinum at some distance on either side of the field of laser annealing was found. This shows the stresses and thermal expansion of the film.

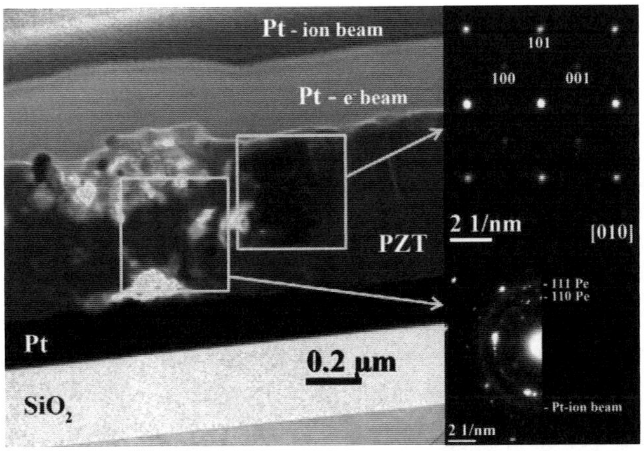

Figure 3. Brightfield TEM image of microstructure incision (left) and corresponding diffraction images of different areas: perovskite polycrystal (bottom), monocrystal [010] (top). Highlights in PZT layer correspons to the pores.

TEM showed that the crystallization region is a hemispherical region centered on the film surface. Big (500 nm) and small (50 nm) perovskite crystals are presented. Some microstructures have pyrochlore and platinum inclusions in PZT layer.

4. Conclusions

Thus, we have proposed a method for the local crystallization of perovskite microstructures in amorphous PZT film by femtosecond laser radiation with monitoring of crystallization kinetics by optical second harmonic intensity. SHG technique helped to identify features of crystallization: its explosive nature in the initial stage of crystallization, as well as a transition to a self-sustaining mode at the termination of the annealing. It is confirmed that for obtaining microstructures with smooth spatial profile of the perovskite phase in addition to the previously detected narrow band power density [14], annealing times range is also narrow. Increasing of the annealing time leads to the formation of ring structures with defective perovskite center, which is easily detected using SHG microscopy. The proposed technique significantly increases the controllability of the process of laser annealing and can be used not only for ferroelectrics, but also for a wide range of materials that crystallizes into noncentrosymmetric phase. This work was supported by the Ministry of Education and Science of Russian Federation (project № 11.144.2014)

5. References

[1] Vorotilov K A, Sigov A S, Romanov A A and Mashevich P R 2010 *Nanomaterials and Nanostructures - XXI Century* **1** pp 45-53.
[2] Shen I Y, Cao G Z, Chia-Che Wu and Cheng-Chun Lee 2006 *Ferroelectrics* **342** p 15-34
[3] Ai-Dong Li, Di Wu, Hui-Qin Ling, Mu Wang, Zhiguo Liu, Naiben Ming 2002 *Journal of Crystal Growth* **235** p 394-400
[4] Khmelenin D N, Zhigalina O M, Vorotilov K A and Lebo I G 2012 *Sol. St. Phys.* **54** 5 p 939-941
[5] Firsova N Yu, Mishina E D, Sigov A S, Senkevich S V, Pronin I P, Kholkin A, Bdikin I and Yuzyuk Yu I 2012 *Ferroelectics* **433** 1 p 164-169
[6] Pronin I P, Kaptelov E Yu, Senkevich S V, Klimov V A, Zaitseva N V, Shaplygina T A, Pronin V P, Kukushkin S A 2010 *Sol. St. Phys.* **52** 1 p 124-128

[7] Huang Z, Zhang Q and Whatmore R W *Journal of Applied Physics* **85** 10 p 7355
[8] Hu H, Peng C J and Krupanidhi S B 1993 *Thin Solid Films* **223** pp 327-333
[9] Tentilova I Yu, Kaptelov E Yu, Pronin I P and Ugolkov V L 2012 *Inorganic Materials* **48** 11 pp 1136–1140
[10] Pronin I P, Kaptelov E Yu, Senkevich S V, Klimov V A, Zaitseva N V, Shaplygin T A, Pronin V P and Kukushkin S A 2010 *Physics of the Solid State* **52** 1 pp 132–136
[11] Emelyanov V I, Panin I M 1993 *Appl. Phys.* **57** pp 561-566
[12] Kukushkin S A, Tentilova I Yu, Pronin I P 2012 *Sol. St. Phys* **54** 3 pp 571-575
[13] Kukushkin S A and Osipov A V 2014 *J. Phys. D: Appl. Phys.* **47** 313001
[14] Firsova N Yu, Elshin A S, Marchenkova M A, Bolotov A K, Ivanov M S, Pronin I P, Senkevich S V, Kiselev D A, Mishina E D 2014 *Journal of Nano and Microsystem Technique* **7** pp 43-46

The efficiency of UV LEDs based on GaN/AlGaN heterostructures

A S Evseenkov[1], S A Tarasov[1], S Yu Kurin[2], A S Usikov[3,4], B P Papchenko[3], H Helava[4], Yu N Makarov[2,4] and A V Solomonov[1]

[1]Saint-Petersburg Electrotechnical University "LETI", 5 Prof. Popov Str., 197376 St. Petersburg, Russia
[2]Nitride Crystals Group Ltd., pr. Engel'sa 27, St. Petersburg 194156, Russia
[3]University ITMO, Kronverkskiy pr. 49, St. Petersburg 197101, Russia
[4]Nitride Crystals Inc., 181 E Industry Court, Suite B, Deer Park, NY 11729, USA

E-mail: as.evseenkov@gmail.com

Abstract. The UV LED GaN/AlGaN heterostructures obtained by HVPE approach were investigated. It was shown that the peak wavelength of UV LEDs was in the range of 360-380 nm with FWHM of 10-13 nm. At operating current of 20 mA, the active region temperature T_j was 43°C, the output optical power and efficiency - 1.14 mW and 1.46%, respectively. It was shown that the use of HVPE method allowed to achieve a high degree of structural perfection of epitaxial structures.

1. Introduction

In modern optoelectronic industry in order to create LED structures, the compositions based on solid solutions of semiconductor nitrides are used. Their most important characteristic is the conversion efficiency of electrical energy into optical radiation. The main problems of UV LEDs are low quality of epitaxial heterostructures [1, 2], low light extraction index [3], and self-heating effect. Addressing these problems will allow us to create high-quality GaN/AlGaN heterostructures for such promising areas as photoelectrochemical hydrogen generation, photopolymerization, medicine.

The main factor limiting the light extraction from the chip and reducing the efficiency of LEDs is total internal reflection [4, 5] at the interface of materials with high and low optical density. The effective way to solve this issue is introducing into the structure the surfaces that scatter light. When creating a specific shape roughness the light extraction index increases due to a significant increase in the critical angle. At present time, the texturing of the substrates and epitaxial layers is widely used [6]. The texturing can be created using a variety of methods: the technique of wet etching in combination with the Laser-lift-off technology [7], creation of microoptical elements on a sapphire substrate [8], micromachining [9] etc.

The experimental study of the optical and electrical properties of UV LEDs based on GaN/AlGaN [10-12] heterostructures has been carried out. Special attention was paid to investigation of the electroluminescence spectra at different forward currents and temperatures and determination of optical power and luminescence efficiency.

2. Experimental results

The main objective of the experiment was to investigate the structural perfection and features of the distribution of impurities in the UV LED heterostructures grown by HVPE [13], and determine their most important operating parameters, including output optical power, the temperature of the active region and efficiency [14, 15].

The design of UV LEDs based on GaN/AlGaN heterostructures is shown in Fig. 1. Counting from the surface of the sapphire substrate (0001), the heterostructure includes the following epitaxial layers: nucleation layer, stress control layers, electron emitter, undoped active region, electron-blocking layer, p-contact layer. The total thickness of the above-mentioned layers is about 5 μm. Planar chips were fabricated with dimensions of 0.31 × 0.31 mm^2 and then packaged. The Ti/Al and Ni/Au metal compositions were used as n-and p-type electrodes, respectively.

Good quality of the structures was confirmed [16] by XRD method. The FWHM of the active layer peak on XRD rocking curve obtained in a symmetric reflection (0002) was less than 400 arcsec. The threading dislocation density measured by AFM varied from $8 \cdot 10^7$ to $9 \cdot 10^8$ cm^{-2}, which is typical for thin epitaxial nitride layers grown on sapphire substrates. The wavelength of PL maximum was in the range of 359.5–362 nm, the average PL FWHM was 11 nm. The position of the PL peak wavelength and FWHM of the main peak were almost identical for different areas on the wafer surface. The results showed that the use of rapid and low-cost HVPE method allowed us to achieve a high degree of structural perfection of epitaxial structures.

When studying the operating parameters of LEDs we focused on the investigation of spectral (Fig. 2), power-current (P-I) (Fig. 3) and I-V characteristics, the measurement of power and efficiency of the luminescence, as well as the study of the impact on them of the direct current and ambient temperature. The investigated structures demonstrated high enough output optical power. At 20 mA current it was 1.15 mW and reached a maximum of 4.2 mW at 120 mA. A deviation of the P-I characteristics from the linear dependence was observed at current values above 30 mA. The LEDs showed a working capacity in the continuous operation mode up to the values of the direct current of 140 mA.

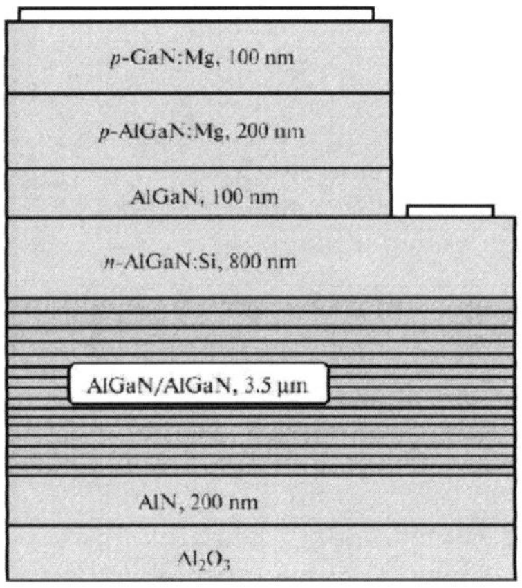

Figure 1. Schematic of the precise laser machining system.

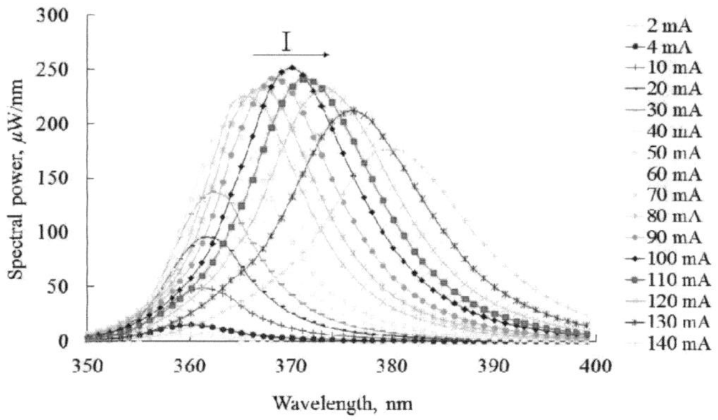

Figure 2. Electroluminescence spectra of UV LEDs at different values of direct current.

Depending on the wavelength of the maximum of the luminescence spectrum on the current value (Fig. 2) one can see that the shift of the spectral characteristics to long-wave region accelerates with an increase in current. This effect is due to a rapid increase in the temperature of the crystal. In general, the peak wavelength is seen to shift by 20 nm with increasing of the current from 2 mA to 140 mA

(Fig. 4). Special attention was paid to the influence of direct current on the temperature of the active region of the structure (Fig. 5). The experimental data indicate that the p-n-junction temperature T_j at 20 mA is 43°C, which is higher than the values of blue emitters based on InGaN. At the current of 60 mA T_j exceeded 100°C and then the temperature continued to increase significantly. A comparison dependencies of wavelength and temperature on the current leads to a conclusion that self-heating is a one of the main process that reduces the output optical power of the structures. It was determined that the thermal resistance of the samples at I_f was equal to 250°C/W.

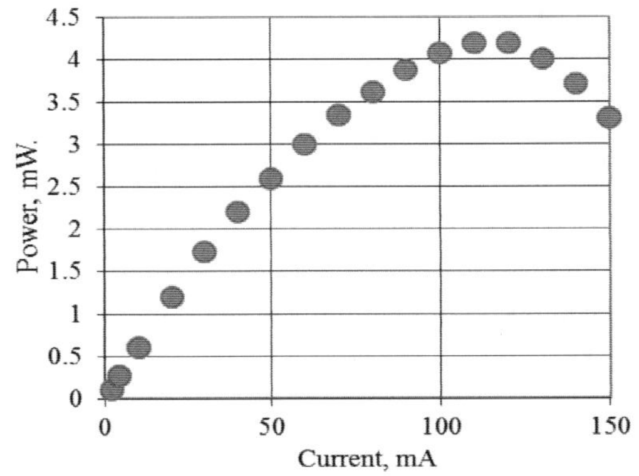

Figure 3. The power-current characteristics of UV LEDs.

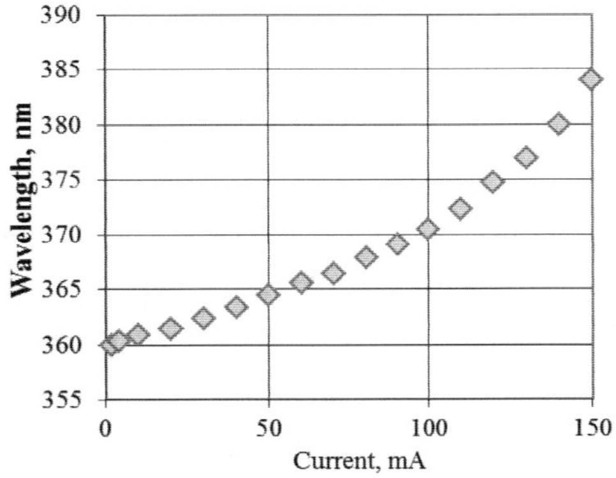

Figure 4. The dependence of the wavelength at the maximum of the spectral characteristics on the direct current.

Maximum wall-plug efficiency of packaged UV LED chips was equal to 1.5% at If (Fig. 6). At 100 mA the efficiency decreased by about 2 times.

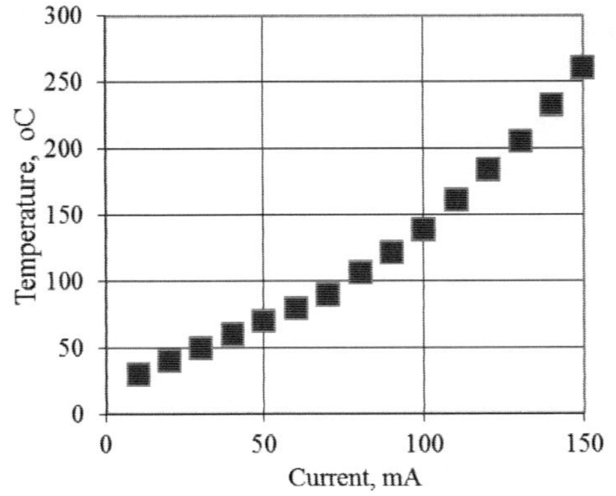

Figure 5. The dependence of the temperature of the UV LED active region on the direct current.

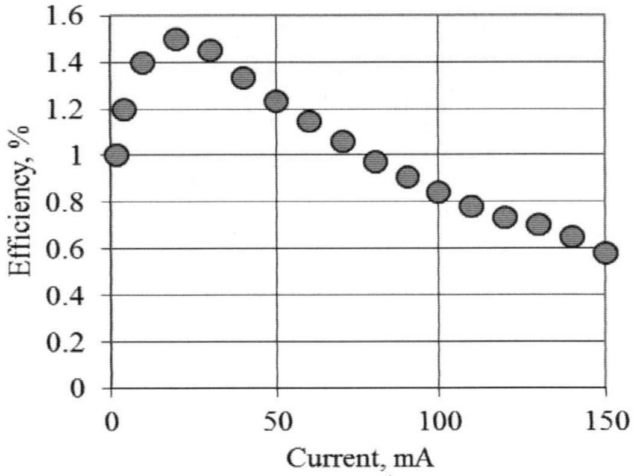

Figure 6. The dependence of the efficiency of the UV LED on the direct current active region on the direct current.

3. Conclusions

The UV LED heterostructures were grown by HVPE approach. The FWHM of the active layer peak on XRD rocking curve obtained in a symmetric reflection (0002) was less than 400 arcsec. The threading dislocation density measured by AFM varied from $8 \cdot 10^7$ to 9 10^8 cm^{-2}. PL measurements revealed a uniform distribution of peak wavelength and FWHM across the wafer. The results showed that the use of HVPE method allowed us to achieve a high degree of structural perfection of epitaxial structures.

The experiment showed that the peak wavelength of UV LEDs was in the range of 360-380 nm with FWHM of 10-13 nm. At operating current of 20 mA the active region temperature Tj is 43°C, the output optical power - 1.14 mW. At 100 mA the efficiency decreased by about 2 times.

Acknowledgments
The work was supported by the Ministry of Education of the Russian Federation in the framework of the project of the state task in the field of scientific activity, project number 16.1307.2014K.

References

[1] JE.F. Schubert 2006 Light Emitting Diodes New York, Cambridge University Press
[2] M Kneissl et al 2011. *Semicond. Sci. Technol* **26** 01403
[3] F I Manyakhin et al, 2010 *Izv. Vyshh. Uchebn. Zaved* **50** 54
[4] J K Kim et al 2008 *Appl Phys Lett* **93** 221111
[5] J H Son et al 2012 *Advanced Materials* **24** 2259-2262
[6] C Rooman et al 2005 *Photonics Technology Letters* 2649-2651
[7] A Deinega et al 2011 *J.Opt.Soc. Am.* A **28** 5
[8] V Osincky et al 2003 *FIP* **1** 94.
[9] K D Lee et al 2010 *Semiconductor lighting* **5** 22.
[10] E A Menkovich, S A Tarasov, I A Lamkin 2012 *Functional Materials* **19** 233-237.
[11] I Lamkin, S Tarasov 2013 *Journal of Physics: Conference Series* **461** 012025
[12] E A Menkovich, A V Solomonov, S A Tarasov, P A Yurgin 2014 *Functional materials* **21** 186-189
[13] S Kurin, A Antipov, I Barash, A Roenkov, H Helava, S Tarasov, E Menkovich, I Lamkin, Yu Makarov 2013 *Physica status solidi (c)* **10** 289–293
[14] E A Menkovich, S A Tarasov, I A Lamkin, S Yu Kurin, A A Antipov, et al 2013 *Journal of Physics: Conference Series* **461** 012027
[15] A V Solomonov, S A Tarasov, E A Men'kovich, I A Lamkin, S Yu Kurin, A A Antipov et al 2014 *Semiconductors* **48** 245–250
[16] E A Menkovich, S A Tarasov, I A Lamkin, A V Solomonov, S Yu Kurin, A A Antipov, I S Barash, A D Roenkov, A S Usikov, H I Helava and Yu N Makarov 2014 *Journal of Physics: Conference Series* **541** 012054

Combined EPR and ODMR study of Ce^{3+} optical emitters in yttrium aluminium garnet

A S Gurin[1], D D Kramushchenko[1], Yu A Uspenskaya[1], G R Asatryan[1], A G Petrosyan[2], D O Tolmachev[1], N G Romanov[1], P G Baranov[1,3]
[1] Ioffe Physical-Technical Institute of the Russian Academy of Sciences, 194021 St. Petersburg, Russia
[2] Institute for Physical Research, National Academy of Sciences of Armenia, 0203 Ashtarak-2, Armenia
[3] St. Petersburg State Polytechnical University, 195251 St. Petersburg, Russia

E-mail sasha.gurin@mail.ioffe.ru

Abstract. Electron paramagnetic resonance (EPR) was detected in the ground state of Ce^{3+} ions in YAG:Ce single crystals by monitoring the intensity of photoluminescence under non-resonant excitation. The photoluminescence intensity was found to increase strongly in magnetic field at liquid helium temperature. The optically detected magnetic resonance (ODMR) effect was of the order of percents and corresponded to resonance decrease in the luminescence intensity. ODMR correlated with the conventional EPR spectra of Ce^{3+} ions, which occupy the dodecahedral Y^{3+} sites in the YAG crystal lattice. The increase of the PL intensity in magnetic field was concluded to be caused by Boltzmann distribution between spin sublevels of the ground state of Ce^{3+} ions. The ODMR tend to equalize the populations of levels in resonance and results to a decrease of PL intensity. Different types of Ce^{3+} centres characterized by the presence of nearby antisite defects were observed, which should be taken into account in applications of single ion spectroscopy based applications of YAG:Ce.

1. Introduction

Rare-earth doped crystals with the garnet structure play an important role in quantum electronics. Interest in garnet crystals with the Ce^{3+} activator has significantly increased in recent years due to the increasing potential for their use as high-performance scintillators in many fields, including positron emission tomography imaging [1, 2] and high energy physics [3]. Ce^{3+} doped phosphors with the garnet structure are widely used in white light-emitting diodes [4].

Recent publications on the application of garnet crystals doped with rare earth elements for quantum computing open up new possibilities in this direction. It was declared that rare-earth-doped crystals are excellent hardware for quantum storage of optical information [5]. Coherent properties of rare-earth single-spin qubits in yttrium aluminium garnets (YAG) have been demonstrated [6]. Hyperfine coupling to aluminium nuclear spins suggests that cerium electron spins can be exploited as an interface between photons and long-lived nuclear spin memory. Additional functionality of these materials is added by their wave-guiding properties allowing for on-chip photonic networks. Combined with high brightness of Ce^{3+} emission and a possibility of creating photonic circuits out of the host material, this makes cerium spins an interesting option for integrated quantum photonics.

Obviously, progress in these fields depends in many respects on the degree of understanding of the spectroscopic properties of the considered crystals and, on this basis, the development of methods for producing crystals with desired parameters. One of the most powerful direct methods to investigate the properties of materials is electron paramagnetic resonance (EPR) [7, 8], which makes it possible to determine the chemical and charge states of an impurity centre, its local symmetry, the composition of the nearest environment, the structure of energy levels, the specific features of the interaction with the crystal lattice, etc.

For single spin manipulation one needs to use optically detected magnetic resonance (ODMR) techniques [9, 10]. ODMR is "trigger detection" in that the absorption of a resonance microwave photon triggers a change in emission (absorption) of an optical photon due to the selective feeding of the magnetic sublevels. A giant gain in sensitivity appears since the energy of an optical quantum is by several orders of magnitude higher as compared with a microwave one, and it becomes possible to detect a very small number of spins down to a single spin. ODMR of Ce^{3+} ions in garnets was studied via the magnetic circular dichroism (MCD) in absorption [11, 12]. In Ref. 12 EPR of Ce^{3+} ions in $Y_3Al_5O_{12}$ has been observed via MCD and the MCD bands at 227, 270, 338 nm and at 458.5 nm have been definitely attributed to Ce^{3+}.

In the present paper, we report results of joint ODMR and EPR study of YAG:Ce single crystals. EPR of the ground state of Ce^{3+} ions was detected by monitoring the photoluminescence intensity and correlated with conventional EPR spectra. Several types of Ce^{3+} centres characterized by the presence of nearby defects were observed by EPR.

2. Experimental details

In this work, we investigated $Y_3Al_5O_{12}$ single crystals doped with cerium at concentrations of 0.1 to 0.2 at %. The crystals were grown from the melt by vertical directed crystallization [13, 14] in an Ar/H_2 atmosphere with the use of molybdenum containers and seed crystals oriented along the crystallographic axis [001]. The EPR studies were performed on the samples with specified activator concentrations which were cut from the regions without facet growth forms and light-scattering inclusions in the form of rectangular parallelepipeds $1.5 \times 2 \times 5$ mm^3 in size.

The EPR experiments were carried out by using X-band (9.3 GHz) JEOL EPR spectrometer equipped with a helium gas-flow cryostat, which was manufactured in the laboratory and provided temperature control within the range of 4–300 K. The EPR spectra of a residual paramagnetic impurity of Mo^{3+} were used to control the sample orientation in the magnetic field. It was found that the long sample edge (the rotation axis) was turned by 24^0 from the crystal axis [100] and rotation about it corresponded to polar angles $\varphi = 24°$ and $\theta = 0°$ to 360°.

Photoluminescence (PL) spectra were measured at a temperature between 1.8 and 300 K. A deuterium arc lamp as well as a semiconductor laser ($\lambda = 405$ nm) were used as excitation sources. A photomultiplier in combination with a grating monochromator was used for light detection.

ODMR measurements were performed on ODMR spectrometer operating in Q-band (35 GHz) providing the magnetic field up to 4.5 T and microwave power up to 0.5 W. The sample was placed in the centre of the cylindrical microwave cavity H_{011} with optical access for excitation and emission light. ODMR spectra were recorded in the temperature range between 1.8 and 4.2 K by monitoring the intensity of luminescence. No modulation of the microwaves was applied because of long spin–lattice relaxation times of the investigated paramagnetic centres.

3. Experimental results and discussion

Figure 1 shows the PL spectrum of YAG:Ce (0.1%) single crystal recorded at a temperature of 1.8 K. The broadband phonon-assisted emission of the Ce^{3+} ions is observed, a sharp zero-phonon line at 489 nm [5] characteristic for Ce^{3+} ions is not visible in the figure because of its low intensity and not high enough resolution. The laser line (405 nm) which was used for the excitation of PL is depicted in the left part of Figure.

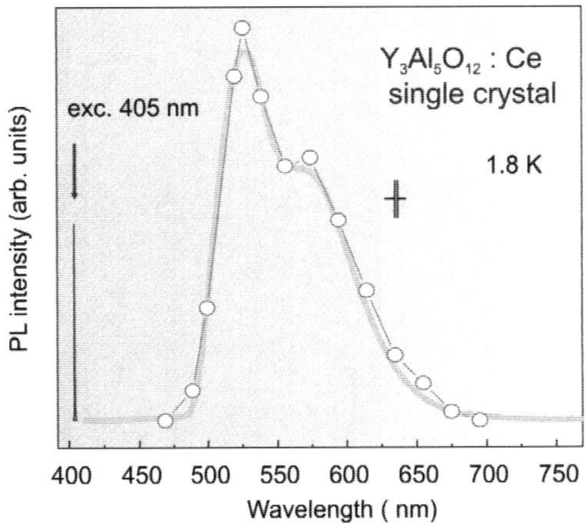

Figure 1. The photoluminescence spectrum of the Ce^{3+} ions in YAG:Ce (0.1%) single crystal. The laser line (405 nm) which was used for excitation of PL is marked by an arrow. Open circles display the dependence of the Ce^{3+} ODMR signal amplitudes on the detection wavelength.

In magnetic field at a temperature of 1.8 K, the intensity of the total PL of the YAG:Ce single crystal was found to increase. Application of 35 GHz microwaves resulted in appearance of a number of anisotropic resonance signals (ODMR signals) corresponding to a decrease in PL intensity as one can see in Figure 2 which shows the magnetic field dependences of the relative PL intensity in the YAG:Ce single crystal at $\lambda = 570$ nm for several orientations of the sample in magnetic field. The dips on these dependences correspond to EPR in the ground state of Ce^{3+} ions. The dependence of the ODMR amplitude on the detection wavelength is shown in Figure 1 by open circles. It coincides with the PL spectrum of Ce^{3+} ions.

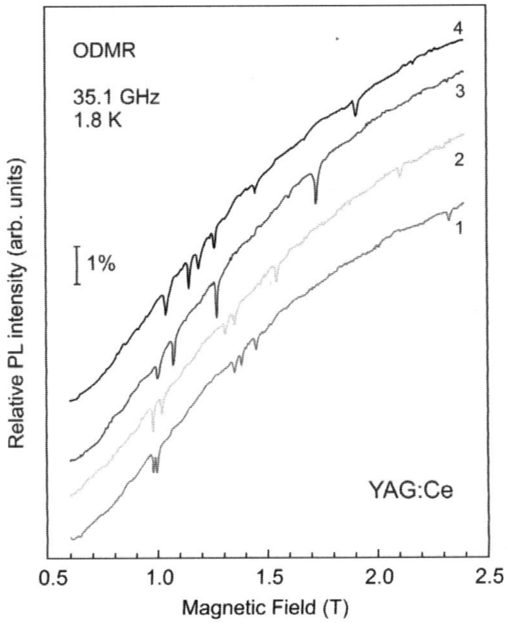

Figure 2. The magnetic field dependences of the relative PL intensity at a wavelength of 570 nm recorded in the YAG:Ce single crystal at a temperature of T = 1.8 K in the presence of 35.1 GHz microwave field. The curves 1 to 4 correspond to different orientations of the crystal in magnetic field: spherical coordinates are $\varphi = 24°$ and $\theta = 85°$ (1), 75° (2), 65° (3), and 55° (4).

Figure 3 shows the angular variations of the 35.1 GHz ODMR (a) and 9.3 GHz EPR (b) spectra recorded in the YAG:Ce single crystal with rotation of the sample corresponding to $\varphi = 24°$ and $\theta = -5° \div 95°$. Top axes display the effective g-factor values in the same scale for (a) and (b). In the EPR spectra (b), lines from Mo^{3+} impurity ions were also detected in addition to Ce^{3+}.

Cerium has only even isotopes with a zero nuclear magnetic moment (I = 0). The angular dependences of the EPR spectra are well described by the spin Hamiltonian of orthorhombic symmetry in the form

$$\hat{H} = \mu_B \boldsymbol{S} \cdot \hat{g} \cdot \boldsymbol{B}, \tag{1}$$

where μ_B is the Bohr magneton, $\boldsymbol{S}$ is the effective spin (S = 1/2), $\boldsymbol{B}$ is the vector of the external magnetic field, and $\hat{g}$ is the g-factor tensor.

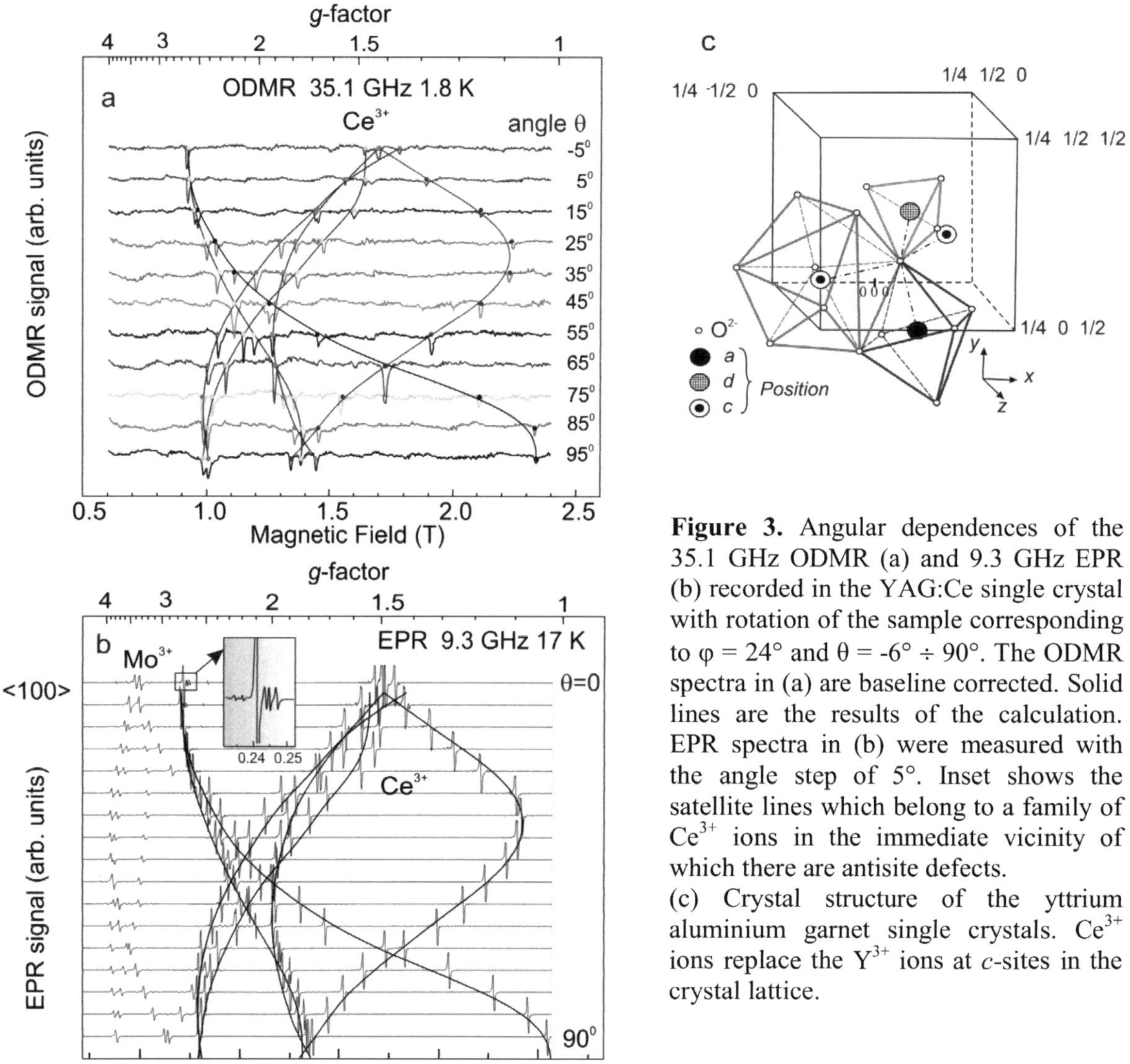

Figure 3. Angular dependences of the 35.1 GHz ODMR (a) and 9.3 GHz EPR (b) recorded in the YAG:Ce single crystal with rotation of the sample corresponding to φ = 24° and θ = -6° ÷ 90°. The ODMR spectra in (a) are baseline corrected. Solid lines are the results of the calculation. EPR spectra in (b) were measured with the angle step of 5°. Inset shows the satellite lines which belong to a family of Ce^{3+} ions in the immediate vicinity of which there are antisite defects.
(c) Crystal structure of the yttrium aluminium garnet single crystals. Ce^{3+} ions replace the Y^{3+} ions at c-sites in the crystal lattice.

In the YAG single crystals, the rare-earth ions occupy as a rule dodecahedral sites of the crystal lattice, thus replacing the Y^{3+} ions. In this position (c-sites), the Y^{3+} ions are coordinated by eight oxygen ions with local symmetry D_2 as shown in Figure 3 (c) and form 6 nonequivalent positions. The principal directions of the local magnetic axes of the Ce^{3+} ion are oriented in such a way that the x axes are along one of the crystallographic directions <001> and the directions of the axes y and z coincide with the <011> directions. The corresponding Euler angles for one centre in the dodecahedral position are as follows: α = 90°, β = 135° and γ = 90°. The orientations of other five Ce^{3+} centres are obtained by the symmetry operations in the YAG crystal lattice. From the experimental orientation dependences of the EPR line positions, the following parameters of the g-tensor have been obtained: g_x = 2.74 ±0.05, g_y = 1.87 ± 0.05, and g_z = 0.91 ± 0.05 in agreement with the data reported in [15]. The calculations were performed with a special computer program developed by V. Grachev [16].

In the EPR spectra of YAG : Ce single crystals, along with the main intense lines of Ce^{3+} ions located in the *c*-sites, additional lines of approximately two orders of magnitude lower intensity were observed as shown in the inset in Figure 3 (b). These lines have similar angular dependences and can be attributed to Ce^{3+} centres the immediate vicinity of which is distorted by a defect. From the analysis of the angular and temperature dependences of EPR spectra, as well as the average values of the *g*-factors the satellite lines have been ascribed to the Ce^{3+} with nearby antisite defects [17].

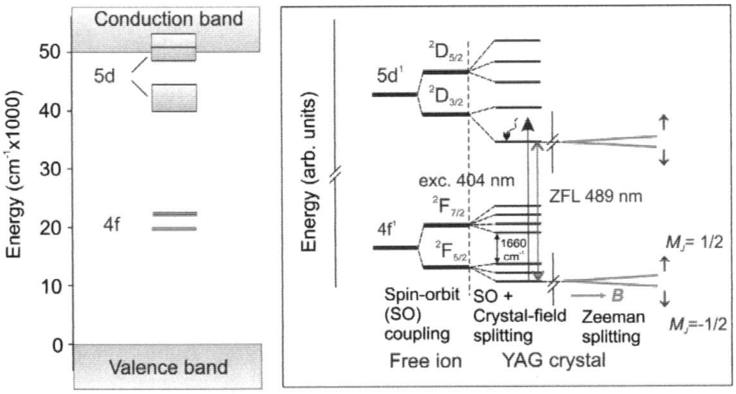

Figure 4. Positions of the Ce^{3+} ion states relative to the conduction and valence bands of the YAG crystal (left panel) and the electronic energy level structure of Ce^{3+} ion in a YAG crystal (right panel). The Zeeman splitting of the lowest 4*f* Kramer's doublet for the ground state and the lowest 5*d* Kramer's doublet for the excited state are depicted.

Figure 4 presents the electronic level structure of Ce^{3+} ion in a YAG crystal [5, 18, 19]. The positions of the Ce^{3+} ion states relative to the conduction and valence bands are shown in the left panel. The configuration of Ce^{3+} is $4f^1$ so the ground term is ^2F. The spin-orbit coupling splits this term into $^2F_{5/2}$ and $^2F_{7/2}$ sub-terms. The separation between these levels is about 2200 cm^{-1} for the free ion. The energy levels of the lowest sub-term of the Ce^{3+} ion are $^2F_{5/2}$. In a crystal field, the sixfold degeneracy will be lifted (the number of levels is given by $2J+1$). The separation between the adjacent *J* levels due to the spin-orbit coupling exceeds the splitting of the ground-state levels due to the crystal field by at least an order of magnitude. Since the crystal fields will be slightly different for the different environment of cerium ions, there will be a difference in the parameters of the anisotropic *g*-factor for Ce^{3+} centres with defects in the immediate vicinity. In the YAG single crystals, the above disturbances of the regular lattice can manifest themselves as extrinsic (non-equivalent) substitutions of Y^{3+} ions for Al^{3+} ions in octahedral sites (*a*-sites) or as substitutions of Al^{3+} ions for Y^{3+} ions in the c-sites, so called "antisite" defects. The formation of such defects is possible within the limits of 4–6% during the growth of single crystals from the melt.

As was recently shown rare-earth-doped YAG crystals are excellent hardware for quantum storage of optical information. Coherent properties of single rare-earth spin qubits were demonstrated, where ODMR of a single Ce^{3+} ion in YAG ceramic crystals was recorded via the intensity of photoluminescence. To force the creation of a nonequilibrium population of spin sublevels in the ground state of cerium ions resonant circularly polarized optical excitation was used in [5], which changed the emission intensity of Ce^{3+} determined by thermalized populations of the ground-state spin sublevels. This happened because the strongest transition between the ground $4f^1$ and the optically excited $5d^1$ spin doublets under σ^+ circularly polarized excitation is the a spin flip transition $4f^1$(spin down) $\leftrightarrow$ $5d^1$ (spin up). The other 3 transitions are almost 3 orders of magnitude weaker. However, the decay back to the ground state is equally probable. This results in efficient pumping of the ion into $4f^1$ spin-up state and a reduction of the intensity of the phonon-assisted Ce^{3+}emission [5].

In our experiments, excitation of PL was non-resonant and was not specifically polarized but we did observe an increase in the emission intensity with magnetic field and resonance decreases of the order of a percent at magnetic fields corresponding to EPR in the ground state of Ce^{3+} ions. It can be assumed that there is a link between the ground state spin polarization and he PL intensity. The resonant microwave transitions, which tend to equalize the Boltzmann populations of the ground state spin sublevels can result in a drop of the luminescence intensity. Further studies are under way which should clarify the details of the mechanism which could couple the spin polarization of the ground state and the emission intensity and exclude the unintentional polarization effects. ODMR of excited state of Ce^{3+} did not observed in our experiments, probably because of short lifetime and short spin relaxation time in excited state.

4. Conclusions

An increase of the photoluminescence intensity with magnetic field and anisotropic ODMR signals corresponding to resonant decreases of the PL intensity were found in YAG:Ce single crystals at liquid helium temperature. The ODMR spectra correlated with the conventional EPR spectra of the ground state of Ce^{3+} ions, which substitute for the Y^{3+} ions and occupy the dodecahedral sites (c-sites) in the crystal lattice. In our experiments, excitation of the Ce^{3+} luminescence was non-resonant and was not specifically polarized but we did observe the effect of the ground state spin polarization of on the emission intensity. These effects are definitely connected with the Boltzmann population distribution in the ground state of cerium ions but further studies are required to clarify the details of the mechanism which couples the ground state spin polarization and the emission intensity.

Several different types of Ce^{3+} centres with slightly different parameters, which are characterized by the presence of nearby defects (antisite defects) were observed by EPR. This should be taken into account in applications using single Ce^{3+} ions in YAG.

Acknowledgments

This work has been supported by Russian Science Foundation under Agreement #14-12-00859 and Russian Foundation for Basic Research under grant # 13-02-00821

5. References

[1] Autrata R, Schauer P, Kvapil Jos and Kvapil J 1978 *J. Phys. E: Sci. Instrum.* **11** 707
[2] Moszynski M, Ludziewski T,Wolski D, Klamra W and Norlin L O 1994 *Nucl. Instrum. Methods Phys. Res., Sect.* A **345** 461
[3] Dujardin C, Mancini C, Amans D, Ledoux G, Abler D, Auffray E, Lecoq P, Perrodin D, Ovanesyan K and Petrosyan A 2010. *J. Appl. Phys.* **108** 013510
[4] Schlotter P, Schmidt R, and Schneider J 1997 *J. Appl. Phys.* **A 64** 417
[5] Siyushev P, Xia K, Reuter R, Jamali M, Zhao N, Yang N, Duan C, Kukharchyk N, Wieck A D, Kolesov R and Wrachtrup J 2014 *Nature Communications* **5** 3895
[6] Kolesov R, Xia K, Reuter R, Jamali M, Stohr R, Inal T, Siyushev P, Wrachtrup J *Phys. Rev. Lett.* 2013 **111** 120502
[7] Al'tshuler S A and Kozyrev B M 1974 *Electron Paramagnetic Resonance in Compounds of Transition Elements, 2nd ed.* (Wiley, New York)
[8] Abragam A and Bleaney B 1970 *Electron Paramagnetic Resonance of Transition Ions* (Oxford University Press, Oxford)
[9] S. Geschwind 1972 Electron Paramagnetic Resonance, ed. S. Geschwind (Plenum, New York)
[10] Spaeth J-M Niklas J R, Bartram R H, 1992 Structural Analysis of the Point Defects in Solids (Springer-Verlag, Berlin-Heidelberg)
[11] Reyher H-J, Faust B, Sugg B, Rupp R and Ackermann L 1997 *J. Phys.: Condens. Matter* **9** 9065
[12] Reyher H-J, Hausfeld N, Pape M, Baur J and Schneider J 1999 *Solid State Commun* **110** 345
[13] Chernov A A, Givargizov E I, Bagdasarov Kh S, Kuznetsov V A, Dem'yanets L N and Lobachev A N 1994 in: Modern Crystallography (Ed. Vainshtein B K, Springer-Verlag, Berlin)
[14] Petrosyan A G, 1994 *J. Crystal Growth* **139** 372
[15] H. R. Lewis, J. Appl. Phys. 37 (2), 739 (1966).
[16] Grachev V View EPR/ENDOR program http://www.physics.montana.edu/faculty/grachev
[17] Asatryan G R, Kramushchenko D D, Uspenskaya Yu A, Baranov P G, Petrosyan A G, 2014 *Physics of the Solid State* **56** 1150
[18] Hamilton D S, Gayen S K, Pogatshnik G J, Ghen R D,1989 *Phys. Rev. B* **39** 8807
[19] Rotman S R 1992 *Phys. Status Solidi A* **132** K61

Journal of Physics: Conference Series **661** (2015) 012040 doi:10.1088/1742-6596/661/1/012040

Selectivity control of photosensitivity of Ag-GaP and Ag-AlGaN structures

I A Lamkin[1], S A Tarasov[1], A V Solomonov[1], M Y Andreev[1], S Yu Kurin[2]

[1]Saint-Petersburg Electrotechnical University "LETI", Prof. Popova 5, St. Petersburg 197376, Russia
[2]GaN-Crystals Ltd., 27 Engels Ave, St. Petersburg 194156, Russia,

E-mail: ialamkin@mail.ru

Abstract. Design, growth and studies of photosensitive structures based on Ag-GaP and Ag-$Al_xGa_{1-x}N$ contacts are reported. Methods for structure selectivity control, which allow changing the sensitivity spectrum half-width in a range of 11-210 nm were worked out. By varying the metal layer thickness, a set of Ag-GaP short-wavelength photodetectors (PD) was fabricated. The set includes PDs from broadband (spectrum half-width $\Delta\lambda$=210 nm, sensitivity S_I = 0,19 A/W) to visible-blind ($\Delta\lambda$=15 nm, S_I = 0,034 A/W). The use of Ag-$Al_xGa_{1-x}N$ structures provided increased sensitivity (S_I = 0,071 A/W) and $\Delta\lambda$ reduced to 11 nm due to special selection of solid solution composition.

1. Introduction

One of the key objectives of modern optoelectronics is development of photodetectors. In many cases the necessity occurs to develop a wavelength-selective narrowband photodetector. Applying broadband photodetectors in combination with optical filters is a regular solution. However such filters significantly worsen device parameters, in particular leading to reduced sensitivity, an increase of dimensions and cost. Besides, it is often a challenge to adjust the filters to the required band pass. The problem is especially important in designing short-wavelength photodetectors. Therefore, development of selective ultraviolet (UV) photodetectors with the structure providing the selectivity effect without the use of additional optical filters is a relevant problem.

Existing silicon-based photodiodes [3] show high sensitivity in the visible range of the spectrum due to a narrow band gap of the semiconductor. The use of GaP with a wider band gap allows eliminating part of spurious signals [4]. The most perspective are wide-bandgap $Al_xGa_{1-x}N$ solid solutions [5, 6] with a band gap that is wide enough to eliminate sensitivity in the visible portion of the spectrum. In this work, the results of development and research of methods for selectivity control of the photodetectors with a possibility of varying their sensitivity within both short- and long-wavelength regions of the spectrum are presented.

2. Sample and experimental technique

The structures on the basis of gallium phosphide and solid solutions of aluminium gallium nitride have been investigated. Sulfur doped *n*-GaP epitaxial layers with a free carrier concentration of $(0.5\ldots5)\cdot10^{16}$ cm^{-3} were grown by vapour phase epitaxy. Schottky barriers were formed by vacuum thermal evaporation of silver on the semiconductor surface preliminarily cleaned in tetrachloromethane environment. The ohmic contacts were formed by Au deposition in accordance with the standard technique [7].

$Al_xGa_{1-x}N$ epitaxial layers were grown by hydride vapour phase epitaxy (HVPE) technique [8]. The top doped n-$Al_xGa_{1-x}N$ layer had a carrier concentration of 10^{16} cm^{-3}. Metal layers were deposited by vacuum thermal evaporation at a residual gas pressure of 10^{-5} Torr. Ag was used as a rectifying contact material. Al with Ti underlayer were used to form Ohmic contacts. Prior to metal deposition the structures were cleaned by potassium hydroxide (KOH) during 2 min and washed with distilled water. For good metal adhesion the structures were heated to a temperature of 300°C during deposition.

Structure characteristics were measured using a set-up based on a monochromator MDR-3 with a diffraction grating and a xenon lamp as a light source. The set-up allows measuring spectral characteristics in a range of 200-2500 nm and current-voltage characteristic in a range of 10^{-15} A - 1 A.

3. Experimental results and discussions

In this work, a research based on two basic selectivity control methods for the sensitivity spectrum of photodetector structures is presented. One is based on using selective properties of a Schottky contact metal layer (silver layer), while in the other method, the selectivity is controlled by varying the composition of a semiconductor epitaxial layer.

Silver is a unique material for fabrication selective ultraviolet photodetectors. The presence of a narrow spectral window around a wavelength of 322 nm allowed realizing so-called visible-blind photodetectors which are not sensitive to the visible light. It can be carried out through making a junction between silver and wide bang gap semiconductor. The parameter which mostly affects the selectivity of such photodetectors is the silver layer thickness.

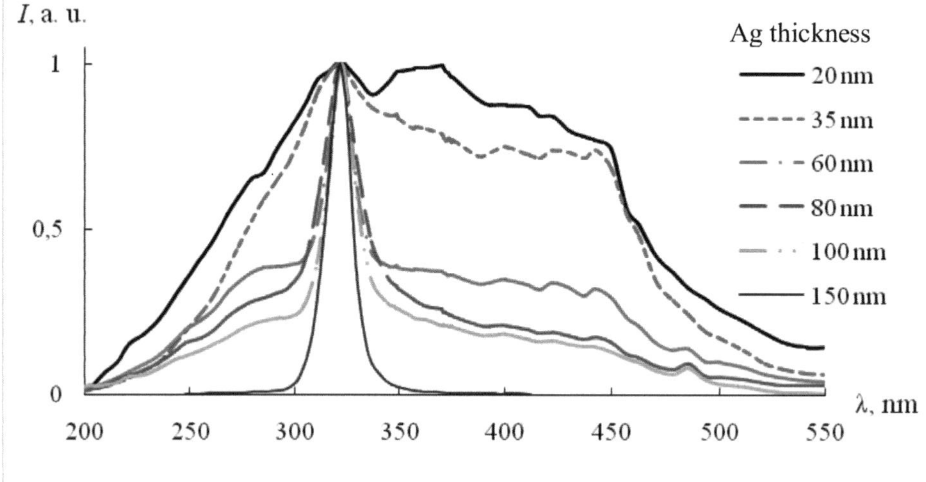

Figure 1. Spectral characteristics of Ag- GaP photosensitive structures with different selectivity degree

A set of photosensitive structures based on gallium phosphide and nitride semiconductor solid solutions with different selectivity degree was fabricated. By using thin silver layers (less than 20 nm), broadband short-wave photodetectors with the sensitivity range of 200-600 nm were designed. The sensitivity at the maximum was 0.19 A/W, the value of the spectrum half-width was greater than 210 nm. The reverse current value at a voltage of 1 V did not exceed 10^{-14} A enabling to reach good detectivity.

Varying the silver layer thickness allowed to control the selectivity degree of spectral characteristics as shown in Fig. 1. It is shown that increasing the layer thickness leads to more rapid decreasing of sensitivity in both short- and long-wave ranges of the spectrum than at the maximum of

322 nm. At 150 nm thickness of the Ag layer a visible-blind UV photodetector with the sensitivity maximum at 322 nm and a half-width of 15 nm. The sensitivity was 0.034 A/W.

To increase the photosensitivity within the UV spectrum range, structures based on direct band gap solid solutions $Al_xGa_{1-x}N$ were investigated. The use of $Al_xGa_{1-x}N$ solid solutions allows not only to improve the sensitivity, but eliminate spurious signals in the long-wave region of the spectrum and reduce the spectrum half-width. Photodetector structures based on Schottky barriers Ag-$Al_{0,08}Ga_{0,92}N$ were fabricated. Such photodetectors have the long wavelength edge of photosensitivity less than 350 nm. It complies with requirements to visible-blind photodetectors. The half-width was in a range of 15-40 nm depending on the Ag layer thickness varied from 15 to 150 nm.

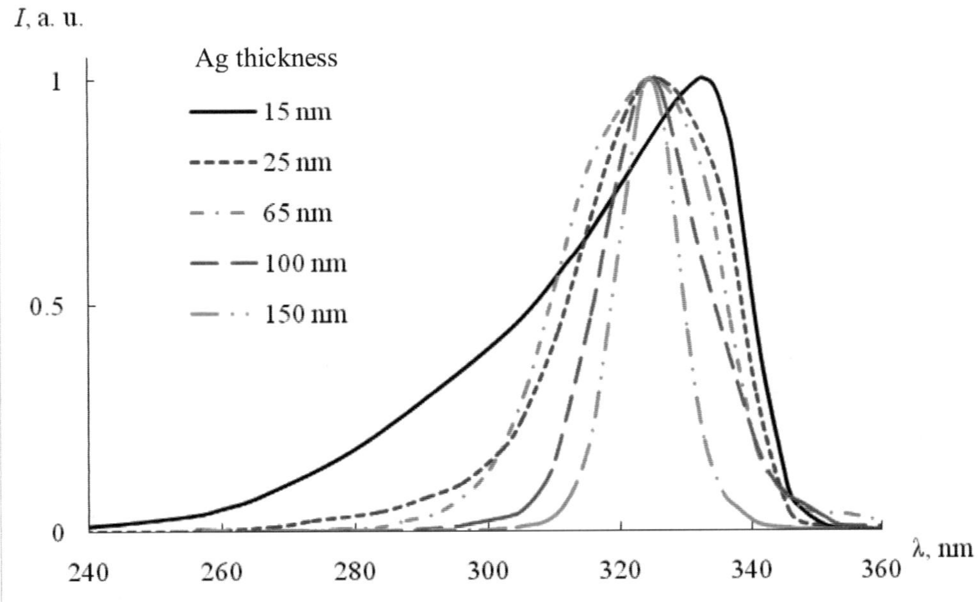

Figure 2. Spectral characteristics of Ag-$Al_{0,08}Ga_{0,92}N$ photosensitive structures with different selectivity degree

The photosensitivity spectrum is mainly determined by the absorption spectrum of a semiconductor. The photodiode sensitivity peak is within a range of maximum photoabsorption which corresponds to the photons with energy more than the band gap. The limit of photodiode sensitivity is defined by the theoretical limit of quantum efficiency η_{max}, that is, when every absorbed photon leads to formation of an electron-hole pair. The theoretical limit of quantum efficiency is shown in Fig. 3 by the dashed line.

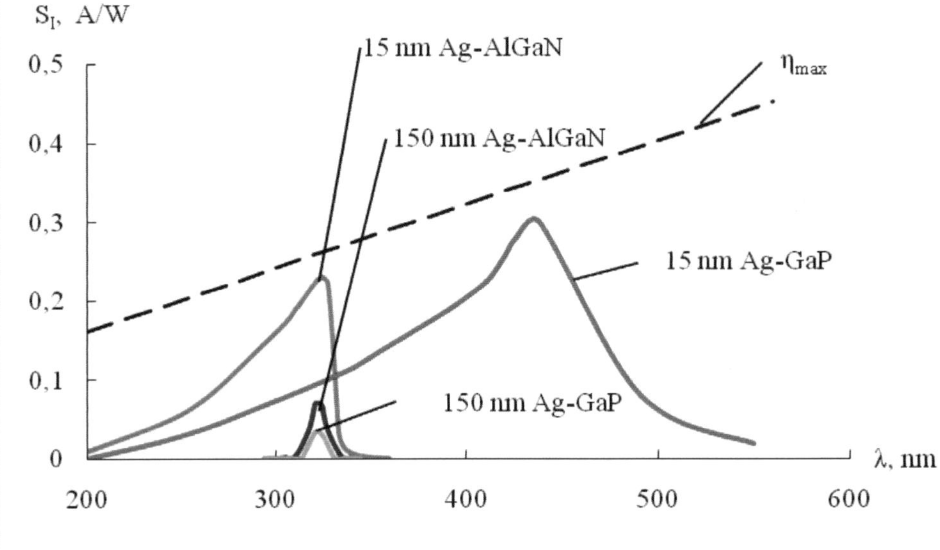

Figure 3. Theoretical spectral characteristics of photodiodes based on silver layers

In a real Schottky barrier photodiode, the cutoff wavelength on the short-wavelength side strongly depends on the surface recombination velocity. The higher the surface recombination velocity, the less the quantum efficiency of a photodetector and the sharper decline in the short-wavelength range of the spectral characteristic is observed. For a gallium phosphide photodetector in particular, the efficiency at a wavelength of 322 nm decreases by almost 3 times with respect to the efficiency maximum at a wavelength of 420 nm (Fig. 3). Since the band gap of a $Al_xGa_{1-x}N$ solid solution exceeds the GaP band gap, the efficiency decline occurs at significantly shorter wavelengths. As a result, the quantum efficiency of a photodetector with $Al_xGa_{1-x}N$ epitaxial layer at a wavelength of 322 nm is higher than that of GaP photodetector, that is, a selective $Al_xGa_{1-x}N$-based Schottky barrier photodiode will have higher sensitivity then a Ag-GaP-based photodiode.

The use of $Al_xGa_{1-x}N$ solid solution makes it possible not only to increase the sensitivity of silver based photodetector structures but additionally control their selectivity. Appropriate choice of solid solution composition shifts the red limit of photoeffect which narrows the sensitivity spectrum of a photodetector into longer wavelength range. For this purpose Ag-$Al_{0,15}Ga_{0,85}N$ structures were created with spectral characteristics of photosensitivity shown in Fig.4. The sensitivity spectrum of the structure with Ag layer thickness of 15 nm is shown by the solid line. In this case the selectivity effect barely reveals itself, and the spectrum half-width is more than 40 nm. On the right side the sensitivity is limited by the solid solution band gap, which corresponds to a wavelength of 340 nm. When increasing the Ag layer thickness, the sensitivity in the short-wavelength range dramatically reduces due to absorption in metal, and blue limit of photoeffect shifts into a longer wavelengths range. At the same time the right side of the spectrum remains unchanged since the solid solution composition remains permanent. It allowed reaching higher degree of photosensitivity selectivity. The half-width of sensitivity obtained for Ag thickness of 150 nm was less than 11 nm.

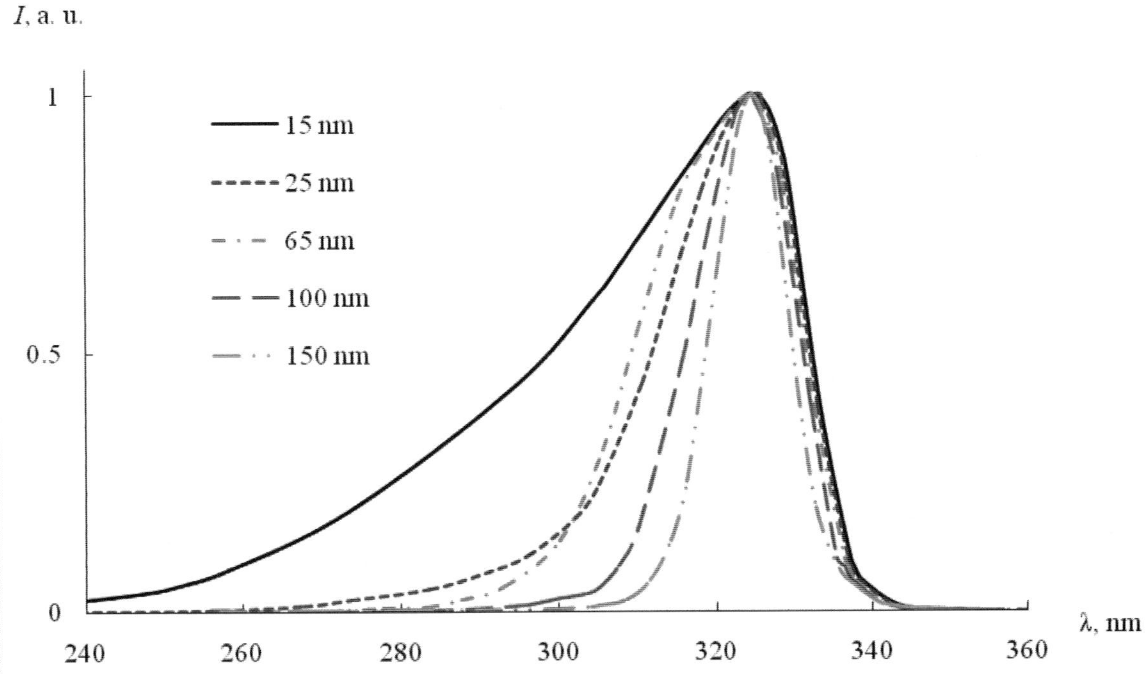

Figure 4. Spectral characteristics of Ag-$Al_{0,2}Ga_{0,8}N$ photosensitive structures with different selectivity degree

Besides, altering solid solution composition to $Al_{0,2}Ga_{0,8}N$ resulted in a sensitivity increase of a selective photodiode as the long wavelength limit of photosensitivity was shifted into a range of shorter wavelengths by increasing the Al fraction in the solid solution of the epitaxial layer. At that, the maximum absorption of the solid solution coincided with the Ag spectral window position.

4. Conclusion

In this work, photosensitive structures with different degree of spectral characteristics selectivity on the basis of gallium phosphide and aluminium gallium nitride solid solutions have been created and investigated. Two methods of structure selectivity control which allow changing the sensitivity spectrum half-width in a range of 11-210 nm were worked out: by using selective properties of the Schottky contact metal layer (silver layer) and varying the composition of the semiconductor epitaxial layer.

By using thin silver layers (less than 20 nm) broadband shortwave photodetectors based on Ag-GaP structures with the sensitivity range of 200-600 nm have been created. The sensitivity at the maximum was 0,19 A/W, the value of the spectrum half-width was more than 210 nm. It is shown that increasing the layer thickness leads to more rapid decreasing of sensitivity in both short wave and long wave range of the spectrum than at the maximum at 322 nm. With the Ag layer thickness of 150 nm a visible-blind ultraviolet photodetector with the sensitivity peak of 322 nm and 15 nm half-width. The sensitivity was 0,034 A/W.

The structures based on Ag-$Al_{0,08}Ga_{0,92}N$ Schottky barriers have been manufactured in order to improve photosensitivity in the UV range and eliminate spurious signals in a long wavelength range. It allowed to realize photodetectors which have a long wavelength edge of photosensitivity less than 350 nm. The half-width was in a range of 15-40 nm depending on the Ag layer thickness varied from 15 to 150 nm. The proper choice of $Al_xGa_{1-x}N$ solid solution composition provided increased photoresponse with further reducing the half-width to 11 nm by matching the peaks of the silver transmittance spectrum and the structure sensitivity spectrum. The obtained parameters complie with requirements to visible-blind photodetectors. The sensitivity was 0,071 A/W.

The work was supported by the Ministry of Education of the Russian Federation in the framework of the project of the state task in the field of scientific activity, project number 16.1307.2014K.

References

[1] Menkovich E A, Tarasov S A, Lamkin I A Luminescence of nanostructures based on semiconductor nitrides *Functional Materials* vol. 19, № 2, p. 233, 2012.

[2] Kurin S Yu, Antipov A A, Roenkov A D, Barash I S, Helava H I , Menkovich E A, Tarasov S A, Lamkin I A, Shmidt N M, Makarov Yu N UV LEDs for high-current operation *Journal of Physics: Conference Series* vol. 461. p. 012028, 2013.

[3] Chu-Hsuan Lin, Chee Wee Liu, Metal-Insulator-Semiconductor Photodetectors *Sensors* vol. 10, p. 8797, 2010.

[4] Pikhtin A N, Tarasov S A, Kloth B Ag-GaP Schottky photodiodes for UV sensors *IEEE Transactions on Electron Devices* vol. 50, № 1, p. 215, 2003.

[5] Lamkin I, Tarasov S Ultraviolet photodiodes based on AlGaN solid solutions *Journal of Physics: Conference Series* vol. 461, p. 012025, 2013.

[6] C. J. Collins, T. Li, D. J. H. Lambert et al Selective regrowth of $Al_{0.30}Ga_{0.70}N$ *p–i–n* photodiodes *Appl. Phys. Lett.* vol. 77, p. 2810, 2000.

[7] Pikhtin A N, Tarasov S A, Kloth B New values of the Ag-n-GaP potential barrier *Technical Physics Letters* vol. 28, № 10, p. 872-873, 2002.

[8] Kurin S, Antipov A, Barash I, Roenkov A, Helava H, Tarasov S, Menkovich E, Lamkin I, Makarov Yu. CHVPE growth of AlGaN-based UV LEDS *Physica Status Solidi (C) Current Topics in Solid State Physics* vol. 10, № 3, p. 289-293, 2013.

Journal of Physics: Conference Series **661** (2015) 012041 doi:10.1088/1742-6596/661/1/012041

Stereoscopic 3-dimensional spectral imaging systems based on a single acousto-optical tunable filter

A S Machikhin[1], V E Pozhar [1]
[1] Laboratory of acousto-optic spectrometry, Scientific and Technological Center of Unique Instrumentation of Russian Academy of Sciences, 15 Butlerova, Moscow, 117342 Russia

E-mail: aalexanderr@mail.ru

Abstract. The novel technique for 3-dimensional (3D) spectral imaging is proposed. It is based on simultaneous spectral acousto-optic filtration of a pair of stereoscopic convergent light beams via single acoustic wave. A few promising configurations of such acousto-optic tunable filters (AOTFs) is proposed, discussed and compared. It is shown that these elements may be ultra-compact and PC controlled modules, ready to be integrated into many existing schemes of stereoscopic 3D imagers. The experiment carried out with the help of single-AOTF-based spectral stereomicroscope demonstrates the efficiency of proposed technique.

1. Introduction

Imaging acousto-optic tunable filters (AOTFs), in which dynamic ultrasound grating provides effective spectral filtration of light, nowadays have a lot of applications in spectroscopy [1-3]. These filters have no moving elements and provide fast (less than 1 ms) arbitrary spectral addressing in the wide range (up to 400 nm), rather high spectral (up to 0.1 nm) and spatial (up to 1000 resolved elements along each direction) resolution. Among the most promising applications are microscopy, aerial photography, machine vision, where the spectral filtration in the particular wavelengths produces contrast patterns of physical, chemical and other properties of the inspected objects. The multi- and hyperspectral imaging reveals important additional features in comparison with the ordinary color visualization.

In particular, two-dimensional (2D) spectral imaging is used for 3D structure reconstruction by means of spectral-domain optical coherent tomography thus providing information about the location and the shape of the inner elements in the object of interest [4]. At the same time, there are many optical 3D imaging techniques (stereomicroscopy, 3D television) which do not use spectral filtration. And they obviously can be more effective with use of the spectral characterization.

General outline of this tendency is illustrated in Fig. 1. Addition of any new dimension to the imaging system $I(x,y)$ provides essentially new capabilities: either 3D imaging $I(x,y,z)$ or spectral imaging $I(x,y,\lambda)$. And incorporation of both coordinates (z and λ) provides even more information. In Fig. 1 two possible approaches to the development of 3D spectral imagers are shown: departing from 2D spectral systems or from 3D systems. In the first way, it is necessary just to combine two ordinary spectral imagers into the stereo system. But probably the most direct way is an integration of filters

[1] To whom any correspondence should be addressed.

into the existing 3D system. In this way, for example, stereomicroscopes are often supplied with a set of color glass filters. However, it is not enough for precise spectrum characterization.

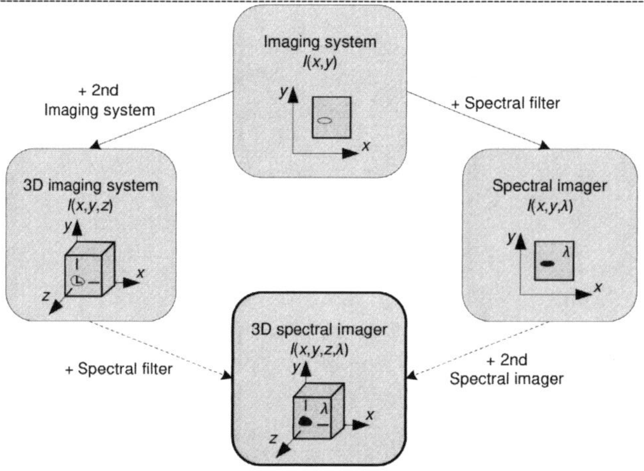

Figure 1. Evolution of optical systems from an imaging system to 3D spectral imager.

Utilization of an AOTF may solve this problem. However, unlike glass filters, the integration of AOTF into the system encounters some difficulties. Available AOTFs cannot be used for these purposes without significant upgrade. One needs either rather small AOTFs to insert them in both channels of the conventional stereoscopic system or very large AOTF with the input aperture filtering two parallel spatially separated light beams. Moreover, a pair of small AOTFs must be identical to ensure the same spectral dependence $\lambda(f)$ on ultrasound frequency f. For a large AOTF, there should be provided identical conditions of light acousto-optic interaction of both stereoscopic beams. The divergence and attenuation of ultrasonic wave may lead to significant difference of image distortions between two channels.

Thus, this straightforward approach for the development of 3D spectral imagers is not ideal. Therefore, it is necessary to analyze other ways to create and implement specialized AOTFs which are capable to filter two stereoscopic beams in the same volume by the same ultrasound column. This requirement is non-trivial as the Bragg condition of diffraction unambiguously associates the incident angle of light beam, selected wavelength, ultrasound parameters and aberration characteristics [5]. To find an appreciable solution of the described problem, it is necessary to analyze the whole optical system including the principles of diffracted stereoscopic beams detection. Classical stereoscopic systems form stereoscopic images directly at operator's eyes for visual observation or project these images on the screen for the observation via special glasses [6]. Two stereoscopic beams must be distinguishable at least by one characteristic of light: incident angle Θ, polarization p, wavelength λ. And also they can be separated in time t. In each of these modes detection principles is different: while the angle-separated beams can be observed directly, the beams differed in polarization or wavelength can be detected via polarization or spectral glasses accordingly. And for the time-divided beams synchronously switched glasses are used.

Ultrasound-controlled AO cell is capable to operate any of these physical characteristics (Fig. 2). Consider the incident light beam propagating in the direction **m** and characterized with ordinary («o») or extraordinary («e») linear polarization p. The wavelength of diffracted light λ is defined by the ultrasound frequency f and the incident angle θ. The polarization p' of the output beam and its propagation direction $\mathbf{m}'$ (and accordingly output angle θ') are determined by the crystal shape and the ultrasound characteristics. The intensity of diffracted beam $I'(t)$ repeats the time variations of the ultrasound power $P(t)$. Therefore, AOTF changes one or few parameters of the light beam and may be used in each operation mode.

It should be noted, however, that one of the mentioned above methods (spectral) of the stereopair separation is not applicable for 3D spectral imaging because it requires identical wavelengths λ for both stereoscopic beams after AO diffraction. Three other methods are examined in this paper.

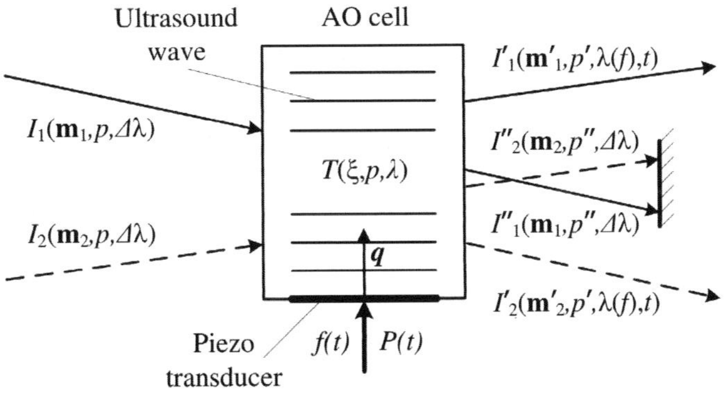

Figure 2. Principle diagram of the two-beam diffraction via the ultrasonic wave. I, $\mathbf{m}$, p, λ – intensity, orientation, polarization and wavelength of the optical beam; P, $\mathbf{q}$, f – power, wave vector and frequency of the ultrasound wave

2. Configurations of single AOTF-based 3D spectral imagers

2.1 Angular separation of stereo images

For this method the wave vectors $\mathbf{k} = (2\pi\mathbf{m}/\lambda)$ of two stereoscopic light beams must have the same amplitude but different directions θ_1 and θ_2. It is realizable by means of anisotropic diffraction (Fig. 3b). Additional condition is "wide-aperture" filtering [7] which leads to the single solution (i.e. single configuration) for each incident angle θ. The difference between the angles θ_1-θ_2 may vary from 0 to the maximal value defined by the refractive indexes of the AO cell $\Delta n = n_e - n_o$. So, one can choose the angle θ_1-θ_2 so that both beams provide necessary separation angle after diffraction. Optical scheme for this method is shown in Fig. 3a.

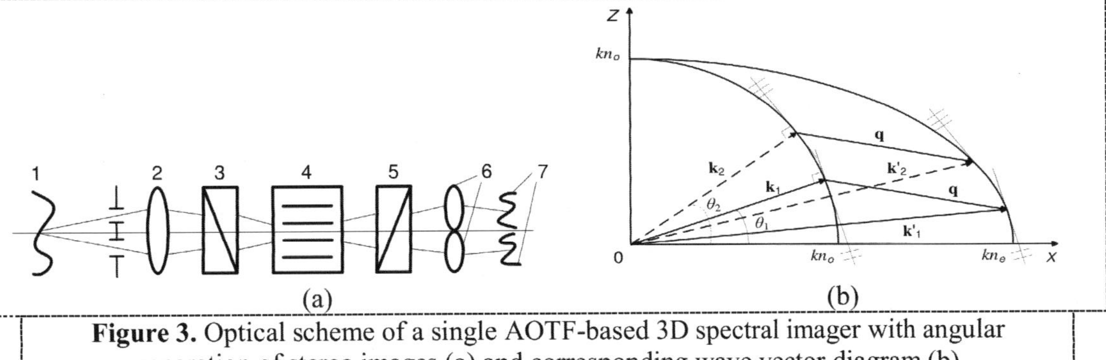

Figure 3. Optical scheme of a single AOTF-based 3D spectral imager with angular separation of stereo images (a) and corresponding wave vector diagram (b)

2.2 Time separation of stereo images

Stereoscopic systems with time separation of images use the single image sensor while two images are alternatively demonstrated. An appropriate AOTF-based optical spectral imager is presented in Fig. 4.

Journal of Physics: Conference Series **661** (2015) 012041 doi:10.1088/1742-6596/661/1/012041

Two channels are formed with input optomechanical system comprising two slits 8, two static mirrors 9 and bi-position mirror 10 (Fig. 4a,b). The basic path of the system is the same for both channels: objective 2, polarizers 3,5, AO cell 4, output lens 6 and photodetector 7. For visual perception, time-synchronous glasses are used. The main advantage of the scheme is that most elements are used in both channels. It allows the development of devices with relatively simple and compact design.
Time separation may be implemented also by the use of specialized AOTF. It is possible to provide such incident angles θ_1 and θ_2 that corresponding frequencies of ultrasound f_1 and f_2 will differ greatly (Fig. 4c,d). Then, filtration of both beams may be done by varying f. Changing the frequencies f_1 and f_2 but keeping the ratio f_2/f_1 other wavelengths may be selected. It may be done in a particular spectral range out of which one acoustic frequency matches two different wavelengths λ in the stereoscopic channels. To limit the range it is necessary to insert additional band-pass filter. Changing this filter it is possible to change the spectral range.

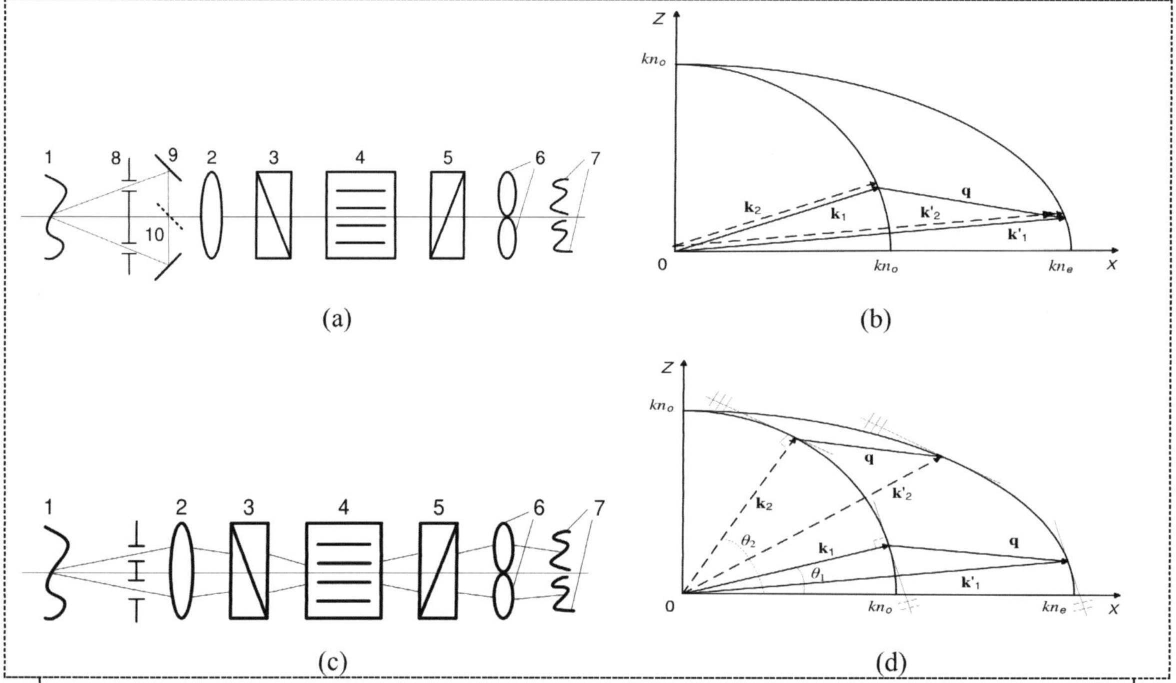

Figure 4. Optical schemes of single AOTF-based 3D spectral imager with time separation of stereo images (a,c) and corresponding wave vector diagrams (b,d)

2.3 Polarization separation of stereo images

Imaging AOTFs use the change of polarization mode for spectral selection the light. This fact can be exploited in the separation of stereoscopic beams. If two incident beams ($\mathbf{k}_1$, $\mathbf{k}_2$) have different polarizations: ordinary («o») and extraordinary («e»), then diffracted beams ($\mathbf{k'}_1$, $\mathbf{k'}_2$) also differ in polarization (Fig. 5b) and can be separated by polarizing glasses.
However, two conditions must be fulfilled: first, the diffraction of both beams must meet wide-aperture condition [7] and, second, both beams must be diffracted via the same acoustic wave $\mathbf{q}$. These two requirements can be completed with an extra condition, assuming that both diffracted beams propagate in the same direction. As it follows from the analysis of formulae describing wide-aperture diffraction [8], for each angle of diffraction θ_{dif} there are single values of incident angles θ_1, θ_2 and ultrasound direction angle γ. Then, one can define the optimum angles [9] corresponding to the maximum value of stereo parallax (θ_2-θ_1):

$$\theta_1^{opt} = \mathrm{arcctg}\left(\left(n_e/n_o\right)^2\right),$$

$$\theta_2^{opt} = \mathrm{arctg}\left(\left(n_e/n_o\right)^2\right).$$

The principle implementation of this scheme is shown in Fig. 5a.

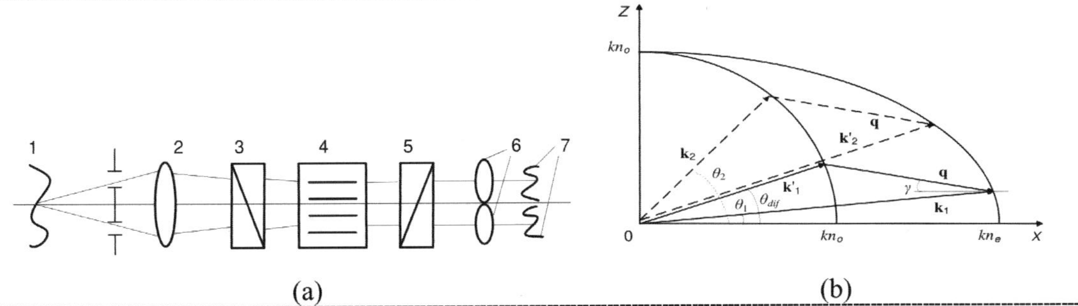

Figure 5. Optical scheme of single AOTF-based 3D spectral imager with polarization separation of stereo images (a) and corresponding wave vectors diagram (b)

3. Example. AOTF-based stereo microscope

To prove the effectiveness of single-AOTF 3D spectral imagers the device based on stereomicroscope was developed and tested. AOTF with two entrance pupils was integrated into the optical scheme of Abbe stereomicroscope after zoom lenses where the light beams in both channels are parallel to the optical axis [10]. It provides the identical geometry of AO interaction for both beams.

The wavelength to be selected is assigned by the user via PC. Controller board, radiofrequency generator and two amplifiers are integrated into AOTF. Illumination from the inspected object is collected by the microscope objective, is transmitted through set of zoom lens. After AO filtration spectral image is formed by tube lenses and may be observed through the oculars or focused on the detector of video camera.

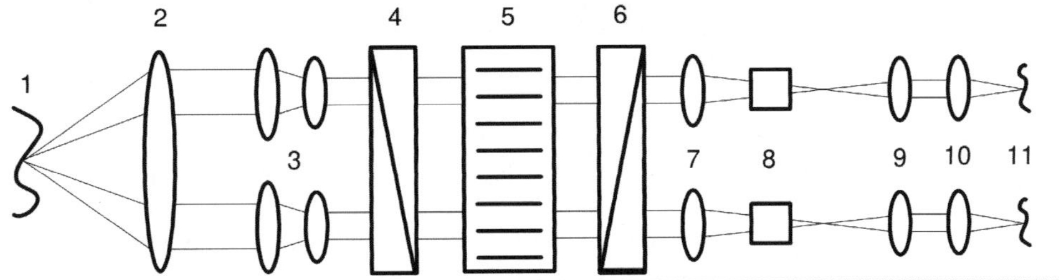

Figure 6. Stereoscopic spectrometer based on Abbe microscope
a) optical scheme: 1 – inspected object, 2 – common main objective, 3 – stereo zoom lens, 4,6 – polarizers, 5 – AO cell, 7 – tube lenses, 8 – image inversion, 9 – eyepieces, 10 – objectives of digital cameras, 11 – registered images

In Fig. 7a the stereoscopic images I and II obtained by the device described are shown. The 3D structure of blood smear reconstructed by means of digital image processing [11] shows vividly an erythrocyte among all other cells (Fig. 7b). Such a 3D imaging system may be useful in biology – for contrast selection of 3D elements in various objects, medicine – for cytometry, gemology – for discovery and localization of foreign inclusions, research of plasma – for visualization of chemical composition and temperature distribution and many other applications.

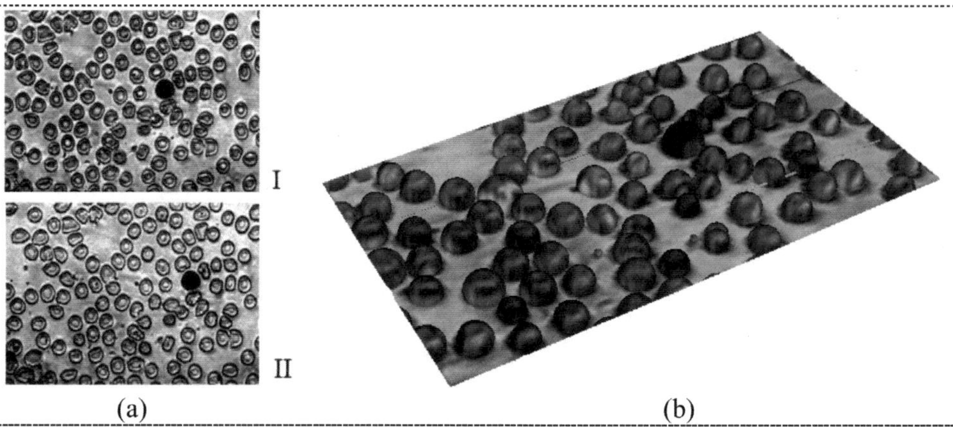

Figure 7. Stereoscopic images (a) and 3D reconstruction (b) of blood smear obtained by stereoscopic spectrometer based on AO monochromator and microscope (λ = 600 nm)

4. Conclusion

For the first time, the implementation of single-AOTF-based stereoscopic 3D spectral imaging is analyzed. It is shown that there are several AOTF configurations applicable for spectral filtration of two stereopair beams. All the schemes are divided into three classes based on angle, time and polarization separation. The experiment with use of the single dual-channel AOTF shows the principal efficiency of AOTF-based 3D spectral imaging.

In future, 3D spectral imagers based on proposed AOTF configurations may be widely used for 3D image reconstruction in microscopy, film industry and other fields.

5. References

[1] Glenar D A Hillman J J Saif B Bergstrakh J 1994 Acousto-optic imaging spectropolarimetry for remote sensing *Appl. Opt.* **33(31)** 7412-24

[2] Bouhifd M Whelan M Aprahamian M 2005 Fluorescence imaging spectroscopy utilizing acousto-optic tunable filters *Proc. SPIE* **5826** 185-93

[3] Dekemper E Loodts N Van Opstal B Maes J Vanhellemont F Mateshvili N Franssens G Pieroux D Bingen C Robert C De Vos L Aballea L Fussen D 2012 Tunable acousto-optic spectral imager for atmospheric composition measurements in the visible spectral domain *Appl. Opt.* **51(25)** 6259-67

[4] Viskovatykh A V Machikhin A S Pozhar V E Pustovoit V I Viskovatykh D A 2014 Combined optical-coherence and spectral microscopy based on tunable acousto-optic filters of images *Tech. Phys. Lett.* **40(2)** 157-60

[5] Pozhar V E Machihin A S 2012 Image aberrations caused by light diffraction via ultrasonic waves in uniaxial crystals *Appl. Opt.* **51(19)** 4513-9

[6] Ezhov V A Studentsov S A 2005 Volume (or stereoscopic) images on the screens of standard computer and television displays. *Proc. SPIE.* **5821** 102-16

[7] Chang I C 1975 Analysis of non-collinear acousto-optic filters *Electron. Lett.* **11(25)** 617-8

[8] Machikhin A S Pozhar V E 2010 Image transfer under wide-angle acousto-optic interaction *Quantum Electronics* **40(9)** 837-41

[9] Machikhin A S Pozhar V E 2014 Obtaining spectral stereo images with electron spectral tuning and polarization separation *Tech. Phys. Lett.* **40(9)** 823-6

[10] Pozhar V E Machihin A S AOTF-based 3D spectral imaging system 2012 *AIP Conf. Proc.* 1433 65-67

[11] Gorevoy A V Machikhin A S 2014 Method to improve accuracy of the geometrical parameters measurement in stereoscopic AOTF-based spectral imagers *J. of Physics: Conf. Series* **584** 012004

Journal of Physics: Conference Series **661** (2015) 012042 doi:10.1088/1742-6596/661/1/012042

Surface modes of big clusters and resonances in generalized LC–model of metal-dielectric nanocomposites

N A Olekhno[1,2], **Y M Beltukov**[1,3] **and D A Parshin**[2]

[1]Ioffe Physico-Technical Institute, 194021 Saint Petersburg, Russia
[2]Saint Petersburg State Polytechnical University, 195251 Saint Petersburg, Russia
[3]Université Montpellier II, CNRS, Montpellier 34095, France

E-mail: [1]`olekhnon@gmail.com`

Abstract. Random impedance networks with inductive L and capacitive C bonds have been widely addressed in a literature to describe properties of disordered metal-dielectric nanocomposites. In this paper we show that networks with single bonds in the form of parallel oscillatory LC–circuits and capacitances C are more appropriate model for metal-dielectric composite in optical frequency range. Resonant spectrum of such networks demonstrates absence of resonances at frequencies higher than the plasma frequency of metal ω_p. Eigenmodes of big ordered clusters are studied. Their resonances are shown to be surface modes having the form of multipoles of various order. It is in agreement with theory of surface plasmons in metallic nanowires. The idea for metal-metal and metal-semiconductor nanocomposites with different plasma frequencies is put forward.

1. Introduction

Disordered nanocomposites consisting of small ($\leq$ 30 nm) granules of noble metal (Ag,Au) incorporated into dielectric matrix attracting considerable attention due to their interesting optical properties. First of all, it is giant fluctuations of local electric fields in composite which lead to Surface-Enchanced Raman scattering (SERS) [1]. This behavior is caused by surface plasmon resonances in metallic grains which allow to concentrate the energy of electromagnetic wave in volumes much smaller than λ^3, where λ is a wavelength of incident radiation. The main feature for SERS is the presence of "hot spots" – regions in composite, where local electric field is much higher than those in the other parts of the composite. Another interesting properties are anomalous optical absorption, higher harmonic generation and other nonlinear effects.

Optical properties of disordered nanocomposites can be adequately modeled by electric networks consisting of inductances L and capacitances C, randomly distributed in a lattice [2]. Inductive bonds L here model metallic grains, while capacitances C model dielectric bridges between them. This representation follows from discretization of Maxwell equations for given composite in quasistatic approximation.

Random LC–networks have a wide resonant spectrum with complicated structure [3]. Local electric fields in such networks exhibit giant fluctuations which have multifractal structure [3]. Multifractality is a result of dipole–dipole interaction between resonant clusters which can be considered as harmonic oscillators [2, 4, 5].

2. Approximation of metal-dielectric nanocomposites by random LC-networks

Maxwell equations for a composite have a form

$$\operatorname{rot}\mathbf{E} = -\frac{1}{c}\frac{\partial \mathbf{B}}{\partial t}, \tag{1}$$

$$\operatorname{rot}\mathbf{H} = \frac{4\pi}{c}\mathbf{j} + \frac{1}{c}\frac{\partial \mathbf{D}}{\partial t}, \tag{2}$$

$$\operatorname{div}\mathbf{D} = 4\pi\rho, \tag{3}$$

$$\operatorname{div}\mathbf{B} = 0. \tag{4}$$

For a non-magnetic composite permeability $\mu = 1$ and $\mathbf{B} = \mathbf{H}$. Let us assume that there are no exterior local charges in the composite, so $\rho = 0$ and $\operatorname{div}\mathbf{D} = 0$. In the quasistatic approximation the magnetic field vanishes, hence $\operatorname{rot}\mathbf{E} = 0$ and we can introduce an electric potential $\mathbf{E} = -\operatorname{grad}\varphi$. Also from $\operatorname{rot}\mathbf{H} = 0$ we get the equation

$$\mathbf{j} = -\frac{1}{4\pi}\frac{\partial \mathbf{D}}{\partial t} \tag{5}$$

where $\mathbf{D} = \mathbf{D_0}\exp(-i\omega t)$. The electric induction $\mathbf{D}$ is related with the electric field $\mathbf{E}$ by the equation $\mathbf{D} = \varepsilon(\omega)\mathbf{E}$. The permittivity of dielectric ε_d can be considered as a constant ε in the optical range of frequencies. The permittivity of metal ε_m in Drude model without dissipation reads

$$\varepsilon_m(\omega) = 1 - \frac{\omega_p^2}{\omega^2}. \tag{6}$$

This relation is valid for metals where $\omega_p\tau \gg 1$ in the frequency range $\omega\tau \gg 1$.

From Eq. (5), we get the expression for the displacement current density in dielectric

$$\mathbf{j} = -i\omega\frac{\varepsilon_d}{4\pi}\operatorname{grad}\varphi. \tag{7}$$

It is equivalent to the relation $I = YU$ between the current $I = ja^2$ and the voltage $U = \varphi_1 - \varphi_2$ in a cubic lattice of capacitances with small lattice constant a. Here $Y = -i\omega C_d$ is the admittance of the capacitor with the capacitance $C_d = \varepsilon a/4\pi$. On the other hand, in metals making use of Eq. (6) we have

$$\mathbf{j} = -\left(i\omega\frac{1}{4\pi} + \frac{\omega_p^2}{i\omega 4\pi}\right)\operatorname{grad}\varphi \tag{8}$$

which corresponds to a cubic lattice of parallel LC–circuits which has the admittance $Y = -i\omega C_m - 1/i\omega L_m$ with the inductance $L_m = 4\pi c^2/(a\omega_p^2)$ and the capacitance $C_m = a/4\pi$. The resonant frequency $c/\sqrt{L_m C_m}$ of this circuits is the plasma frequency ω_p. In the low-frequency range $\omega \ll \omega_p$ the parallel LC–bonds can be replaced by a single inductances L because $i\omega C_m \ll ic^2/\omega L_m$.

As shown in our previous work [6], Kirchhoff equations for LC–network can be reduced to a generalized eigenvalue problem

$$\sum_j K_{ij}\varphi_j - \omega^2\sum_j C_{ij}\varphi_j = 0. \tag{9}$$

Here φ_i are potentials at the lattice sites enumarated by index i, K_{ij} is the matrix of inverse inductances $1/L_{ij}$ and C_{ij} is the matrix of capacitances between sites i and j. If sites i and j

are not connected by inductance or capacitor, corresponding matrix elements K_{ij} or C_{ij} is zero. Diagonal elements of both matrices are

$$K_{ii} = -\sum_{j \neq i} K_{ij}, \quad C_{ii} = -\sum_{j \neq i} C_{ij}. \tag{10}$$

The resonant frequencies ω of the system can be calculated from the equation

$$\det\left(K - \omega^2 C\right) = 0. \tag{11}$$

3. Generalizations of LC–model

Disordered nanocomposites can be adequately modeled by electric lattice with randomly distributed inductances L and capacitances C [2, 3, 9]. The inductances and the capacitances qualitatively correspond to randomly positioned metal grains and dielectric bridges between them. We will consider an LC–network with topology of two-dimensional square lattice, whose sites are connected with bonds which are randomly chosen to be inductances L or capacitances C with probabilities p and $1-p$ respectively. We will refer these probabilities as concentrations of L– and C–entries. Such planar networks correspond to a composite consisting from long metallic rods, randomly positioned in dielectric but parallel to each other.

As was mentioned above, networks with purely L– and C–bonds can be used to model metal-dielectric composites in the low-frequency range $\omega \ll \omega_p$. However this model gives a significant number of resonant frequencies $\omega \approx \omega_p$ and even $\omega \gg \omega_p$ (Figure 1). Therefore it is necessary to study generalized networks with randomly positioned bonds in the form of parallel LC–circuits and capacitances C.

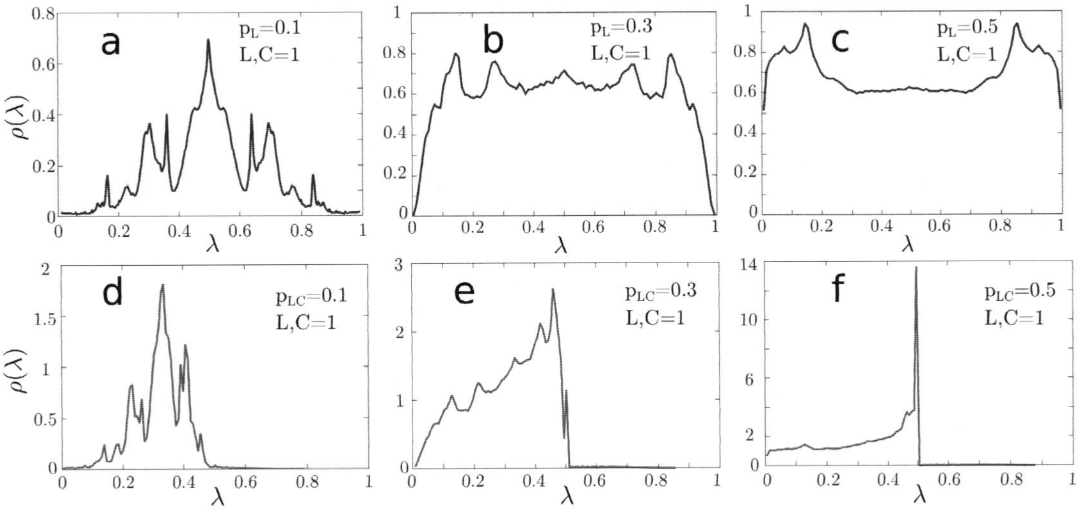

Figure 1. Spectral density of resonances $\rho(\lambda)$ for $2d$ LC–networks with various concentration p of L–bonds (a,b,c) and LC–C–networks with various concentration of parallel LC–circuits (d,e,f). Numerical computation for networks of size 20 x 20 sites with averaging over 20 000 realizations.

Let us introduce the reduced resonant frequencies [2, 3]

$$\lambda = \frac{(\omega/\omega_0)^2}{1+(\omega/\omega_0)^2}, \tag{12}$$

which maps the frequency spectrum $0 < \omega < \infty$ to the interval $[0, 1]$ for λ. It is convenient and allows to investigate some special types of symmetry.

The Figure 1 shows the comparison of the spectral density of resonances (SDR) $\rho(\lambda)$ for random networks with L– and LC–bonds in a lattice of capacitances C for various values of the probability p. The spectral density of resonances for networks with L- bonds is taken from [3]. The spectral density of resonances $\rho(\lambda)$ for various concentrations of LC–bonds p is shown in Figure 1d-f. In the case of low concentrations p (Figure 1d) the shape of $\rho(\lambda)$ resembles that for networks with L– and C–bonds. The redshift of the central frequency from $\lambda = 1/2$ to $\lambda = 1/3$ arises because the frequency of a single "metallic" bond with L,$C = 1$ in capacitive lattice $C = 1$ is now

$$\omega_0^{LC} = \frac{c}{\sqrt{2LC}} = \frac{\omega_0^L}{\sqrt{2}} \tag{13}$$

instead of $\omega_0^L = c/\sqrt{LC} = 1$ for LC–networks. When concentration p is high, the spectrum looks completely diferent (Figure 1f).

One can note that there are no resonant frequencies higher than ω_p ($\lambda = 0.5$) in such random networks with parallel LC circuits. It is in agreement with the fact that metal in the Drude model becomes transparent to incident radiation in frequency range $\omega > \omega_p$ and thus has no absorption. Experimental observations of metal-dielectric nanocomposites show presence of weak absorption peaks with frequencies $\omega > \omega_p$, but since it is a result of nonlinear effects they cannot appear in our linear model. High peak at $\lambda = 0.5$ ($\omega = \omega_p$) which arise near the percolation threshold corresponds to a volume plasmon in metal and is formed by eigenfrequencies of parallel LC–circuits.

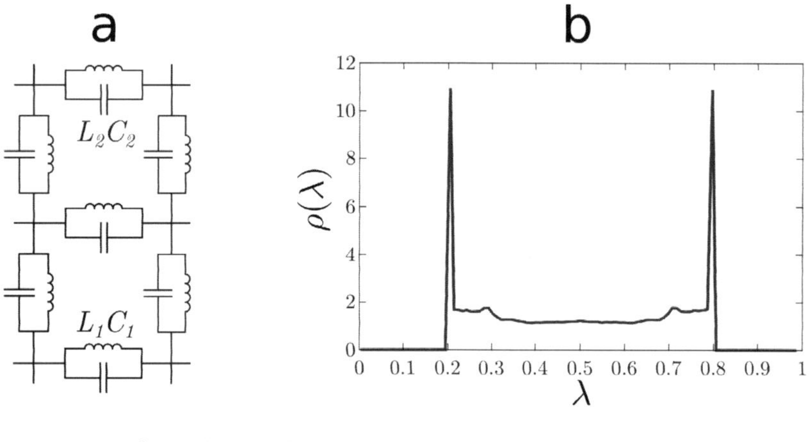

Figure 2. Example of random $2d$ network with two types of parallel LC circuits and corresponding spectral density of resonances $\rho(\lambda)$ for network at the percolation threshold $p_{LC1} = p_{LC2} = 0.5$. Eigenfrequency of parallel LC–bonds of the first type $\omega_{p1} = 1/\sqrt{L_1C_1} = 0.5$ ($\lambda = 0.2$), for the second type $\omega_{p2} = 1/\sqrt{L_2C_2} = 2$ ($\lambda = 0.8$). Numerical computation for networks of size 20 x 20 sites with averaging over 20 000 realizations.

We can introduce more general model in the form of random network consisting from two types of parallel LC–circuits with two different resonant frequencies $\omega_{01} = c/\sqrt{L_1C_1}$ and $\omega_{02} = 1/\sqrt{L_2C_2}$ (see Figure 2a). Such a network corresponds to a mixture of two metals with different plasma frequencies ω_{p_1} and ω_{p_2} or to a composite consisting of metallic grains in the matrix of n–type degenerate semiconductor, which plasma frequency

$$\omega_p^{sc} = \sqrt{\frac{4\pi n e^2}{m}} \tag{14}$$

which depends on the concentration of free carriers n and can be much smaller than those of metal.

It is well seen from the figure that all resonances are positioned in the band between ω_{p1} and ω_{p2} and there are no higher and lower frequencies (Figure 2b). It is in a full agreement with surface plasmon resonance condition, which requires the permittivity of grains ε_g and permittivity of background medium ε_b to be of the opposite signs. In the frequency range $\omega < \omega_{p1}$ both permittivities are negative, while for $\omega > \omega_{p2}$ both ε_g,$\varepsilon_b > 0$ and we have a non-resonant mixture of two dielectrics.

One can see presence of two peaks at the edges of the resonant spectrum (Figure 2a). They have a form of δ-function like singularities and arise close to the percolation threshold. In $2d$ square networks with two types of parallel LC circuits it is obvious that at least one of the singularities exists for any p since percolation threshold $p_C = 0.5$.

4. Resonances of big clusters

Resonances in disordered LC–networks and resonances in ordered linear [2], staircase [9] and fractal [2] inductive clusters in capacitive lattice have been studied in the literature. However resonances in closed dense inductive clusters in the form of different geometrical figures (squares, rectangles, circles etc.) with big sizes have not been studied yet. Such configurations in two-dimensional networks correspond to metal nanowires of different cross-section placed into a dielectric medium (or vacuum). We will examine such a configurations in the framework of the generalized model with parallel LC–circuits and C–bonds considering square clusters from parallel LC–circuits as an example. Such cluster of size $l \times l$ (for $l = 2$) is shown on Figure 3a.

As can be seen from eigenmode potential distribution, which is given by coresponding eigenvector in generalized eigenvalue problem (9), some of the modes have a form of surface oscillations (Figure 3b-d). Number of these surface modes n_s is equal to the number of LC–bonds at the border between cluster and capacitive lattice, i.e. $4l$. Other modes have a structure of volume oscillations at plasma frequency. Therefore the total numder of resonances is equal to the total number of oscillatory LC–bonds in the cluster.

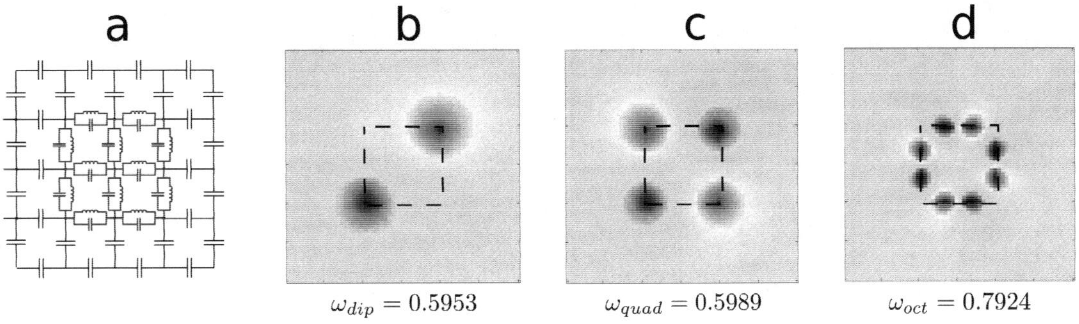

Figure 3. Example of square LC–cluster of size l=2 in lattice of capacitors C (a) and surface modes in square cluster of size l=20 in form of multipoles of various order with corresponding resonant frequencies (b,c,d)

In a 2d lattice the surface modes have a form of planar multipoles of the order m beginning with $m = 1$ (dipole) and higher. The resonant frequency increases with the order of the multipole (Figure 3). It is in a good agreement with classical theory of surface plasmons in metallic nanowires [7]. Some of these surface modes (for example, dipole mode) are two times degenerated due to their symmetry.

It is interesting to note that in the case of purely-inductive clusters (made only from L bonds) all resonant modes have a form of similar surface modes (see Figure 3b-d) with different frequencies. But there are no volume modes in the cluster, hence the total number of resonances in this case $n_{res} = 4l$.

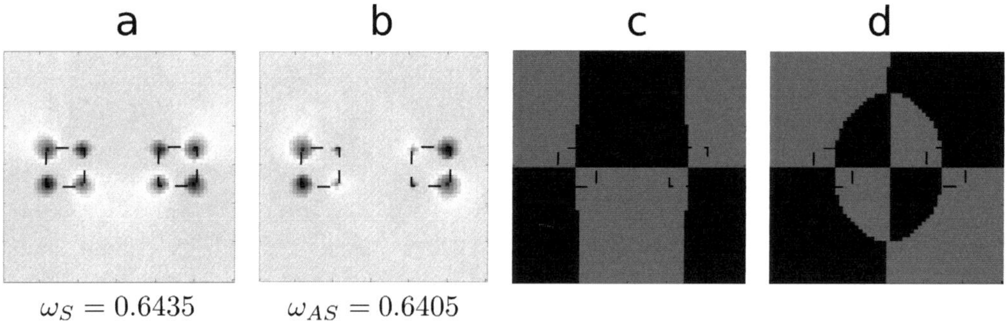

Figure 4. Symmetric (a) and antisymmetric (b) collective eigenmodes in system with two square LC–clusters of size l=10, positioned at distance 9 bonds between them and corresponding potential signatures (c,d)

In the system of two identical dense clusters, placed at some distance from each other the initial multipole modes split on symmetric (Figure 4a) and antisymmetric one (Figure 4b) with different frequencies ω_S and ω_{AS} and thus became non-degenerate due to their interaction. One can see that frequency of the symmetric mode ω_S is higher than of the antisymmetric one ω_{AS}, while usually we have $\omega_{AS} > \omega_S$. This can be understood from signatures of the potential distribution in far field (Figure 4c,d). One can see that the symmetric collective eigenmode has a form of multipole of the higher order than the antisymmetric one and therefore has higher eigenfrequency.

5. Conclusion

More general network description of metal-dielectric nanocomposites in the form of networks with LC– and C–bonds has been proposed. We have shown that spectral properties of such networks is more similar to the spectrum of metal-dielectric nanocomposites than that of networks with purely-inductive and capacitive bonds. Surface modes in dense clusters have a form of multipoles of various order, which is in agreement with theory of surface plasmon resonances in metallic nanowires. The model for metal-metal and metal-semiconductor composites with different plasma frequencies is also proposed.

We are very grateful to Prof. V. I. Kozub for many fruitful discussions. One of the authors (YMB) thanks Dynasty Foundation for financial support.

References

[1] Brouers F, Blacher S, Lagarkov A N, Sarychev A K, Gadenne P and Shalaev V M 1997 *Phys. Rev. B* **55** 13234
[2] Clerc J P, Giraud G, Luck J M and Robin Th 1996 *J. Phys. A* **29** 4781
[3] Jonckheere Th and Luck J M 1998 *J. Phys. A* **31** 3687
[4] Levitov L S 1989 *Europhys. Lett.* **9** 83
[5] Parshin D A and Schober H R 1998 *Phys. Rev. B* **57** 10232
[6] Olekhno N A, Beltukov Y M and Parshin D A 2014 *J. Phys. Conf. Ser.* **541** 012075
[7] Klimov V V, Nanoplasmonics, 2009
[8] Clerc J P, Giraud G, Laugier J M and Luck J M 1990 *Adv. Phys.* **39** 191
[9] Schafer S, Raymond L and Albinet G, 2005 *Eur. Phys. J. B* **43** 81

Nonlinear elastic waves in layered bars made of different materials

A V Belashov[1,2], **G V Dreiden**[1], **A M Samsonov**[1], **I V Semenova**[1]
[1] Ioffe Physical Technical Institute, 26, Polytekhnicheskaya, St. Petersburg, 194021, Russia
[2] National Research Univ. of Information Technologies, Mechanics and Optics, Kronverkskij pr. 49, St. Petersburg, 197101, Russia

E-mail: irina.semenova@mail.ioffe.ru

Abstract. The paper presents experimental observation of strain solitary waves in layered waveguides with layers made of two glassy polymers, PMMA and polystyrene. It is shown that in delaminated structures solitons propagate in each layer independently with parameters typical for waves in each material. In bars with layers bonded with the glassy adhesive a combined wave is formed, propagating as a single one in both layers and exhibiting soliton properties.

1. Introduction

A wide variety of laminated composite materials are used nowadays for various applications, including aerospace and automotive industries. Laminated structural elements may be produced either of the same material or of different ones. Robust coupling of materials with different adhesive and mechanical characteristics is always a complicated task; and further in-service monitoring of their structural integrity becomes an important and challenging problem.

Recently we demonstrated the feasibility of strain solitary waves as a candidate approach for NDT of layered structural elements made of the same material [1]. This approach may be further applied to address the problem of delamination detection in structures made of different materials. In this paper we present experimental results on strain soliton evolution in bonded and delaminated bars with layers made of two glassy polymers, polystyrene (PS) and PMMA. Note that peculiarities of strain soliton formation in these materials and its behaviour in uniform waveguides made of them were studied in detail in our previous papers (see e.g. [1,2] and references therein).

2. Experimental arrangement

Experiments on generation and observation of bulk strain solitons in layered bars were performed using the experimental setup applied in our former research on nonlinear bulk elastic waves in various waveguides (see e.g. [1,2]). The experimental technique is based on laser generation and subsequent holographic recording of the waves under study. The strain soliton is formed in the uniform PMMA rod from the initial shock wave generated in water nearby its input cross section via laser evaporation of the metallic foil (see Fig. 1). The rod (10 mm in diameter, 50 mm long) is bonded to the two-layered bar (10x10 mm cross section) with layers (5x10 mm cross section) made of PS and PMMA. The two waveguides are used in experiments: one with delaminated (clenched) layers, and the other with layers perfectly bonded by the ethylcyanoacrylate (CA), “Super-Moment” adhesive. The initial 'rod' part of the waveguide provides optimal conditions for a soliton formation and ensures that it is definitely the soliton entering the layered part under study. Note that our previous experiments showed that the CA adhesive is close in linear elasticity to that of PMMA and PS (see Table 1 for elastic characteristics of both polymers and CA adhesive) and does not introduce any significant distortions in strain soliton behaviour ([1]). Wave patterns in the bar were recorded using the technique of classical 'analog' holographic interferometry, providing observation of bulk density waves in an optically transparent waveguide via wave-induced local variations of the refractive index of the waveguide.

Table 1. Elastic parameters of PS, PMMA and CA adhesive.

	Density	**Young modulus**	**Poisson ratio**	**3rd order elastic moduli (Murnaghan)**			**Nonlinearity coefficient**	**Sound velocity (rod)**
	ρ g/cm³	E GPa	ν	l GPa	m GPa	n GPa	β GPa	C_r m/s
PS	1.06	4.03	0.34	-18.9	-13.3	-10.0	-35.3	1870
PMMA	1.16	5.27	0.34	-10.9	-7.7	-1.4	-15.9	2080
CA	1.05	2.76						

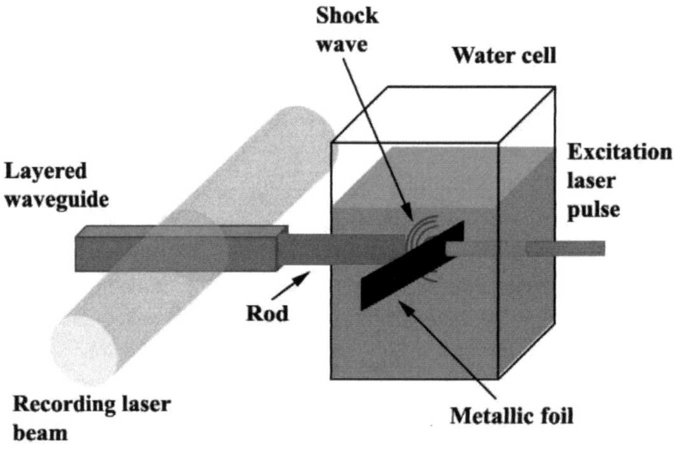

Figure 1. Experimental layout

3. Interferogram processing

The traditional interferogram processing procedure includes measurements of fringe shifts with consecutive calculations of wave amplitude and length. Various defects in interference fringes, discontinuities in particular, complicate or even preclude automatic fringe tracking and interferogram processing by means of known algorithms. We have recently demonstrated ([3]) the possibility to process classical holographic interferograms in finite width fringes using algorithms developed for reconstruction of off-axis digital holograms (e.g. [4-6]) and specifically those sustainable to fringe discontinuities, e.g. [6]. This latter algorithm was applied in the present work for processing of interferograms demonstrating strain soliton evolution in layered bars. The results obtained are shown in Figs. 2 and 3.

The soliton parameters can be directly obtained from the fringe shifts on interferograms or from the reconstructed arrays of phase shift values in the following way. In classical case the soliton amplitude can be calculated from the fringe shift by the formula:

$$A = \frac{\Delta K \lambda}{h(n_1 - 1)(1 - \nu)} \tag{1}$$

where ΔK is the fringe shift, λ is the recording laser wavelength (0,694 μm), n_1 is the refractive index of the material, h is the waveguide thickness along the recording laser beam (10 mm), ν is the Poisson ratio. Since $\Delta K = \varphi / 2\pi$ (where φ is phase), we obtain for the case of digital hologram reconstruction algorithm:

$$A = \frac{\varphi \lambda}{2\pi h(n_1 - 1)(1 - \nu)} \tag{2}$$

4. Results and discussion

The experiments were performed using two specimen in the form of two-layered waveguides with layers made of PMMA and PS. Layers of the first specimen were completely delaminated and clenched together. Layers in the second specimen were bonded by the CA adhesive. In both configurations the layered part of the waveguide was bonded to the initial part (PMMA rod) by the CA adhesive.

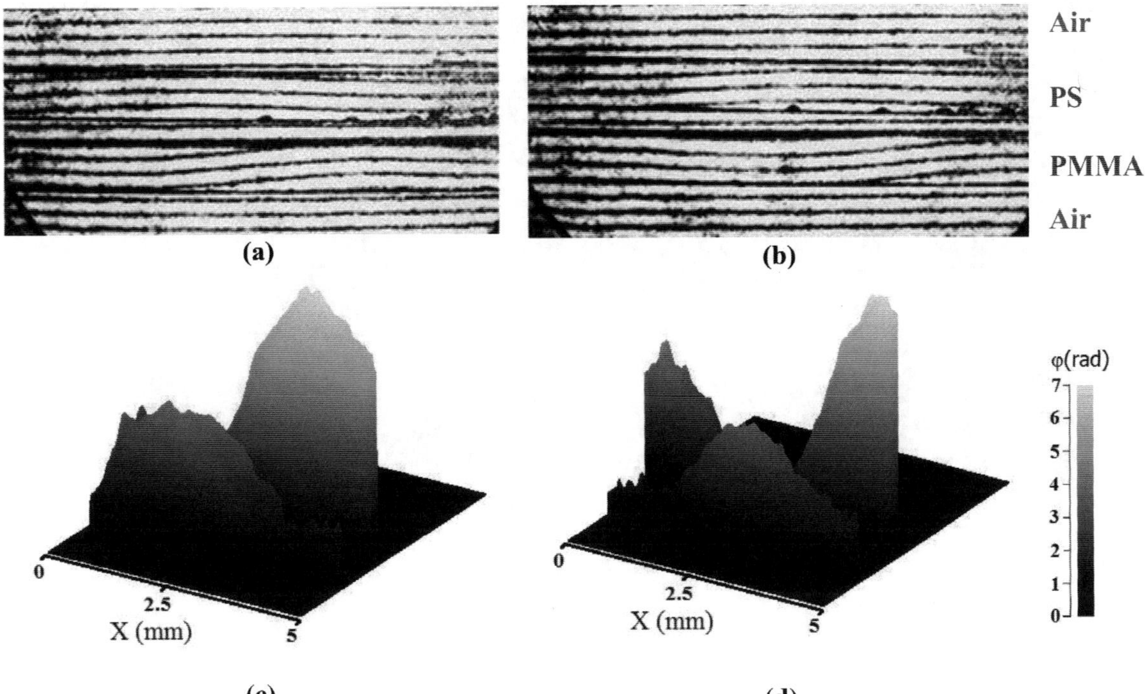

Figure 2. Holographic interferograms (a, b) and results of their processing (c, d) for the strain solitons in a bar with delaminated layers made of PS and PMMA at the distance 20-70 mm from the layered part input. Waves move from left to right. The upper layers (PS) on the interferograms (a, b) correspond to the front curves on the processed phase distributions (c, d).

Figure 2 presents interferograms of wave patterns and results of their processing in a waveguide with delaminated layers at the same distance from the input of the layered part of the waveguide (20-70 mm) but at two succeeding time moments. Obviously wave patterns in the layers differ quite noticeably, solitons in the materials propagate with different velocities, characteristic for each material. Their amplitudes are remarkably different. Moreover in the PMMA layer one can observe the formation of a soliton train, which is in close correspondence with our previous experiments and developed theory [2]. We can conclude that in this case the initial soliton enters two separate layers of PS and PMMA and then two individual solitons propagate along them independently. Due to the difference in velocities (1870 m/s in PS and 2080 m/s in PMMA) the soliton in the upper layer (PS) lags behind the soliton in the lower layer (PMMA). The soliton amplitude in the PS layer is much lower than that in the PMMA layer. This difference is due to two factors. Firstly, at the transverse junction of the PMMA rod and PS layer a part of the soliton energy is reflected back due to the difference in acoustic impedances of the materials. And secondly, it is a consequence of the difference in refractive indices of the materials (1.49 for PMMA and 1.59 for PS) resulting in a slightly different fringe shift for the same density variation (see eq. 1).

The wave patterns shown in Figure 3, where the PS and PMMA layers are bonded with the CA adhesive, are quite different: a combined solitary wave is formed propagating in the layered structure with the resultant velocity, amplitude and shape. Note that the waveguide area covered by the two interferograms in Fig. 3 (a,c) comprises 100 mm which equals to three soliton widths and would be sufficient to detect soliton separation in the layers, if any.

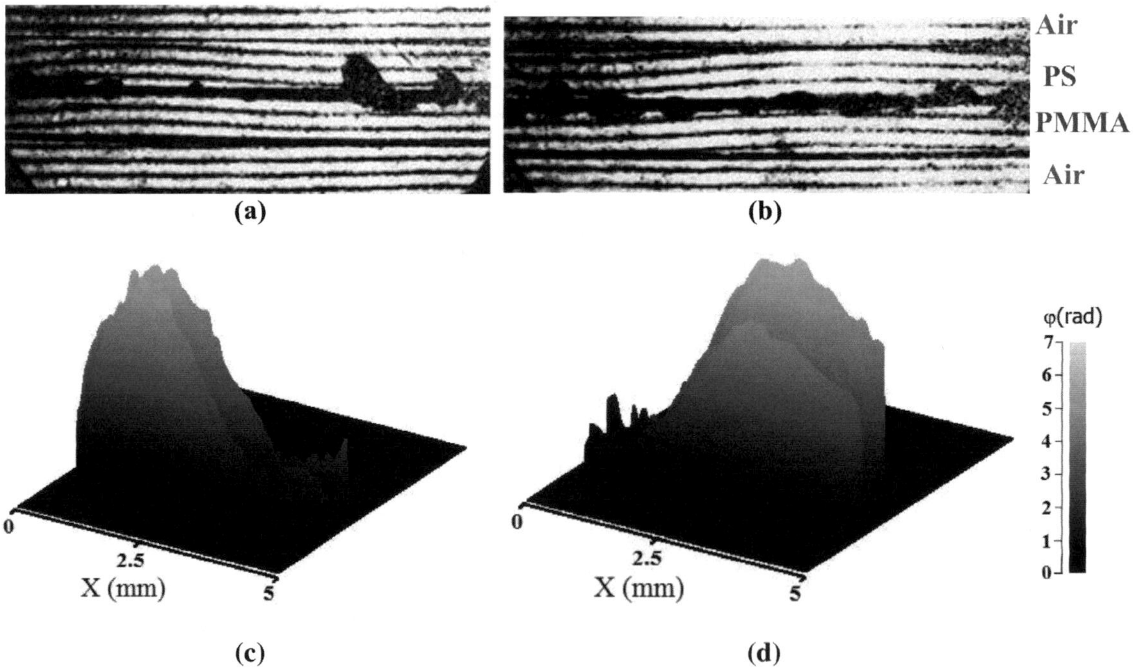

Figure 3. Holographic interferograms (a, b) and results of their processing (c, d) for the strain soliton in a bar with bonded layers made of PS and PMMA at the distance 20-70 mm (a, c) and 70-120 mm (b, d) from the layered part input. Waves move from left to right. The upper layers (PS) on the interferograms (a, b) correspond to the front curves on the processed phase distributions (c, d).

5. Conclusions

Thus it was shown that wave patterns representing strain soliton behaviour in bonded and delaminated waveguides with layers made of different materials are substantially distinctive. In delaminated layers solitons propagate independently exhibiting parameters specific for each material. In CA-bonded layers a single combined soliton is formed which propagates in the layered waveguide as a whole with some resultant parameters. The effect observed can be applied for in-operation delamination detection in layered structural elements.

Acknowledgments

The work was performed under the financial support from Russian Science Foundation (Project # 14-12-00342). AV Belashov acknowledges the governmental financial support for leading universities of Russian Federation (grant # 074-U01).

References

[1] Dreiden G V, Khusnutdinova K R, Samsonov A M, Semenova I V 2012 *J. Appl. Phys.* **112, 6,** 063516
[2] Dreiden G V, Khusnutdinova K R, Samsonov A M, Semenova I V 2010 *J. Appl. Phys.* **107, 3,** 034909
[3] Belashov A V, Petrov N V, Semenova I V 2015 *Tech. Phys. Lett.* (in press)
[4] Verrier N, Atlan M 2011 *Appl. Opt.* **50**, H136-H146
[5] Tahara T, Awatsuji Y, Shimozato Y, Kakue T, Nishio K, Ura S, Kubota T, Matoba O 2011, *Opt. Lett.* 36, 3254-3256
[6] Liebling M, Blu T, Unser M 2004 *JOSA A* **21, 3,** 367-377

Holographic detection of nonradiative transitions in oxygen molecules: digital and classical approach

D M Beltukova[1,3], **A V Belashov**[1,2], **N V Petrov**[2], **I V Semenova**[1], **O S Vasyutinskii**[1]

[1] Ioffe Physical Technical Institute, 26, Polytekhnicheskaya, St. Petersburg, 194021, Russia

[2] National Research Univ. of Information Technologies, Mechanics and Optics, Kronverkskij pr. 49, St. Petersburg, 197101, Russia

[3] St.Petersburg State Polytekhnic University; 29, Polytekhnicheskaya, St.Petersburg, 195251, Russia

E-mail: irina.semenova@mail.ioffe.ru

Abstract. The paper presents a novel approach to detect and monitor nonradiative transitions in singlet oxygen molecules in water by means of holographic recording of thermal disturbances. The approach is realized experimentally using classical and digital holographic techniques. Advantages and disadvantages of each of the techniques are considered.

1. Introduction

Singlet oxygen, the oxygen molecule in its first excited state $a^1\Delta_g$, is one of the active oxygen species which is of paramount importance for a wide variety of biophysical and biochemical processes in nature and living organisms in particular. In medicine it is a key agent in photodynamic therapy (PDT) used for treatment of malignant tumors, some pathogenic diseases, dermatopathies etc.

The most commonly used method for singlet oxygen formation is its photosensitized generation. The detection of singlet oxygen is usually based on the processes associated with its deactivation. The radiative transition from the excited singlet to the ground triplet state of the oxygen molecule $a^1\Delta_g \rightarrow X^3\Sigma_g$ is strongly forbidden by several selection rules that results in the extremely low phosphorescence intensity at 1270 nm. Nonradiative transitions play a key role in the deactivation of singlet oxygen [1]. The nonradiative deactivation of excited oxygen molecules via surrounding molecules causes temperature gradients which may be detected by the techniques sensitive to refractive index variations. Photoacoustic and photorefractive approaches have already been applied for detection of deactivation processes in singlet oxygen ([see e.g. 2]) and have demonstrated their efficiency in providing important information on the process efficiency, quantum yield and temporal characteristics. In the present paper singlet oxygen was detected by means of two holographic techniques based on the recording of phase distributions of a probe laser beam. These techniques are widely used nowadays to obtain information on spatial distributions of refractive index variations caused by different processes.

Various methods of phase reconstruction are developed by now. The classical approach is based on the recording and subsequent optical reconstruction of holograms and holographic interferograms. Some methods are based on the application of specific devices, such as Shack-Hartmann wavefront sensors. Another approach utilizes the wavefront reconstruction from digital holograms. Each of the

Journal of Physics: Conference Series **661** (2015) 012044 doi:10.1088/1742-6596/661/1/012044

approaches exhibit advantages and disadvantages defining their application domains. In this paper we present results of wavefront reconstruction using classical and digital (DH) holographic techniques along with their comparison.

2. Experimental arrangement

In the experiments singlet oxygen was generated in an aqueous solution of Radachlorin photosensitizer [3] by the 50 mW laser diode emitting at 405 nm in the absorption band of Radachlorin. Laser radiation was directed vertically onto the solution surface in the water cell with the excitation area of about 1 mm in diameter (Fig. 1, a). The recording was performed in the perpendicular direction using pulsed Ruby laser (694 nm, 20 ns) in the case of classical arrangement (Fig. 1, b) and CW HeNe laser (633 nm) in digital recording arrangement (Fig. 1, c). In the latter case the Videoscan 205 digital camera with 4.65x4.65 μm pixel size was used for hologram recording.

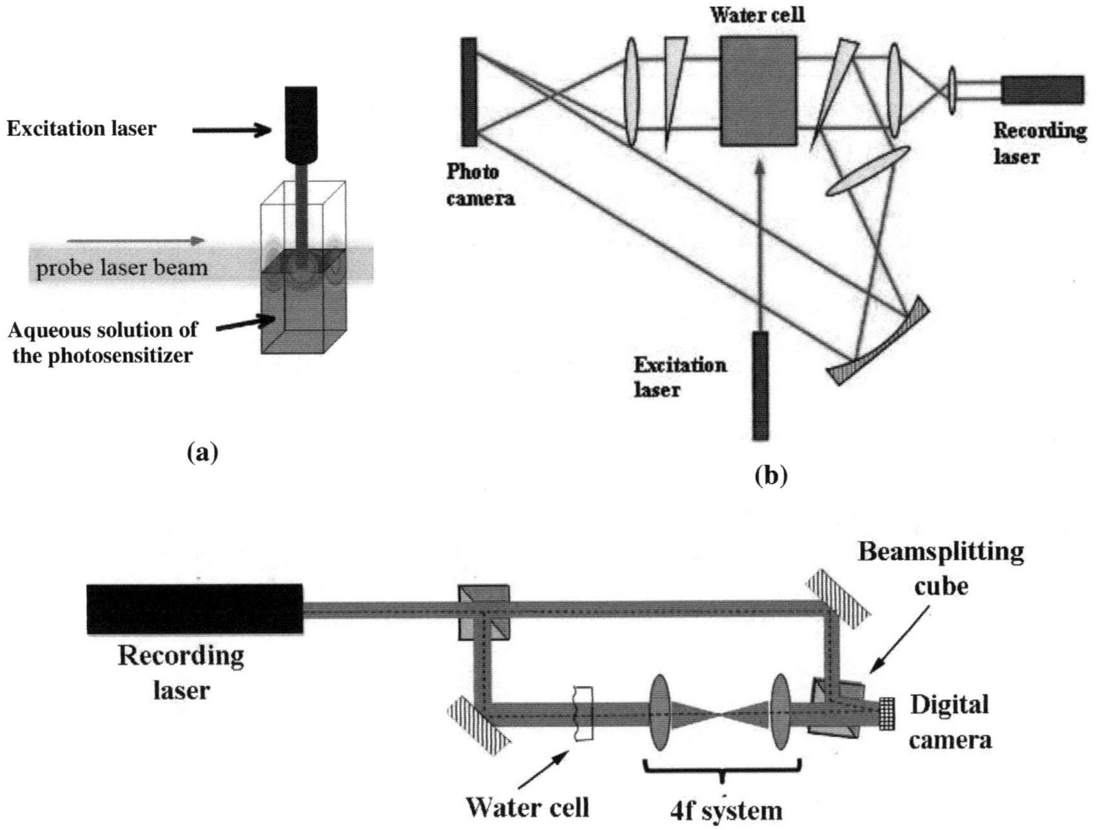

Figure 1. Schematics of the experimental setups. Excitation channel layout (a), classical holographic interferometry arrangement (b), digital holography arrangement (c).

Phase reconstruction from classically recorded holographic interferograms was performed using the "Fringe Analyzer" software, Mathcad 14 had been used for temperature mapping. As a result of the processing temperature maps of the recorded areas were reconstructed. The phase retrieval from digitally recorded holograms was performed using the two-step complex-wave retrieval algorithm developed in [4]. Note that the recently suggested modification of this algorithm [5] provides data retrieval from even highly noisy or modulated by speckle structures digital holograms.

The major distinction of DH arrangement from the classical approach is a relatively strict constraint on the angle formed by the object and reference beams which defines the interference fringe width. From one hand this angle must be smaller than that defined by the Nyquist theorem. On the other hand many reconstruction algorithms (e.g. [4, 6] impose a requirement on the reference wave phase: in the recording plane it should vary much faster than that of the object wave. This condition holds when the angle between the reference and object waves is sufficiently big (usually a few tenths of degree). Another specific aspect of DH is the necessity to reconstruct two digital holograms recorded before and in the presence of the disturbance. The difference of phase distributions of object waves recorded by these holograms represents the sought phase retardation induced by the disturbance.

Note that digital cameras also pose some limitations to experimental realizations. First of all modern digital cameras can not provide frame rate sufficient for recording processes in nanoscale time frame with high spatial resolution. Also due to their high sensitivity a presence of strong extraneous radiation which may result from physical processes occurring with the object under study may hinder hologram recording. In the classical realization such spurious radiation may worsen the quality of each hologram. However at the reconstruction stage only object waves corresponding to the coherent recording radiation are formed, and the quality of the resulting optically reconstructed interferogram does not deteriorate much.

3. Data processing

The processing procedure of classically recorded interferograms includes measurements of fringe shifts with consecutive calculations of phase, refractive index and finally of temperature variations. In the digital approach phase distributions are obtained first from each hologram and then the resulting distribution of phase retardation is calculated as a difference of the two phase distributions taken at the desired time moments. Temperature variations are then calculated from these distributions.

Note that in both realizations integral phase retardation distributions along the recording beam path are obtained from the recorded holograms. To obtain a phase distribution in some cross section an inverse problem is to be solved. In the case of cylindrical symmetry the inverse Abel transform is usually applied. For more correct reconstruction of the initial distribution in a discrete case as many as possible readings should be used for calculations. In DH their number is defined by the pixel size of the recording photosensitive array. In classical holographic interferometry, regardless of the initial higher resolution, only a limited number of readings can be taken which is due to much lower density of the resulting interference fringes.

4. Results and discussion

Figure 2 (a-c) presents holographic interferograms of thermal disturbances induced in 1% Radachlorin solution in water by 405-nm laser excitation of 10s, 15s and 20s duration, respectively. Interferograms were recorded in the classical holographic arrangement. The temperature maps obtained as a result of their processing are shown in Fig. 2 (d-f). Figure 3 presents temperature maps obtained from digitally recorded holograms for 0.17%, 0.35% and 0.7% Radachlorin solution in water at 5s, 10s and 15s excitation durations. The comparison of the results obtained using these two techniques shows a good correspondence of the data retrieved by the two holographic approaches.

Note that the released heat is directly connected to the number of excited oxygen molecules, giving thus an opportunity to obtain information on the spatial distribution of oxygen molecules in the medium under study.

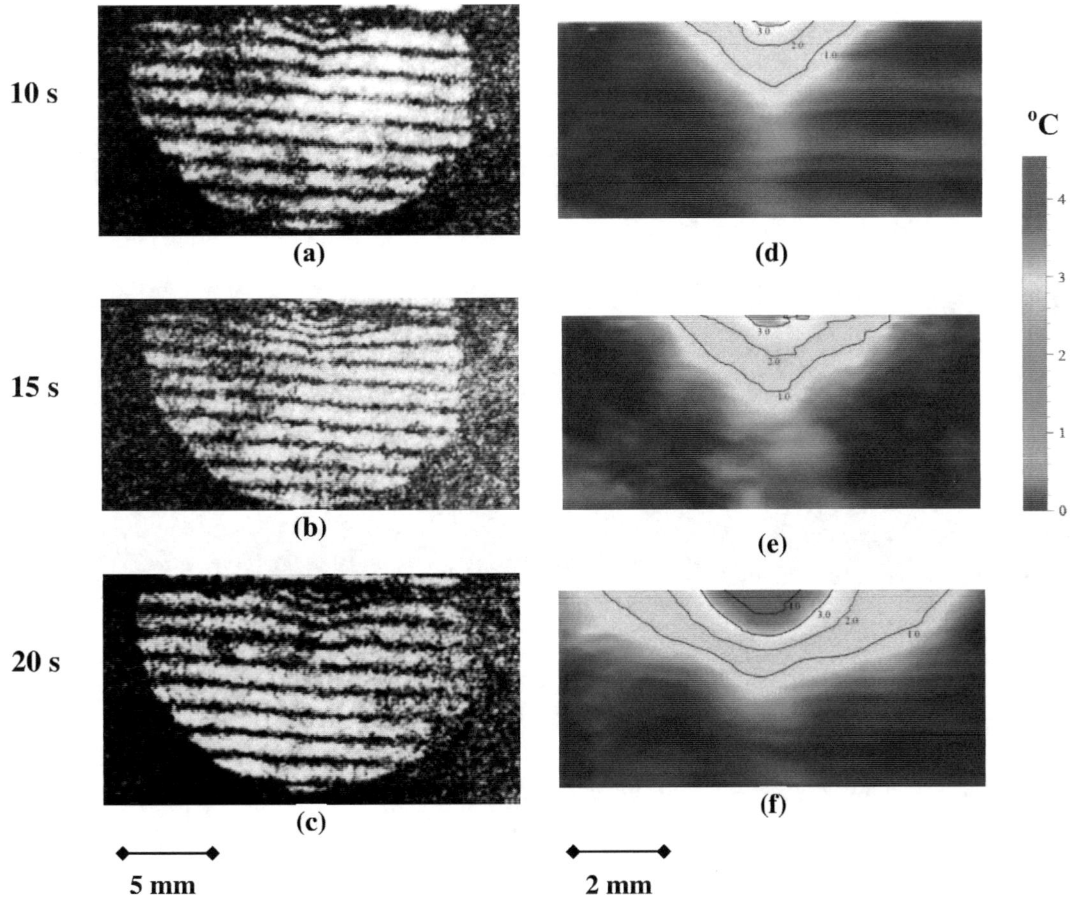

Figure 2. Holographic interferograms (a-c) and corresponding magnified fragments of reconstructed temperature maps (d-f) obtained in 1% aqueous solution of Radachlorin under excitation of 405-nm laser radiation, excitation duration is indicated in the left column.

5. Conclusions

Thus it was shown that both the classical and digital holographic approaches provide reliable (and similar) data on the process under study. Classical holographic interferometry is more suitable to retrieve information on fast developing processes since conventional holographic materials allow hologram recording with pulsed lasers on nano- or even picosecond time scale. Digital holography on the other hand, although providing recording on microsecond time scale, allows one to monitor the process evolution and provides an opportunity to compare phase distributions recorded at any selected time moments.

Mention also that generally speaking the suggested approach is demonstrated to be promising for recording and monitoring of nonradiative deactivation processes in optically transparent media.

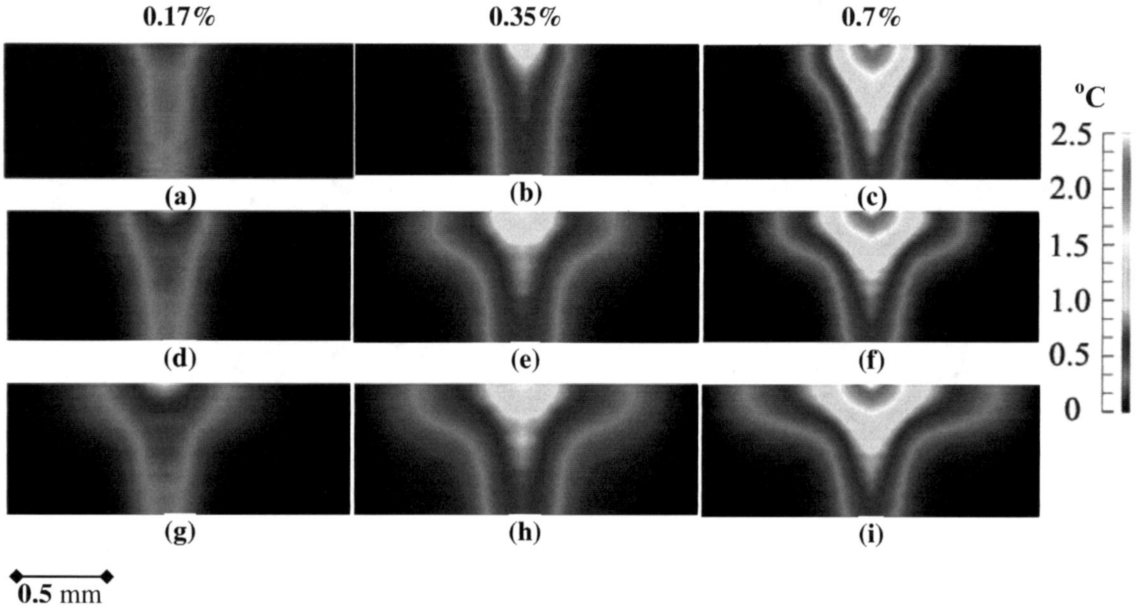

Figure 3. Temperature maps reconstructed from a set of digital holograms recorded at various concentrations of the photosensitizer solution (indicated over each column) and various excitation durations: 5 s – first row, 10 s – second row and 15 s – third row.

Acknowledgments
The work was performed under the financing from Russian Science Foundation (Project # 14-13-00266). AV Belashov acknowledges the governmental financial support for leading universities of Russian Federation (grant # 074-U01). The authors thank RadaPharma company for rendering Radachlorin® photosensitizer for the experiments, and M.V.Volkov (University ITMO) for the "Fringe Analyzer" software.

References

[1] Schweitzer C, Schmidt R 2003 *Chem. Rev.* **103**, 1685–1757
[2] Braslavsky S E, Heibel G E 1992 *Chem. Rev.* **92** 1381-1410
[3] http://www.radapharma.ru/radahlorin.php
[4] Liebling M, Blu T, Unser M 2004 *JOSA A* **21, 3,** 367-377
[5] Belashov A V, Petrov N V, Semenova I V 2014 *Opt. Express* **22**, 28363-28376
[6] Kerr D, Santoyo F M, Tyrer J R 1990 *JOSA A* **7**, 820-826

Journal of Physics: Conference Series **661** (2015) 012045 doi:10.1088/1742-6596/661/1/012045

Generation of multiplexed harmonics at interaction of multi-harmonic light beams in dielectric media with induced plasma.

S.A. Stumpf[1]

ITMO University, 49 Kronverksky ave., Saint-Petersburg, 197101 Russia.

E-mail: stumpf@rain.ifmo.ru

Abstract. A significant number of papers investigate a propagation of intense two-color beam through air in conditions when inharmonic oscillations of valence electrons convert to a plasma nonlinearity. This paper extends the scope of analysis, examining more complex combinations of synchronized beams propagating through dielectric media: triple beam and a beam combined with quazi-static field. It is shown that adding a third harmonic to a mix of first and second ones does not show principal changes in interaction picture: all phenomena appear to be described well before. But adding a static field to an interaction scene shows new peculiarities in redistribution of pulse spectrum energy to infra-red wing of optical range.

1. Introduction.

A generation of spectral supercontinuum by a pair of synchronized few-cycle pulses, including a generation of radiation in terahertz range (THz), is an object of complex and thorough studies this decade [1-5]. Fine tuning of generation procedure, opening a way to cost-effective and efficient generation of long-wave radiation (up to 1 mm), requires an accurate and applicable theory, combining a description of optical processes in dielectric media with analysis of high-energetic processes such as plasma induction. We derived the theory several years ago [6]. This theory gives an excellent correlation between computed data and experimental results of observation of quazi-periodic dependence of generated THz radiation power on interacting pulses spatial shift.

Noticing a fast progress of experimental set-ups, allowing generation and combination of multiple synchronized optic pulses, we found out that theoretically it is possible to increase an efficiency of THz generation by adding pulses of tripled basic wavelength to an interaction scene. So the present paper shows some results of examining such multiwave high-intense femtosecond pulse propagation through dielectric media with plasma excitation. The additional feature of this study is an investigation of high-intense few-cycle pulse interaction with media at presence of quasi-static electric field, causing a separate ponderomotive dynamics of free electrons induced by input pulse.

2. A model for few-cycle pulse evolution in isotropic dielectric media with excited free electrons.

The model for strong few-cycle pulse dynamics in transparent isotropic dielectric media we deduced in [6] based on density matrix for three-band approximation of media energy structure. We assume the higher conductivity subband as a quazi-free motion state for valence electrons, thus it describes an induced photo-plasma. Other assumptions are: an electric field of input wave is essentially smaller than a field of ionized atom on a Bohr radius $E_g = 10^9 V/m$, i.e. we consider pulse intensities $I << 10^{17} W/cm^2$. This allows use of habitual energetic model of a dielectric with quasi-discrete energy state structure. We consider dielectric medium as transparent in a wide range, assuming that the spectrum of input and propagating pulse is very wide but does not exceed the media transparency range. This assumption limits pulse duration to several femtoseconds that stands for only few oscillations of optical field. We assume also linear component of polarization response notably exceeding the nonlinear one, that is met, for example, in fused silica at pulse intensities up to $I \sim 5 \cdot 10^{14} W/cm^2$. Thus we analyze only electronic nonlinearity, neglecting other mechanisms in few-cycle field because of their inertia characteristic times cardinally exceed pulse length [7, 8].

The result has a form of system of waveform evolution equation and dynamic material ones, considering inertial part of field-cube electron nonlinearity as well as inertial plasma nonlinearity induced in pulse strong field. Neglecting a self-reflected wave, as grounded in [9, 10], we combine this system to united evolution equation (1):

$$\begin{cases} \dfrac{\partial E}{\partial z} - a\dfrac{\partial^3 E}{\partial \tau^3} + g\dfrac{\partial E^3}{\partial \tau} + \dfrac{\partial}{\partial \tau}\left(g^{(1)} E \left(\dfrac{\partial E}{\partial \tau}\right)^2 + g^{(2)} E^2 \dfrac{\partial^2 E}{\partial \tau^2} \right) + \dfrac{2\pi}{cn_0} j = 0 \\ \dfrac{\partial \rho}{\partial \tau} + \dfrac{\rho}{\tau_p} = \alpha E^2 \\ \dfrac{\partial}{\partial \tau} j + \dfrac{j}{\tau_c} = \beta \rho E^3, \end{cases} \tag{1}$$

where E is pulse field electrical component, z is a spatial coordinate collinear to pulse propagation direction, $\tau = t - \frac{n_0}{c} z$, c – vacuum light speed, n_0 and a characterize a dependence of linear refraction index n on a frequency ω

$$n(\omega) = n_0 + ca\omega^2, \tag{2}$$

which can be easily refined, when necessary, with higher order addictives, by expanding (3) with higher order field time derivatives [10], j is induced electric circuit density, ρ is a density of excited (to lower conductivity band) dipoles; factors g, $g^{(1)}$, $g^{(2)}$ describe nonlinearity refraction index of dielectric media

$$n_2(\omega) = n_2^0 + A\omega^2, \tag{3}$$

where $n_2^0 = \frac{3}{2} cg + \frac{1}{2} \tau_{12} cg^{(1)}$, $A = \frac{1}{2} c(g^{(1)} - 3g^{(2)})$, $\alpha = \frac{1}{2n_0} \frac{|p_{23}|^2}{T_{21}(\hbar\omega_{32})^2}$, $\beta = \frac{e^2}{m_e^* c} \cdot \frac{n_0^2 - 1}{\hbar\omega_{21}}$ characterizes inertial plasma nonlinearity of dielectric media (see [6]), p_{ik} are scalar values of dipole moments of interstate transitions $i \to k$, ω_{ik} – frequencies according to these transition energy, T_{ik} are interstate relaxation times and τ_{ik} are state excitation relaxation times, m_e^* is an effective electron mass in conductivity band, $\sigma = 2\tau_c^{-1}$ is defined by average collision relaxation time τ_c of free electrons, which has an order of tens femtoseconds for most dielectrics.

This model was applied to an investigation of field structure of generated IR and THz radiation in an experiment of bi-harmonic beams propagation in air held in ITMO University [6]. A dependence of generation efficiency on filament (generated by pulses interaction in air) length is acquired and examined. It is shown that the dependence is periodical by filament lengths. We gave some recommendations on optimization of experimental setup for acquiring a most intensive THz radiation: an output power can be significantly increased by controlling a filament length (by focusing system parameters). It was also shown that integral intensity of generated pulse in range 25.5 μm – 0.2 mm depends periodically on filament length, and this THz pulse appears as few-cycle.

3. An interaction of multi-harmonic input pulse with dielectric media at photoplasma induction conditions.

Figures 1-2 illustrate results of numerical simulations of multi-harmonic few-cycle pulse represented as

$$E(t) = E_0 \exp(-(tt_u^{-1})^2)\sin(\omega_0 t) + E_1 \exp(-(tt_u^{-1})^2)\sin(2\omega_0 t + \varphi_1) + E_2 \exp(-(tt_u^{-1})^2)\sin(3\omega_0 t + \varphi_2), \quad (4)$$

where E_0 – initial electric field amplitude, t_u – input pulse duration, ω_0 – central wavelength, E_1, E_2 are amplitudes of synchronized derivative harmonics (they are functions of E_0, determined by a scheme of experimental set-up), φ_1, φ_2 are relative phase shifts accordingly. Modeled media is chosen to be an atmosphere air, so we use the following values for media properties: a linear refraction index $n_0 = 1.00027$, nonlinear refraction index $n_2 = 5.6 \cdot 10^{-19} cm^2 / W$, $\rho = 1.225 \cdot 10^{-3}\, g/cm^3$, $\tau_c = 65\ fs$, $\tau_p = 150\ fs$ ([11, 12, 13]).

Figure 1 shows the extrabroadening of input pulse spectrum (dotted line) in a wavelength range of $350 - 950 nm$, at pulse properties: duration $t_u = 10 fs$, central wavelength $\lambda_0 = 2\pi c/\omega_0 = 800 nm$, input peak intensity $I = 1 \cdot 10^{14} W / cm^2$, propagation distance $z = 0.32 mm$, $E_1 = 0.5 E_0$, $E_2 = 0$. To illustrate the role and importance of ionization mechanisms, we compare the modelled spectrum (solid line) with the one obtained by modeling the same circumstances and propagation regime, except plasma nonlinearity (dashed line). It is notable that plasma nonlinearity causes a drastic broadening of pulse spectrum into a long-wave range, that was examined in a set of previous our publications [14, 15, 16]. We see that this effect is mostly caused by plasma nonlinearity, as reference dashed line (effect without plasma) shows almost no generation in long-wave wing.

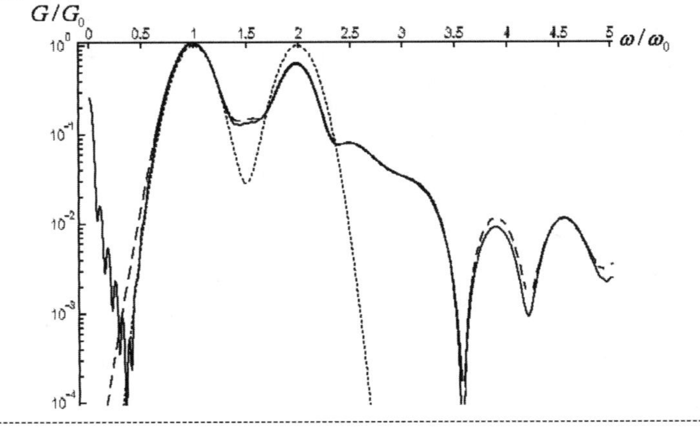

Figure 1. Spectral density of energy of electric field for pulse (4) with $E_2 = 0$ propagating with plasma nonlinearity (solid line), compared with the one in case of plasma neglecting (dashed) and initial spectrum (dotted). G_0 stands for maximum of spectral density.

Figure 2 illustrates adding the third harmonic to the interaction scheme: $E_2 = 0.5 E_0$, all other properties are the same as above. The most noticeable thing is that no principal addition is caused to an interaction scheme by such addition: because a generation of third harmonic is generic and well-known effect of inertia-less nonlinearity mechanisms, the photo-induced plasma makes no principal contribution in the whole process. All differences betweed examined (solid) and reference (dotted) plots are included into Figure 1 and caused by second harmonic participation in interaction process.

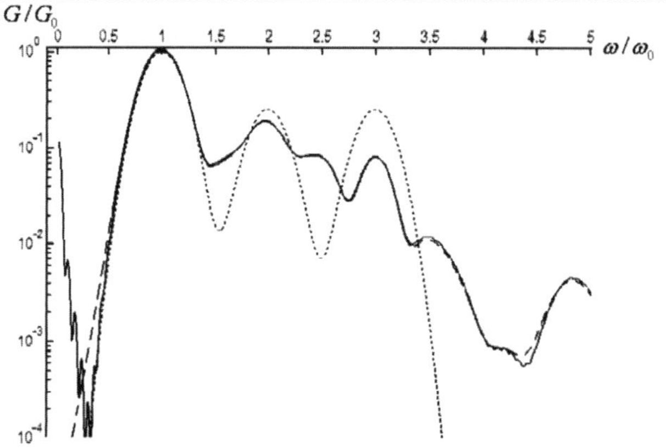

Figure 2. Spectral density of energy of electric field for pulse (4) with $E_2 = 0.5E_0$ propagating with plasma nonlinearity (solid line), compared with the one in case of plasma neglecting (dashed) and initial spectrum (dotted). G_0 stands for maximum of spectral density.

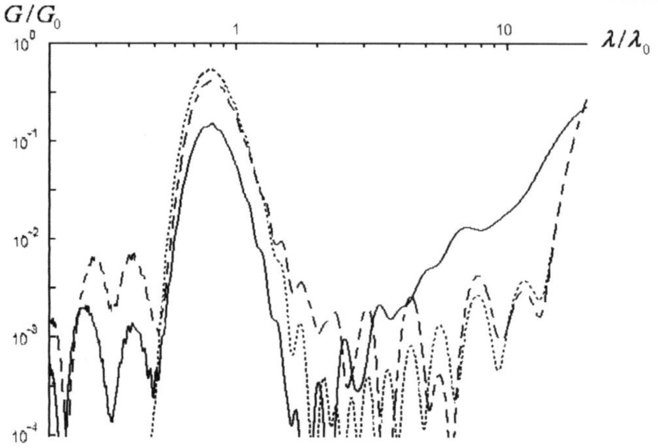

Figure 3. Spectral density of energy of electric field for pulse (4) with $E_2 = 0.5E_0$ propagating with plasma nonlinearity (solid line), compared with the one in case of plasma neglecting (dashed) and initial spectrum (dotted). G_0 stands for maximum of spectral density.

Figure 3 shows a result of a propagation of high-intense few-cycle pulse in dielectric media at presence of quazi-static electric field:

$$E(t) = E_0 \exp(-(t/t_u)^2)\sin(\omega_0 t) + E_1 \exp(-(t/10t_u)^2), \tag{5}$$

It is noticeable that the most visible effect of such propagation is a notable redistribution of pulse energy to low-wave wing of optical spectrum (for better view, exponential abscissa is presented on a plot for wavelength). In experiment, the situation can be modelled by adding an electric condenser with charging contour to a focusing point of input pulse. It is expectable to obtain wide-band radiation in IR-spectrum, caused by an interaction of quazi-free electrons induced by optical pulse, static and oscillating electric field.

4. Conclusion.

The study reports an application of dynamic interaction model to simulate a propagation of complex multi-harmonic high-intense femtosecond pulse through dielectric media, under circumstances initiating an excitation of free electrons in media volume. We show that a propagation of plasma-inducing pulse through dielectric at presence of quazi-static field is accompanied by a notable redistribution of pulse energy to low-wave wing of optical spectrum. That gives a recommendation how to set up an experiment for obtaining wide-band radiation in IR-spectrum, caused by an interaction of quazi-free electrons induced by optical pulse, static and oscillating electric field. The model shows its definite practical value in prospective application to experiment planning and experimental data processing.

References

[1] J. Dai, J. Liu, X. C. Zhang 2011 *IEEE J. Sel. Top. Quantum Electron.* **17** 183–90
[2] N. Yang, H.-W. Du 2014 *OSA Opt. Expr.* **22** 25494–99
[3] K. Kim, J. Glownia, A. Taylor, G. Rodriguez 2012 *IEEE J. Quantum Electron.* **48** 797–805
[4] A. V. Borodin, N. A. Panov, O. G. Kosareva, V. A. Andreeva, M. N. Esaulkov, V. A. Makarov, A. P. Shkurinov, S.L. Chin, X.-C. Zhang 2013 *Opt. Lett.* **38** 1906–08
[5] Andreev A A, Bespalov V G, Gorodetsky A A et. al. 2009 *Opt. Spectr.* **107** 569-76
[6] Stumpf S A, Korolev A A, Kozlov S A 2007 *Izv. RAN, Ser. Phys.* **71** 158-61
[7] Berkovsky A N, Kozlov S A, Shpolyansky Yu A 2004 *Topics in coherent and nonlinear optics*, 170-88
[8] Husakou A V, Herrmann J 2001 *Phys. Rev. Lett.* **87** 203901
[9] Kozlov S A, Sazonov S V 1997 *JETP* **111** 404–18
[10] Bespalov V G, Kozlov S A, Shpolyansky Yu A, Walmsley I A 2002 *Phys.Rev.A.* **66** 0138111
[11] Reference Tables on Physical Constants 1991 1232
[12] Kandidov V P, Kosareva O G, Tamarov M P, Brode A, Chin S 1999 *Quant. Electron* **29** 73-77
[13] Sudrie L, Couairon A, Franco M, Lamouroux B, Prade B, Tzortzakis S, Mysyrowicz A 2002 *Phys. Rev. Lett.* **89** 186601.
[14] Shtumpf S A, Kozlov S A, Korolev A A 2013 *Journ. Phys. Conf. Ser.* **461** 012045.
[15] Shtumpf S A, Shpolyansky Y A, Korolev A A, Kozlov S A 2013 *Journ.Optic.Techn.* **80** 269-73
[16] Shtumpf S A, Korolev A A, Kozlov S A 2007 *Journ.Optic.Techn.* **74** 723-25

Analytical estimations for statistical characteristics of Casimir field detected by indirect measurement scheme

Sheremetyev V.O., Trifanov A.I.
ITMO University, Kronverksky, 49, Saint-Petersburg, 197101, Russia
E-mail: vlad.sheremetiev@gmail.com, alextrifanov@gmail.com

Abstract. We investigate the process of Casimir field indirect photodetection. The protocol of measurement is implemented via interaction between two-level atom and cavity. We evaluated basic statistical characteristics of conditional state of Casimir field .

1. Introduction

Dynamical Casimir Effect (DCE) is the process of photon creation from vacuum due to the cavity boundary movement or medium physical parameters changing (e.g. dielectric constant). This effect is based on the parametric process of vacuum fluctuations amplification [1, 2, 3].

Earlier the main difficulty of DCE experimental verification was the very high frequency variation of moving boundaries [4]. Experimental confirmations of DCE was obtained by means of the superconductive scheme [5]. Here authors realized the idea of imitation of boundary moving [6, 7]).

In this paper, we investigate the process of electromagnetic field photodetection, generated from DCE. In particular, we are interested in indirect measurement scheme implemented via interaction between two-level atom and cavity electromagnetic field. To analyze statistical properties of DCE field we use a two-level atom-pointer passing through the cavity. After the interaction atomic state is detected in ionizing chamber. Examination of detection result allows us to determine properties of cavity field conditional state. Through repeating the interaction many times we can investigate the quantum trajectory for the field state using atomic state detection statistics. Using this ideas we evaluated analytically basic statistical characteristics of conditional state of DCE field .

2. The model

A composite mirror changes its reflection properties (under intermittent laser light irradiation) in empty cavity for periodic movement imitation. It is used for field generation in GHz frequency range [7]. Nonadiabatic change of cavity spectrum produces real photons with frequencies determined by parametric resonance conditions. We assume that the main contribution to photon production is given by squeezing process and relativistic effect of intermode interaction due to mirror's accelerated movement is neglected. We investigate the generate regime of photon creation which allows us to reduce the problem to interaction of atom with single mode of intracavity electromagnetic field. We describe the interaction between the atom and the field by the following

Hamiltonian [8]:

$$\widehat{H} = x\widehat{a}^{\dagger}\widehat{a} + \frac{\Delta + x}{2}\widehat{\sigma}_z + i\frac{\varepsilon\nu}{4}\left(\widehat{a}^{\dagger 2} - \widehat{a}^2\right) + g\left(\widehat{a}^{\dagger}\widehat{\sigma}_- + \widehat{a}\widehat{\sigma}_+\right), \tag{1}$$

where $\widehat{a}$ is the field annihilation operator; g is the atom–field coupling constant (assumed real); $\varepsilon\nu$ is the effectiveness of generation process; x is the resonance shift between modulation frequency and cavity frequency; Δ is the atom-field detuning; $\widehat{\sigma}_- = |g\rangle\langle e| = \widehat{\sigma}_+^{\dagger}$ are Pauli operators, $|g\rangle$ and $|e\rangle$ are atomic ground and excited states; $\widehat{\sigma}_z = |e\rangle\langle e| - |g\rangle\langle g|$.

Applying Bogolyubov transformation of (1) with squeezing operator $\widehat{S} = \exp\left[\frac{1}{2}\left(\beta^*\widehat{a}^2 - \beta\widehat{a}^{\dagger 2}\right)\right]$, where

$$\beta = re^{i\theta}, \quad r = \frac{1}{2}\operatorname{arctanh}\frac{\varepsilon\nu}{2x}, \quad \theta = \pi/2, \tag{2}$$

we get the following form of Hamiltonian $\widehat{H}_B = \widehat{S}^{\dagger}\widehat{H}\widehat{S}$:

$$\widehat{H}_B = \mu\left(\widehat{a}^{\dagger}\widehat{a} + \frac{1}{2}\right) + \frac{\alpha}{2}\widehat{\sigma}_z + \\ + g\left[\left(\widehat{a}\cosh(r) - i\widehat{a}^{\dagger}\sinh(r)\right)\widehat{\sigma}_+ + \left(\widehat{a}^{\dagger}\cosh(r) + i\widehat{a}\sinh(r)\right)\widehat{\sigma}_-\right]. \tag{3}$$

Here $\alpha = \Delta + x$ and

$$\mu = \frac{\sqrt{4x^2 - (\varepsilon\nu)^2}}{2}.$$

From (2) condition of validity of this unitary transformation was obtained: $|\varepsilon\nu/2x| < 1$.

Also we assume resonance regime $|\mu - \alpha| \ll \mu, \alpha$ in system (3). Writing (3) in interaction picture with operators $\exp\left(i\mu(\widehat{a}^{\dagger}\widehat{a} + \frac{1}{2})t\right)$ and $\exp\left(i\alpha\frac{\widehat{\sigma}_z}{2}t\right)$ and applying RWA we reach the following Hamiltonian in the form of Jaynes-Cummings:

$$H_{J-C} = g\cosh(r)(\widehat{a}\widehat{\sigma}_+e^{it(\mu-\alpha)} + \widehat{a}^{\dagger}\widehat{\sigma}_-e^{-it(\mu+\alpha)}). \tag{4}$$

From here we can obtain analytical expressions for operator of vacuum state conditional evolution. It can be done by expanding evolution operator in basis of atomic states. Then we solve system of operator-valued differential equation and get expressions for case of the atom prepared in its ground state and detected in excited state. This result for operator of vacuum state conditional evolution is given by

$$U_{ge}(t) = -i\frac{(g\cosh(r))\,e^{-i\delta t/2}}{\sqrt{\delta^2/4 + (g\cosh(r))^2 aa^{\dagger}}}\sin\left(t\sqrt{\delta^2/4 + (g\cosh(r))^2 aa^{\dagger}}\right)a, \tag{5}$$

where $\delta = \mu - \alpha$.

Operators U_{ge} have the following physical interpretation: they map initial state $|\varphi(0)\rangle$ of intracavity field onto the final state $|\varphi(t)\rangle$, conditioned by detection result e. This final state however is not normalized and full transformation is

$$|\varphi_e(t)\rangle = \frac{U_{ge}(t)\,|\varphi(0)\rangle}{P_e(t)}, \quad P_e(t) = \langle\varphi_e(t) \mid \varphi_e(t)\rangle. \tag{6}$$

Here $P_e(t)$ is the probability to detect the atom in state $|e\rangle$ after interaction lasted time t.

we will consider the system under resonance conditions $\delta = 0$

3. Analytical results

From (3-5) we can evaluate basic statistical properties of DCE field quantum state, specifically: probability to detect atom in its excited state

$$P_e(t) = \langle \phi_e(t)|\phi_e(t)\rangle = \sum_m |\langle 0|S|m\rangle|^2 \sin^2\left(gt\sqrt{m}\right); \tag{7}$$

mean number of produced photons

$$\langle \widehat{n}(t)\rangle = \langle \phi_e(t)|\widehat{a}^\dagger \widehat{a}|\phi_e(t)\rangle = \sum_m |\langle 0|S|m\rangle|^2 \sin^2\left(gt\sqrt{m}\right)(m-1); \tag{8}$$

probability distribution for mean photon number

$$P_e(n) = |\langle 0|\widehat{S}^\dagger U_{ge}\widehat{S}|n\rangle|^2 = \left|\sum_m \langle 0|S^\dagger|m\rangle \frac{\sin^2\left(gt\sqrt{m}\right)}{\sqrt{m}}\sqrt{m+1}\langle m+1|S|n\rangle\right|^2; \tag{9}$$

dispersions of field quadratures

$$\langle(\Delta\widehat{X})^2\rangle = \frac{1}{2}\left[\sum_m \langle 0|S^\dagger|m\rangle M\langle m|S|0\rangle + \sum_m |\langle 0|S|m\rangle|^2 \sin^2\left(gt\sqrt{m}\right)(2m)\right], \tag{10}$$

where $M = \frac{\sin\left(gt\sqrt{m+2}\right)}{\sqrt{m+2}}\frac{\sin\left(gt\sqrt{m}\right)}{\sqrt{m}}\left(m\sqrt{(m+1)(m+2)} + (m-2)\sqrt{m(m-1)}\right);$

$$\langle(\Delta\widehat{P})^2\rangle = -\frac{1}{2}\left[\sum_m \langle 0|S^\dagger|m\rangle M\langle m|S|0\rangle - \sum_m |\langle 0|S|m\rangle|^2 \sin^2\left(gt\sqrt{m}\right)(2m)\right], \tag{11}$$

where

$$\langle n|S|m\rangle = \frac{\left(e^{i\theta}\tanh r\right)^n}{\sqrt{2}^n\sqrt{n!}\cosh r}\exp\frac{e^{-i\theta}m^2\tanh r - m^2}{2}H\left(n, \frac{me^{-i\frac{\theta}{2}}}{\sqrt{2\cosh r\sinh r}}\right). \tag{12}$$

Then we present the results followed from the numerical analysis and comparison of evolution generated by Hamiltonian (1) and expressions $(7-11)$. The system parameters were chosen in such a way to meet conditions which gives by transformations leading from (1) to (4). Also we have chosen parameters so as to reach the full resonance ($|\mu - \alpha| = 0$). Namely, the following values of parameters are used:

$$\varepsilon\nu = 1.37, \quad \Delta = -0.29, \quad g = 0.7, \tag{13}$$

then $x = (-((\varepsilon\nu)^2 + 4\Delta^2))/8/\Delta = 0.954$.

In first figure the probability to detect atom in excited state is shown: we see what probabilities corresponding with (1) and (4) are close in start of interaction between atom and field. In second figure we show the mean cavity photon number. These quantities, conditioned by detection result, are close in the same period of time. For the instant of time $gt \approx 1$ the mean photon number probability distributions are presented in Fig.2. These distributions contain the nonzero probability only for odd numbers of photons. The matter is that in the absence of atom the intracavity field is in squeezed vacuum state (which has the distribution with only even photons). When atom is found in its excited state, the photon number in cavity decreases by one. We can see that probabilities to find the certain number of photons are almost identical.

In Fig.3 field quadratures are shown. These quantities describe the nonclassicallity of the generated intracavity field interacted with atom during certain time interval. From here it is easy to see that statistical properties of the field change dramatically [9] due to the interaction with atom. In the origin of this process quadrature of momentum decreases to the values, corresponded to squeezed state. Quadratures of momentums and quadratures of coordinates are well coincide in the certain period of time.

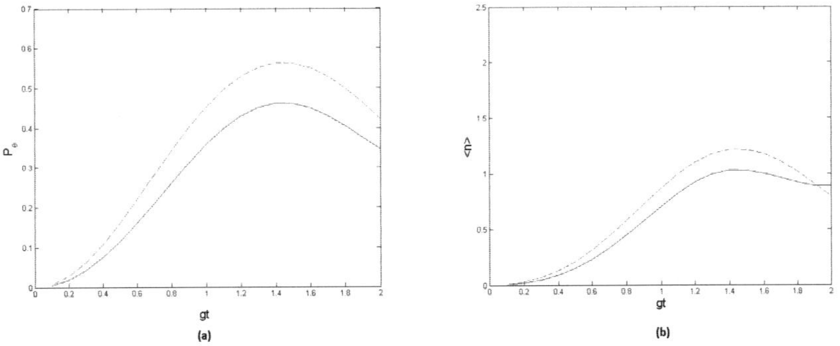

Figure 1. Probability **(a)** and mean photon number **(b)** as functions of dimensionless time gt; dash line corresponds to analytical analysis, solid line corresponds to numerical analysis

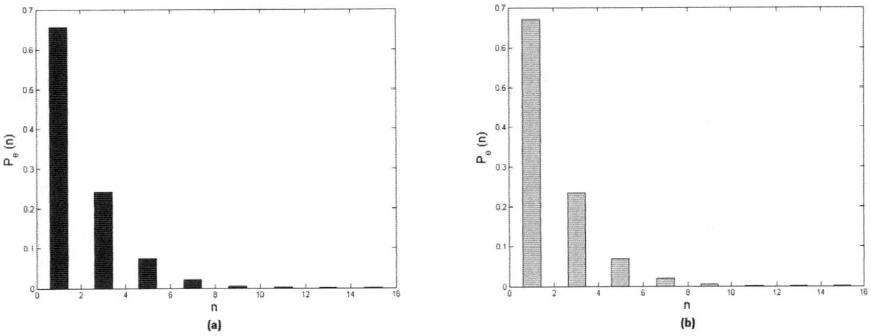

Figure 2. Photon number probability distributions for $gt \approx 1$; **(a)** corresponds to numerical analysis, **(b)** corresponds to analytical analysis

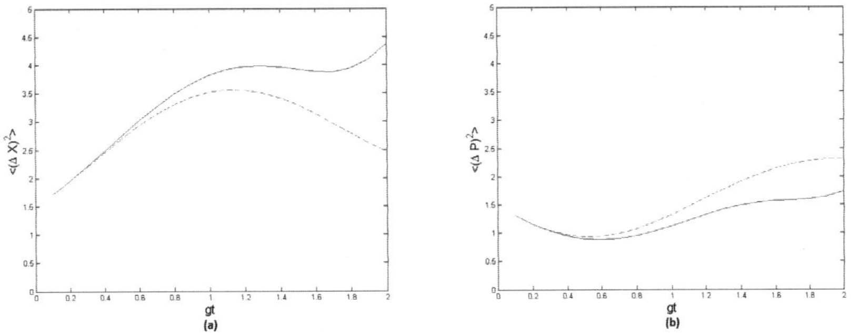

Figure 3. Field quadratures: $\langle(\Delta X)^2\rangle$ **(a)** , $\langle(\Delta P)^2\rangle$ **(b)** ; dash line corresponds to analytical analysis, solid line corresponds to numerical analysis

4. Conclusion

As a result, we have evaluated basic statistical properties of DCE field quantum state. These expressions are used to research evolution of electromagnetic field state and, specifically, to demonstrate non-classical properties of Casimir field.

Acknowledgments

This work was partially financially supported by the Government of the Russian Federation (grant 074-U01), by Ministry of Education and Science of the Russian Federation (GOSZADANIE 2014/190, Project 14.Z50.31.0031 and ZADANIE No. 1.754.2014/K), by grant of Russian Foundation for Basic Researches and grant of the President of Russia (MK-2736.2015.2).

References

[1] Law C K 1994 *Phys. Rev. A* **49** 433-537
[2] Dodonov V V and Klimov A B 1996 *Phys. Rev. A* **53** 2664-2682
[3] Schützhold R, Plunien G and Soff G 1998 *Phys. Rev. A* **57** 2311-2318
[4] Dodonov V V 2010 *Physica Scripta* **82** 038105
[5] Wilson C M et all 2011 *Nature* **479** 376-379
[6] Yablonovitch E 1989 *Phys. Rev. Lett.* **62** 1742-1745
[7] Braggio C et all 2005 *Europhys. Lett.* **70** (6) 754–760
[8] Dodonov V V 2012 *Phys. Rev. A* **85** 2311-2318
[9] Dodonov V V 2011 *Physics Letters A* **375** 4261-4267
[10] Trifanov A I and Miroshnichenko G P 2013 *Nanosystems: Phys., Chem., Math.* **4** (5) 635-647
[11] Clerk A A 2010 *Rev. Mod. Phys.* **82** 1155-1208
[12] Nation P D 2012 *Rev. Mod. Phys.* **84**
[13] Lamoreaux S K 1997 *Phys. Rev. Lett.* **78** 5-8
[14] Sheremetyev V O, Trifanov A I and Trifanova E S 2014 *J. Phys.: Conf. Series* **541** 012105

Noise compensation in a Fabry-Perot-based displacement sensor operating at picometer-level resolution

Nikolai A. Ushakov, Leonid B. Liokumovich, Alexandr A. Markvart

Department of Radiophysics, St. Petersburg State Polytechnical University, Polytechnicheskaya, 29, 195251, St. Petersburg, Russia

n.ushakoff@spbstu.ru

Abstract. An approach for compensating the influence of the interrogator noises on the readings of the extrinsic Fabry-Perot interferometric displacement sensor is proposed. With the use of a reference interferometer and special processing, the sensor resolution was about twofold improved and comprised 12-25 pm instead of 15-46 pm for a wide range of EFPI baselines.

1. Introduction

During the last two decades a great progress in manufacture and implementation of the fiber optic sensors based on the extrinsic Fabry-Perot interferometers (EFPI) [1] has been achieved by the academic institutions and commercial companies. Such sensors demonstrate a great dynamic measurement range and a high resolution [2], [3]. Sensors of a great variety of physical quantities have been designed and implemented. The most commonly used EFPI baseline demodulation approaches are the white-light interferometry, using a scanning readout interferometer [4], [5] and wavelength-domain interferometry, in which the measurement and further processing of the interrogated interferometer spectral function is used to measure the interferometer OPD [6]–[9]. The best attained baseline resolution was 14-15 picometers [10], estimated as doubled standard deviation of the measured baseline fluctuations. The analysis of the noise sources and the light propagation inside the EFPI cavity was made in a paper [10], dedicated to resolution limits analysis for single EFPI sensors.

An idea of compensating hardware noises in interferometric sensing systems had been proposed long ago, mainly dedicated to reducing the laser frequency noise influence. In a paper [11] the idea of noise reduction with a compensating interferometer was realized for extrinsic Fabry-Perot sensors, interrogated by means of the wavelength-domain interferometry. As a result, the sensor resolution was tenfold improved (from 700 nm to 70 nm).

However, the approach proposed in [11] is suitable only for suppression of the fluctuations, induced by environmental influences and isn't able to suppress OPD noises, induced by fluctuations in the acquisition hardware. The goal of the current paper is therefore to develop an approach of OPD resolution enhancement by means of reducing the influence of hardware noises on the OPD readings. This will provide an ability to either use cheaper interrogation units with poorer characteristics or to improve the resolution of high end displacement sensors.

Throughout this paper we consider the case of registering the spectrum of the light reflected from the sensor, which is the most common case for both single-sensor and multiplexed systems. The spectral function of a low-finesse Fabry-Perot interferometer is a sum of quasi-static and quasi-harmonic components, and is expressed as $S_{FP}(L, \lambda) = S_0(L, \lambda) + S_M \cdot S(L, \lambda)$, where [10]:

$$S(L, \lambda) = \cos[4\pi nL/\lambda + \gamma_R(L, \lambda)], \quad (1)$$

$$S_M = 2\sqrt{R_1 R_2 \cdot \eta(L)}, \qquad \eta(L) = \frac{(\pi n w_0^2)^2}{L^2\lambda^2 + (\pi n w_0^2)^2}, \quad (2)$$

$$\gamma_R(L, \lambda) = \psi_R + \varphi_R = \operatorname{atan}\left[L/z_R \cdot \left(4L^2/z_R^2 + 3\right)\right] + \varphi_R, \quad (3)$$

and an approximation of Gaussian profile was applied to the fiber mode and the beam inside the interferometer. $\eta(L)$ is a coupling coefficient of a light beam, irradiated by a fiber mode and travelled a distance $2L$ back to a fiber mode [10]; L is the interferometer baseline; w_0 is an effective radius of a mode at the output of the first fiber; λ is the light wavelength; n is the refractive index of the media between the mirrors; $z_R = \pi n w_0^2/\lambda$ – Rayleigh length of the beam; the argument additive $\gamma_R(L, \lambda)$ contains a phase shift ψ_R, induced by the light diffraction inside the cavity, and a phase φ_R, induced by the mirrors (typically for dielectric mirrors $\varphi_R = \pi$). The equations above are valid for the case of parallel mirrors.

One of the most attractive approaches for estimating the baseline L from the registered spectrum is to approximate its variable component $S'(\lambda)$ with analytical expression (2) by means of least-squares fitting. Such fitting returns the global minimum of the residual norm, given by expression

$$R(L) = \|S'(\lambda) - S(L, \lambda)\| = \sqrt{\sum_i [S'_i - S_i(L)]^2}, \quad (4)$$

where $S'_i = S'(\lambda_i)$, $S_i(L) = S(\lambda_i, L)$, $\lambda_i = \lambda_0 + i \cdot \Delta$, Δ is the step between the spectral points, $i = -(M-1)/2, \ldots (M-1)/2$, M is the number of points in digitized spectrum (for the current notation M must be odd, which corresponds to the utilized interrogator and the performed simulations).

2. Noise mechanisms

An extensive study of single EFPI displacement sensors resolution limits with wavelength-scanning interrogation was done in [10]. It was shown that the main noise sources are:

1. Absolute wavelength scale shift $\Delta\lambda_0$, determined by fluctuations of the triggering of the scanning start, $\sigma_{\Delta\lambda} = \text{stdev}\{\Delta\lambda_0\}$.
2. Jitter of the wavelength points $\delta\lambda_i$, caused by the fluctuations of the signal sampling moments, $\sigma_{\delta\lambda} = \text{stdev}\{\delta\lambda\}$.
3. Additive noises δs_i, produced by the photo registering units, by the light source intensity noises, etc. $\sigma_s = \text{stdev}\{\delta s\}$.

These mechanisms will result in distortion of the registered interferometer spectral function $S'(\lambda)$. Therefore, the spectrum approximation procedure gives an erroneous result, denoted throughout the paper as L_a. When considering a vector of consequently measured baseline values, its standard deviation σ_{La} can be calculated. Generally, σ_{La} is used as a quantitative characteristic of a sensor resolution, which is approximated as $2 \cdot \sigma_{La}$.

The first mechanism provides the shift of the measured interferometer spectrum, inherently shifting the displacement sensor readings as follows

$$\delta L \approx -\Delta\lambda_0 \cdot L_0/\lambda_0. \quad (5)$$

The jitter of spectral points during interrogation produces the distortion of the measured spectral function $S'(\lambda)$. This distortion can be interpreted as an additive noise with some variance. The resulting signal-to-noise ratio SNR_{JIT} can easily be estimated by simple trigonometric derivations [10]:

$$\text{SNR}_{\text{JIT}} = \frac{S_M^2/2}{\left(\sigma_{\delta\lambda}\, S_M\, 4\pi n L_0/\lambda_0^2\right)^2} = \frac{2\lambda_0^4}{\left(8\pi n L_0\, \sigma_{\delta\lambda}\right)^2}. \quad (6)$$

Considering the third mechanism, one has to take into account that generally the noise variance can depend on the mean optical power P_I, incident to the photodetector (shot noise level and laser intensity noise influence are strongly related to the mean power level). The dependency can be adequately approximated by a power function as follows

$$\sigma_s = aP_I^b, \tag{7}$$

which, for an EFPI, produces the signal-to-noise ratio of the spectrum, written as follows

$$\mathrm{SNR}_s = \frac{S_m^2/2}{\sigma_s^2} = \frac{2P_0^{2-2b}}{a^2}\frac{R_1R_2^*}{\left(R_1 + R_2^*\right)^{2b}}, \tag{8}$$

P_0 is the optical power irradiated by the light source; $R_{1,2}^* = R_{1,2}\cdot\eta$ are effective mirrors reflectivities, taking into account light losses due to divergence of a non-guided beam inside the cavity. The parameters a and b must be obtained explicitly for a given experimental setup.

The influence of the additive noises on the standard deviation of the baseline measurement can be found either by numeric simulation, or analytically using the Cramer-Rao bound [12]. It can be shown that for the approximation approach [3] the corresponding relation can be written as

$$\sigma_L(\mathrm{SNR}) = C\cdot\mathrm{SNR}^{-1/2}, \tag{9}$$

where C is within the interval $(9 \div 11)\cdot 10^{-4}\mu\mathrm{m}^{-1}$ for different parameters of the baseline measurement approach [3].

Finally, the expression for the baseline standard deviation can be obtained by combining expressions (5) and (9) and taking into account the variance summation rule:

$$\sigma_L = \sqrt{\left(\frac{L}{\lambda_0}\right)^2\cdot\sigma_{\Delta\lambda}^2 + \frac{C^2}{\mathrm{SNR}_s} + \frac{C^2}{\mathrm{SNR}_{JIT}}}. \tag{10}$$

Substituting (2), (6) and (8) to (10), one will be able to obtain the final explicit expression, which isn't done due to excessive bulkiness.

3. Baseline noises statistics

Throughout the paper the following noise compensation scheme will be assumed: two interferometers with similar parameters are used for the measurement, one of them is exposed to the target perturbation (which is the actual measurand) and will be referred to as sensing or signal interferometer, and the second one is isolated from any environmental changes and will be referred to as reference interferometer. The interferometer baselines will be denoted L_S and L_R, respectively. Both these interferometers will be assumed to be interrogated by the same tunable laser, while their spectral functions will be registered by independent (although, similar) photodetectors.

Considering the task of noise compensation, one needs to determine the likelihood of the two interferometers' measured baselines fluctuations, which can be done by means of correlation coefficient. Therefore, in order to study the ability of noise cancellation, one needs to study the behavior of the correlation coefficient C_{BF} between the registered baseline values of reference and sensing interferometers with respect to the mechanisms, mentioned in the previous section.

As can be seen from expression (5), the absolute shift of the wavelength scale produces an error, proportional to the baseline value, therefore, its influence will be correlated for the signal and reference interferometers with arbitrarily different baselines.

Considering the impact of individual spectral points jitter, one has to take into account that it is indirect in nature: jitter itself produces equivalent additive noises, which, in turn, affect the accuracy of the approximation procedure [3]. If the sensing and the reference interferometers have equal baselines $L_S \approx L_R$, the shapes of their spectral functions will be the same, resulting in identical noise patterns, produced by the jitter. This will provide equal baseline measurement errors for the both interferometers and $C_{BF} = 1$. Recall the properties of the spectral function $S'(\lambda_i)$, particularly that it is quasi-harmonic with period close to $\lambda_0/2$, one can expect secondary maximums of correlation in cases $|L_S - L_R| = \lambda_0/2$ and $|L_S - L_R| = \lambda_0/4$ (when interferometers spectral functions are nearly inversed). On the other hand, for arbitrarily different baselines L_S and L_R, the spectral functions will be uncorrelated, producing sufficiently different noise patterns, which will result in uncorrelated baseline noises. The dependency of the produced baseline noises correlation on the difference of interferometer baselines can be found by means of the numeric calculations.

Considering the additive noises (the third mechanism), one needs to take into account their two main sources – laser intensity noises, equal for the both interferometers and the photodetector noises, which are produced by different devices, and hence, are uncorrelated, producing uncorrelated errors. Laser intensity noises, being the same for the measurement and reference interferometers, will produce identical errors in case of close baseline values $L_S \approx L_R$ and $|L_S - L_R| = \lambda_0/2$. In case of $|L_S - L_R| = \lambda_0/4$ the influence of laser intensity noises will be inverse for signal and reference interferometers and will produce anti-correlation $C_{BF}(\lambda_0/4) = -1$. For arbitrarily different baselines the baseline fluctuations will be uncorrelated, $C_{BF} = 0$. Dependency $C_{BF}(L_S - L_R)$ for a particular setup will depend on the relation of the above mentioned noise sources.

In order to study the exact dependency of the signal and reference interferometers' baseline noises correlation C_{BF} on the baseline difference, a numeric simulation has been performed. Two spectral functions $S(L_{S,R}, \lambda)$ were calculated according to (1), the difference $\delta L = L_S - L_R$ was varied within the interval $0 \div 0.8$ μm, the values $L_{S,R}$ themselves were around 200 μm. The rest parameters were chosen close to the experimental: wavelength scanning range $1510 \div 1590$ nm, interval between spectral points $\Delta = 4$ pm, resulting in $M = 20001$ points in spectrum; wavelength jitter stdev $\sigma_{\delta\lambda} = 1$ pm; scale shift stdev $\sigma_{\Delta\lambda} \approx 0.05$ pm; laser output optical power $P_0 \approx 0.06$ mW; RIN in full frequency band 60 dB (in case of ~ 1 MHz photodetector band this corresponds to practical 120dB/√Hz RIN spectral density); photodetector noise equivalent power (NEP) 30nW in the full frequency band ($3\cdot10^{-11}$ W/√Hz for ~ 1 MHz photodetector band). For all baseline combinations L_S, L_R an ensemble of N=100 realizations of noises were calculated for better statistical validity of the simulation results. The correlation coefficients C_{BF} of the resulting vectors of the baseline fluctuations were calculated. In figure 1 the simulated dependency $C_{BF}(\delta L)$ is shown as a solid line.

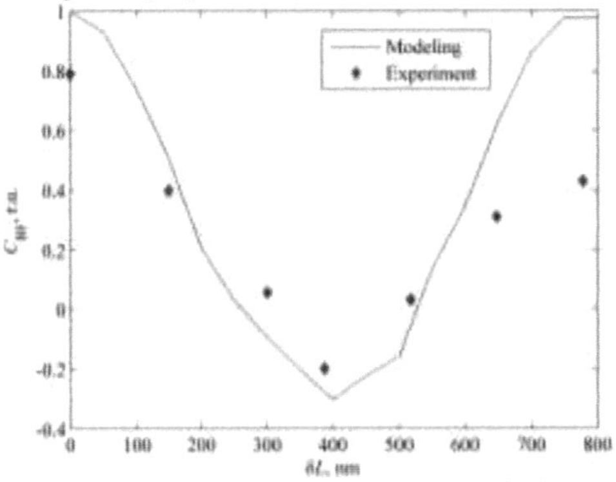

Figure 1. Dependencies of the baseline fluctuations correlation coefficient on the baseline difference for all noise mechanisms: modelling (line) and experimental (points).

From this dependence it can be predicted that for baselines difference less than 50 nm the correlation is still enough for noise compensation.

4. Experimental noise compensation

In order to support the theoretical results, an experimental study of EFPI displacement sensor resolution was carried out. Spectral measurements were performed using the optical sensor interrogator NI PXIe 4844, utilizing a tunable laser with SMF-28 single-mode fiber output. Spectrometer parameters are the following: scanning range [1.51; 1.59] μm; wavelength jitter stdev $\sigma_{\delta\lambda} = 1$ pm; optical power $P_0 \approx 0.06$ mW; scale shift stdev $\sigma_{\Delta\lambda} \approx 0.05$ pm; spectral points stepping $\Delta = 4$ pm; number of spectral points $M = 20001$; spectrum acquisition rate about 1 Hz, spectrum acquisition time 0.035 s.

The sensing and reference interferometers were formed by the two ends of SMF-28 fiber packaged with PC connectors, fixed in a standard mating sleeve. The air gaps L_S and L_R between the fiber ends was varied from ~100 μm up to ~800 μm with steps ~100 μm by the use of Standa 7TF2 translation stages. The experimental setup is schematically shown in figure 2.

The aim of the first performed experiment was to verify the correlation properties of the baseline fluctuations, predicted by means of numeric modeling. As in the modeling, the baselines of the both interferometers were set to ~ 200 μm, the signal interferometer baseline was fixed, while the reference interferometer baseline was scanned with step ~100 nm. In figure 1 the experimental relation $C_{BF}(\delta L)$ (dots) is compared with the simulated dependency $C_{BF}(\delta L)$ (solid line). Not exact correspondence of simulated and experimental dependencies can be due to some discrepancy of experimental setup parameters and those implied in the simulation. Also, a lower level of correlation is observed in experiment, therefore, limiting the admissible baseline discrepancy to somewhat 10 nm.

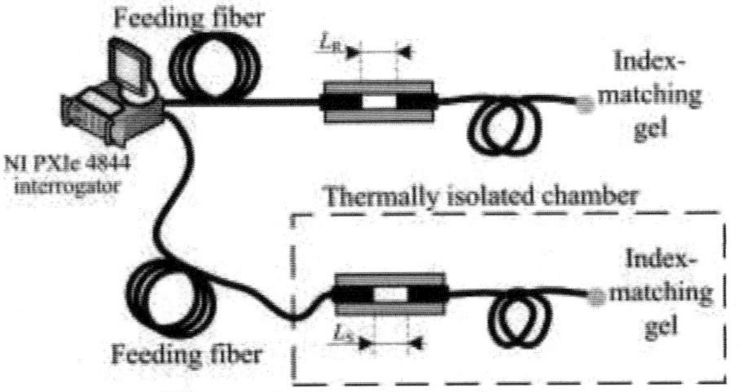

Figure 2. Experimental setup.

After that the noise compensation possibilities were tested. In order to compensate the parasitic baseline fluctuations, the baselines of both the signal and reference interferometers L_S and L_R were measured, after that the deviations of the reference interferometer baseline from the initial value were subtracted from the sensing interferometer baseline values:

$$L_{SC}(t) = L_S(t) - [L_R(t) - L_R(0)]. \quad (11)$$

The baselines of both interferometers were set nearly equal with accuracy better than 1 nm. For each baseline value the measurements were performed for about 5 minutes, resulting in 300 measured points, with respect to them the standard deviation was calculated. Then the baselines were changed and measurements were repeated for another L_S and L_R values. The dependencies of standard deviations of a single sensor readings L_S and noise-suppressed double sensor readings L_{SC} on the baseline value are shown in figure 3. Estimation of the measured baseline standard deviation calculated according to (10) is also presented as a solid curve for reference.

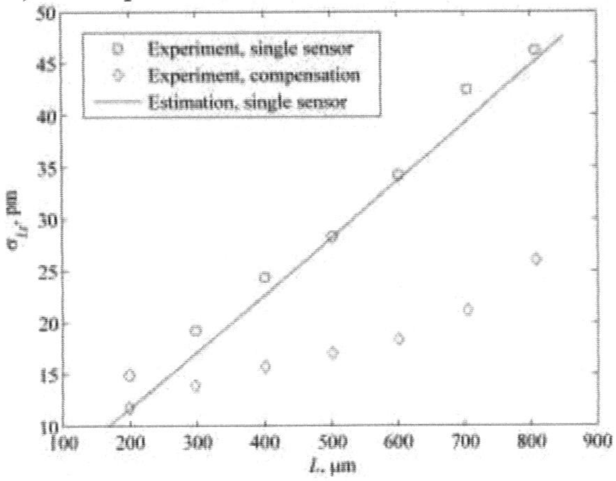

Figure 3. Dependencies of measured baselines fluctuations' standard deviations on the mean baseline value.

As can be seen in figure 3, in the performed experiment the effort of the noise compensation exceeded 1.5 for baselines greater than 400 μm and reached 2 for $L_S \approx L_R \approx 700$ μm.

5. Conclusion

In the current paper an original approach for eliminating the influence of the interrogating unit noises on the measured value of interferometer baseline is proposed. Parasitic baseline fluctuations, induced by unideal interrogator operation are estimated by means of additional reference interferometer, isolated from the environment, which had the same parameters as the sensing interferometer. After that the fluctuations of the reference interferometer measured baseline were subtracted from the measured baseline of the signal interferometer. As a result, the standard deviation of the signal interferometer measured baseline value was about twofold decreased.

Acknowledgments

The work was supported by the program State Tasks for Higher Educational Institutions (projects no. 3.1446.2014K and 2014/184).

References

[1] K. Chang, Z. Fang, K. K. Chin, R. Qu, and H. Cai, *Fundamentals of Optical Fiber Sensors*. Hoboken: John Wiley & Sons, 2012.

[2] W. Wang and F. Li, "Large-range liquid level sensor based on an optical fibre extrinsic Fabry–Perot interferometer," *Opt. Lasers Eng.*, 52, 201–205, 2014.

[3] N. A. Ushakov, L. B. Liokumovich, and A. Medvedev, "EFPI signal processing method providing picometer-level resolution in cavity length measurement," in *Proc. SPIE*, 8789, 87890Y (2013).

[4] J. Sirkis and C.-C. Chang, "Multiplexed optical fiber sensors using a single Fabry-Perot resonator for phase modulation," *J. Light. Technol.*, 14, 1653–1663, (1996).

[5] V. Bhatia, K. A. Murphy, R. O. Claus, M. E. Jones, J. L. Grace, T. A. Tran, and J. A. Greene, "Optical fibre based absolute extrinsic Fabry-Perot interferometric sensing system," *Meas. Sci. Technol.*, 7, 58–61, (1996).

[6] M. Han, Y. Zhang, F. Shen, G. R. Pickrell, and A. Wang, "Signal-processing algorithm for white-light optical fiber extrinsic Fabry-Perot interferometric sensors.," *Opt. Lett.*, 29, 1736–1738, (2004).

[7] F. Shen and A. Wang, "Frequency-estimation-based signal-processing algorithm for white-light optical fiber Fabry-Perot interferometers.," *Appl. Opt.*, 44, 5206–5214, (2005).

[8] Z. Wang and Y. Jiang, "Wavenumber scanning-based Fourier transform white-light interferometry.," *Appl. Opt.*, 51, 5512–5516, (2012).

[9] X. Zhou and Q. Yu, "Wide-Range Displacement Sensor Based on Fiber-Optic Fabry–Perot Interferometer for Subnanometer Measurement," *IEEE Sens. J.*, 11, 1602–1606, (2011).

[10] N. A. Ushakov and L. B. Liokumovich, "Resolution limits of extrinsic Fabry–Perot interferometric displacement sensors utilizing wavelength scanning interrogation," *Appl. Opt.*, 53, 5092–5099, (2014).

[11] Y. Jiang and C. Tang, "A high-resolution technique for strain measurement using an extrinsic Fabry–Perot interferometer (EFPI) and a compensating EFPI," *Meas. Sci. Technol.*, 19, 065304, (2008).

[12] H. Fu and P. Y. Kam, "MAP / ML Estimation of the Frequency and Phase of a Single Sinusoid in Noise," *IEEE Trans. Signal Process.*, 55, 834–845, (2007).

Journal of Physics: Conference Series **661** (2015) 012048 doi:10.1088/1742-6596/661/1/012048

Dual-tunable multiferroic active ring filter for microwave photonic oscillators

V V Vitko, A A Nikitin, A B Ustinov and B A Kalinikos

St. Petersburg Electrotechnical University, St. Petersburg, 197376 Russia

E-mail: vitaliy.vitko@gmail.com

Abstract. A theoretical model of a microwave active ring filter based on a ferrite–ferroelectric layered structure serving as a waveguide for spin-electromagnetic waves is developed. An experimental prototype of the device is fabricated and characterized. The device is implemented as an active-ring resonator with a microwave amplifier and a ferrite-ferroelectric delay line. The resonance properties of this system are studied theoretically and experimentally. The results show dual control of central frequency of the filter with magnetic and electric fields. An effective Q-factor of 50 000 and tuning by 5 MHz with an electric field are achieved at 8 GHz.

Recent years an increased interest for fabrication and investigation of new composite and structured magnetic materials is observed (see e.g. [1]). Spin-electromagnetic waves (SEWs) propagating in artificial multiferroics composed in the form of ferrite and ferroelectric multilayers have a good potential for application in microwave electronics, including filters, phase shifters, etc. [2]. A distinctive feature of SEWs is that their spectrum can be electronically tuned with both electric and magnetic fields. Magnetic tuning is made through variation of the microwave magnetic permeability of the ferrite layer with variation of an external magnetic field. Electrical tuning is accomplished through the variation of the dielectric permittivity of the ferroelectric layer with the application of an electric field.

One of the promising multiferroic devices based on artificial multiferroic structure is an active ring resonator [3]. This device can be used as a *tunable* active ring filter for microwave photonic oscillators. Microwave photonic oscillators typically consist of optical and radiofrequency paths. The radiofrequency path usually contains a frequency selective element (microwave filter) based on a resonator having fixed resonance frequency [4]. In order to obtain frequency tunable microwave photonic oscillator a high-Q tunable microwave filter should be used. An active ring resonator based on ferrite-ferroelectric structure is good candidate as a frequency selective element of such oscillator due to high Q-factor [3, 5]. The most demanding applications are aerospace and avionics industries where the environment is harsh and the compactness, size and power consumption of a component are critical to the system design.

The purpose of this work is theoretical and experimental investigation of active ring filter performance characteristics. The layered structure consisted of yttrium iron garnet (YIG) and barium strontium titanate (BST) was used in the filter as a waveguide for SEWs. Dispersion characteristics of the SEWs are controlled by both bias magnetic and electric fields. Therefore, the central frequency of the filter can be controlled with both electric and magnetic fields.

A schematic view of the investigated filter is shown in Figure 1. The active ring contains a multiferroic delay line described by a complex propagation constant K=k-iα (where k is a wave number and α is a damping constant) and an amplifier characterized with a gain coefficient g_0. Directional couplers are used for microwave signal input and output. A theoretical model of the active ring filter was developed using the SEW dispersion law [6]. A complex transmission coefficient was derived as a ratio of the input signal amplitude to the output signal amplitude, which was a result of superposition of infinite number of decaying waves circulating in the ring. Thus, formulae for amplitude-frequency characteristic and phase-frequency characteristic were obtained.

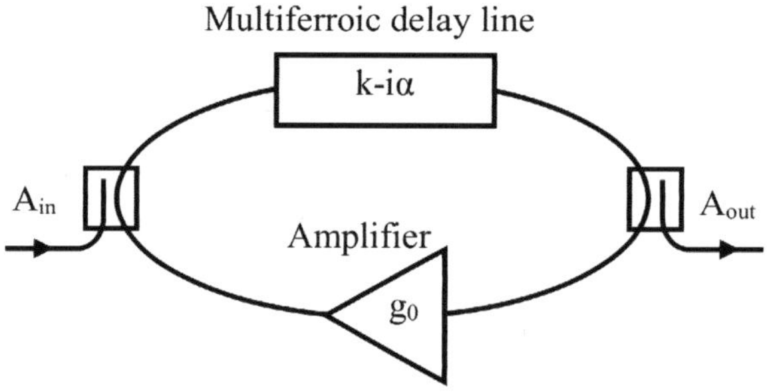

Figure 1. A diagram of the multiferroic active ring filter.

The trial calculations were carried out for a multiferroic bilayer consisting of a 15-μm-thick YIG film and a 500-μm-thick BST slab. A ring length was taken to be equal to a SEW propagation distance of 5 mm. A standard for YIG saturation magnetization M=1750 G was used. A BST dielectric permittivity was taken ε=1500. A bias magnetic field H was 1500 Oe. Figure 2 shows results of the numerical simulation of the amplitude-frequency characteristic of the filter. Well pronounced peaks correspond to the resonant frequencies of the ring.

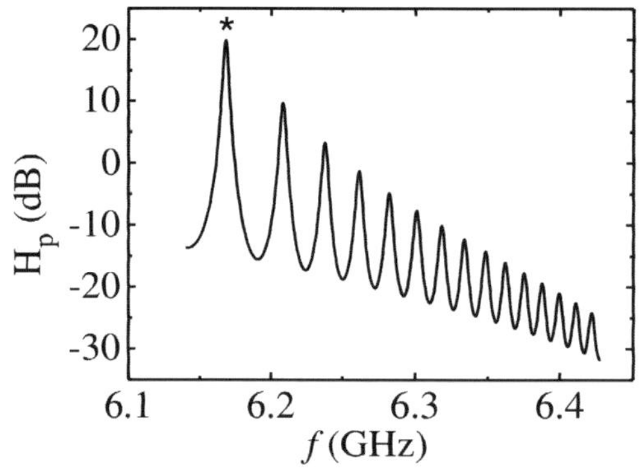

Figure 2. Amplitude-frequency characteristic of the active ring filter.

Let us analyze a pass band of the filter corresponding to the lowest-frequency resonant peak marked by asterisk in Figure 2. Figure 3(a) shows a frequency tuning of the central frequency with the change in ε. Note that it is due to SEW spectrum shift shown in Figure 3(b). We also studied magnetic field tuning. Variation of magnetic field from 1500 Oe to 2500 Oe changed the resonant frequency from 6.17 GHz to 9.13 GHz. Therefore, dual tuning of the multiferroic active ring filter was demonstrated.

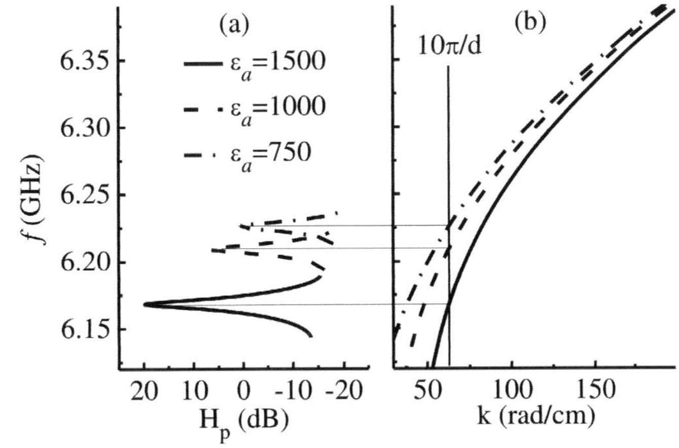

Figure 3. Fragments of theoretically calculated amplitude-frequency characteristics for lowest resonant peak of the active ring filter (a) and SEW dispersion curves (b) for different values of dielectric permittivity.

Experimental measurements were performed for an active ring filter structure similar to that reported in [3]. In consistent with figure 1, the main elements of the filter are a ferrite-ferroelectric delay line, a microwave amplifier, a variable attenuator, and the microwave input and output directional couplers. The ferrite-ferroelectric structure utilized in the delay line was composed of a 5.5 μm thick, 2 mm wide, and 40 mm long single-crystal YIG film and a 500 μm thick, 10 mm wide, and 5 mm long ceramic BST slab. One surface of the slab was covered by 50 nm thick chromium electrode. Such a thickness for the chromium electrode was much smaller than the skin depth for the operating frequencies. A copper electrode with thickness of 5 μm covered the other surface of the BST slab. Two 50 μm wide and 2 mm long short circuited microstrip antennas were used in order to excite and receive waves in the layered structure. The distance between the antennas was taken to be 8 mm. The antennas were fed by the microstrip transmission lines of 50 Ω characteristic impedance. The microstrip parts were fabricated by conventional photolithography on grounded alumina substrates with thickness of 500 μm. The YIG film stripe was positioned over the antennas with the YIG film side in contact with the antennas. The BST slab was pressed from below to the surface of the YIG film so that the chromium covered side was in contact with the film.

In the experiments ε was changed through application of electric field to BST. Figure 4 shows comparison of typical experimental and theoretical data for saturation magnetization M=1815 G of YIG film and external magnetic field H=2017 Oe. In this case centre frequencies were about 7.8 GHz and theoretical data were calculated for parameters used in experiment. The detailed study shows that an effective electric field control of the active ring filter central frequencies of the pass band takes place for relatively thick YIG films.

In agreement with the theory of hybrid spin-electromagnetic waves [6], the shift in central frequencies of the pass band is dependent on the eigenmode number n of the ring. Namely, the shift of central frequencies of the pass band decreased for the modes with higher frequencies, i.e., for higher values of n. The maximum shift of about 5 MHz was obtained for the lowest mode of the ring. At the same time, the maximum Q-factor of about 40 000 for G=−1.5 dB is measured. This mode demonstrated a frequency shift of 2.9 MHz. Approaching of G close to zero resulted in an increase in Q-factor up to 50 000. With proper choice for the geometry of the layered structure, in particular, the thicknesses of YIG and BST components, it is possible to increase the electric field tuning bandwidth of the resonator.

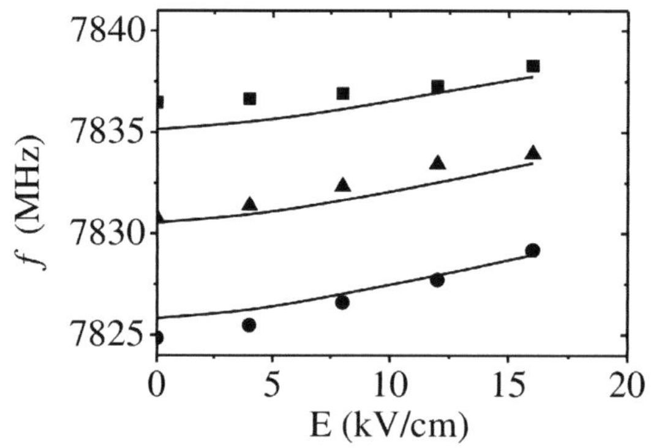

Figure 4. Central frequencies of the pass bands of active ring filter as a function of electric field (signs – experiment, solid lines – theory).

In conclusion, we developed a theoretical model of a microwave active ring filter based on a ferrite–ferroelectric layered structure. The results show a dual control of central frequency of the filter with magnetic and electric fields. An electric field tuning by 5 MHz and Q-factor up to 50 000 were obtained for operating frequencies close to 8 GHz. The investigated tunable active ring filter is very promising for application as a frequency selective element of tunable microwave photonic oscillators.

The work was supported in part by the Russian Foundation for Basic Research and by Ministry of Education and Science of Russian Federation (Project "Goszadanie").

References

[1] Vaz C A F, Staub U 2013 *Journal of Materials Chemistry C.* **1** № 41 6731
[2] Ozgur U, Alivov Y and Morkoc H 2009 *J. Mater. Sci.: Mater. Electron.* **20** 911
[3] Ustinov A B, Srinivasan G and Kalinikos B A 2008 *Appl. Phys. Lett.* **92** 193512
[4] Capmany J, Novak D 2007 Nature Photonics **1** (6) 319-330
[5] Demidov V E, Kalinikos B A, Kovshikov N G, Edenhofer P 1999 *Electronics Letters* **35** (21) 1856
[6] Demidov V E, Kalinikos B A, Edenhofer P 2002 *J. Appl. Phys.* **91** 10007

Microwave filter based on Lamb modes for optoelectronic generator

V V Vitko[1,2], A A Nikitin[1], A V Kondrashov[1,2], A A Nikitin[1,2], A B Ustinov[1], P. Yu. Belyavskiy[1], B A Kalinikos[1] and J E Butler[3]

[1]St. Petersburg Electrotechnical University, St. Petersburg, 197376 Russia
[2]Institute of Applied Physics of the Russian Academy of Science, Nizhny Novgorod, 603950 Russia
[3]Cubic Carbon Ceramics, Huntingtown, MD 20639, USA

E-mail: vitaliy.vitko@gmail.com

Abstract. Experimental results for narrowband filter based on yttrium iron garnet film epitaxially grown on gadolinium gallium garnet substrate have been shown. The principle of operation of the filter is based on excitation of Lamb modes in the substrate. We demonstrated also that the use of single crystal diamond as a substrate will significantly reduce the phase noise of the designed optoelectronic microwave generator.

One of the advanced devices in microwave photonics [1] is a microwave generator. It belongs to a new class of high-stability oscillators that are capable to work at the frequency ranging from hundreds of megahertz up to several hundreds gigahertz [2].

The main advantage of the microwave generator in comparison with a traditional microwave generator is a possibility to compromise between the frequency tuning enhancement and phase noise minimization [3]. There are designs of the optoelectronic microwave generators with very low phase noise about -160 dBc/Hz at 10 kHz offset frequency [4-8]. The maximum frequency of generation is limited by the pass-band of optoelectronic components that have currently lying in the range of hundreds of gigahertz [4]. Well known that such performance characteristics are difficult to achieve by the ordinary microwave generators due to the fundamental limitations.

The schematic view of an optoelectronic microwave generator is shown in Figure 1. The generator has a ring circuit that consisted of the radiofrequency and optical paths [4]. Usually the optical path consists of a light-emitting diode, an electro-optical modulator, an optical fiber, and a photodetector. The main elements of the radiofrequency path are a microwave amplifier and a microwave filter. The microwave filter plays a key role in the circuit providing the required operating parameters of the optoelectronic generator.

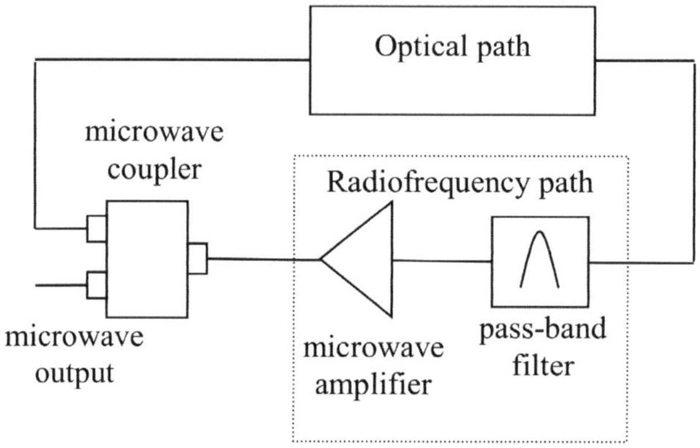

Figure 1. A block diagram of an optoelectronic generator.

Consider now the principle of operation of the ring circuit. The diode emits an optical carrier signal. Electro-optical modulator performs an amplitude-modulation of the optical signal by the microwave signal. That signal propagates through the optical fiber to the photodiode [8]. The optical fiber delays the signal. The delay time is determined by the length of optical fiber. An increase in time delay leads to an increase in the phase-frequency characteristic slope decreasing the phase noise of the generator. It was shown in [8] that the increase of optical fiber length from 100 m to 1000 m reduces the phase noise from -95 dBc/Hz to -107.57 dBc/Hz at 10 kHz offset frequency (Figure 2).

The photodiode receives the optical signal and demodulates it into the microwave signal. After demodulation the microwave signal propagates through the radiofrequency path. The microwave filter provides frequency selectivity and determines the generation frequency. In this case phase balance condition is

$$\varphi_{opt} + \varphi_e = 2\pi n, \tag{1}$$

where φ_{opt} is a phase shift in the optical fiber, φ_e is a phase shift in electrical circuits, and n is a number of generation mode. In fact the optical fiber length significantly exceeds the length of the electrical circuits in radiofrequency path, i.e. phase shift in radiofrequency path circuits can be neglected $\varphi_e \approx 0$.

Time delay determines generation frequency by the phase balance condition generation frequencies become closer by increasing of time delay. Therefore, the requirement for narrow band microwave filter considerably increases. The microwave amplifier is used for compensation of loss. The equidistant spectrum of generation frequencies that carried out the phase balance condition is shown in Figure 3. Note that, the suppression level of the remaining harmonics determines the lowest possible phase noise of developed microwave generator.

It should be noted that the frequency of generated signal can be tuned in accordance with the bandwidth tuning of the tunable microwave filter. For example, the frequency tuning of the generated signal in the range from 6 to 12 GHz with a phase noise level of -128 dBc/Hz (at 10 kHz offset frequency) was demonstrated in [9]. The similar tuning can be achieved by the band-pass filter based on a film of yttrium iron garnet (YIG). The YIG-film filter operates due to the phenomenon of spin wave propagation.

The actual problem is to develop a high-Q band-pass microwave filter. The required high-Q (narrow-band) microwave filter can be realized using magnetoacoustic interaction. This interaction results in excitation of high-Q acoustic oscillations (Lamb modes) by the spin waves propagating in single-crystal YIG films [10 – 14].

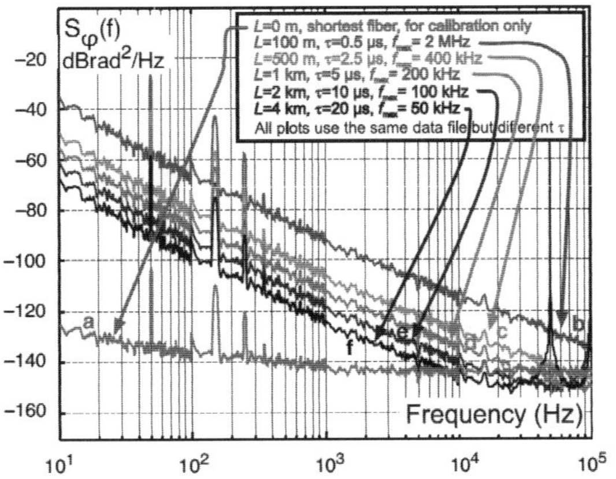

Figure 2. Dependence of the phase noise for different lengths of optical fiber [9].

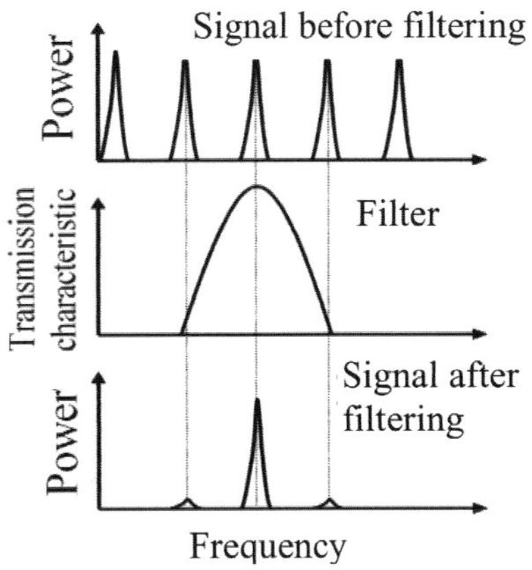

Figure 3. Equidistant spectra of optoelectronic generator; transmission characteristic of the microwave filter and the microwave signal after filtering.

Let us consider the operation and design of the band-pass magnetoacoustic filter for application as a frequency control element in optoelectronic generators. The experimental prototype used for the investigation of the magnetoacoustic oscillations is shown in Figure 4. The prototype was consisted of a 500-μm-thick single-crystal gadolinium gallium garnet (GGG) substrate with the 5.7-μm-thick YIG films epitaxially grown on the both sides of the substrates. The YIG film had a width of 2 mm, length of 30 mm, and a saturation magnetization of 1750 G. Ferromagnetic resonance linewidth was 0.5 Oe at 5 GHz frequency. The microwave microstrip antennas with length of 2 mm and width of 50 μm have been formed on alumina substrates metallized from back side. Thickness of the substrate was of 0.5 mm. This structure was magnetized with external magnetic field of 1938 Oe directed perpendicular to the surface of the YIG film that is the condition for the excitation of forward volume spin waves.

The antenna shown in Figure 4 as "Port 1" was exploited as the microwave input. It excited forward volume spin waves in YIG film. Antenna served as "Port 2" received the forward volume spin waves. It was the microwave output. The distance between the "Port 1" and "Port 2" antennas was 4 mm. Antenna denoted as a "Port 3" was used for the reception of spin waves excited from the other side of the GGG substrate. This antenna was positioned of 3 mm distance from the antenna "Port 1".

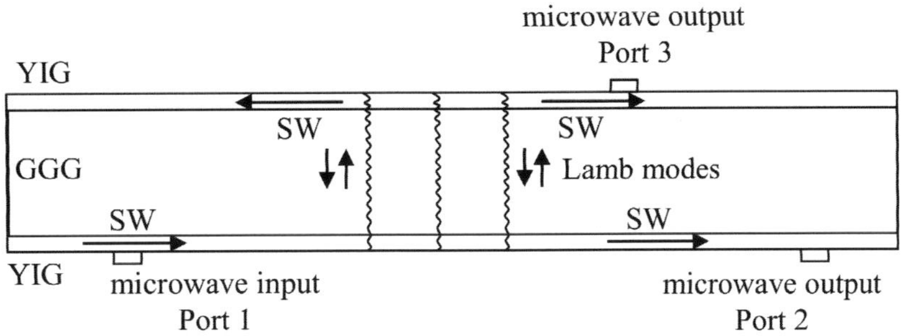

Figure 4. Prototype for the study of magneto-oscillations in the structure of the YIG / GGG / YIG.

Measured transmission characteristics S_{21} and S_{31} are shown in Figure 5. Dips on the S_{21} characteristic correspond to the excitation of Lamb modes by spin waves propagated along the "bottom" YIG film [12]. In turn, these Lamb modes propagated across the substrate GGG excite the forward volume spin waves in the "upper" YIG film. Note that the dips at amplitude frequency characteristic $S_{21}(f)$ are equidistant that confirms acoustic nature of these excitations.

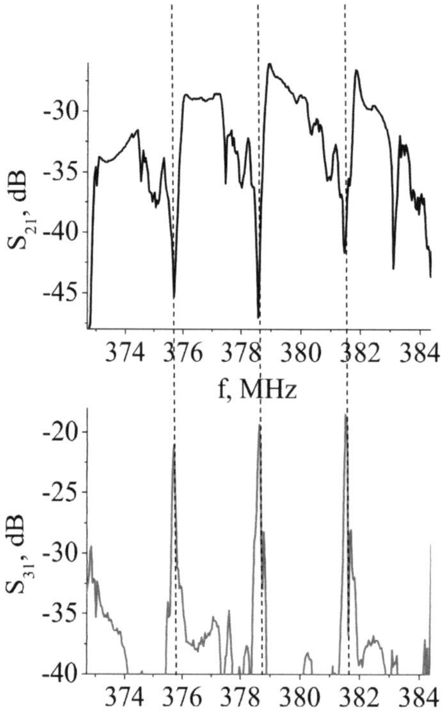

Figure 5. Experimental transmission characteristics S_{21} and S_{31}.

The S_{31} characteristic corresponds to the transmission characteristic of the developed narrowband filter. The quality factor Q is about 5000. Note that the obtained value of the Q-factor exceeds 5-10 times Q-factor of ordinary YIG-filter based on phenomenon of ferromagnetic resonance. One way to further increase of Q-factor is using instead GGG another material with better acoustic properties such as diamond. The acoustic properties of diamond and GGG are shown in Table 1.

Table 1.

	Diamond [15]	GGG [16-17]
Young's modulus	1050 GPa	222 GPa
Poisson's ratio	0.1	0.28
Speed of sound	17500 m/s	3521 m/s

One can see that Young's modulus of diamond higher than of GGG. Therefore the use of single-crystal diamond substrate instead of GGG allows to increase the Q-factor [18]. It will significantly reduce the phase noise of the designed optoelectronic microwave generator.

In conclusion, the design of band-pass magnetoacoustic filter based on Lamb modes for application as a frequency control element in optoelectronic generators was considered. The experimental prototype was fabricated and experimental transmission characteristics were measured. The Q-factor

Journal of Physics: Conference Series **661** (2015) 012049 doi:10.1088/1742-6596/661/1/012049

was about 5000. It was shown that the use of single crystal diamond as a substrate will significantly reduce the phase noise of the designed optoelectronic microwave generator.

The work was supported in part by the Russian Foundation for Basic Research, Russian Science Foundation, Grant of the President of the Russian Federation for young scientists and PhDs (MK-6229.2015.8) and by Act 220 of the Russian Government (Agreement No. 14.B25.31.0021 with host institution IAP RAS).

References

[1] Capmany J 2007 *Nature photonics* **1** 319
[2] Belkin M E and Sigov A S 2009 *Journal of Communications Technology and Electronics* **54** No 8 855
[3] Seeds A J and Williams K J 2006 *OSA Journal of Lightwave Technology* **241** No 12 4628
[4] Yao X S and Maleki L 1994 *Electronics Letters* **30** 1525
[5] Larger L, Udaltsov V S and Poinsot S 2005 *Journal of Optical Technology* **72** Issue 5 378
[6] Yao X S 2002 *Cambridge University Press* 255
[7] Oewaves, Inc URL: http://www.oewaves.com/advanced-oeo-sp-1171610647
[8] Hayata A, Bacoua A, Rissonsa A and Molliera J C 2008 *Proceedings of International Topics Meeting on Microwave Photonics MPW/APMP* 98
[9] Volyanskiy K, Cussey J, Tavernier H, Salzenstein P, Sauvage G, Larger L and Rubiola E 2008 *J. Opt. Soc. Am. B* **25**, No. 12 2140
[10] Wood D L and Remeika J R 1967 *J. Appl. Phys.* **38** No 3 1038-1045
[11] Kozhevnikov A V, Nikitov S A and Filimonov Yu A 1997 *J. de Physique* **7** 401
[12] Camley R E 1979 *J. Appl. Phys.* **50** No 8 5272
[13] Gulyaev Yu V, Zil'berman P E, Kazakov G T, Sysoev V G, Tikhonov V V, Filimonov Yu A, Nam B P and Khe A S 1981 *JETP Lett.* **34** No. 9 477
[14] Zilberman P E; Kazakov G T and Tikhonov V V 1985 *Soviet Technical Physics Letters* **11** 319
[15] Diamond materials 2008 URL: http://www.diamond-materials.com/downloads/cvd_diamond_booklet.pdf
[16] Pardavi-horvath M 1984 *Journal of materials science* **19** 1159
[17] Kleszczewski Z and Bodzenta J 1988 *Phys. stat.* sol. (b). **146** 467
[18] Biderman V L 1980 Teorija mehanicheskih kolebanij (Theory of mechanical oscillations). *Moscow: Vyshaja shkola (in Russian)* 408

Theoretical investigation of nonlinear damping and nonlinear phase shift of spin-electromagnetic waves propagating in infinite multiferroics at sub-terahertz frequencies

I A Ustinova, M A Cherkasskii, A B Ustinov and B A Kalinikos

Department of Physical Electronics and Technology, St.Petersburg Electrotechnical University, 5 Popov Street, St.Petersburg 197376, Russia

E-mail: ustinovairin@yahoo.com

Abstract. The nonlinear phase shift and nonlinear damping of spin-electromagnetic waves were theoretically studied for the first time in sub-terahertz frequency range in infinite homogeneous longitudinal magnetized multiferroics. The research was based on the solution of the Ginzburg-Landau equation. It is shown that the saturation of the phase shift occurs due to the nonlinear damping if the nonlinear damping coefficients exceed $\upsilon_1=10^8$ s^{-1} and $\upsilon_2=10^9$ s^{-1}.

Recently an interest to study new microwave structured and composite magnetic materials has been increased. One class of such materials is multiferroics which combine both electric and magnetic wave nonlinearities [1-6]. In particular, the envelope solitons of spin-electromagnetic waves (SEWs) in infinite multiferroic medium was studied theoretically [5]. A nonlinear microwave phase shifter based on a planar multiferroic composite was fabricated and investigated experimentally in [6]. A survey of the literature shows that the nonlinear properties of the SEWs are studied not enough. Some examples of the issues which are addressed by the researchers are nonlinear damping and nonlinear phase shift of the high-power spin waves propagating in thin magnetic films [7] as well as their application for development of microwave devices such as nonlinear phase shifters [8, 9], a nonlinear directional coupler [7], and a nonlinear interferometer [10].

The aim of the present work is a theoretical investigation of nonlinear damping and nonlinear phase shift of the high power SEWs propagating in longitudinally magnetized infinite multiferroic medium. The study was carried out in two stages. In the first stage, the nonlinear dispersion characteristics were numerically simulated and analyzed with the use of the nonlinear dispersion equation derived in the work [5]. In the second stage, the SEWs propagation was simulated numerically in a stable nonlinear regime. The stable nonlinear regime is due to nonlinear four-wave interaction processes for which enrichment of spectrum of microwave signal carried by SEWs does not take place. Nonlinear damping and nonlinear phase shift of SEWs belong to the four-wave processes. Therefore, numerical simulation of these processes was based on the nonlinear evolutional Ginzburg-Landau (GL) equation [8].

A dispersion law for the case of linear propagation of SEWs in longitudinally magnetized medium is well known (see, e.g., [11]). A nonlinear dispersion law was derived in [5]. The electric and magnetic wave nonlinearities were taken into account in the linear dispersion law. The relationship between the electric field vector and the variable magnetization vector was derived from Maxwell's equations. It has the following form: $\vec{e} = \omega \cdot \mu \cdot \vec{k} \times \vec{m} \cdot (k_0^2 - k^2)^{-1}$. Substituting this expression into the linear dispersion law and introducing the normalized variable magnetization amplitude as $|U| = |\vec{m}| \cdot (2^{1/2} \cdot \vec{M}_0)^{-1}$ we get the nonlinear dispersion law of SEWs:

$$k^2 = \frac{\varepsilon_0^2 \cdot \varepsilon_L^3 \cdot \mu_0 \cdot \omega^2 \cdot \left(\omega \pm \omega_H \pm \omega_M \cdot \left(1-|U|^2\right)\right)}{\varepsilon_0 \cdot \varepsilon_L^2 \cdot (\omega \pm \omega_H) - 2 \cdot |U|^2 \cdot \mu_0 \cdot N_\varepsilon \cdot \left(\omega \pm \omega_H \pm \omega_M \cdot \left(1-|U|^2\right)\right) \cdot M_0^2}, \quad (1)$$

where k is a wave number, ω is a frequency of SEWs, $\varepsilon_0 = 8.85 \cdot 10^{-12}$ F/m is permittivity of vacuum, $\varepsilon_L = 19$ is relative permittivity, $\mu_0 = 4\pi \cdot 10^{-7}$ H/m is permeability of vacuum, $\omega_H = |g|\mu_0 H$, $\omega_M = |g|\mu_0 M$, $|g|$ is module of the gyromagnetic ratio for the electron spin, H is a value of the static magnetic field, M_0 is static magnetization in the absence of waves, $N_\varepsilon = -10^{-10}$ is a coefficient of dielectric nonlinearity. In the expression (1) the upper signs correspond to waves with left circular polarization, and the bottom signs correspond to waves with right circular polarization. Hereinafter we will be interested only in waves with right circular polarization.

The expression (1) is correct only for an ideal medium in which there are no losses. In fact, the propagation of waves is inevitably accompanied by their damping. The losses were taken into account as in [8]:

$$\alpha_+ = \frac{\omega}{c} \cdot \left(\frac{\varepsilon_L}{2}\right)^{1/2} \cdot \left[\left(\mu_+'^2 + \mu_+''^2\right)^{1/2} - \mu_+'\right]^{1/2}, \quad (2)$$

where α is a damping constant, μ_+' and μ_+'' are the real and imaginary parts of the effective scalar magnetic permeability for the circularly polarized wave, c is light's speed in vacuum.

The calculation was carried out for the homogeneous multiferroic medium having parameters corresponding to Al-substituted barium hexaferrite $BaAl_2Fe_{10}O_{19}$ [12]: saturation magnetization $M_0 = 147$ kA/m, the uniaxial magnetocrystalline anisotropy field $H_A = 2677$ kA / m.

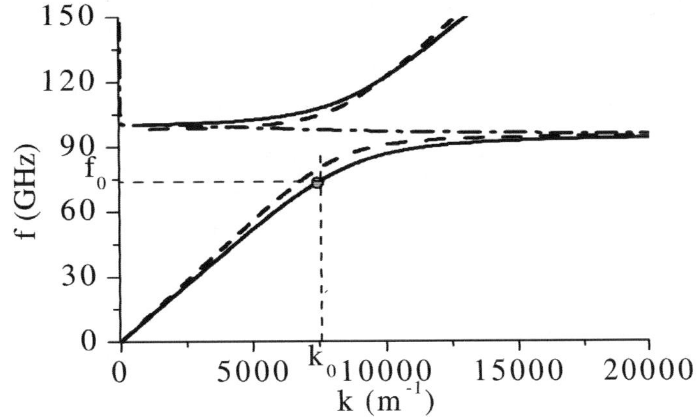

Figure 1. The spectrum and the losses of SEWs. Solid curves correspond to the spectrum of linear SEWs. Dashed curves correspond to the spectrum of nonlinear SEWs for $|U|^2$=0.4. Dash-dotted curve shows SEW dissipation.

The linear and nonlinear dispersion laws and the dependence of losses for waves in the hexaferrite are shown in figure 1. The "operating point" for further calculations was chosen to be in the region of strong dispersion. It has the following parameters: k_0=7550 rad/m, $f_0=7.6\cdot 10^{10}$ Hz. The losses were α=0.957 at this point. Note, that for the hexaferrite one can find a point of intersection of linear and nonlinear dispersion laws at k=10025 rad/m and $f=1.2\cdot 10^{11}$ Hz. This corresponds to the change in the type of nonlinearity. Spectrum exists in the sub-terahertz frequency range.

Further, the theoretical investigation of nonlinear properties of the SEWs in a stable regime of propagation was carried out. It was based on the dispersion law with the use of the method of envelopes [13]. The nonlinear GL equation was obtained under this method [7]:

$$(dU/dt)+V_g\cdot(dU/dz)-i\cdot(D/2)\cdot(d^2U/dz^2)+(\nu_1+i\cdot N)\cdot|U|^2\cdot U+\nu_2\cdot|U|^4\cdot U=-\eta\cdot U \quad (3)$$

It has the following coefficients: $V_g=\partial\omega/\partial k$ is group velocity, $N=\partial\omega/\partial|U|^2$ is nonlinear self-interaction coefficient, $D=d^2\omega/dk^2$ is dispersion coefficient, η is linear damping coefficient (which equals α), υ_1 and υ_2 are cubic and quintic damping coefficients. Note that the nonlinear coefficient is a function of both magnetic and electric field components of the SEWs in the model.

We assume that the coefficient D equals zero for the continuous wave signal, as well as for the pulsed signal when the microwave pulses have a relatively long duration (more than 1 μs). In this case the time-independent analytic solution of GL equation (3) for the wave envelope amplitude U(z) is described by the transcendental equations below. Introducing the notation $G=4\eta\nu_2-\nu_1^2$, these equations can be written in the following form:

$$\frac{z\eta}{V_g}=-\ln\left|\frac{U(z)}{U_0}\right|+\frac{1}{4}\ln\left|\frac{\nu_2U^4(z)+\nu_1U^2(z)+\eta}{\nu_2U_0^4+\nu_1U_0^2+\eta}\right|+ \\ +\frac{\nu_1}{2G^{1/2}}\left[\tan^{-1}\left(\frac{2\nu_2U^2(z)+\nu_1}{G^{1/2}}\right)-\tan^{-1}\left(\frac{2\nu_2U_0^2+\nu_1}{G^{1/2}}\right)\right] \quad (4)$$

for $4\eta\nu_2>\nu_1^2$,

$$\frac{z\eta}{V_g}=-\ln\left|\frac{U(z)}{U_0}\right|+\frac{1}{4}\ln\left|\frac{\nu_2U^4(z)+\nu_1U^2(z)+\eta}{\nu_2U_0^4+\nu_1U_0^2+\eta}\right|- \\ -\frac{\nu_1}{4(-G)^{1/2}}\ln\left|\frac{\left(2\nu_2U^2(z)+\nu_1+(-G)^{1/2}\right)\left(2\nu_2U_0^2+\nu_1-(-G)^{-1/2}\right)}{\left(2\nu_2U^2(z)+\nu_1-(-G)^{1/2}\right)\left(2\nu_2U_0^2+\nu_1+(-G)^{1/2}\right)}\right| \quad (5)$$

for $4\eta\nu_2<\nu_1^2$, and

$$\frac{z\eta}{V_g}=\frac{1}{2}\ln\left|\frac{U_0^2(\nu_1U^2(z)+2\eta)}{U^2(z)(\nu_1U_0^2+2\eta)}\right|-\frac{\eta}{\nu_1U^2(z)+2\eta}+\frac{\eta}{\nu_1U_0^2+2\eta} \quad (6)$$

for $4\eta\nu_2=\nu_1^2$. In this equations U_0 is the amplitude of SEWs at z=0. The solution of the GL equation for phase shift has the following form:

$$\begin{cases} \varphi_{NL}(U) = -(N / V_g)\left[z(U)\cdot U^2 + \int_U^{U_0} z(U')\cdot 2U'dU' \right] \\ z = z(U) \end{cases} . \quad (7)$$

The function z(U) is defined by relations (4)-(6).

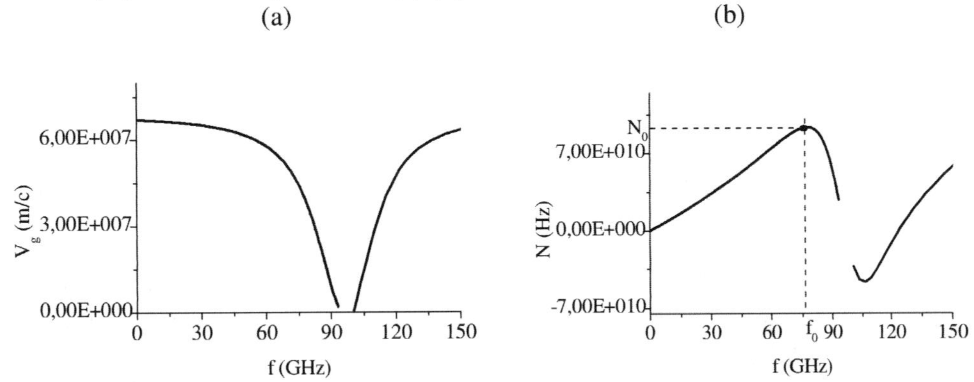

Figure 2. Group velocity (a) and nonlinear coefficient (b) as a function of frequency

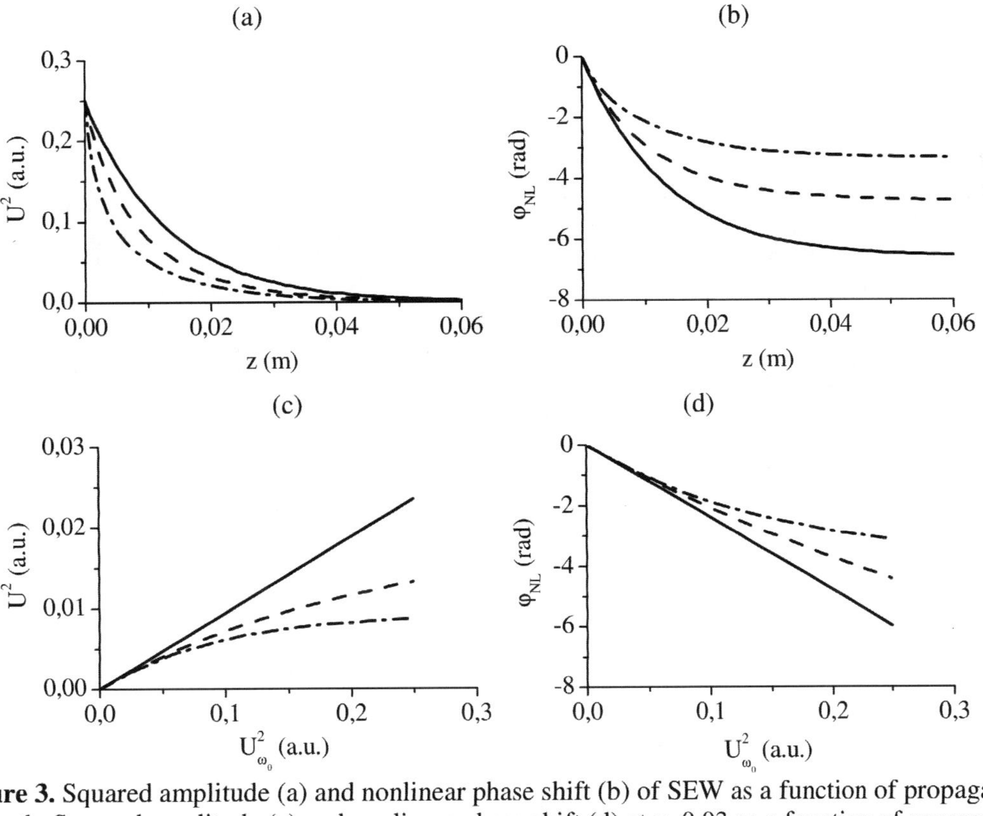

Figure 3. Squared amplitude (a) and nonlinear phase shift (b) of SEW as a function of propagation path. Squared amplitude (c) and nonlinear phase shift (d) at z=0.03 as a function of squared amplitude at z=0.

The results of the numerical simulation of group velocity and nonlinear coefficient as a function of frequency are shown in figure 2. It is seen that $V_g = 4.6\cdot 10^7$ m/c and $N = 9.3\cdot 10^{10}$ s^{-1} correspond to frequency f_0 = 76 GHz.

Figure 3 shows the typical dependences of $U^2(z), \varphi_{NL}(z), U^2(U_{\omega 0}^2), \varphi_{NL}(U_{\omega 0}^2)$ calculated by equations (4)-(7). Solid lines are drown for nonlinear coefficients υ_1=0 s^{-1}, υ_2=0 s^{-1}, dashed lines are drown for $\upsilon_1=6\cdot 10^9$ s^{-1}, υ_2=0 s^{-1}, dash-dotted lines are drown for $\upsilon_1=6\cdot 10^9$ s^{-1}, $\upsilon_2=9\cdot 10^{10}$ s^{-1}. Characteristics of $U^2(U_{\omega_0}^2)$ and $\varphi_{NL}(U_{\omega_0}^2)$ were calculated for z=0.03 m. As is seen from the figure 3 (b) the nonlinear phase shift as a function of distance demonstrates the saturation even for the case of absence of the nonlinear damping. Occurrence of the nonlinear damping (see figure 3 (a)) reduces the saturation level. Figure 3 (c) shows saturation of wave amplitude with an increase in excitation amplitude U_0 in the case of nonlinear damping. Similar saturation takes place for nonlinear phase shift shown in Figure 3 (d). Detailed analysis of the theoretical data obtained for various values of υ_1 and υ_2 shows that the saturation of the phase shift occurs due to the nonlinear damping if the nonlinear damping coefficients exceed $\upsilon_1=10^8$ s^{-1} and $\upsilon_2=10^9$ s^{-1}.

In conclusion, the results of our investigation show that the presence of nonlinear damping leads to a strong decrease in amplitude and to saturation in the nonlinear phase shift of SEWs. The features of the nonlinear damping of the SEWs obtained in this work could be used for creating of new sub-terahertz multiferroic devices. Thus, the examined materials are perspective for using in devices such as a phase shifter and a power limiter.

Acknowledgement
The work was supported in part by the Russian Foundation for Basic Research, grant of President of Russian Federation, and by Ministry of Education and Science of Russian Federation (Project "Goszadanie").

References
[1] Vaz C A F 2012 *J.Phys.: Condens. Matter.* **24** 333201
[2] Grishin S V, Beginin E N, Morozova M A, Sharaevskii Yu P and Nikitov S A 2014 *J. Appl. Phys.* **115** 053908
[3] Burdin D A, Chashin D V, Ekonomov N A, Fetisov L Y, Fetisov Y K, Sreenivasulub G and Srinivasan G 2014 *Journal of Magnetism and Magnetic Materials.* **358-359** 98
[4] Sadovnikov A V, Bublikov K V, Beginin E N and Nikitov N A 2014 *Journal of Communications Technology and Electronics.* **59** 914
[5] Cherkasskii M A and Kalinikos B A 2013 *Tech. Phys. Lett.* **39** 182
[6] Ustinov A B, Kalinikos B A and Srinivasan G 2014 *Appl. Phys. Lett.* **104** 052911
[7] Ustinov A B and Kalinikos B A 2006 *Appl. Phys. Lett.* **89** 172511
[8] Ustinov A B and Kalinikos B A 2008 *Appl. Phys. Lett.* **93** 102504
[9] Kuanr B K, Anderson N R, Celinski Z J and Camley R E 2015 *IEEE Magnetics Letters* **6** Article № 3500304
[10] Ustinov A B and Kalinikos B A 2007 *Appl. Phys. Lett.* **90** 252510
[11] Lax B and Button K J 1962 *Microwave ferrites and ferrimagnetics* (New York: McGraw-Hill)
[12] Ustinov A B, Tatarenko A S, Srinivasan G and Balbashov A M 2009 *J. Appl. Phys.* **105** 023908
[13] Karpman V I 1975 *Nonlinear waves in dispersive media* (London: Pergamon Press)

Generation of the pure spin current *via* Auger recombination in Rashba quantum wells

A N Afanasiev, A A Greshnov and G G Zegrya

A F Ioffe Physico-Technical Institute, 26 Politekhnicheskaya Street, St Petersburg 194021, Russian Federation

E-mail: afanasiev.an@mail.ru

Abstract. We propose new non-optical mechanism of the pure spin current generation *via* Auger recombination in the quantum wells with Rashba spin-orbit coupling. It is shown that such process is allowed due to interference between the Coulomb matrix elements corresponding to two different transitions during Auger recombination, leading to non-diagonal transversal components of the spin current tensor, $J_R = J^{xy} = -J^{yx}$. In the limit of low temperatures the total spin current is proportional to the Rashba constant γ_R, spin relaxation time τ_s and the third power of both the concentration n, inverse quantum well width a^{-1} and E_g^{-2}. Estimations show that typical magnitude of the generated spin current by this way is much greater than the ones obtained using intraband optical excitation mechanisms and comparable with spin currents driven by interband optical excitation, provided that the quantum well bandgap is narrow enough ($E_g < 0.5$ eV).

1. Introduction

Intensive development of the modern spin physics of low-dimensional semiconductor structures is primarily associated with the idea of creation of electronics based on carriers spin. One of the main goals of research in this area is the invention of an effective mechanism for creation of spin currents. The flow of electrical current may be accompanied by the transfer of spin if carriers are polarized. But it is more interesting to consider situation when directed motion of spin polarized carriers leads to zero charge current i.e. a pure spin current flows. Spin polarization in semiconductor structures can be created using spin-orbit splitting of bands in inversion-asymmetric materials. In two-dimensional structures spin splitting of conduction band is linear with respect to in-plane wave vector. This can be described by effective Rashba hamiltonian [1] related to structural asymmetry of confining potential. In case of [001] quantum well growth direction Rashba term is given by

$$H_R = \gamma_R(\sigma_x q_y - \sigma_y q_x) , \tag{1}$$

with corresponding dispersion relation

$$E_\mu(\mathbf{q}) = \frac{\hbar^2 q^2}{2m^*} + \mu\gamma_R q , \tag{2}$$

where q is the in-plane component of wave vector, γ_R is the Rashba constant describing splitting value, $\mu = \pm1$ is the spin index, $\sigma-$ Pauli matrixes. Rashba spin splitting leads to opposite

Journal of Physics: Conference Series **661** (2015) 012051 doi:10.1088/1742-6596/661/1/012051

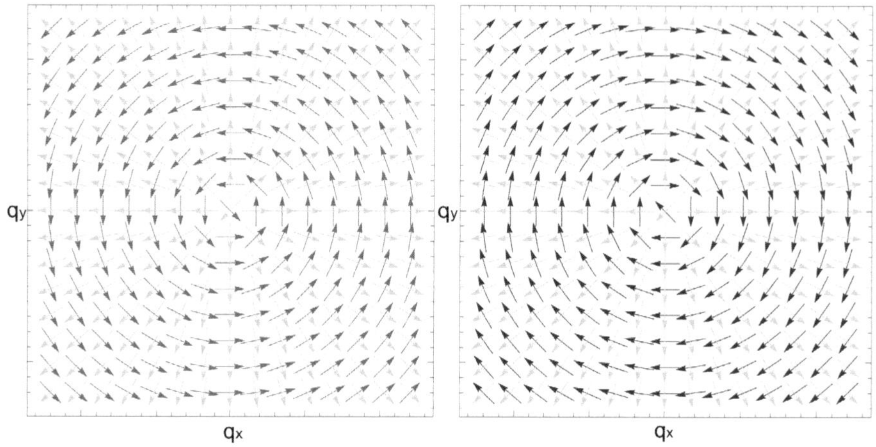

Figure 1. Spin and velocity vector fields in wave vector space, obtained from effective Hamiltonian (1). Green arrows correspond to velocity field, red and blue– to spin fields for energy branches (2) $\mu = 1$, $\mu = -1$ respectively.

motion of carriers with opposite spin direction (see Fig.1). In equilibrium, contributions to spin current from the energy branches (2) compensate each other and the total pure spin current flow vanishes.

Most proposed spin currents generation methods are based on optical excitation of carriers. In this paper non-equilibrium distribution required for generation of pure spin currents is excited by non-optical method. Consider a junction with the quantum well. Applying bias causes carrier injection into the well followed by Auger recombination. We consider the most probable type of Auger recombination in quantum wells– so-called CHCC– process [2] illustrated on Fig. 2. There are two electrons 1 and 2 in the initial state on one quantization level in the quantum well. One of them recombines with heavy hole 3 and transfers energy to another electron by the Coulomb interaction, which is excited to state 4 in the continuous spectrum with an energy near bandgap.

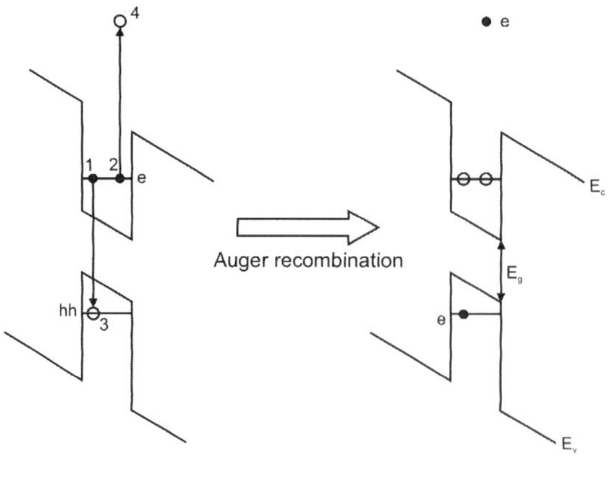

Figure 2. Generation of spin current-carrying states due to Auger recombination.

In two dimensional structures Auger recombination is significantly enhanced since energy threshold is removed due to lack of momentum conservation in growth direction. Spatial confinement results in modification of E_g functional dependence of the rate of Auger recombination in comparison with the bulk semiconductor. In quantum wells Auger recombination rate contains E_g^{-n} term [2],[3] while for bulk it is proportional to $\exp(-E_g/kT)$. We consider the quantum well with narrow bandgap and high carrier concentration. This let us neglect other types of recombination in compare with Auger recombination.

High-energy electrons 4 are weekly influenced by well potential and hence $q-$linear spin splitting of their states is small. This, together with the rapid spin relaxation of heavy holes let us assume that the final states 3 and 4 do not contribute to the creation of spin polarization. Therefore, the spin current is carried by released states in the Fermi distribution of electrons in the quantum well (" holes " in the Fermi sea of electrons), which are generated in pairs during each elementary act of CHCC– process of Auger recombination (see Fig.2).

2. Carriers state and Auger recombination rate

In order to describe Auger recombination it is crucial to take into account mixing of the valence and conduction bands states to each other. Since the spin relaxation processes are rapidly accelerated with increasing of temperature we restrict ourselves to the limit of low temperatures $T \to 0$. For sake of simplicity we consider carrier concentration n satisfying the relation

$$q \le q_f \ll \frac{\pi}{a} \ll k_g = \sqrt{3}\frac{E_g}{P}\,, \tag{3}$$

where q is the in-plane wave vector, a is the well width, $q_f = \sqrt{2\pi n}$ is the Fermi wave vector, E_g is the bandgap, P is the Kane matrix element. Thus, electrons and holes occupy only the ground level of size quantization.

To obtain carriers states, first we consider the symmetric well without spin splitting and further modify the obtained functions with respect to Rashba spin-orbit coupling. Heavy holes in symmetric well are described by Luttinger model [5] for infinitely deep well, which gives

$$\psi^{(3)}_{\mathbf{q},\uparrow\downarrow}(z) = \sqrt{\frac{2}{a}}\begin{pmatrix} \frac{\sqrt{2}}{\pi}\frac{P}{E_g}q_+ \\ 0 \\ \cos\frac{\pi z}{a} \\ -i\frac{\sqrt{3}}{2}\frac{q_+a}{\pi}\left(-\sin\frac{\pi z}{a}+\frac{2z}{a}\right) \\ 0 \\ 0 \end{pmatrix},\ \sqrt{\frac{2}{a}}\begin{pmatrix} 0 \\ -\frac{\sqrt{2}}{\pi}\frac{P}{E_g}q_- \\ 0 \\ 0 \\ i\frac{\sqrt{3}}{2}\frac{q_-a}{\pi}\left(-\sin\frac{\pi z}{a}+\frac{2z}{a}\right) \\ \cos\frac{\pi z}{a} \end{pmatrix}. \tag{4}$$

Electrons are described by Kane model [4]. Thus, we obtain the following wave functions for electrons in quantum well

$$\psi_{\mathbf{q},\uparrow,\downarrow} = \sqrt{\frac{2}{a}}\begin{pmatrix} \cos\left(\frac{\pi z}{a}\right) \\ 0 \\ -\frac{1}{\sqrt{2}}\frac{P}{E_g}q_-\cos\left(\frac{\pi z}{a}\right) \\ \sqrt{\frac{2}{3}}\frac{iP}{E_g}\frac{\pi}{a}\sin\left(\frac{\pi z}{a}\right) \\ \frac{1}{\sqrt{6}}\frac{P}{E_g}q_+\cos\left(\frac{\pi z}{a}\right) \\ 0 \end{pmatrix},\ \sqrt{\frac{2}{a}}\begin{pmatrix} 0 \\ \cos\left(\frac{\pi z}{a}\right) \\ 0 \\ -\frac{1}{\sqrt{6}}\frac{P}{E_g}q_-\cos\left(\frac{\pi z}{a}\right) \\ \sqrt{\frac{2}{3}}\frac{iP}{E_g}\frac{\pi}{a}\sin\left(\frac{\pi z}{a}\right) \\ \frac{1}{\sqrt{2}}\frac{P}{E_g}q_+\cos\left(\frac{\pi z}{a}\right) \end{pmatrix}, \tag{5}$$

where P – Kane matrix element, $q_\pm = q_x \pm iq_y$, z is growth direction. We describe spin splitting of electron states in the well by mixing the multi-band functions (5) corresponding to different spins with the coefficients obtained from the single-band model (1).

$$\psi^{(e)}_\xi(z) = \frac{\psi_{\mathbf{q},\uparrow}(z) + i\mu e^{i\varphi}\psi_{\mathbf{q},\downarrow}(z)}{\sqrt{2}}\,, \tag{6}$$

where $\mathbf{q} = (q\cos\varphi, q\sin\varphi, 0)$, φ– angle of in-plane wave vector with respect to x axis, $\xi = \{\mathbf{q}, \mu\}$ is the set of the quantum numbers.

The wave functions (4), (6) are used to calculate the rate of Auger recombination. Auger recombination probability per unit time is calculated by the Fermi golden rule.

$$P_{i\to f} = \frac{2\pi}{\hbar} \mid M_{fi} \mid^2 \delta(E_f - E_i) \ , \tag{7}$$

where

$$M_{fi} = \langle \psi^{(f)}_{\xi_3,\xi_4,k_4}(\mathbf{r}_1, \mathbf{r}_2) \left| \frac{e^2}{\varkappa |\mathbf{r}_1 - \mathbf{r}_2|} \right| \psi^{(i)}_{\xi_2,\xi_1}(\mathbf{r}_1, \mathbf{r}_2) \rangle \tag{8}$$

is the Coulomb matrix element. Antisymmetry requirement of two-particle wave functions describing the initial and final states results in two terms in the matrix element of Auger transition. This leads to contribution containing their interference

$$|M_{fi}|^2 = |M_I|^2 + |M_{II}|^2 + 2Re(M_I^* M_{II}) \tag{9}$$

It can be shown that matrix element of electron-electron interaction under assumptions (3) reduces to product of overlap integrals of the initial and final single particle states

$$M_{I,II} = \frac{2\pi e^2}{\varkappa \, q_{I,II}} \langle \psi^{(e)}_{\xi_{2,1}} | \psi^{(4)}_{\xi_4,k_4} \rangle \langle \psi^{(e)}_{\xi_{1,2}} | \psi^{(3)}_{\xi_3} \rangle \ \delta_{\mathbf{q}_{1,2} - \mathbf{q}_3, \mathbf{q}_{2,1} - \mathbf{q}_4} \ , \tag{10}$$

here $q_I = |\mathbf{q}_1 - \mathbf{q}_3|$, $q_{II} = |\mathbf{q}_2 - \mathbf{q}_3|$– transferred in-plane wave vector.

3. Spin current derivation

Operator of the spin current is given by product of velocity and spin operators

$$j^{\alpha\beta} = v_\alpha s_\beta \ . \tag{11}$$

In case of Rashba spin splitting there are only off-diagonal non-zero average values of the components of the spin current operator tensor

$$j^{xy}(\xi) = -\frac{1}{2}\left(\frac{\gamma_R}{\hbar} + \mu \sin^2(\varphi)\frac{\hbar q}{m^*}\right) \ . \tag{12}$$

The corresponding components of the macroscopic spin current is calculated in the approximation of the relaxation time

$$J_R = -\tau_s \sum_{\xi_1,\xi_2} P(\xi_1,\xi_2) f_0(\xi_1) f_0(\xi_2) [j(\xi_1) + j(\xi_2)] \ , \tag{13}$$

where τ_s is the spin relaxation time, $P(\xi_1, \xi_2)$is the rate of Auger recombination for fixed states of electrons in the well, i.e. spin current carriers generation rate

$$P(\xi_1,\xi_2) = \sum_{\xi_3,\xi_4,k_4} P(\xi_1,\xi_2,\xi_3,\xi_4,k_4) f_0^h(\mathbf{q}_3) \ . \tag{14}$$

After summation over in-plane wave vectors and spin indexes, final analytical expression for the spin current in this system can be obtained

$$J_R = \tau_s \frac{\gamma_R}{\hbar} \frac{8}{3\sqrt{3}} \frac{e^4 \pi^4}{\varkappa^2 \hbar a^3} \frac{P^5}{E_g^6} q_f^6 \frac{I_e}{(2\pi)^3} \ . \tag{15}$$

I_e is the dimensionless integral given by

$$I_e = -\int_0^1 \frac{d^2q_3}{(2\pi)^2} \int_0^{2\pi}\int_0^{2\pi} \frac{d\varphi_1 d\varphi_2}{(2\pi)^4} \int_0^1 q_1 dq_1 \left\{ \int_0^1 q_2 dq_2 \cos(arg(\alpha)) - \right.$$

$$\left. - 2\left[q_1 \sin^2(\varphi_1)\cos\left(arg(\alpha) + \varphi_2 - \varphi_1\right) + \sin^2(\varphi_2)\cos(arg(\alpha))\right]_{q_2=1} \right\} \quad (16)$$

where $arg(\alpha)$ is the argument of complex value $\alpha = (q_3^+ - q_2^+)(q_3^- - q_1^-)$. This expression makes it clear, that both the Fermi surface and Fermi sea contribute to spin current. Only interference term $2Re(M_I^* M_{II})$ in the square of the matrix element (8) give contribution to I_e.

4. Discussion and conclusion

In order to analyze results it is more convenient to extract from (15) spin current unit parameter and rewrite analytical expression for spin current in low temperature regime in the form

$$J_R = \frac{8\pi^4 I_e}{3\sqrt{3}} J_0 \left(\frac{e^2/\varkappa a}{\hbar/\tau_s}\right)^2 (na^2)^3 \frac{\gamma_R P^5}{E_g^6 a^6}, \quad (17)$$

where $J_0 = (\tau_s a)^{-1}$ is the "elementary" spin current. The final result strongly depends on band structure parameters as P^5 and E_g^{-6} which is consequence of generation *via* Auger recombination. Since Auger recombination is three-particle process, the value of generated spin current has cubic dependance on carrier concentration.

Estimation of the generated spin current (17) using the following set of parameters– $E_g = 0.4eV$, $P = 7\cdot 10^{-7} eV\cdot cm$, $\tau_s = 10^{-9} s$, $\gamma_R = 10^{-10} eV\cdot cm$, $a = 10\ nm$, $\varkappa = 10$, $n = 10^{11} cm^{-2}$– gives $J_R \simeq 10^{18} cm^{-1}s^{-1}$. For comparison, in case of optical excitation with $I = 1kW/cm^2$ and the same parameters quantum well photoionization mechanism [6] gives $J \sim 10^{10} cm^{-1}s^{-1}$, while interband absorption of light mechanism [7] – $J \sim 10^{16} cm^{-1}s^{-1}$. Estimations show that typical magnitude of the generated spin current by Auger recombination is much greater than the ones obtained using intraband optical excitation mechanisms and comparable with spin currents driven by interband optical excitation, provided that the quantum well bandgap is narrow enough. Unlike mechanisms using optical excitation, in the considering structure it is sufficient to apply voltage to the junction to create spin currents. This is more convenient for integration into the semiconductor schemes.

Acknowledgments

This work was supported by "Dynasty" Foundation and Russian Federation President support program of leading scientific schools (NSh-5062.2014.2).

References

[1] Bychkov Y A and Rashba E I 1984 *JETP Lett.* **39** 78
[2] Zegrya G G and Polkovnikov A S 1998 *JETP* **86** 815
[3] Dyakonov M I and Kachorovskii V Yu 1994 *Phys. Rev.* B **49** 17130
[4] Kane E O 1957 *J. Phys. Chem. Solids* **1** 249
[5] Luttinger J M 1956 *Phys. Rev.* **102** 1030
[6] Utesov O I, Zegrya G G and Greshnov A A 2010 *JETP Lett.* **92** 33
[7] Ivchenko E L and Tarasenko S A 2008 *Semicond. Sci. Technol.* **23** 114007

Effect of subband nonparabolicity on optical properties of InSbAs/AlSb deep quantum well heterostructures

N. V. Pavlov, G. G. Zegrya

Ioffe Institute, 194021, Russia, St.-Petersburg, Politekhnicheskaya St., 26

E-mail: pavlovnv@mail.ru

Abstract. Optical properties of heterostructures with deep quantum wells have been studied in the framework of four-band Kane mode permitting a nonparabolic energy spectrum of charge carriers to be taken into account. The system $AlSb/InAs_{0.84}Sb_{0.16}/AlSb$ was used as an example. It is established that the nonparabolicity weakly influences the overlap integral between s-and p-states, but notably increase the state density and optical absorption coefficient in the conduction band.

1. Introduction

Heterostructures with deep quantum wells that use InAsSb solid solutions as the active region can be considered as promising materials for infrared optoelectronics [1, 2]. These compounds are characterized by the bandgap width E_g and effective electron mass m_c that are minimum among A_3B_5 semiconductors [3, 4]. There is here a significant nonparabolicity in the energy spectrum of charge carriers. This necessarily leads to adjusting substantial corrections to the energies of dimensional quantization levels as compared to those observed under the parabolic dispersion law, even for the ground state of conduction band. The purpose of present investigation is to calculate the optical absorption coefficient and radiation recombination rate for intraband optical transitions between various subbands of dimensional quantization with and without nonparabolicity in the energy spectrum of $AlSb/InAs_{0.84}Sb_{0.16}/AlSb$ heterostructure with deep quantum wells. The nonparabolicity in the energy spectrum is described with a four-band Kane model.

2. Basic Equations

For the charge carriers, we used the wave functions of the type $\psi = \Psi_s \,|\, s\rangle + \mathbf{\Psi} \,|\, \mathbf{p}\rangle$, where $|\, s\rangle$ and $|\, \mathbf{p}\rangle$ are the Bloch functions of s- and p-type, and Ψ_s and $\mathbf{\Psi}$ are the spinors. Near the Γ-point, the spherical approximation Kane equations are given by the following expressions:

$$(E_C - E)\Psi_s - iP\nabla\mathbf{\Psi} = 0,$$

$$(E_V - \delta - E)\mathbf{\Psi} - iP\nabla\Psi_s + \frac{\hbar^2}{2m_0}(\gamma_1 + 4\gamma_2)\nabla(\nabla\mathbf{\Psi})$$

$$-\frac{\hbar^2}{2m_0}(\gamma_1 - 2\gamma_2)[\nabla\times[\nabla\times\mathbf{\Psi}]] + i\delta[\boldsymbol{\sigma}\times\mathbf{\Psi}] = 0. \qquad (1)$$

Here E_C and E_V are the energies of conduction and valence band edges, P is the Kane matrix element, $\delta=\Delta_{so}/3$ is the spin-orbit splitting constant, m_0 is the free electron mass, γ_1 and γ_2 are generalized Lattinger parameters, and $\boldsymbol{\sigma}=(\sigma_x, \sigma_y, \sigma_z)$ are the Pauli matrices.

The nonparabolicity in the energy spectra are described by the following expressions:

$$E_c = \frac{\hbar^2(k_c^2+q^2)}{2m_c},$$

$$E_{hh} = -E_g - \frac{\hbar^2(k_h^2+q^2)}{2m_{hh}},$$

$$E_{lh} = -E_g - \frac{\hbar^2(k_l^2+q^2)}{2m_{lh}}, \qquad (2)$$

where E_c, E_{hh} and E_{lh} are the energies of electrons, heavy and light holes, k_c, k_h and k_l are the quantized wave vector components of electrons, heavy and light holes, m_h and m_l are effective masses of heavy and light holes.

The energy spectra with regard to their nonparabolic features are written for electrons and light holes as:

$$P^2(k_c^2+q^2) = \frac{E_c(E_c+E_g)(E_c+E_g+3\delta)}{E_c+E_g+2\delta},$$

$$E_{lh} = -E_g - \frac{3}{2}\delta - \frac{\hbar^2(k_l^2+q^2)}{4}(m_l^{-1}+m_h^{-1}) + \sqrt{2\delta^2 + \left(\frac{\delta}{2} - \frac{\hbar^2(k_l^2+q^2)}{4}(m_l^{-1}-m_h^{-1})\right)^2}, \qquad (3)$$

where $m_l^{-1} = -\frac{2P^2}{\hbar^2 E_{lh}} + m_0^{-1}(\gamma_1+4\gamma_2)$ and $m_h^{-1} = m_0^{-1}(\gamma_1-2\gamma_2)$.

The optical absorption coefficient α as a function of transition frequency ω is given by the following expression [6]:

$$\alpha(\omega) = \sum_{i,j} \frac{4\pi}{\sqrt{\kappa_0}} \frac{e^2}{\hbar c} \frac{1}{a\hbar\omega} \int q\mathrm{d}q P_{ij}^2 \delta(E_i - E_j - \hbar\omega), \qquad (4)$$

where the indices i and j refer to various subbands in the conduction and valence bands, κ_0 is a static dielectric permittivity, $P_{ij}^2 = 2\left|P_{ij}^x\right|^2 + \left|P_{ij}^{\|}\right|^2$, and $\boldsymbol{P}_{ij}$ is the matrix element of optical transition in the four-band Kane model which is defined as $\boldsymbol{P}_{ij} = iP\int(\overline{\Psi_{si}}\boldsymbol{\Psi}_j + \overline{\Psi_{sj}}\boldsymbol{\Psi}_i)\mathrm{d}x$.

The radiative recombination rate of quantum wells for non-degenerated electrons can be calculated using the following formula [7]:

$$R_{ph} = \frac{4\kappa_\infty}{\sqrt{\kappa_0}} \frac{e^2}{\hbar c} \frac{1}{\hbar^3 c^2} \sum_{i,j} \left|P_{ij}(q_T)\right|^2 (E_i(q_T) - E_j(q_T)) \frac{n_i p}{N_j} \exp\left(-\frac{\varepsilon_j}{k_B T}\right), \qquad (5)$$

where κ_∞ is a high frequency permittivity, $q_T = \frac{\sqrt{2k_B m_c}}{\hbar}$ is the wave vector of thermal motion, T is absolute temperature, $N_j = \frac{m_j k_B T}{\pi\hbar^2}$ is an effective density of states in the valence band, n and p are the carrier concentrations, and ε_j is the distance between the subband j and ground level in the valence band.

The radiative recombination time is presented by the following equation [7]:

$$\tau_{phij} = \frac{\sqrt{\kappa_0}}{4\pi\kappa_\infty} \frac{\hbar c}{e^2} \frac{\hbar c^2 m_j k_B T}{p\left|P_{ij}(q_T)\right|^2 (E_i(q_T) - E_j(q_T))} \exp(\frac{\varepsilon_j}{k_B T}). \quad (6)$$

In the case of strong degeneracy, we should use q_F instead of q_T in formulas (5) and (6).

3. Results

The energies and quantized wave vector components calculated with and without nonparabolicity of energy spectrum with the quantum well width *of* a=10 *nm* are presented in Table 1.

Table 1. Energies and quantized wave vector components for various levels of dimensional quantization in $AlSb/InAs_{0.84}Sb_{0.16}/AlSb$ heterostructure with the quantum well width of a=10 nm as calculated with and without allowance for the energy spectrum nonparabolicity.

Dimensional quantization level	Without nonparabolicity		With nonparabolicity	
	Quantized wave vector component k, 10^6 cm^{-1}	Energy level E, meV	Quantized wave vector component k, 10^6 cm^{-1}	Energy level E, meV
c1	2.78	163	2.456	95
c2	5.481	632	5.206	294
c3	7.458	1170	8.051	501
hh1	2.66	6	2.66	6
hh2	5.3	26	5.3	26
lh1	1.606	46	1.377	20

The effective masses of charge carriers related to the dimensional quantization level and calculated with the four-band Kane model, are presented in Table 2.

Table 2. Charge carrier effective masses of various dimensional quantization levels in the $AlSb/InAs_{0.84}Sb_{0.16}/AlSb$ heterostructure with the quantum well width of a=5 and 10 nm as calculated with and without regard to the nonparabolicity in the energy spectrum.

Dimensional quantization level	a=5 nm		a=10 nm	
	Energy level E, meV	$\frac{m_i}{m_{c0}}$	Energy level E, meV	$\frac{m_i}{m_{c0}}$
c1	220	1.78	95	1.341
c2	612	3.104	294	2
c3	1010	4.4	501	2.728
c4			720	3.46
c5			938	4.181
c6			1150	4.784
lh1	114	2.267	20	1.339

As can be seen, the allowance for the nonparabolicity in the energy spectrum significantly influences the results of calculation for the positions of dimensional quantization levels and the carrier effective mass and, therefore, the density of states function.

The frequency dependences of a total optical absorption coefficient as calculated with and without allowance for nonparabolicity in the energy spectrum are presented in Fig. 1.

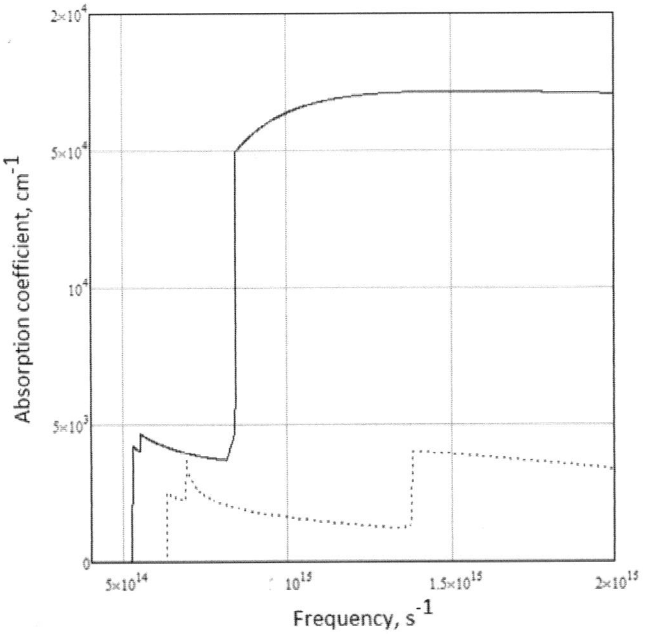

Fig. 1. Dependence of total optical absorption coefficient α on the optical transition frequency ω in the AlSb/InAs$_{0.84}$Sb$_{0.16}$/AlSb heterostructure with the quantum well width of a=10 nm as calculated in the framework of the parabolic spectrum model (dashed curve) and with allowance for the nonparabolicity in this spectrum, in the framework of four-band Kane model (solid curve).

The radiation recombination time for various optical transitions as calculated in cases of non-degenerated and strongly degenerate electrons, is given in Table 3.

Table 3. Radiative recombination time τ_{phij} for various optical transitions in AlSb/InAs$_{0.84}$Sb$_{0.16}$/AlSb heterostructure with quantum well width of a=10 nm, T=300 K, and p=10^{12} cm^{-2} as calculated with and without nonparabolicity in the energy spectrum in the cases of non-degenerated and strongly degenerate electrons (n=5×10^{12} cm^{-2}).

Transition	Without nonparabolicity		With nonparabolicity	
	Non-degenerate electron case τ_{phij}, s	Strongly degenerate case τ_{phij}, s	Non-degenerate electron case τ_{phij}, s	Strongly degenerate case τ_{phij}, s
c1-hh1	$8.1\cdot10^{-9}$	$7.1\cdot10^{-9}$	$1.2\cdot10^{-8}$	$1.5\cdot10^{-8}$
c2-hh1	$2.5\cdot10^{-8}$		$2.2\cdot10^{-8}$	
c3-hh1	$4.3\cdot10^{-7}$		$9.6\cdot10^{-5}$	
c1-hh2	$4.5\cdot10^{-7}$	$2.7\cdot10^{-8}$	$2.9\cdot10^{-7}$	$7.6\cdot10^{-8}$
c2-hh2	$3.6\cdot10^{-9}$		$1.1\cdot10^{-8}$	
c1-lh1	$1.02\cdot10^{-7}$	$1.8\cdot10^{-7}$	$1.1\cdot10^{-7}$	$2\cdot10^{-7}$
c2-lh1	$5.8\cdot10^{-7}$		$4.6\cdot10^{-7}$	

As can be seen, the allowance for the nonparabolicity in the spectrum for non-degenerate electrons leads to the radiative recombination time that increases for the transitions with coinciding parity of the initial and final states and to this time decreasing for the transitions between the states of different parity.

4. Conclusions

It has been thus calculated that the nonparabolicity in the energy spectrum in heterostructures with deep quantum wells significantly influences the energy levels of dimensional quantization, leading to a significant increase in the radiation absorption coefficient, especially for the optical transitions with participation of high-energy electrons.

This work was partly supported by Grant of Science School #5062.2014.2.

References

[1] Zegrya G G 1997 *Mid-infrared strained diode lasers* (Antimonide-Related Strained Layer Heterostructures) ed Manasreh M O (Gordon & Breach Science Publishers, Amsterdam) p 273

[2] Mikhailova M P, Danilov L V *et al* 2013 *Superlinear luminescence and enhancement of optical power in GaSb-based heterostructures with high conduction band offsets and nanostructures with deep quantum wells* (The Wonder of Nanotechnology: Quantum Optoelectronic Devices and Applications) edited by Razeghi M *et al* (SPIE Press, Bellingham, WA) p 105

[3] Vurgaftman I, Meyer J R, Ram-Mohan L R, J. 2001 *Appl. Phys.* **89** 5815

[4] Cripps S A, Hosea T J C, Krier A *et al* 2007 *Appl. Phys. Lett.* **90** 172106

[5] Asryan L V, Gun'ko N A, Polkovnikov A S, Zegrya G G, Suris R A, Lau P, and Makino T 2000 *Semicond. Sci. Technol.* **15** 1132

[6] Zegrya G G and Polkovnikov A S 1998 *J. Exp. Theor. Phys.* **86** 815

[7] Pavlov N V, Zegrya G G 2014 *Semicond.* **48** 1185

Journal of Physics: Conference Series **661** (2015) 012053 doi:10.1088/1742-6596/661/1/012053

Development of methods for orderly growth of nanowires

R R Reznik[123], K P Kotlyar[2], A I Khrebtov[16], Yu B Samsonenko[134], I P Soshnikov[134], V Dyakonov[78], U M Zadiranov[3], E M Tankelevskaya[3] and D A Kudryashov[1], D S Shevchuk[1] and G E Cirlin[12345],

[1]St-Petersburg Academic University – Nanotechnology Research and Education Centre RAS, Khlopina 8/3, 194021, St-Petersburg, Russia

[2]St-Petersburg State Polytechnical University – Nanotechnology, Polytechnicheskaya 29, 195251, St-Petersburg, Russia

[3]Ioffe Physical Technical Institute RAS, Politekhnicheskaya 26, 194021, St-Petersburg, Russia

[4]Institute for Analytical Instrumentation RAS, Rizhsky 26, 190103, St-Petersburg, Russia

[5]Fock Institute of Physics,St-Petersburg State University,198504 Petrodvorets, Russia

[6]Vavilov State Optical Institute, 199034, Birzhevaya line 12, St-Petersburg, Russia

[7]ZAE Bayern, Energieeffizienz Am Galgenberg 87, 97074 Würzburg, Germany

[8]University of Wuerzburg, Sanderring 2, 97070, Würzburg, Germany

E-mail: moment92@mail.ru

Abstract. Method of manufacturing substrates for self-catalyst/free-catalyst ordered growth of the nanowires has been developed. Experiments show the possibility of autocatalytic growth of ordered GaAs NWs on the substrates produced during the research.

1. Introduction

Nanowires are crystalline structures the length of which considerably exceeds the diameter which is from 8 to 300 nm. Nowadays the interest in nanowires stems from the prospects of their use in many applications, such as: chemical and biological highly sensitive sensors [1-4], emission cathodes and cantilevers for STM / AFM, fuel cells and batteries [5], in optoelectronics and onolactone, in electromechanical devices (tensor sensors, microphones, piezoresistors, nanobodies).

The creation of device structures based on nanowires requires the development of methods and technologies for the synthesis of ordered semiconductor NWs and research of their features. We emphasize that approach developed by many researchers based on the creation of orderly catalytic growth centers of NWs [6, 7.8, 9] has a number of disadvantages, including the possibility of growth centers' shifting to their complete removal from the surface, branching of NWs, uncontrolled growth in other directions , uncontrolled alloying of catalyst by the material etc. Therefore, the work is dedicated to the research aimed at the development of autocatalytic and catalyst free methods of growth of ordered arrays of NWs in materials A^3B^5.

2. The essence of the method.

The aim of the work is to determine the possibility of producing A^3B^5 structures with ordered free-catalyst or self-catalytic NWs through inhibitory layer with ordered windows of growth. The initial goal is the formation of holes in the inhibitory ordered SiO_2 layers by electron lithography.

Polished wafers Si (111) brand KDB-10 with SiO_2 layer about 50-100 nm thick previously synthesized by magnetron sputtering and / or thermal oxidation were used as the samples.

Electron resist layer 672 ARP (PMMA-polymethylmethacrylate K 950) was applied with centrifugation on the surface of the samples and after that dried for 15 minutes at T = 90C. Then the sample was exposed in the electron microscope Zeiss SUPRA 25 and subsequently developed in special solutions. The next step was etching with argon ion beam at energy of 0.5 keV to form lithographic windows in the SiO_2 layer and removal of residual polymer. At the final stage MBE growth of GaAs NWs was carried out using self-catalyst method [10, 11].

3. Results and discussion.

The sample surface after electron-beam lithography is presented in figure 1. The image shows three groups (2x10)of arrays produced with lithography at accelerating voltages of 10, 15 and 20 kV.

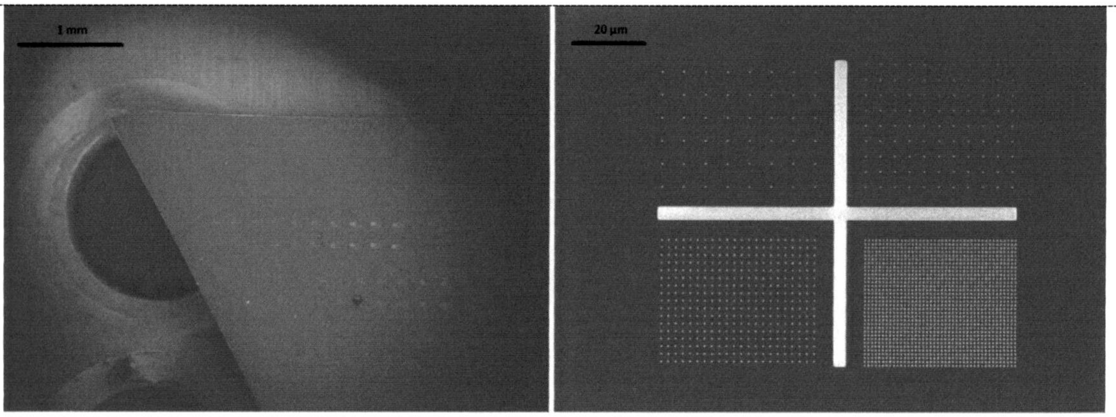

Figure 1. Review electron microscopic image in geometry top view of the sample surface after carrying out lithography.

When the exposure time is less than the minimum response time of EBW, ticks appear instead of points due to the finite time of operation of the system control beam in the electron microscope (figure 2). This fact means that we are limited in the choice of the values of the beam current and exposure time below.

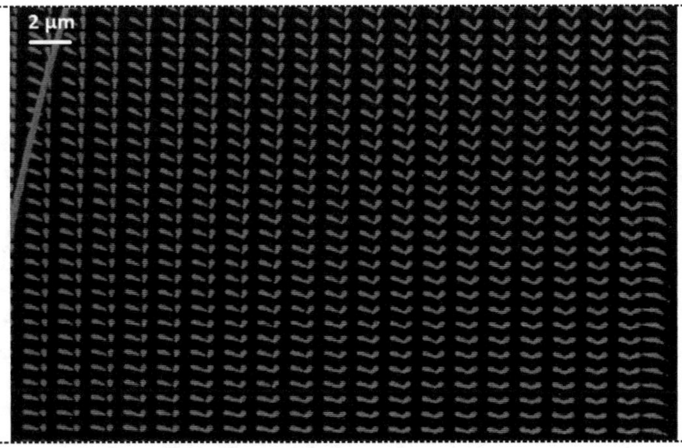

Figure 2. Electron microscope image in geometry top view of the array of slots in the resist during the exposure time greater than the minimum response time ELS. Lithography was performed at an accelerating voltage of 20kV.

Figure 3 shows the dependence of the diameter of the "window" in resist on the distance between adjacent "windows" at different accelerating voltages. By increasing the step up to two microns the diameter of the element decreases. Such dependence of the size on the step can be explained by the influence of proximity effects when the step is less than 2 microns. Lithographic elements of all sizes, used to analyze the results, were measured in areas with a step of 2, 4 and 6 microns.

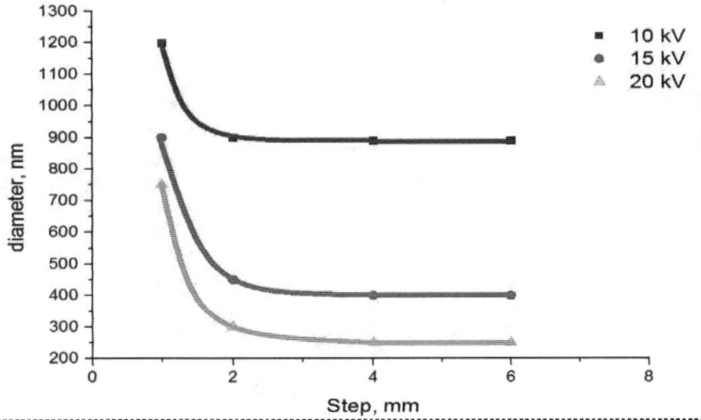

Figure 3. The dependence of the diameter of the "window" in the resist on the step between adjacent elements at accelerating voltages of 10, 15 and 20 kV and a dose of 600.

Dependence of the diameter of the "window" on the dose is shown in figure 4. At high doses (> 100 mC / cm^2) a "burning" of the resist can be seen where the electron beam enters into the sample. It should be noted that in following experiments on the growth of the NWs observed the formation of holes where the resist had been burnt.

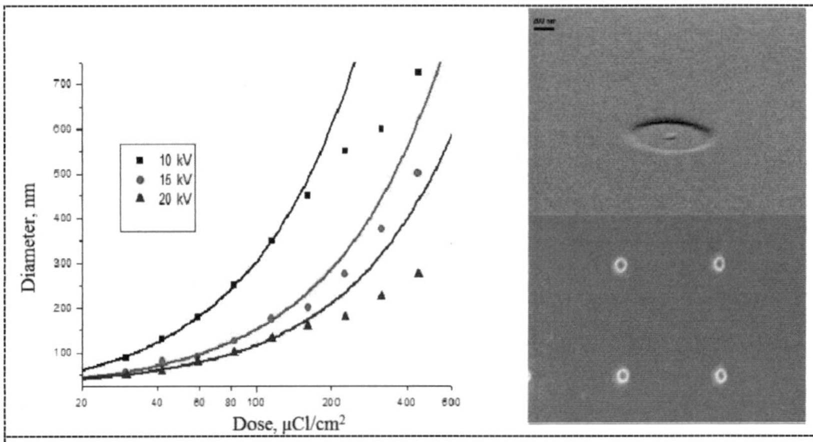

Figure 4. The dependence of the diameter of the "window" on the dose for a single layer of resist. Lithography was performed at accelerating voltages of 10, 15 and 20 kV and Electron microscope image in geometry top view of the array of slots in the resist, there is the burning of the resist. The exposure was conducted at an accelerating voltage of 20 kV and a beam current of~1 nA.

The result ion etching is represented in figure 5. Analysis of the characteristic size of the bitmap window showed that the minimum stable diameter Windows in the SiO_2 layer is about 40 nm.

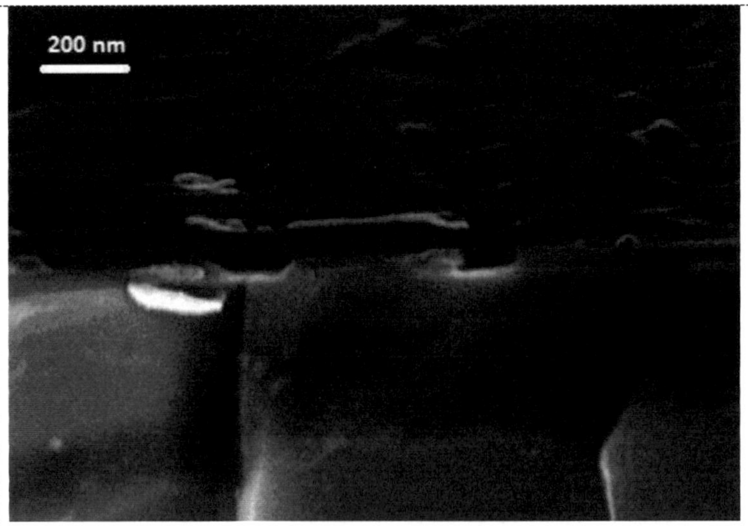

Figure 5. Electron microscope image in ISO array of "Windows" in the SiO_2 layer on the Si substrate after ion etching

There is SiO_2 flaking from the silicon surface in a number of cases after etching on the samples with SiO_2 layer was applied by magnetron sputtering (figure 6). This happens due to the lack of adhesion between SiO_2 layer and the silicon substrate during magnetron sputtering. No flaking of the inhibitory layer is observed on the samples with SiO_2 layer produced by oxidation of the substrate.

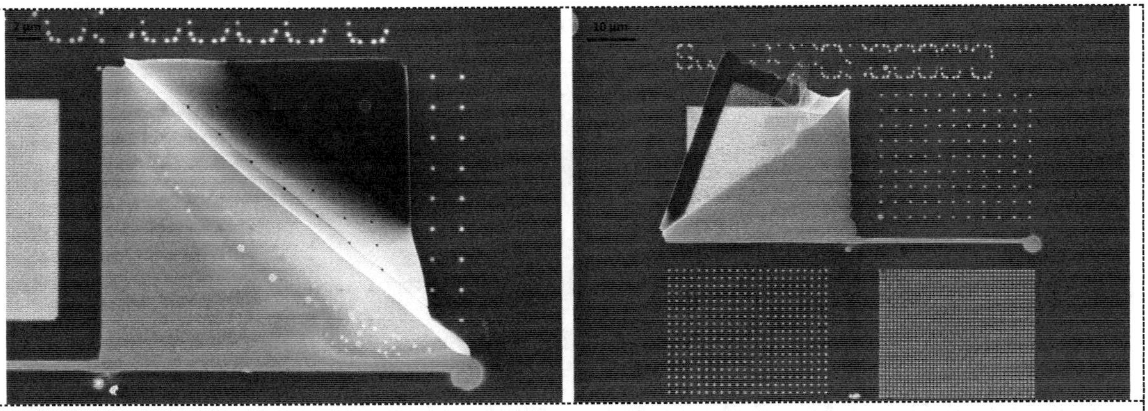

Figure 6. Electron microscope image in geometry top view of the array of slots in the SiO_2 layer obtained by precipitation with magnetron sputtering on a substrate of Si(111) after processing and ion etching.

Preliminary experiments on the growth of GaAs NWs are carried out in this work. The Example of the sample with surface lithographic structure in the SiO_2 layer is shown in figure 7. NW is 1000 nm high and 30 nm in diameter. The growth of individual NWs with low density is observed off the field of lithographic picture, the arrangement of these NWs is disordered and random and apparently due to insufficient purity of the process.

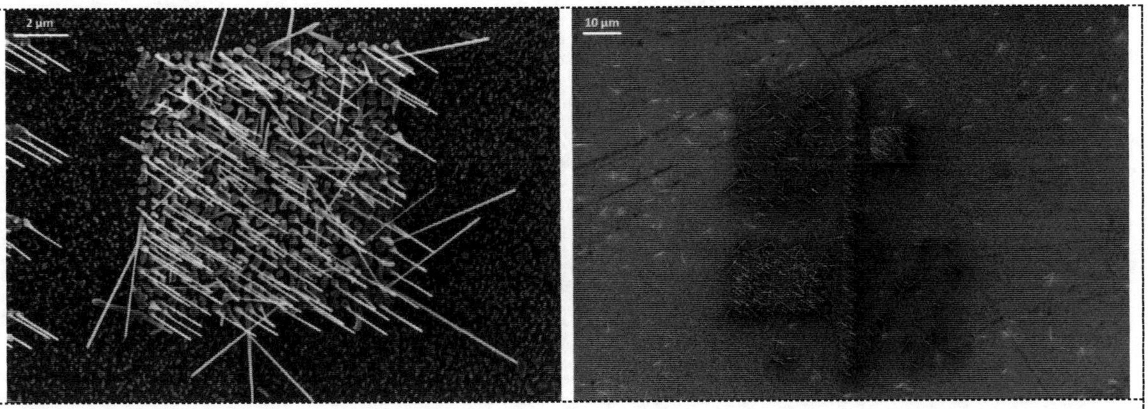

Figure 7. Electron microscope image in geometry top view of ordered arrays of GaAs NW on substrates with a SiO_2 layer with a lithographic Windows.

4. Conclusions.

The Dependence of the lithographic windows size on the main parameters (energy, beam current, the exposure step between elements, the thickness of the resist layer and inhibitory layer, peculiarities of the application of the inhibitory layer, etc.) are studied in the work. It is shown that if the distance between the elements is less than 2 microns, it is necessary to consider the impact of proximity effects. At high doses (> 100 mC / cm^2) "burning" of the resist can be seen where the electron beam enters the sample. The optimal conditions for the formation of lithographic windows with dimensions less than 50 nm are defined. The features of ion etching in the inhibitory SiO_2 layer are searched. The possibility of the formation of the windows with the characteristic sizes from 40 to 250 nm is demonstrated. The effect of inhibitory layer adhesion on the formation of lithographic substrates for growth is also researched.

The experiments show the possibility of autocatalytic growth of orderly GaAs NWs on the substrates produced during the research. The necessity of systematic research of growth conditions using the chosen method to get higher quality (homogeneous) arrays of NWs is evident.

Acknowledgements

Financial support from Russian Science Foundation grant No 14-12-00393 is acknowledged.

References

[1] Patolsky F, Zheng G and Lieber C M 2006 *Nature Protocols* **1** 1711
[2] Mathews M, Jansen R, Rijnders G, Lodder J C,and Blank D H A 2009 *Phys. Rev. B* **80** 064408
[3] Man-Fai Ng, Liping Zhou, Shuo-Wang Yang, Li Yun Sim, Tan Vincent B. C., and Ping Wu 2007 . *Phys. Rev. B* **76** 155435
[4] Markussen T, Rurali R, Brandbyge M and Jauho A-P 2006 *Phys. Rev. B* **74** 245313
[5] Chan C K, Peng H, Liu G, McIlwrath K, Zhang X F, Huggins R A and Yi Cui 2008 *Nature Nanotechnology* **3** 31-35
[6] Givargozov E I 2003 *Nature* **11** 20
[7] Cheng G and Moskovits M 2002 *Advanced Materials* **14** 1567
[8] Roest A L, Verheijen M A, Wunnicke O, Serafin S, Wondergem H and Bakkers E P A M 2006 *Nanotechnology* **17** 271
[9] Persson A I, Fröberg L E, Samuelson L and Linke H 2009 *Nanotechnology* **20** 225304
[10] Noborisaka J, Motohisa J and Fukui T 2005 *Appl. Phys. Lett.* **86** 213102
[11] Soshnikov I P, Petrov V A, Proskuryakov U U, Kudryashov D A, Nashchyokin A V, Cirlin G E, Treharne R, Durose K 2013 *Semiconductors* **47** 865

Simulation of a plasma antenna by PIC method

N N Bogachev[1,2]

[1]Moscow State University of Information Technologies, Radio engineering and Electronics, Moscow, Russian Federation

[2]A M Prokhorov General Physics Institute, Russian Academy of Science, Moscow, Russian Federation

E-mail: bgniknik@yandex.ru

Abstract. The numerical model of a plasma asymmetric dipole antenna is studied in this paper. This antenna consists of a dielectric tube with discharge plasma and a metal screen. The plasma in the tube is simulated by Particle-in-cell (PIC) method. The spectra of the electromagnetic field components were studied in the plasma and the near antenna field for various antenna operating modes. The operating modes depend on a ratio of a plasma frequency and a frequency of radiated electromagnetic wave. Langmuir frequency component was found in spectra within plasma for all antenna operating modes. The amplitude of this spectral component is greater than the amplitude of the component at the carrier frequency for non-radiative and non-linear modes. The biggest amplitude of the radiated signal spectrum is at the carrier frequency in the linear mode.

1. Introduction

Plasma antennas with discharge plasma in dielectric tubes are researched by many scientific groups [1-10]. These antennas are very perspective, because they have some advantages such as low radar cross section and fast reconfigurable parameters (a frequency, a radiation pattern, a radiation resistance and etc). However, when we change the antenna parameters, we should use receiving and transmitting modes in which a signal hasn't noticeable non-linear distortions and high noises. Nowadays the researches of the plasma antenna noises and the non-linear distortions of a signal are key trends.

One of the most convenient plasma antenna types is a Plasma Assymetrical Dipole antenna (PAD or "monopole"). The plasma assymetrical dipole antenna consists of a dielectric discharge tube with plasma and a metal screen (see figure 1). A control of PAD parameters can be made by changing a plasma concentration or the length of the plasma column.

A dispersion equation of a surface electromagnetic wave on a plasma cylinder was solved for a cylinder radius r_0=0.5 cm and plasma density values n_e=10^{10}-10^{12} cm^{-3} and the plasma assymetrical dipole was simulated by a Drude model in KARAT code in work [10]. In this work three operation modes of a plasma antenna were found: *non-radiative*, *non-linear* and *linear*. These operation modes depend on a ratio of an electromagnetic frequency f_0 (or $\omega_0=2\pi f_0$) and a plasma frequency f_p (or $\omega_p=2\pi f_p$) and relate with propagation conditions of the surface wave [11-13]. Only a potential surface wave exists along the plasma cylinder and PAD's parameters differ a lot from parameters of the same metal antenna on the frequency f_0 into the non-radiative mode. In the non-linear mode a velocity of surface wave is less than the velocity of light c on the frequency f_0 and the PAD's parameters differ from the metal antenna parameters. In the linear mode the surface wave velocity is close to the velocity of light c on the same frequency and the plasma antenna parameters are almost coincide with the metal antenna parameters. However, the Drude model considers only dielectric properties of the plasma, which are determined by a plasma density and an electron collision frequency.

But an interaction of a propagation electromagnetic wave and plasma can bring non-linear distortions and noises on a transmitting and receiving signal in the different operation modes. A selection of the operating modes with the minimal distortions demands the study of signal spectra of the plasma dipole antenna. In this task the plasma should be studied as a collection of particles. The Particle-in-Cell (PIC) method allows studying plasma as the collection of the particles.

A propagation in plasma antenna and radiation of a Gaussian form pulse is researched in this paper. The plasma asymmetrical dipole antenna parameters are: the length l=4 cm and the inner radius of the discharge tube r= 0.5 cm and the Gaussian form pulse parameters are: the duration τ_i = 15 ns and the carrier frequency f_0=1.7 GHz (ω_{ew0}=1.07·10^{10} rad·s^{-1}). This consideration will allow progress in solving two major problems: understanding of the process of plasma antenna signal radiation in different modes and the effect of plasma particles in the emitted signal.

2. Numerical model

A PIC model of the plasma asymmetric quarter wave dipole antenna have been done in KARAT code [14]. KARAT code is a program for electromagnetic numerical simulation, it uses the Final Differences in Time Domain (FDTD) method for solving Maxwell's equations. A scheme of plasma antenna model is shown in figure 2. From the coaxial cable 1 the electromagnetic wave passes to the plasma in dielectric tube 2, which together with the metal screen 3 forms an asymmetric dipole antenna. Perfect Matching Layer (PML) 4 is located on the borders of the modeling area. The spectra of electromagnetic field components E_r and E_z were recorded inside the plasma at a point number 1 (z = 7 cm, r = 0.5 cm) and in the near antenna field zone at a point number 2 (z = 8 cm, r = 5.8 cm). The point number 2 is located in the near zone, because the size increase of the simulated area endangering a stability of the PIC model. PIC plasma consists of two particle sorts, they are electrons and argon ions.

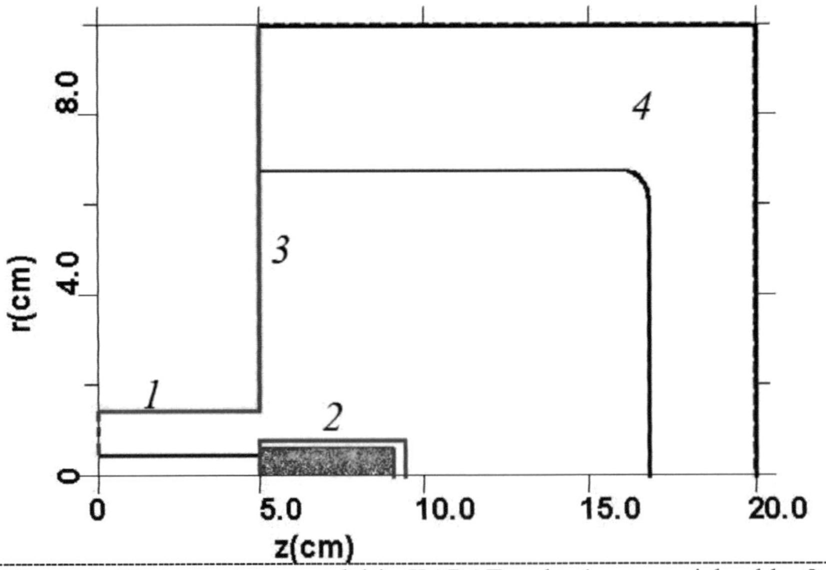

Figure 1. Plasma antenna model in KARAT code: 1 — coaxial cable, 2 — dielectric discharge tube with PIC plasma, 3 — metal screen, 4 – perfect matching layer (PML).

3. Simulation results and discussion

In this chapter I will discuss the spectra of the electromagnetic fields components $E_я$ and E_r, which were obtained by simulation of the plasma asymmetric dipole antenna.

The spectra are shown in figure 2 in the non-radiative mode. Parameters of plasma in this mode are the plasma frequency $f_p = 2 \cdot f_0 = 3.4\,\text{GHz}$ ($\omega_p = 2\omega_{ew0} = 2.14 \cdot 10^{10}$ rad/s) and the plasma density $n_e = 1.4 \cdot 10^{11}\,\text{cm}^{-3}$. Amplitudes of E_z and E_r components at the frequency 1.7 GHz are more than the ones at the frequency 3.4 GHz inside plasma (figure 2a). The amplitude of E_r components at low frequencies is more than the one at the frequency f_0. In point number 2 (Figure 3b) there are a low-frequency noise component, the carrier frequency f_0 and Langmuir frequency f_p in the spectra of E_z and E_r. This antenna operating mode is a *non-radiative,* since the amplitude of the signal in the near field of the antenna is much smaller than in plasma. The amplitude of the low-frequency noise component in the signal is more than the amplitude of the carrier signal frequency f_0. Langmuir frequency f_p coincides with the double harmonic of the carrier frequency $2f_0$, it is resonant for this type of antenna. Therefore, it is highly expressed in the signal spectrum in the points 1 and 2.

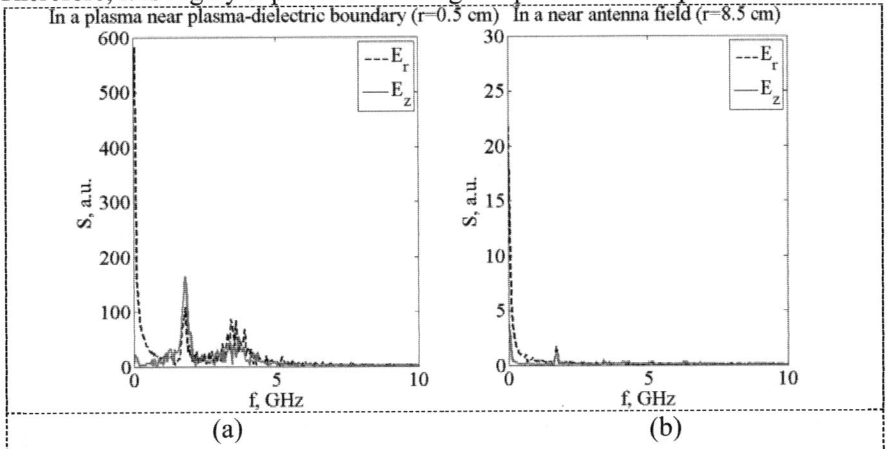

Figure 2. Electromagnetic wave spectra in non-radiative mode in: (a) point 1, (b) point 2.

The spectra of the signal are presented in figure 3 in the non-linear mode. The parameters of plasma in this mode are the plasma frequency $f_p = 5 \cdot f_0 = 8.5\,\text{GHz}$ ($\omega_p = 5\omega_{ew0} = 5.35 \cdot 10^{10}$ rad/s), the plasma density $n_e = 9{,}1 \cdot 10^{11}\,\text{cm}^{-3}$.The spectra of E_z and E_r have noises components between the frequencies f_0 and f_p and the high amplitude double components near the frequency f_p inside plasma (figure 3a). In point number 2 (Figure 3b) the amplitude at the signal frequency f_0 is best visible in spectrum of E_z, but the low-frequency components, the harmonics $2f_0$, $3f_0$ and noises with small amplitude is visible from f_0 till f_p. The amplitude of the low-frequency noise component is the biggest in the spectrum of E_r as for the rest spectrum of E_r is near the same the spectrum of E_z.

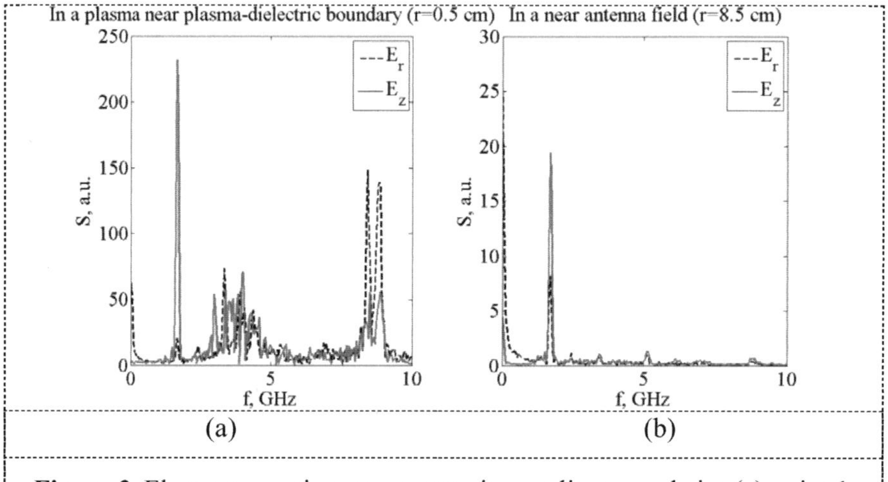

Figure 3. Electromagnetic wave spectra in non-linear mode in: (a) point 1, (b) point 2.

The signal spectra of the linear plasma antenna operation mode are presented in figure 4. The plasma parameters in this mode are the plasma frequency $f_p = 10 \cdot f_0 = 17$ GHz ($\omega_p = 10\omega_{ew0} = 10.7 \cdot 10^{10}$ rad/s) and the plasma density $n_e = 3.6 \cdot 10^{12}$ cm^{-3}. The Langmur frequency component f_p has the biggest amplitude in E_r spectrum inside plasma (Figure 5a), other components have smaller amplitudes than the f_p component. The components of the 6.1 GHz, f_p and f_0 are the best visible in the E_z spectrum, they amplitude are smaller than the amplitude E_r at f_p. This antenna operating mode is a linear because the f_0 spectrum component is the biggest for E_z and E_r signal spectra in the near antenna field (figure 4b). The spectrum component at the frequency f_0 is bigger than 2 f_0, 6.1 GHz, f_p components in figure 4b and f_0 components in figure 3b. Let us note, that the field of the quarter wave dipole antenna has structure of a TM-mode and in the radiation direction of the dipole antenna $E_r|_{r=\infty} = 0$.

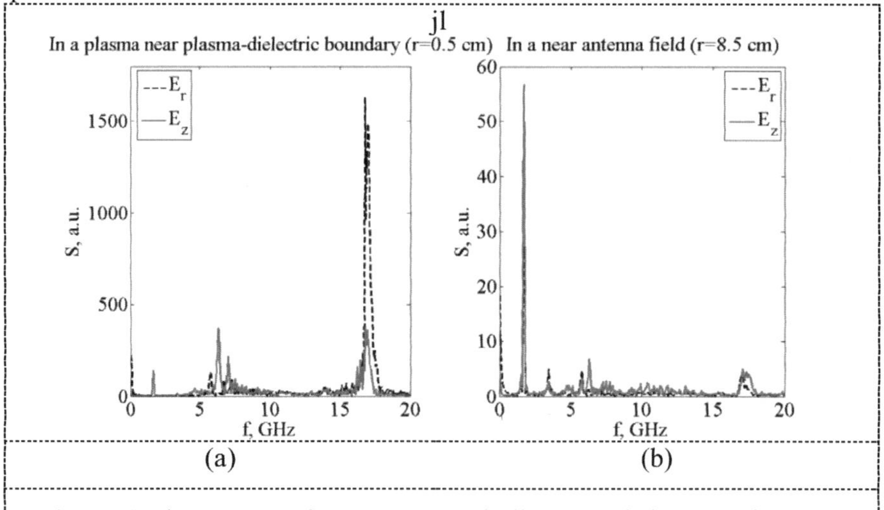

Figure 4. Electromagnetic wave spectra in linear mode in: (a) point 1, (b) point 2.

4. Conclusions

PIC antenna model confirmed the relationship obtained in the Drude model of three plasma antenna operating modes to the terms of propagation of a surface wave [8,9]. It also allowed to study the effect of plasma oscillations in the radiated signal. It was showed in the present work that inside the plasma Langmuir oscillations with the frequency f_p and high-frequency harmonics of the input signal are excited in all antenna operating modes. The field components amplitude at the plasma frequency exceeds the amplitude at the frequency f_0 inside plasma in the non-radiative and non-linear modes. The biggest amplitude of the radiated signal spectrum is at the carrier frequency f_0 in the linear mode, but there is high-frequency noises in the band from f_0 to f_p. Parameters of the plasma antenna are changed in a wide range in linear mode. Energy losses of the input signal are due to the excitation of Langmuir oscillations in the plasma.

5. Acknowledgments

This work was supported by the Russian Foundation for Basic Research (RFBR) project number 14-08-31336 mol_a. The author is grateful to Professor Namik G. Gusein-zade and Ph. D. Irina L. Bogdankevich for discussions and useful comments.

References

[1] Borg G G, Harris J H, Miljak D G and Martin N M 1999 *Appl. Phys. Lett.* **74** 3272.
[2] Rayner J P, Whichello A P, Cheetham A D 2004 *IEEE Trans. on Plasma Science* **32(1)** 269.
[3] Alexeff I, Anderson T, Parameswaran S, et al 2006 *IEEE Trans. on Plasma Science* **34(2)** 166.
[4] Istomin E N, Karfidov D M, Minaev I M, et al 2006 *Plasma Physics Reports* **32(5)** 388.
[5] Liang C, Xu Y and Wang Z 2008 *Chin. Phys. Lett.* **25(10)** 3712.
[6] Minaev I M, Gusein-zade N G and Rukhadze K Z 2010 *Plasma Physics Reports* **36(10)** 914.
[7] Gusein-zade N G, Minaev I M, Rukhadze A A and Rukhadze K Z 2011 *Journal of Communications Technology and Electronics* **56(10)** 1207
[8] Zhao J S, Zhang Z T, Li Y J, et al 2013 *Journal of Physics: Conference Series* **418** 2.
[9] Chen Z, Zhu A and Lv J 2013 *WSEAS Trans. on Com.* **12(2)** 63.
[10] Bogachev N N, Bogdankevich I L, Gusein-zade N G and Sergeichev K F 2015 *Acta Polytechnica* **55(1)** 34.
[11] Moisan M, Shivarova A and Trivelpiece A W 1982 *Plasma Phys.* **24** 1331
[12] A. Aleksandrov, L. Bogdankevich, A. Rukhadze. Principles of Plasma Electrodynamics. Springer Verlach, Heidelberg, 1984.
[13] Kukushkin A V, Rukhadze A A and Rukhadze K Z 2012 *Phys.-Usp.* **55** 1124.
[14] Tarakanov V P, User's Manual for Code KARAT (Springfield: VA) 1992.

Journal of Physics: Conference Series **661** (2015) 012055 doi:10.1088/1742-6596/661/1/012055

Numerical simulation of heat fluxes in a two-temperature plasma at shock tube walls

E A Kuznetsov[1,2], **S A Poniaev**[2,3]

[1] Saint-Petersburg Polytechnical University
[2] Ioffe Physico-Techical Institute
[3] Mozhaikiy military space academy

Abstract. Numerical simulation of a two-temperature three-component Xenon plasma flow is presented. A solver based on the OpenFOAM CFD software package is developed. The heat flux at the shock tube end wall is calculated and compared with experimental data. It is shown that the heat flux due to electrons can be as high as 14% of the total heat flux.

1. Introduction

Supersonic flights give rise to high temperatures near surfaces of flying objects which are caused by a strong gas compression in front of a moving body and heat release due to internal friction in the gas. High temperatures give rise to nonequilibrium chemical properties of the gas flow, i.e., dissociation and recombination of gas molecules, atom ionization, and chemical reactions in the flow and at the flying object surface. Therefore, studies of supersonic gas flows should take into account these processes and variations in gas properties with temperature [1-5].

Particular attention in development of high-speed flying vehicles is paid to the choice of material and construction of the hull which is able to withstand high temperatures [1]. Thus, it is important to have a correct estimate of the heat flux at the flying object surface. To this end, modern mathematical models including the chemistry of gas flows and ionization and recombination of atoms and also modern numerical methods should be used.

The goal of our study was to simulate the inert gas plasma flows which are frequently used in experimental investigations of supersonic flows in shock tubes and to estimate heat fluxes at the shock tube wall and compare them with experimental data [2].

2. Mathematical model

We consider inert gas Xenon the plasma of which contains neutral atoms, electrons, and ions, i.e., we consider a three-component plasma. The electron gas temperature T_e differs from that of heavy particles (atoms and ions) T. The basic processes that determine the concentration of charged particles in the plasma are the electron impact ionization and three-body recombination [6]

$$Xe + e \rightleftharpoons Xe^{+} + e + e. \tag{1}$$

Then the rate of electron concentration variation $\dot{n}_e$ is given by

$$\dot{n}_e = k_i n_e n_a - k_r n_e^3, \tag{2}$$

where k_i and k_r are the ionization and recombination coefficients, respectively, n_e is the electron concentration (which is equal to the ion concentration), and n_a is the atom concentration. The ionization and recombination coefficients are obtained by using a combined consideration of the ionization and recombination kinetics and population of excited states. The recombination coefficient k_r is the function of electron temperature. Fig. 1 shows the dependence of k_r on temperature for Xenon.

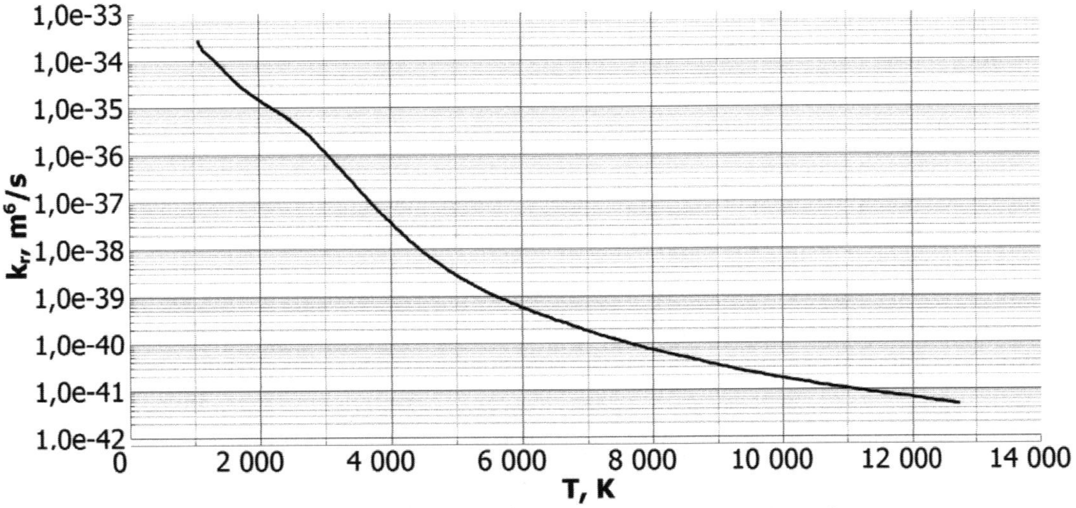

Figure 1. Recombination coefficient for Xenon as a function of temperature

The ionization coefficient k_i is found from the recombination coefficient by using the equilibrium constant K [6]

$$k_i = Kk_r \tag{3}$$

$$K = 1.936 \cdot 10^{22} \cdot T_e^{\frac{3}{2}} \cdot \exp\left(-\frac{140363}{T_e}\right) \tag{4}$$

There are many models which can be used to calculate the average collision frequencies between particles of different types [7]. In our simulation we used the models which seemed most appropriate for our case. We obtained the average frequencies of electron-ion and electron-atom collisions as

$$\nu_{ei} = n_e \frac{4\sqrt{2\pi}}{3} \left(\frac{m_e}{kT_e}\right)^{\frac{3}{2}} \left(\frac{e^2}{4\pi\varepsilon_0 m_e}\right)^2 ln\Lambda \cong 3.64 \cdot 10^{-6} n_e \frac{ln\Lambda}{T_e^{\frac{3}{2}}} \tag{5}$$

$$\Lambda = 1.24 \cdot 10^7 \left(\frac{T_e{}^3}{n_e}\right)^{\frac{1}{2}} \tag{6}$$

$$\nu_{ea} = n_a \frac{4}{3} \left(\frac{8kT_e}{\pi m_e}\right)^{\frac{1}{2}} \sigma_{ea} \tag{7}$$

where m_e is the electron mass, e is the electron charge and σ_{ea} is the collision cross-section given by

$$\sigma_{ea} = a_0^2(-1.82 + 2.2 \cdot 10^4 T_e^{-1} + 3.8 \cdot 10^{-7} T_e^2) \tag{8}$$

where $a_0 = 0.529 \cdot 10^{-10}$ m is the radius of first Bohr orbit.

Each component of the plasma mixture can be regarded as an individual medium that interacts with other components. Each one-component medium is described by a system of equations that express the laws of conservation for the continuum. It is necessary to include into this system of equations the terms that describe the interaction between one-component media. As a result, we have a system of equations for a two-temperature three-component plasma

$$\frac{\partial\rho}{\partial t}+\nabla\cdot\left(\rho\vec{V}\right)=0 \quad (9)$$

$$\frac{\partial n_e}{\partial t}+\nabla\cdot\left(n_e\vec{V}\right)=\dot{n}_e \quad (10)$$

$$\frac{\partial(\rho\vec{V})}{\partial t}+\nabla\cdot\left(\rho\vec{V}\vec{V}+P\right)=0 \quad (11)$$

$$\frac{\partial E}{\partial t}+\nabla\cdot\left(E\vec{V}\right)=-P\nabla\cdot\vec{V}-\nabla\cdot\vec{q}+\dot{n}_e\varepsilon_i+3\frac{m_e}{m_a}kn_e(\nu_{ei}+\nu_{ea})(T_e-T) \quad (12)$$

$$\frac{\partial}{\partial t}\left(\frac{3}{2}p_e\right)+\nabla\left(\frac{5}{2}p_e\vec{V}\right)-\vec{V}\cdot\nabla p_e=-\nabla\cdot\overrightarrow{q_e}-\dot{n}_e\varepsilon_i-3\frac{m_e}{m_a}kn_e(\nu_{ei}+\nu_{ea})(T_e-T) \quad (13)$$

$$\vec{q}=\lambda\nabla T \quad (14)$$

$$\overrightarrow{q_e}=\lambda_e\nabla T_e \quad (15)$$

$$\lambda_e=\frac{2.4}{1+\frac{\nu_{ei}}{\sqrt{2}(\nu_{ei}+\nu_{ea})}}\cdot\frac{k^2n_eT_e}{m_e(\nu_{ei}+\nu_{ea})} \quad (16)$$

where ρ is the flow density, $\vec{V}$ is the flow velocity vector, P is the pressure tensor, E is the flow energy, $\vec{q}$ is the heat flux due to heavy particles, ε_i is the ionization energy, m_a is the atom mass, p_e is the electron pressure, $\overrightarrow{q_e}$ is the heat flux due to electrons, λ is the heat conductivity of the material, and λ_e is the electron heat conductivity.

To solve the system of equations we obtained, we used the free, open source CFD software package OpenFOAM [8]. This allowed us to carry out operations with scalar, vector, and tensor fields in the framework of the finite-volume method.

One of important features of OpenFOAM is the possibility to make one's own utilities and solvers. OpenFOAM includes different solvers which are commonly accepted for the solution of one-temperature gas flows. So our solver for the solution of the two-temperature three-component plasma flow was built by adding additional equations and terms of equations to one of the available solvers for the one-temperature model.

3. Calculation results

In order to compare the results of numerical modulation and experimental data, we calculated the supersonic Xenon flow in the shock tube 10 cm in diameter with a section length of 0.5 m. The Mach number of the shock wave in the shock tube was 6. The calculations were performed for the two-dimensional axisymmetric case. The computational region is presented in Fig.2.

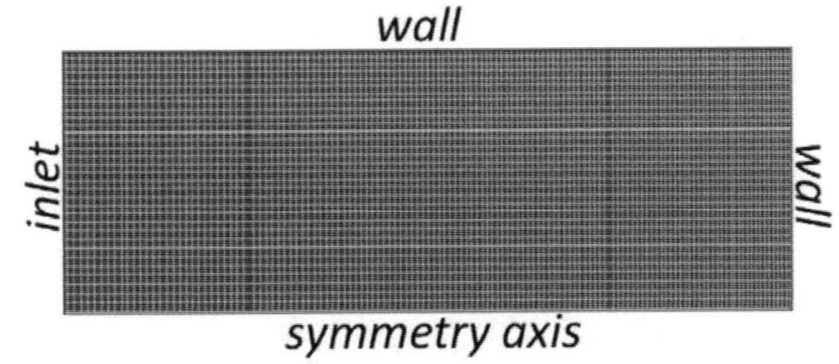

Figure 2. Computation region and mesh

The gas parameters in the low-pressure chamber of the shock tube were as follows: velocity - 0 m/s, gas pressure - 2235 Pa, temperature $T = T_e = 293$ K, and $n_e = 0$ m^{-3}. After the shock wave passed through the gas in the low-pressure chamber and gas ionization occurred, the gas parameters became as follows: velocity - 770 m/s, pressure - 10^5 Pa, $T = T_e = 3552$ K, $n_e = 2.17 \cdot 10^{17}$ m^{-3}.

Figures 3 and 4 show simulated and experimental heat fluxes [9,10] at the shock tube wall, respectively. It can be seen that the results of simulation and experimental data on the heat flux value at the shock tube end wall are in fairly good agreement when the heat flux reaches an approximately stationary level of 1400 kW/m^2. As one can see from Fig.3, there are two heat fluxes: q_T is the heat flux due to heavy plasma components and q_{Te} is the heat flux due to electrons (q is the total heat flux at the wall). The value of q_{Te} reaches approximately 14% of the total heat flux.

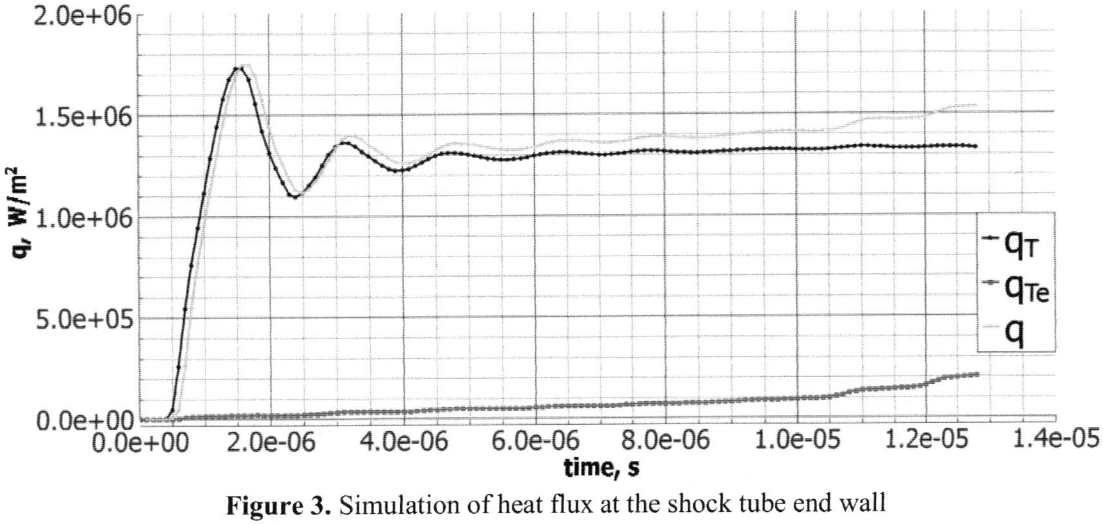

Figure 3. Simulation of heat flux at the shock tube end wall

No satisfactory agreement was found in the simulated and experimental temporal behaviors of the heat flux. This can be explained by the fact that some effects in the high-temperature gas were not taken into account in our simulation, and the processes at the shock tube turned out to be quicker in the simulation than in the experiment. For example, the length of the low-pressure chamber was 10 m and there a boundary layer developed near the walls in the experiment, while the simulation was performed for a 0.5-m long section which could be insufficient for a viscous flow development. It can also be important that in the simulation a change in gas parameters in the shock wave was an instantaneous process, and in reality it took some time to occur.

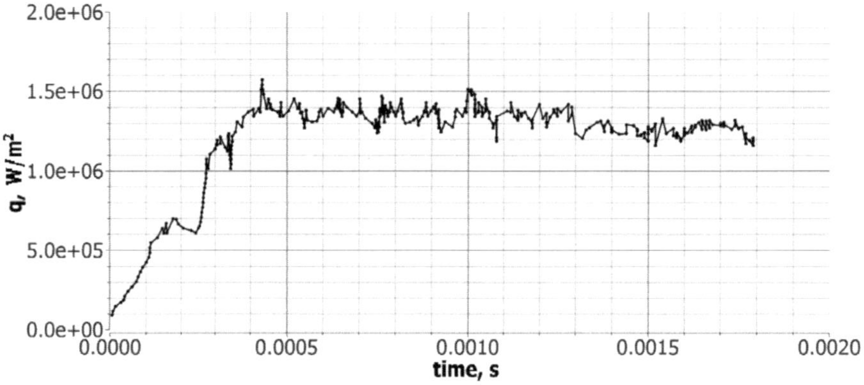

Figure 4. Experimental data on heat flux at the shock tube end wall

Conclusion

Numerical simulation has shown that the solver we have made can be used to estimate pulsed heat fluxes in the flows with different temperatures of heavy plasma particles and electrons. It can be used to calculate heat fluxes at the bodies flowed around by the supersonic flow created in the nozzle connected with the shock tube or arc-heated facilities. Model gases, such as Argon and Xenon, are frequently used in such investigations, and it is necessary to take into account the nonequilibrium ionization state and the two-temperature nature of the plasma flows. As shown above, the heat flux due to electrons can constitute a significant portion of the total heat flux.

References

[1] Edney B E *AIAA Journ.* 1968. V. **6**. N 1. P. 16-24.
[2] Sapochnikov S Z, Mitiakov V Y, Mityakov A V, Measurements of nonstationary heat fluxes by gradient sensors based on single-crystalline anisotropic bismuth, *Technical physics*, 2004, vol. 49, issue 7, pp 920-926
[3] Friedman H S and Fay J A Heat Transfer from Argon and Xenon to the End Wall of a Shock Tube *Physics of Fluids*, Vol. **8**, No. 11, 1965, pp. 1968-1975.
[4] Nowak R J and Yuen M C Heat Transfer to a Hemispherical Body in a Supersonic Argon Plasma *AIAA J.*, Vol. **11**, No. 11, 1973, pp. 1463–1464.
[5] Dillion R E and Nagamatsu H.T. Heat Transfer and Transition Mechanism on a Shock-Tube Wall *AIAA J.*, Vol. **22**, No. 11, 1982, pp. 1524-1528.
[6] Mitchner M and Kruger C, 1973, Partially ionized gases, Wiley, 518 p.
[7] Chernyi G G, Losev S A, Macharet S O, Potapkin B V, Physical and chemical processes in gas dynamics, 2002, Progress in Astronautics and Aeronautics, vol. 196
[8] OpenFOAM: the open source CFD toolbox. – http://www.openfoam.com
[9] Bobashev S V, Mende N P, Popov P A, Reznikov B I, Sakharov V A, Measurements of convective heat fluxes by sensors based on anisotropic thermal elements in pulsed gasdynamic processes, 2012, AIAA Paper 2012-0263
[10] Bobashev S V, Golovachov Y P, Mende N P, Reznikov B I, Popov P A, Sakharov V A, Schmidt A A, Chernyshev A S, Sapozhnikov S Z, Mityakov V Y, Mityakov AV, Application of the gradient heat flux sensor to study pulsed processes in a shock tube, 2008, Tech. Phys., v.53, 12, pp 1634-1635

Transverse and longitudinal vibrations in amorphous silicon

Y M Beltukov[1,2], C Fusco[3], A Tanguy[3] and D A Parshin[4]
[1]A. F. Ioffe Physical-Technical Institute, 194021 Saint-Petersburg, Russia
[2]Université Montpellier II, CNRS, Montpellier 34095, France
[3]Université Lyon 1, F-69622 Villeurbanne Cedex, France
[4]Saint Petersburg State Polytechnical University, 195251 Saint-Petersburg, Russia

Abstract. We show that harmonic vibrations in amorphous silicon can be decomposed to transverse and longitudinal components in all frequency range even in the absence of the well defined wave vector **q**. For this purpose we define the transverse component of the eigenvector with given ω as a component, which does not change the volumes of Voronoi cells around atoms. The longitudinal component is the remaining orthogonal component. We have found the longitudinal and transverse components of the vibrational density of states for numerical model of amorphous silicon. The vibrations are mostly transverse below 7 THz and above 15 THz. In the frequency interval in between the vibrations have a longitudinal nature. Just this sudden transformation of vibrations at 7 THz from almost transverse to almost longitudinal ones explains the prominent peak in the diffusivity of the amorphous silicon just above 7 THz.

1. Introduction

The propagation of vibrational excitations in disordered media is one of the important problems in condensed matter physics. Microscopic nature of these vibrational excitations is still poorly understood, despite that they are responsible for the specific heat, the thermal conductivity and the sound propagation in amorphous materials.

In 1999 Allen and Feldman proposed a simple classification of vibrations in disordered media [1], which was based on the numerical simulations of amorphous silicon. The low-frequency vibrations are longitudinal (LA) or transverse (TA) plane waves (acoustic phonons). The mean free path of phonons decreases rapidly with frequency due to structural or dynamical disorder. At some frequency $\omega_{\rm IR} \approx 4$ THz the mean free path of phonons becomes comparable with its wavelength. It is a so-called Ioffe-Regel criterion. However, the mobility edge $\omega_{\rm loc}$ for vibration localization is much higher then $\omega_{\rm IR}$ what is not the case for electron localization. Vibrations in amorphous materials in this relatively wide frequency range $\omega_{\rm IR} < \omega < \omega_{\rm loc}$ are delocalized but not propagated as phonons. These vibrations were called *diffusons* because they spread by means of diffusion [1, 2].

However, this picture cannot clearly explain the prominent peak (more than 3 times in the value) in the diffusivity of vibrations in amorphous silicon between 7 and 13 THz [1, 3]. Allen and Feldman noticed that the sharp rise of the diffusivity at 7 THz corresponds to the end of TA branch in crystalline Si which is responsible for the local minimum in the vibrational density of states. So the frequency 7 THz marks a point in the spectrum where vibrations change their character from somewhat TA-like, to somewhat LA-like with much larger group velocity. However the notions of transverse and longitudinal vibrations are ill defined for such big frequencies in amorphous silicon.

In this paper we will generalize the notions of transverse and longitudinal vibrations for disordered systems like amorphous silicon. As an example we study numerically the model amorphous silicon (a-Si) system consisting of $N = 32768$ atoms contained in a periodic cubic box of lengths $L_x = L_y = L_z$ of approximately 87 Å. The technical details of the preparation of the a-Si sample have already been presented in Ref. [4]. The Si-Si interaction in the amorphous silicon studied here is well described by the Stillinger-Weber potential [5].

2. Longitudinal and transverse components of displacements

Low-frequency vibrations even in amorphous media are well-defined plane longitudinal and transverse waves. These vibrations can be described in continuous medium approximation. The displacement fields $\mathbf{u}(\mathbf{r})$ for longitudinal (L) and transverse (T) waves have a form

$$\mathbf{u}_{\mathrm{L,T}}(\mathbf{r}) = \mathbf{u}^{(0)}_{\mathrm{L,T}} \exp(i\mathbf{qr}), \tag{1}$$

$$\mathbf{u}^{(0)}_{\mathrm{L}} \parallel \mathbf{q}, \quad \mathbf{u}^{(0)}_{\mathrm{T}} \perp \mathbf{q}. \tag{2}$$

However, this definition contains the phonon wavevector $\mathbf{q}$, which is ill-defined for high-frequency vibrations in amorphous medium. To generalize this approach let us note that

$$\operatorname{div} \mathbf{u}_{\mathrm{T}} = i\mathbf{q}\mathbf{u}_{\mathrm{T}} = 0. \tag{3}$$

The transverse displacement $\mathbf{u}_{\mathrm{T}}(\mathbf{r})$ has zero divergence so it conserves local volumes.

The natural analog of the local volumes in amorphous media are Voronoi cells around each atom. The Voronoi cell $\mathcal{V}_i$, associated with the atom i is the set of all points in the space around this atom whose distance to the atom position $\mathbf{r}_i$ is not greater than their distance to the other atoms $\mathbf{r}_j$ [6]. This type of cells is also known as Wigner-Seitz cells in crystallography for regular crystals. Figure 1 shows a simple example of Voronoi cells for longitudinal and transverse waves in a simple quadratic lattice. A longitudinal wave evidently changes the volumes of Voronoi cells while the transverse wave does not change the volumes despite the change in the form of the cells.

A displacement of atoms in amorphous media may (or may not) change volumes of Voronoi cells. We will call the displacement of atoms $\mathbf{u}_i$ to be transverse if it does not change the volumes of all Voronoi cells. For that let us introduce a matrix A which is responsible for the relative change of the ith Voronoi cell volume V_i under jth atom displacement in the direction α

$$A_{i,j\alpha} = \frac{1}{V_i}\frac{\partial V_i}{\partial r_{j\alpha}}. \tag{4}$$

The explicit formula for the matrix A will be derived in the next section. Using this matrix the displacement of jth atom in the direction α, $u_{j\alpha}$ produces the following relative change of the Voronoi cell volumes

$$\varepsilon_i = \sum_{j\alpha} A_{i,j\alpha} u_{j\alpha}. \tag{5}$$

In the matrix notation this equation reads $\varepsilon = Au$ where A is a rectangular $N \times 3N$ matrix (with N being the number of particles) and u is a displacement vector with $3N$ elements. The matrix A is a discrete analog of the divergence operator.

By definition the transverse component u_{T} of the displacement u is such that $Au_{\mathrm{T}} = 0$, i.e. u_{T} is the projection of the displacement u to the *null space* of the matrix A. The longitudinal

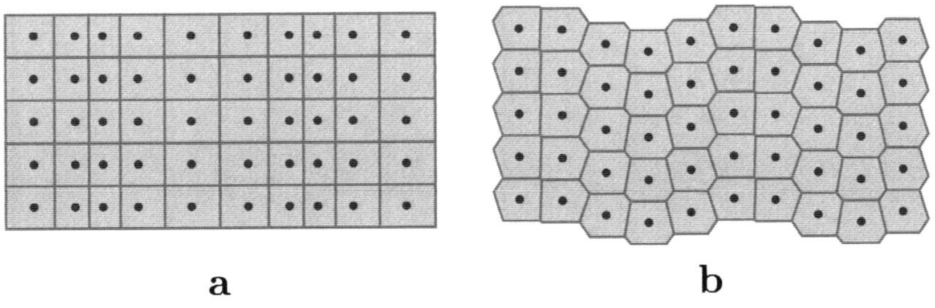

Figure 1. Longitudinal and transverse waves in a simple quadratic lattice.

component $u_{\rm L}$ is a remaining orthogonal component of the displacement field and it is the projection of u to the *row space* of the matrix A. These projections have the following form [7]

$$u_{\rm L} = A^T(AA^T)^{-1}Au, \tag{6}$$

$$u_{\rm T} = u - A^T(AA^T)^{-1}Au. \tag{7}$$

Indeed, $u_{\rm T}$ does not change the Voronoi cell volumes

$$Au_{\rm T} = Au - AA^T(AA^T)^{-1}Au = 0 \tag{8}$$

and $u_{\rm L}$ is orthogonal to $u_{\rm T}$

$$u_{\rm L}^T u_{\rm T} = u^T A^T(AA^T)^{-1}A(u - A^T(AA^T)^{-1}Au) = 0. \tag{9}$$

3. The derivation of the matrix A

Let us show how matrix A can be derived from geometry only. Let we shift only the atom j by the vector $\mathbf{u}$ (Figure 2). The vector $\mathbf{u}$ is small so the Voronoi cells after shifting have almost the same structure, but their facets are slightly shifted and rotated in space. The shifting $\mathbf{u}$ of the atom j can change the volume V_i of a nearest neighbor cell i. By definition the cells i and j are nearest neighbors if they have a common facet $\mathcal{S}_{ij}$.

By definition of the Voronoi cell the facet $\mathcal{S}_{ij}$ lies in the plane $\mathcal{P}_{ij}$, which has the normal $\mathbf{n}_{ij} = \mathbf{r}_{ij}/r_{ij}$, where $\mathbf{r}_{ij} = \mathbf{r}_j - \mathbf{r}_i$ is the vector connecting atoms i and j and go through the point $\mathbf{b}_{ij} = (\mathbf{r}_i + \mathbf{r}_j)/2$. Therefore the equation for this plane reads

$$\mathbf{n}_{ij} \cdot (\mathbf{r} - \mathbf{b}_{ij}) = 0. \tag{10}$$

After shifting of the atom j the new facet $\mathcal{S}'_{ij}$ lies in the plane $\mathcal{P}'_{ij}$ which satisfy the equation

$$\mathbf{n}'_{ij} \cdot (\mathbf{r} - \mathbf{b}'_{ij}) = 0 \tag{11}$$

where $\mathbf{n}'_{ij} = \mathbf{r}'_{ij}/r'_{ij}$, $\mathbf{r}'_{ij} = \mathbf{r}_{ij} + \mathbf{u}$, and $\mathbf{b}'_{ij} = \mathbf{b}_{ij} + \mathbf{u}/2$. The signed distance from an arbitrary point $\mathbf{r}$ to the plane $\mathcal{P}'_{ij}$ is

$$d'_{ij}(\mathbf{r}) = \mathbf{n}'_{ij} \cdot (\mathbf{b}'_{ij} - \mathbf{r}). \tag{12}$$

This distance has the sign "+" if $\mathbf{r}$ lies on the same side to the plane $\mathcal{P}'_{ij}$ as the atom i and the sign "$-$" if $\mathbf{r}$ lies on the opposite side to the plane. If $\mathbf{r}$ lies in the plane $\mathcal{P}'_{ij}$ then the distance

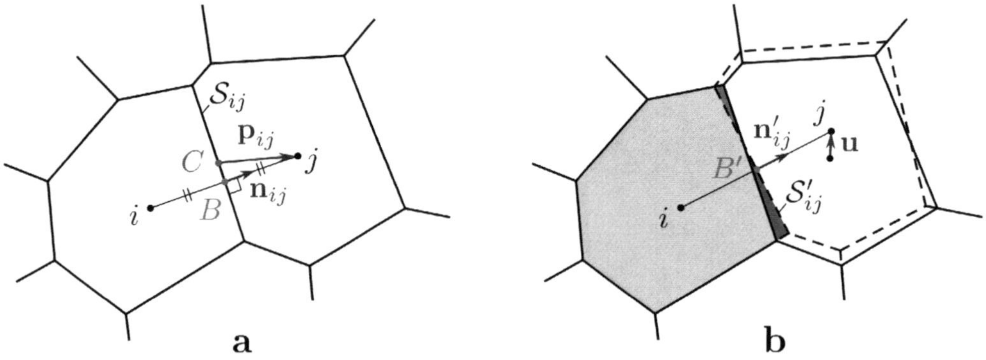

Figure 2. a) Two dimensional example of the Voronoi cells. The points B and C denote the position of $\mathbf{b}_{ij}$ and $\mathbf{c}_{ij}$ respectively. In two dimensional case the point C is the center of the segment $\mathcal{S}_{ij}$. b) Dashed lines shows the Voronoi cells after shifting of the atom j by the vector $\mathbf{u}$. The point B' denotes the position of $\mathbf{b}'_{ij}$. Green and red areas show the increasing and decreasing of the V_i respectively.

$d'_{ij}(\mathbf{r})$ is equal to 0 and we get the Eq. (11). In the linear approximation on $\mathbf{u}$ the change of the volume V_i is the integral of $d'_{ij}(\mathbf{r})$ over the surface of the initial facet $\mathcal{S}_{ij}$

$$\delta V_i = \iint_{\mathbf{r}\in\mathcal{S}_{ij}} \mathbf{n}'_{ij}\cdot(\mathbf{b}'_{ij}-\mathbf{r})\,dS = S_{ij}\mathbf{n}'_{ij}\cdot(\mathbf{b}'_{ij}-\mathbf{c}_{ij}), \tag{13}$$

$$\mathbf{c}_{ij} = \frac{1}{S_{ij}}\iint_{\mathbf{r}\in\mathcal{S}_{ij}} \mathbf{r}\,dS. \tag{14}$$

where S_{ij} and $\mathbf{c}_{ij}$ are correspondingly the area and the centroid of the facet $\mathcal{S}_{ij}$. In the linear approximation on $\mathbf{u}$, the Eq. (13) reads

$$\delta V_i = \frac{S_{ij}}{r_{ij}}\mathbf{p}_{ij}\cdot\mathbf{u}, \quad \mathbf{p}_{ij} = \mathbf{r}_j - \mathbf{c}_{ij}, \tag{15}$$

where we have taken into account that $\mathbf{r}_{ij}\cdot(\mathbf{b}_{ij}-\mathbf{c}_{ij}) = 0$ because $\mathbf{c}_{ij}\in\mathcal{S}_{ij}$.

If all neighboring atoms shift, the change of the volume V_i can be written in the matrix form

$$\delta V_i = \sum_{j\alpha} W_{i,j\alpha}\, u_{j\alpha}. \tag{16}$$

The nondiagonal elements of the matrix W we have already found above

$$W_{i,j\alpha} = \frac{S_{ij}}{r_{ij}}(\mathbf{p}_{ij})_\alpha, \quad i\neq j. \tag{17}$$

The diagonal element $W_{i,i\alpha}$ means the change of the ith Voronoi cell volume under shifting of the ith atom itself. Shifting of the all atoms by the same vector does not change the volumes of the Voronoi cells. Therefore

$$W_{i,i\alpha} = -\sum_{j\neq i} W_{i,j\alpha}. \tag{18}$$

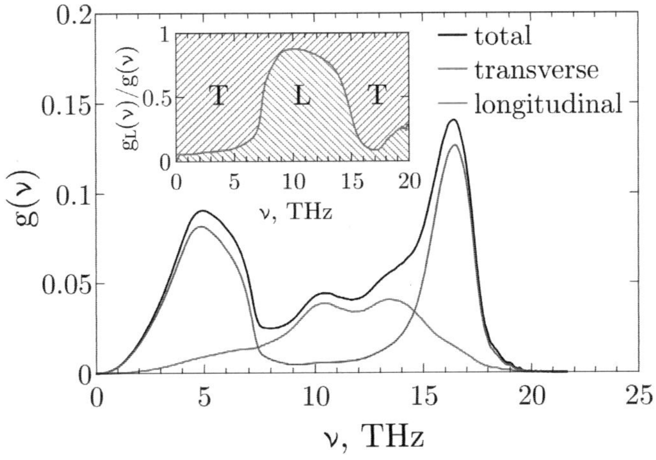

Figure 3. The decomposition of the total vibrational density of states to longitudinal and transverse components. The inset shows the relative number of the longitudinal modes $g_{\mathrm{L}}(\omega)/g(\omega)$ (red line). The relative number of the transverse modes $g_{\mathrm{T}}(\omega)/g(\omega) = 1 - g_{\mathrm{L}}(\omega)/g(\omega)$ is shown by green hatching between red line and the value 1.

After dividing by the volume we finally get the matrix A

$$A_{i,j\alpha} = \frac{1}{V_i} W_{i,j\alpha}. \tag{19}$$

It is notable that finite-elements methods also often use Voronoi cells and have a similar definition for the finite differences for the divergence operator [8].

4. Longitudinal and transverse DOS

The total vibrational density of states can be decomposed to longitudinal and transverse components

$$g(\omega) = \frac{1}{3N} \sum_j \delta(\omega - \omega_j) = g_{\mathrm{L}}(\omega) + g_{\mathrm{T}}(\omega), \tag{20}$$

$$g_{\mathrm{L,T}}(\omega) = \frac{1}{3N} \sum_j \frac{\|u_{\mathrm{L,T}}(\omega_j)\|^2}{\|u(\omega_j)\|^2} \delta(\omega - \omega_j) \tag{21}$$

where ω_j is an eigenfrequency, $u(\omega_j)$ is the corresponding eigenmode, $u_{\mathrm{L}}(\omega_j)$ and $u_{\mathrm{T}}(\omega_j)$ are its transverse and longitudinal components defined by Eqs. (6) and (7). The total vibrational density of states and its longitudinal and transverse components was calculated for the numerical model of a-Si consisting of $N = 32768$ atoms by the Kernel polynomial method (KPM) [9]. The results are shown in the Figure 3. There are three clearly seen regions. The vibrations are mostly transverse below 7 THz and above 15 THz. In the frequency interval in between the vibrations have the longitudinal nature.

The low number of longitudinal vibrations in the low-frequency region is explained by the Debye law, which gives

$$g_{\mathrm{L}}(\omega) \sim \frac{\omega^2}{c_{\mathrm{L}}^3}, \quad g_{\mathrm{T}}(\omega) \sim \frac{2\omega^2}{c_{\mathrm{T}}^3}. \tag{22}$$

In our model the amorphous Si has the longitudinal sound velocity $c_{\rm L} = 7.96$ km/s and the transverse sound velocity $c_{\rm T} = 3.85$ km/s which gives the ratio $g_{\rm L}(\omega)/g_{\rm T}(\omega) = c_{\rm T}^3/2c_{\rm L}^3 = 0.057$ for $\omega \to 0$. This value coincides with the inset in the Figure 3.

The domination of the longitudinal modes between 7 THz and 15 THz in amorphous silicon corresponds to the gap between the upper frequency of TA modes (7.5 THz) and the lower frequency of TO modes (13.9 THz) in crystalline Si [10]. This frequency region in crystalline Si is totally occupied by LA and LO modes without a gap. In the same frequency region the vibrations of amorphous Si have a small transverse component (15–20%). Certainly, there are no optical phonons with well defined wavevector in amorphous silicon due to relatively strong disorder. However, the short-range order of vibrational modes in amorphous phase can be similar to that in crystal phase. At the same time, our definition of the longitudinal and transverse vibrations is local because it is based on the Voronoi cells, which depends on the neighbor atoms only.

The Figure 3 shows a sharp change of the nature of vibrations at 7 THz from almost transverse to almost longitudinal. The longitudinal vibrations correspond to the stretching of the chemical bonds between atoms while the transverse vibrations correspond to the less rigid bond bending and rocking. Therefore longitudinal vibrations can transfer the vibrational energy much faster than transverse vibrations which leads to the sudden rise of the diffusivity at 7 THz observed in Refs. [1, 3]. Furthermore, our additional calculations show that the longitudinal modes still have a well-defined wavevector and a sound velocity up to $\omega_{\rm IR}^{\rm L} \approx 13$ THz unlike the transverse modes, which have the frequency of the Ioffe-Regel crossover $\omega_{\rm IR}^{\rm T} \approx 4$ THz.

5. Conclusion

We have shown that vibrations in amorphous solids can be decomposed to transverse and longitudinal components in a generalized sense. The transverse component does not change the volumes of Voronoi cells around atoms while the longitudinal component is the remaining orthogonal component changing this volume. We show that the rise of the diffusivity at 7 THz in amorphous silicon corresponds to the sharp change of the nature of vibrations from almost transverse to almost longitudinal ones having high sound velocity. This decomposition can be fruitful for the investigation of the correlation function and the diffusivity of transverse and longitudinal components of vibrations in amorphous solids independently from each other.

The 3D-structure of several Voronoi cells in amorphous silicon is available online in supplementary materials.

One of the authors (YMB) thanks the Dynasty Foundation for the financial support. Two authors (DAP and YMB) thanks the University Lyon 1 for hospitality.

References

[1] Allen P B, Feldman J L, Fabian J and Wooten F 1999 *Phil. Mag.* B **79** 1715
[2] Beltukov Y M, Kozub V I and Parshin D A 2013 *Phys. Rev.* B **87** 134203
[3] Feldman J L, Allen P B and Bickham S R 1999 *Phys. Rev.* B **59** 3551
[4] Fusco C, Albaret T and Tanguy A 2010 *Phys. Rev.* E **82** 066116
[5] Stillinger F H and Weber T A 1985 *Phys. Rev.* B **31** 5262
[6] Aurenhammer F 1991 *ACM Computing Surveys* **23** 345
[7] Meyer C D 2000 *Matrix Analysis and Applied Linear Algebra* (SIAM)
[8] Mishev I D 1998 *Numer. Meth. Part. D. E.* **14** 193
[9] Weiße A et al. 2006 *Rev. Mod. Phys.* **78** 275
[10] Tubino R, Piseri L and Zerbi G 1972 *J. Chem. Phys.* **56** 1022

Investigation of spin-electromagnetic wave envelope solitons in a multiferroic layered structure

A B Ustinov[1], A V Kondrashov[1], A A Nikitin[2], M A Cherkasskii[1] and B A Kalinikos[1]
[1]St. Petersburg Electrotechnical University, St. Petersburg, 197376 Russia
[2]ITMO University, St. Petersburg, 197101 Russia

E-mail: kondrashov_av@yahoo.com

Abstract. Hybrid spin-electromagnetic wave (SEW) envelope solitons have been studied both experimentally and theoretically. The solitons were formed during auto-generation of SEWs in an active ring resonator for which the role of the nonlinear dispersive waveguide media was played by a multiferroic layered ferrite-ferroelectric structure. It is demonstrated that the dielectric constant of the structure affects the nonlinear properties of SEWs.

1. Introduction

A nonlinear localized wave packet formed during the propagation of a pulsed excitation as a result of the compensation of the "spreading" effect of the dispersion by the "grouping" effect of the nonlinearity is called an envelope soliton [1]. This is a fundamental physical phenomenon observed in various waveguiding media. The envelope solitons may also be formed via development of a self and induced modulation instabilities during propagation of continuous waves of finite amplitude in a dispersive nonlinear medium [1, 2].

The envelope solitons have been discovered and studied in various media, such as optical fibers, ferromagnetic films, electromagnetic transmission lines, Bose-Einstein condensate, and other media (see, e.g., [1, 3-5]). The soliton phenomena are also being actively studied at present. To last achievements in this field can be attributed the observation of the auto-generation of solitons in metamaterials fabricated on the basis of artificial transmission lines with negative dispersion [6].

In recent years, increased interest has been observed in studies of how magnetic and electric excitations interact in various artificial multiferroic structures [7]. The magnetoelectric interaction in artificial multiferroics can be caused by two effects. One of these is the combination of piezoelectric and magnetostriction effects accompanied by a mechanical deformation of a multiferroic structure in external electric and/or magnetic fields. This is well manifested in artificial structures fabricated from ferromagnetics and piezoelectrics [8]. Another effect is due to the electrodynamic interaction of spin and electromagnetic waves. This effect may give rise to hybrid spin-electromagnetic waves (SEWs) propagating in waveguide multiferroic (ferrite-ferroelectric) structures [9].

Despite that studies of layered multiferroic structures were commenced comparatively long ago [10], nonlinear processes in these structures have been studied insufficiently. For example, the bistability of nonlinear magnetization oscillations in a multiferroic resonator fabricated from a ferromagnetic-piezoelectric planar composite structure was first studied in [11]. The study [12] was concerned with the frequency conversion via a nonlinear magnetoelectric interaction in structures of this kind. The combined wave nonlinearity of SEWs propagating in a multiferroic medium was theoretically examined in [13] and the formation of bright and dark SEW envelope solitons in a ferrite-ferroelectric layered structure was theoretically predicted in [14]. Then experimental observation of dark envelope solitons was reported in [15].

A review of the literature suggests that envelope solitons studied in detail in various waveguide media have been apparently insufficiently analyzed in multiferroics. The goal of our present study was to examine experimentally and theoretically the spin-electromagnetic wave envelope solitons formed in an artificial multiferroic medium, specifically, in a ferrite-ferroelectric layered waveguide structure.

It is known that one of ways to excite envelope solitons is their auto-generation [6, 16-18]. Therefore, we fabricated for experiments an active ring of ordinary type, in which the planar multiferroic structure served as a nonlinear dispersive medium.

The experiments were performed at room temperature. The multiferroic structure comprised a ferrite and a ferroelectric layers (see figure 1). The ferrite layer was fabricated from an epitaxial film of yttrium-iron garnet (YIG) with a thickness of 5.7 μm, width of 2 mm, and length of 40 mm. The ferroelectric layer had the form of a 5×5 mm^2 barium-strontium titanate (BST) slab with a thickness of 500 μm. Two microstrip antennas serving to excite and receive a traveling wave were situated at a distance of 6.7 mm. The antennas were formed by photolithography on alumina substrates. The receiving antenna was connected with the exciting antenna circuit by a feedback circuit constituted by a wideband microwave amplifier and an attenuator controlling the total gain G in the ring.

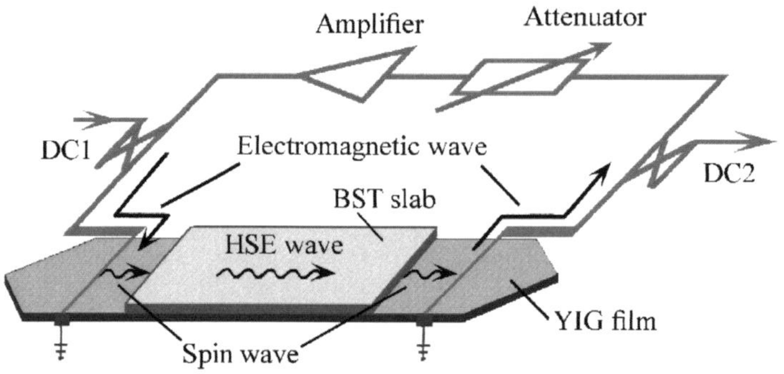

Figure 1. Schematic of the experimental device.

At room temperature, the BST ceramic used in the experiments was in the paraelectric phase. Its permittivity was isotropic and, therefore, the ceramic had no spontaneous polarization. The layered structure was magnetized in its plane perpendicularly to the wave propagation direction. This orientation of the static magnetic field provided conditions for propagation of quasi-surface SEWs [19].

Let us consider the propagation of microwave excitations in the ring (see figure 1). The electromagnetic wave passed through the feedback circuit is converted to the surface spin wave by the input microstrip antenna. This spin wave passes a short distance in the YIG film and enters the multiferroic structure. At the interface with the structure, the spin wave is transformed to a quasi-surface SEW, which then propagates in the YIG-BST structure. Having reached the end of the multiferroic structure, the SEW is converted back to the spin wave, which is received by the output microstrip antenna and converted to the electromagnetic wave. This wave, propagating via the feedback circuit, is amplified and fed to the input microstrip antenna.

In the ring circuit described above, the electromagnetic feedback circuit serves as the active part of the ring resonator, whereas its nonlinearity is mostly determined by the nonlinearity of spin-electromagnetic waves propagating in the multiferroic structure. It is noteworthy that the microwave amplifier had a frequency bandwidth (2-8 GHz) that is wider than that of the multiferroic structure. It is also noteworthy that the propagation time of the microwave carrier signal in the multiferroic structure substantially exceeds its propagation time in the feedback circuit.

The directional couplers DC1 and DC2 with coupling coefficients of -10 dB at the input and output of the waveguide structure were used to input and output the test microwave electromagnetic signal necessary for measuring the ring characteristics below the SEW auto-generation threshold. To observe the shape of the signals generated above this threshold, a small part of the microwave power was extracted from the ring via the output directional coupler.

To control the SEW dispersion by changing the permittivity of the BST slab, chromium electrodes were deposited onto both of its surfaces. The thickness of the metallic electrode situated between the YIG film and BST plate was substantially smaller (~50 nm) than the depth of the skin layer for the microwave signal of the operating gigahertz frequency. This thickness provided an effective hybridization of spin waves localized mostly in the YIG film with electromagnetic waves localized mostly in the BST slab.

According to the theoretical model [20], the eigenmodes of the active ring resonator based on the ferrite-ferroelectric structure have frequencies determined by the phase quantization condition

$$k_{sew}(f)d_{sew}+k_{sw}(f)d_{sw}+\varphi_e=2\pi n, \tag{1}$$

where k_{sew} and k_{sw} are, respectively, the wave numbers of the spin-electromagnetic waves in the YIG-BST structure and spin waves in the YIG film; f is the frequency of waves; d_{sew} and d_{sw} are the wave propagation lengths; φ_e is the microwave signal phase accumulated in the feedback circuit of the ring; and n is an integer.

In contrast to the magnetic-film active ring resonator, the ring resonator based on a multiferroic structure enables a dual control over wave parameters. This is due to the dependence of the group velocity v_{sew} and wave number k_{sew} of spin-electromagnetic waves on both the permeability and permittivity of the multiferroic waveguide. Thus, the eigenmodes of the ring change their frequencies on varying the magnetic (H) and electric (E) bias fields. In this case, the frequency separation between the modes also changes.

Figure 2 shows the transmission coefficient of the ring, measured slightly below the autogeneration threshold at H = 1750 Oe. In these measurements, a low-power (-20 dBm) signal was fed from the microwave generator to the ring input. It can be seen in figure 2 that the transmission coefficient shows a set of narrow peaks corresponding to the eigenmodes of the ring resonator. The experimental results clearly demonstrate the shift of the eignemode frequencies of the ring as a result of the decrease in the dielectric constant of the BST layer upon application of an electric field to this layer. In agreement with the theory [20], the resonance frequencies increase at small k_{sew} (which correspond to lower frequencies) and remain nearly unchanged at large k_{sew} (which correspond to higher frequencies). In particular, the eigenmodes of the ring at frequencies of about 7.0 GHz are shifted by several hundred kilohertz.

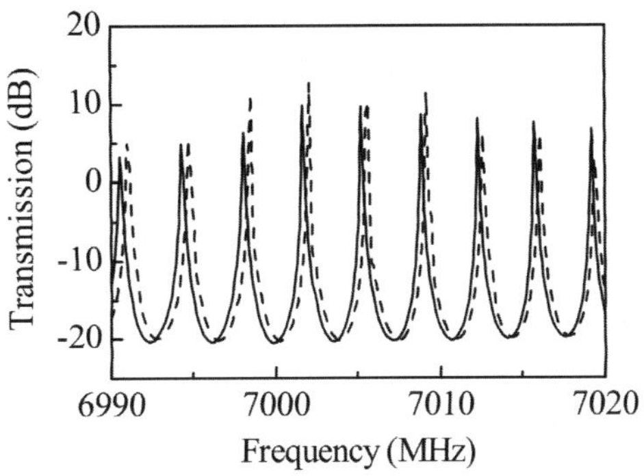

Figure 2. Amplitude-frequency characteristic of the multiferroic ring resonator at bias fields E = 0 (solid line) and E = 12 kV/cm (dashed line).

Experiments demonstrate that the loss during the propagation of SEWs in a multiferroic structure is a comparatively weak function of the applied bias field E. This results in that the auto-generation of a monochromatic wave in the ring appears at different values of the resultant gain G for different values of E. In these experiments, G was conditionally taken to be 0 dB when auto-generation commenced for the field E = 12 kV/cm. Therefore, in the absence of a bias field, the auto-generation appeared at G = 0.2 dB. Periodic sequences of solitons started to be formed at G = 0.25 dB in the absence of a bias field and at G = 0.05 dB in a bias field E = 12 kV/cm.

As expected for the nonlinear dispersive medium with "repulsion" type nonlinearity [21], we observed in our experiments the formation of dark SEW envelope solitons. The upper panels of figure 3 show typical experimental results demonstrating the auto-generation of periodic soliton sequences. The experimentally measured sequences are shown by circles. The solid lines in the same figure represent the results of a numerical simulation, discussed below.

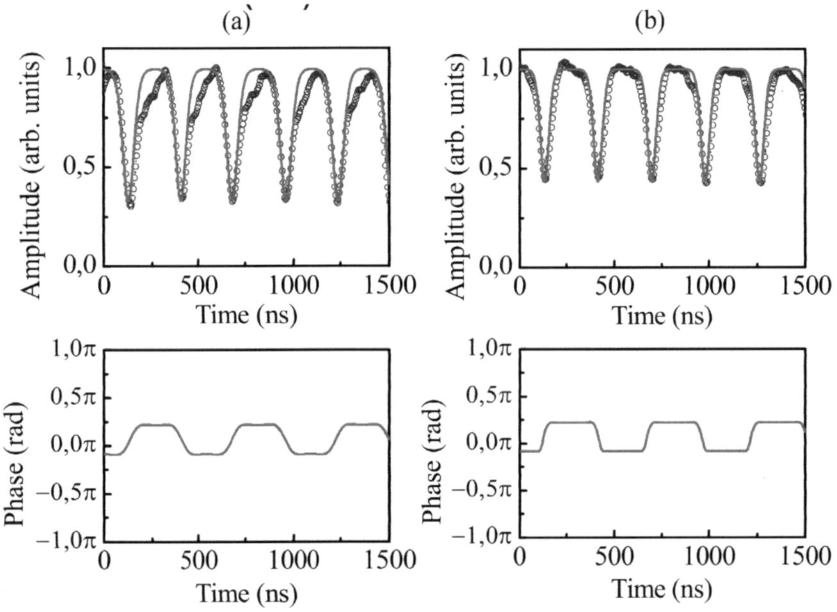

Figure 3. Upper panel: experimental and theoretical profiles of dark spin-electromagnetic wave envelope solitons auto-generated at E = 0 (a) and E = 12 kV/s (b). Lower panel: the corresponding phase profiles.

The experimental data in figure 3(a) were obtained for the gain G = 0.3 dB in the absence of a bias electric field. Under these conditions, dark solitons were formed with duration of 77 ns, repetition period of 273 ns, and carrier frequency of 7000.40 MHz.

Figure 3(b) shows the experimental results obtained at G = 0.1 dB and E = 12 kV/cm. In this case, the soliton repetition period increased to 282 ns because of the decrease in the SEW group velocity. In this case, the soliton duration decreased to 67 ns. Thus, the experimental results demonstrate that the properties of SEW envelope solitons can be controlled by an electric field.

To confirm the solitonic nature of the observed nonlinear wave excitations, we carried out a numerical simulation. Judging from the physics of formation and circulation of solitons in the active ring, they can be attributed to dissipative solitons [22]. Therefore, we chose as the model equation describing the nonlinear dynamics of spin-electromagnetic waves in a ring resonator the Ginsburg--Landau equation

$$i\left[\frac{\partial u}{\partial t}+V_g\frac{\partial u}{\partial x}+(\omega_r-\alpha)u\right]+\frac{D}{2}\frac{\partial^2 u}{\partial x^2}-(N-i\nu)|u|^2 u=0, \quad (2)$$

where u is the slowly varying amplitude of the signal circulating in the ring. The coefficients V_g, D, and N denote, respectively, the group velocity and the dispersion and nonlinear coefficients of the model. The coefficients ω_r and ν have the meaning of the linear and nonlinear relaxation frequencies of spin-electromagnetic waves, and the parameter α reflects the amplification process.

The SEW dispersion law was determined for the experimental structure by using formula (1). In this formula, $k_{sew}(f)$ and $k_{sw}(f)$ were calculated in terms of the theories for SEWs [19] and spin waves [5]. The dependence of the permittivity of the BST slab on the field E was described by the quadratic function $\varepsilon(E) = 2700 - 4.86E^2$, the coefficients of which were found from the best fit of the theoretical and experimental eigenmode frequencies of the ring resonator by using the least-squares method.

As already noted, the results of the numerical simulation of the soliton auto-generation process are shown in figure 3 by solid lines. The simulation was made for the effective values of the coefficients Vg and D, found from the experiment. For E = 0 and E = 12 kV/cm, the values of V_g were, respectively, 2.45×10^4 m/s and 2.38×10^4 m/s, and the values of D were, respectively, -0.686 m^2/(rad s) and -0.551 m^2/(rad s). The values of N were calculated to be, respectively, 1.167×10^{10} rad/s and 1.169×10^{10} rad/s, which is quite reasonable. The solitonic nature of the wave excitations is also confirmed by the occurrence of the fast variation of the envelope phase in the amplitude minimum points of the sequences obtained.

Thus, the results of our calculations confirm the measurement data and support the hybrid-wave nature of the dark envelope solitons auto-generated in the active ring resonator fabricated on the basis of a multiferroic structure.

Acknowledgement
The work at SPbETU was supported in part by Russian Science Foundation. The work at ITMO University was supported by the Government of the Russian Federation (Grant 074-U01).

References
[1] Karpman V I 1975 *Nonlinear waves in dispersive media* (London: Pergamon Press)
[2] Zakharov V E and Ostrovsky L A 2009 *Physika D* **238** 540
[3] Remoissenet M 1999 *Waves Called Solitons: Concepts and experiments* (Berlin: Springer-Verlag).
[4] Kivshar Y S and Agrawal G 2003 *Optical Solitons: From Fibers to Photonic Crystals* (London: Academic Press)
[5] Stancil D D and Prabhakar A 2009 *Spin waves: Theory and applications* (New York: Springer)
[6] Kozyrev A B, Shadrivov I V and Kivshar Y S 2014 *Appl. Phys. Lett.* **104** 084105
[7] Vaz C A F 2012 *J. Phys.: Condens. Matter* **24** 333201
[8] Shastry S, Srinivasan G, Bichurin M I, Petrov V M and Tatarenko A S 2004 *Phys. Rev. B* **70** 064416
[9] Demidov V E, Kalinikos B A and Edenhofer P 2002 *J. Appl. Phys.* **91** 10007
[10] Anfinogenov V B, Verbitskaya T N, Zil'berman P E, Kazakov G T and Tikhonov V V 1986 *Sov. Tech. Phys. Lett.* **12** 185
[11] Ustinov A B, Fetisov Yu K, Lebedev S V and Srinivasan G 2010 *Tech. Phys. Lett.* **36** 166
[12] Fetisov L Y, Fetisov Y K, Sreenivasulu G, Srinivasan G, 2013 *J. Appl. Phys.* **113** 116101
[13] Cherkasskii M A, Kalinikos B A 2013 *Tech. Phys. Lett.* **39** 182
[14] Cherkasskii M A, Kalinikos B A 2013 *JETP Letters* **97** 611
[15] Ustinov A B, Kondrashov A V, Nikitin A A and Kalinikos B A 2014 *Appl. Phys. Lett.* **104** 234101
[16] Kalinikos B A, Kovshikov N G and Patton C E 1997 *Phys. Rev. Lett.* **78** 2827

Journal of Physics: Conference Series **661** (2015) 012057 doi:10.1088/1742-6596/661/1/012057

[17] Beginin E N, Grishin S V and Sharaevsky Yu P 2008 *JETP Lett.* **88** 647
[18] Wu M 2011 Nonlinear Spin Waves in Magnetic Film Feedback Rings *Solid State Physics* **62** 163
[19] Demidov V E, Kalinikos B A, Karmanenko S F, Semenov A A and Edenhofer P 2003 *IEEE Trans. MTT* **51** 2090
[20] Nikitin A A, Ustinov A B, Semenov A A, Kalinikos B A 2012 *Tech. Phys.* **57** 994
[21] Sulem C and Sulem P L 1999 *The Nonlinear Schrödinger Equation: Self-Focusing and Wave Collapse* (New York: Springer-Verlag)
[22] Akhmediev N and Ankiewicz A 2008 *Dissipative solitons: From optics to biology and medicine* (Berlin: Springer)

Spin-wave band-pass filters based on yttrium iron garnet films for tunable microwave photonic oscillators

A B Ustinov, A V Drozdovskii, A A Nikitin and B A Kalinikos

Department of physical electronics and technology, Saint Petersburg Electrotechnical University "LETI", Saint Petersburg, 197376, Russia

E-mail: ustinov_rus@yahoo.com

Abstract. The paper reports on development of tunable band-pass microwave filters for microwave photonic generators. The filters were fabricated with the use of epitaxial yttrium iron garnet films. Principle of operation of the filters was based on excitation, propagation, and reception of spin waves. In order to obtain narrow pass band, the filtering properties of excitation and reception antennas were exploited. The filters demonstrated insertion losses of 2-3 dB, bandwidth of 25-35 MHz, and tuning range of up to 1.5 GHz in the range 3-7 GHz.

Elaboration of microwave signal generators has always been an urgent problem. In recent years, a great interest for development and research of microwave photonic oscillators is evident. The main advantage of the microwave photonic oscillator compared to conventional ones is to overcome the trade-off between the oscillator frequency tuning range and the phase noise level [1]. To date, the microwave photonic oscillators with extremely low phase noise level exist [2-5]. For example, the oscillator developed in work [6] has the phase noise level of -140 dBc/Hz at 10 kHz offset from the carrier. Note that a similar phase noise level is practically unachievable for common microwave generators. Besides in recent years the upper frequency limit of the microwave photonic generators was increased up to hundreds of gigahertz [2,7,8].

A typical microwave photonic oscillator is a ring circuit which includes microwave and optical parts (see figure 1). In the simplest case, the optical part of the generator consists of a laser module, an electro-optic modulator, an optical fiber, and a photo-detector. The main elements of the microwave part include a microwave amplifier and a microwave band-pass filter. As was already mentioned above, an advantage of the microwave photonic generator is a significant reduction of the phase noise level, which depends on the length of the optical fiber. For example, it is shown in work [9] that increase in the fiber length from 100 m to 1000 m reduces the phase noise level from -95 dBc/Hz to -107 dBc/Hz at 10 kHz offset. However, it is clear from the condition of the phase balance that a frequency spectrum of the generator is a series of equally spaced generation frequencies. Thus, the distance between the generation frequencies for the length of the optical fiber of 10 m is about 30 MHz, and for the length of 100 m is 3 MHz. From this follows that in order to select a specific generation frequency it is necessary to use a narrowband band-pass microwave filter. Such a filter must be electronically tunable in order to realize electronic tuning of the generation frequency. Thus, the band-pass microwave filter *plays a key role* in obtaining the desired performance characteristics of the microwave photonic generator.

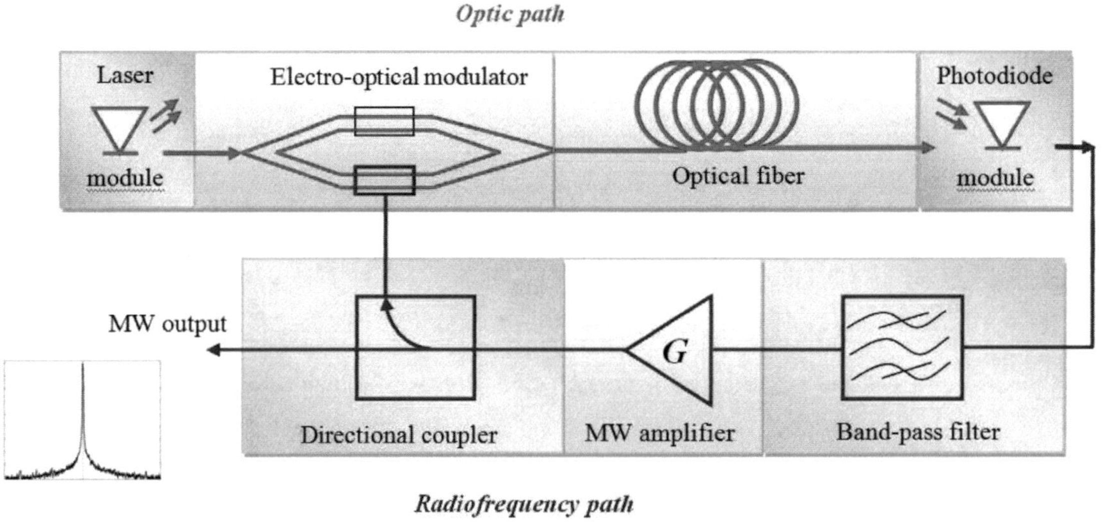

Figure 1. Schematic illustration of microwave photonic oscillator.

It is known from the literature that several types of the filters are commonly used for the frequency filtering in the microwave photonic oscillators. They are cavity resonators [10], Fabry-Perot resonators [10], optical ring resonators [12], as well as whispering gallery mode resonators [6,13]. Actually, a commercial version of the integral microwave photonic generators based on the whispering gallery resonators has been developed by OEwaves, Inc. (www.oewaves.com). The phase noise level of this generator was -108 dBc/Hz at 10 kHz offset. We note that a disadvantage of all above mentioned generators is that they are not tunable.

One of the most effective methods to develop a tunable microwave filter and, therefore, a tunable optoelectronic microwave generator is to utilize for filtering purposes microwave waves travelling in ferromagnetic films and/or in ferrite-ferroelectric structures. Utilization of the travelling waves (spin waves and spin-electromagnetic waves) in the electronically tunable materials for creation of microwave photonic devices was not studied before. On the other hand, ferrite materials (yttrium iron garnet spheres) have been probably used as microwave filters in microwave photonic circuits (see [14,15]). We wrote “probably” because there is no direct information about the types of the microwave filters in the papers [14,15].

Epitaxial single-crystal yttrium iron garnet (YIG) films as well as YIG spheres are used to create microwave devices for many years (see, e.g. [16] and references therein). An advantage of devices based on YIG films is their planar structure allowing to create the filter topology in a single technological cycle. It is more convenient for the production on an industrial scale. Therefore, an elaboration of the tunable YIG-film-based band-pass microwave filters is important task. A survey of the literature and the reviews on the microwave photonic oscillators brings an idea that a further improvement of such oscillators can be achieved with an application of microwave filters tunable in a wide frequency range.

We underline that the tunable YIG-film-based band-pass filters provide not only frequency filtering but a group delay time, which is rather big in comparison with common filters. Due to the principle of operation of the film-based filters the time delay is obtained through travelling of the spin waves. In other words, spin-wave filter has a steep phase-frequency characteristic. Such steepness may be comparable or even bigger than corresponding steepness introduced in the loop circuitry of microwave photonic oscillator by an optical fiber. It is clear physically that this spin-wave effect should lead to a reduction of phase noise level of a microwave photonic generator.

This work reports on development of the YIG-film band-pass filters as frequency selective elements of microwave photonic oscillators. The main goal of the work was elaboration of the filter design providing small insertion loss for microwave signal in relatively narrow pass band because a phase noise of the oscillators is determined by the loss in addition to the time delay. It was also important to achieve relatively broad frequency tuning range.

The filter was designed as a delay line with excitation and reception antennas. An yttrium iron garnet film was utilized in the structure as a spin-wave waveguide. The YIG film was grown on gadolinium gallium garnet substrate by method of liquid phase epitaxy. The film had a thickness of 9 μm and a saturation magnetization $4\pi M_s$ of 1750 G. The film was magnetized in its plane so as to provide a propagation of the surface spin waves. In order to obtain narrow pass band, the filtering properties of excitation and reception antennas were exploited. For this the original topology of the antennas was designed. The experimental prototype was positioned between the poles of electromagnet for characterization.

Amplitude-frequency characteristics (AFCs) of the filter measured for different values of bias magnetic field H are shown in figure 2. Typical frequency dependences of the insertion loss S_{21} and the return loss S_{11} for $H = 1250$ Oe are shown in figure 3. The measurement results show that the frequency tuning range of the filter was of 5140-6320 MHz. The tuning was provided with variation of the bias field in the range of 1150-1530 Oe. In this frequency range the minimum insertion losses were of 2.4-2.9 dB (see figure 4) and return losses were more than 20 dB. The filter bandwidth measured at -3 dB from minimum insertion loss was 28-34 MHz. The steepness of the slopes of AFC from outside the pass band was varied in the range of 0.7-3.8 dB/MHz. The delay time was of 22 ns.

The experimental filter prototype did not use any additional elements that increase the isolation between input and output. Therefore, the out-of-band attenuation was relatively low, in the 35-40 dB. Low-frequency and high-frequency parasitic pass bands are more than 20 dB below the minimum insertion loss.

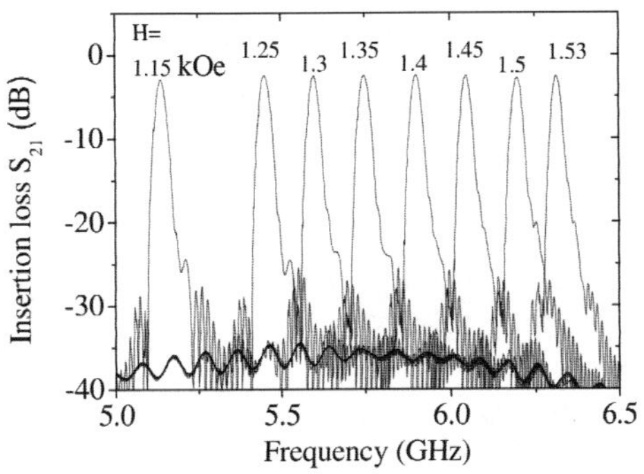

Figure 2. Amplitude-frequency characteristics of the YIG-film filter for $H = 1150$-1530 Oe.

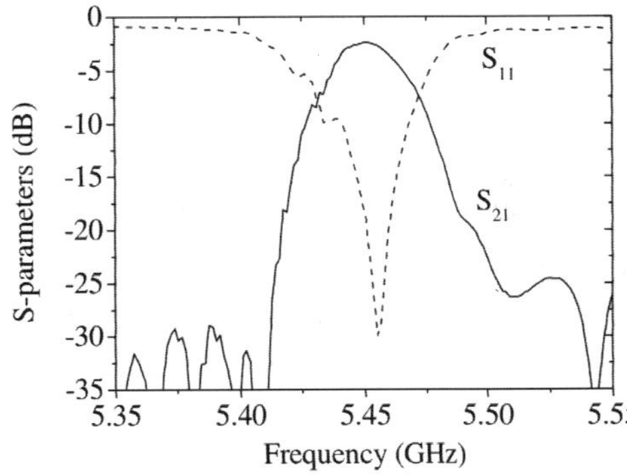

Figure 3. S-parameters of the filter for $H = 1250$ Oe.

It is known that YIG-film filters designed for operation at microwave power of tens to hundreds of milliwatts have a low-frequency limit of tuning. This is due to the existence of low-frequency three-magnon decay of spin waves at such power levels [18]. For YIG films with the saturation magnetization of 1750 G the spin waves decay below 3.2 GHz frequency in the entire spectrum of spin waves. In the range of 3.2-4.9 GHz spin waves demonstrate the decay only in high-frequency part of the surface spin wave spectrum. Therefore, such films are typically used in filters for frequencies above 4.5 GHz. To design the filter for lower frequencies, it is appropriate to use substituted epitaxial YIG films having reduced saturation magnetization. For such films a boundary of the three-magnon decay occurs at the lower frequencies.

Figure 5 shows AFCs of the filter developed with the substituted YIG film having a saturation magnetization of 800 G. The bias field variation in the range of 710-930 Oe resulted in the tuning of the centre frequency in the range of 3-3.6 GHz. The filter bandwidth was of 35 MHz. To increase the isolation between the input and output antennas, special metal screens were used. The delay time was of 29 ns.

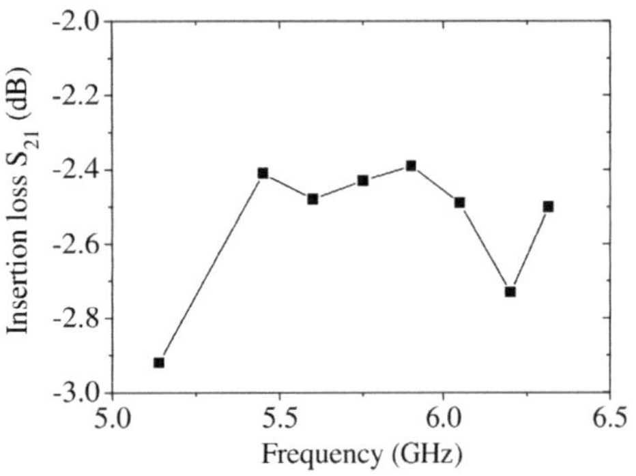

Figure 4. Minimum insertion loss versus frequency characteristic.

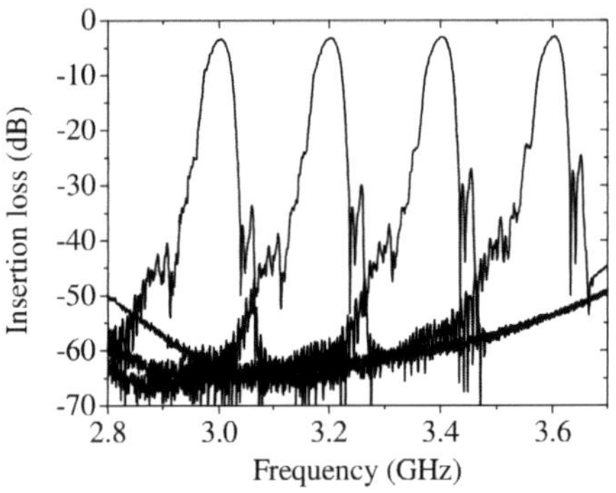

Figure 5. Amplitude-frequency characteristics of the filter made with substituted YIG film for H = 710-930 Oe.

In conclusion, the developed tunable microwave band-pass filters based on epitaxial YIG films demonstrated relatively low insertion loss on the order of 2-3 dB and relatively narrow bandwidth on the order of 30 MHz. These values look favorable in comparison with already existing YIG-film filters [17-21]. The results show that our developed filters have great potential for application. They can be used for development of *tunable* microwave photonic devices, in particular, oscillators. The filters can find conventional application also in microwave equipment.

Acknowledgments
This work was supported in part by the Russian Foundation for Basic Research, Grant of President of the Russian Federation, and by Ministry of Education and Science of the Russian Federation (Project "Goszadanie").

References
[1] Yao X S, Maleki L 1994 Electron. Lett. **30** 1525
[2] Larger L, Udaltsov V S and Poinsot S 2005 J. Opt. Technol. **72** 378
[3] Yao X S 2002 Opto-electronic Oscillators. In book: RF Photonic Technology in Optical Fiber Links. Ed. by W. S. C. Chang (Cambridge University Press) p. 255-292
[4] Product Brief. Advanced Opto-Electronic Oscillator (OEO). Ultra-Low Phase Noise Microwave Signal Source Module. Oewaves, Inc. 2. – URL: www.oewaves.com/advanced-oeo-sp-1171610647
[5] Belkin M, Loparev A 2010 Zhurnal "Elektronika: NTB" **6** 62 (in Russian).
[6] Matsko A B, et al. 2003 Journal of modern optics **50** 2523
[7] Hindle F, Mouret G, Eliet S, Guinet M, Cuisset A, Bocquet R, Rovera D 2011 Appl. Phys. B **104** 763
[8] Qasymeh M, Li W, Yao J 2011 IEEE Trans. MTT **59** 2987
[9] Hayata A, Bacoua A, Rissonsa A, Molliera J.-C 2008 Proceedings of International Topics Meeting on Microwave Photonics MPW/APMP **1** 98
[10] Romisch S, et al. 1999 Proceedings of the 1999 Joint Meeting of the European **2** 561
[11] Ozdur I, et al. 2010 J. Lightwave Technol. **28** 3100
[12] Zhu D, Pan S, Ben D, 2012 IEEE Photon. Technol. Lett. **24** 194
[13] Volyanskiy K, Salzenstein P, Tavernier H, Pogurmirskiy M, Chembo Y K, Larger L 2010 Optics Express **18** 22358
[14] Eliyahu D, Maleki L, 2003 Microwave Symposium Digest **3** 2185
[15] Zhu D, Pan S, Ben D 2012 IEEE Photon. Technol. Lett. **24** 194

[16] Stancil D D and Prabhakar A 2009 Spin waves: Theory and applications (New York: Springer)
[17] Adam J D 1985 IEEE Trans. on UFFC **33** 157
[18] Ishikawa Y, Nishikawa T, Okada T, Shinmura S, Kamado Y, Kanaya F, Wakino K 1990 IEEE MTT-S International Microwave Symposium Digest **1** 143
[19] Popina S M, Chehetin A V 2011 Microwave and Telecommunication Technology (CriMiCo) 21th International Crimean Conference **1** 643
[20] Manuilov S A, Fors R, Khartsev S I, Grishin A M 2009 J. Appl. Phys. **105** 033917
[21] Yang G.-M, Lou J, Wu J, Liu M, Wen G, Jin Y, Sun N X 2011 IEEE MTT-S International Microwave Symposium Digest**, DOI**: 10.1109/MWSYM.2011.5972897

Analysis of chaotic bright spin-wave soliton trains in magnetic films

A B Ustinov[1], A V Kondrashov[1] and A A Nikitin[2]

[1]St. Petersburg Electrotechnical University, St. Petersburg, 197376 Russia
[2]ITMO University, St. Petersburg, 197101 Russia

E-mail: ustinov_rus@yahoo.com

Abstract. We have experimentally studied a chaotic dynamics of bright spin-wave soliton trains. Thin yttrium iron garnet films having pinned surface spins were used for the experiments providing propagation of highly dispersive dipole-exchange spin waves. Chaotic soliton trains were excited in the frequency band around low-frequency part of dipole gap of spin wave spectrum. Analysis of fractal dimension, embedding dimension, and Lyapunov exponents was carried out based on measured time profiles of chaotic soliton sequences. We find that the fractal dimension and Lyapunov exponents are weak function of carrier frequency around dipole gap whereas embedding dimension is almost constant.

1. Introduction

Envelope solitons are formed in the nonlinear dispersive waveguiding media with pulsed excitation if the dispersion spreading of a nonlinear wave packet is compensated by the media nonlinearity. Solitons are also formed with a monochromatic wave excitation through development of the spontaneous modulation instability (SMI) [1,2].

Among other media, high-quality magnetic films, such as single-crystal yttrium iron garnet (YIG) films, were proven to be excellent objects for experiments on nonlinear wave phenomena with microwave spin waves. We underline, that it was the YIG films where a formation of the stationary soliton trains from a monochromatic spin wave (SW) through development of the spontaneous modulation instability (SMI) has been discovered and studied [3-5].

Theoretically, the soliton as a stationary solution of the nonlinear Schrodinger (NLS) equation happens to be stable. Experimentally, the soliton is also usually considered as a stable excitation which preserves its shape and speed during propagation. At the same time, using an optical mode-locked laser, Bolton and Acton [6] have demonstrated in 2000 year a chaotic behavior of solitons which represented pulsations in amplitude of optical pulses circulating in the optical cavity. After that considerable amount of theoretical work was devoted to study of possible chaotic behavior of solitons in the different physical systems [7- 13]. The concept of “dissipative solitons” [14,15] resolves the contradiction.

At present, studies on chaotic wave excitations in waveguiding media and auto-oscillators are of great interest both for basic research and for their possible applications in advanced optical and microwave communication technology [16-18]. Recently, the chaotic behavior of solitons have been observed experimentally for spin waves propagating in magnetic films [19]. It is resulted in a new wave of interest for both theoretical and experimental studies of chaotic dynamics of solitons [20-33].

Purpose of this work is analysis of the chaotic dynamics of solitons formed with monochromatic excitation of dipole-exchange spin waves through simultaneous development of the SMI and the chain of spin-wave parametric interactions in an yttrium iron garnet (YIG) film.

2. Experimental set-up

A 3.1-μm-thick, 1.5-mm-wide, and 30-mm-long YIG film waveguide was used for the experiments. The film was grown on a (111) plane single-crystal gadolinium gallium garnet substrate by the method of liquid phase epitaxy. The YIG film had a ferromagnetic resonance line-width of 0.4 Oe at 3 GHz, and a saturation magnetization $4\pi M_s$ of 1750 G. The YIG film waveguide was magnetized by external constant bias field H_e of 2525 Oe directed perpendicular to the film plane. Note, that the YIG film is assumed to be magnetically isotropic and an internal bias field value calculated as $H_i = H_e - 4\pi M_s$ was 775 Oe that was in good agreement with measurements. The perpendicular magnetization provided a forward volume spin wave (FVSW) propagation in the waveguide.

The film had pinned surface spins and the excitations generated in the experiments were dipole-exchange spin waves demonstrating strong decrease in their group velocity at frequencies close to spin-wave resonances. The spin wave spectrum has so-called "dipole gaps" at these frequencies (see the figure 1). Similar to the paper [19], it is the existence of the dipole gaps that is the key condition for formation of the chaotic soliton trains.

As is well known the pinning of the surface spins arise from surface anisotropy (see e.g. [34]). Physical reason consists in the following. Fields affecting on magnetic moments in the vicinity of the interface between magnetic and non-magnetic materials are differ from those inside the magnetic material. If we do not want to investigate in details magnetic oscillations in these fields, we should take into account their influence on magnetic oscillations inside the ferromagnetic film with an introduction of the boundary conditions (assuming the thickness of the surface interface layer negligibly small). Using this approach it is possible to estimate frequencies of spin-wave resonances (see. e.g. [35]) which coincide with high precision with frequencies of dipole gaps of the spin-wave spectra (calculated with theory published in [36]) shown in figure 1(a) and with frequencies of dips in the amplitude-frequency characteristic shown in figure 1(b). Technological reasons of pinning of surface spins for magnetic film sample used in our experiments are out of the present research.

The spin waves in the YIG film waveguide were excited and detected by conventional microstrip transducers separated by a distance of 4.5 mm. The transducers had a width of 40 μm and a length of 3 mm.

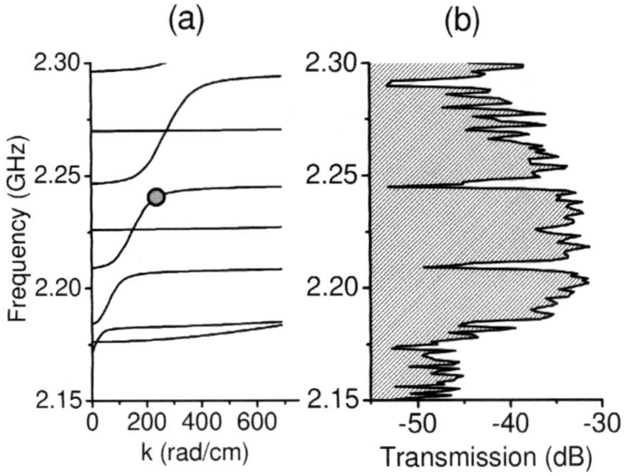

Figure 1. Dipole-exchange spin wave spectrum (a) and amplitude-frequency characteristic (b) of the YIG-film structure

3. Experimental results and discussion

In the experiments a highly monochromatic signal was applied from a microwave source to the input microstrip transducer so as to excite a continuous wave (CW) spin wave in the YIG-film waveguide. The spin waves after propagation the distance of 4.5 mm were detected by the output transducer and then monitored with a microwave spectrum analyzer and a real-time digital oscilloscope having a high sample rate and a long acquisition. It allowed us to record the waveforms during the long time for further mathematical processing. During the measurements the CW input power P_{in} was systematically varied from 0.1 to 40 mW. The CW frequency was preset in the SW excitation range, for the given experimental condition from the FWSW cutoff frequency of 2174 MHz to 2300 MHz at which output signal power level became small and comparable with noise level (see the figure 1).

A formation of the chaotic solitons was found at frequencies around 2247 MHz for the input power level P_{in} around 35 mW. Detailed analysis of the shape of the observed nonlinear excitations was carried out by the method described in our previous work [19] in which chaotic spin-wave soliton trains were observed for the first time. This analysis confirmed solitonic nature of the chaotic pulses.

Figure 2 shows representative data obtained for the case when the monochromatic SW with power of 15.3 dBm at frequency 2247.3 MHz was excited in the waveguide. It is clear that peak power, period, and width of the solitons were chaotically varied.

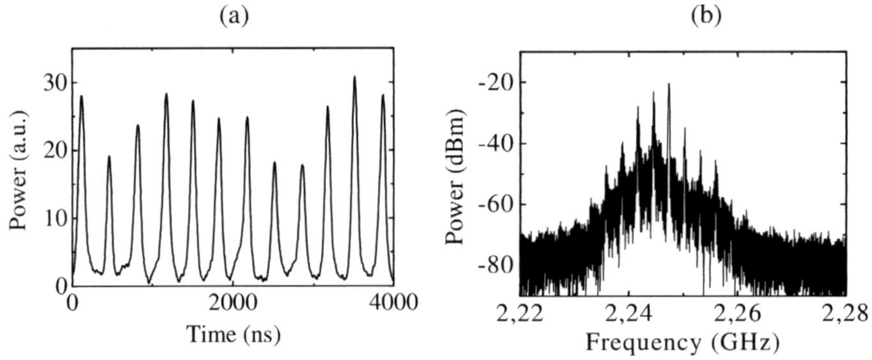

Figure 2. Chaotic bright soliton train (a) and its frequency spectrum (b) at carrier frequency f = 2247.3 MHz.

The observed nonlinear dynamics of the soliton trains contained elements of both regular motion and chaos simultaneously. The regular motion was due to the development of the SMI of the spin waves excited in the film, whereas the chaotic motion was arisen by the development of the chain of parametric interactions (CPI) among the generated SW. It is clearly confirmed by the power-frequency spectrum of the chaotic solitons (see the figure 2(b)), which demonstrate elements of both SMI in the form of the well pronounced equidistant separate harmonics and CPI in the form of the pronounced continuous noise-like pedestal. The carrier frequency around which formation of the chaotic soliton trains took place is shown in the figure 1(a) by circle.

Numerical estimation of the obtained chaotic signal parameters for different values of input signal frequency was done. First multidimensional phase portraits for all frequencies were reconstructed. Conventional method of time delay was used [37]. Dynamic of the spin-system in the phase space is represented by a phase trajectories. Phase trajectory point coordinates were obtained using long waveforms as follows

$$\boldsymbol{x}(t) = \{U(t), U(t+\tau), ..., U(t+(d-1)\tau)\},$$

where τ is time delay, d is a phase space dimension and $U(t)$ is the amplitude of analyzed signal. A form of phase trajectories is defined by an attractor which attracts trajectories of phase space. Regular dynamic phase portrait is defined by so-called regular attractor and chaotic dynamics is defined by a so-called strange attractor. The 2D phase portrait of the observed solitonic chaos is shown in figure 3. According to the paper [37] the phase portrait obtained with the above mentioned method allows to define correctly such attractor parameters as fractal dimensions, minimum embedding dimensions, Lyapunov exponents etc. The values of the chaotic signal parameters allows to define whether the signal chaotic or noisy.

The values of the fractal dimension and minimum embedding dimension were calculated using the standard Grassberger-Procaccia method [38,39]. First, correlation dimensions for several phase space dimensions d were determined. Correlation dimension determines topology of the attractor projection into the phase space with certain dimension d. Due to increase of the phase space the form of attractor projection is encloses to the form of original attractor that leads to an increase in correlation dimension up to some saturation level. The saturation value is the fractal dimension of the attractor D_f. The value

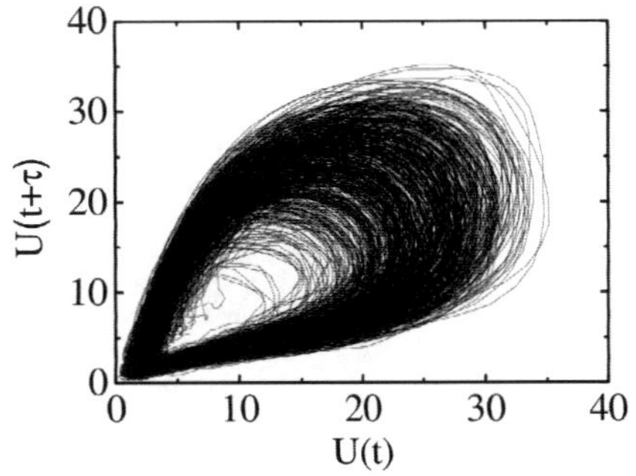

Figure 3. Attractor of the chaotic solitons at carrier frequency f = 2247.3 MHz.

of the phase space dimension for which saturation begins is a minimum embedding dimension. Fractal dimension defines topology of the whole attractor. The attractor corresponding to the monochromatic signal is a stable point and its fractal dimension is zero. The attractor of a periodic signal is a limit cycle. Its fractal dimension equals 1. A noise signal phase trajectory fills up the phase space densely, therefore corresponding fractal dimension is infinite [40]. Dynamical chaos is characterized by appearance of the strange attractor in the phase space. Fractal dimension of the strange attractor is finite and no integer. Minimum embedding dimension defines a degrees of freedom quantity of the investigated spin system. Its value is proportional to required equation quantity for the system modeling.

Using described methods, values of fractal dimensions and minimum embedding dimensions for some particular frequencies of input signal were calculated. The results of calculation are represented in the table 1. The values of D_f and d demonstrate that in addition to regular soliton motion the chaotic modulation of the soliton shape takes place. Calculation results show that in the range of input frequency from 2246.0 MHz to 2247.7 MHz an increase in the frequency leads to fractal dimension weak increase from 5.9 to 6.2. It means that the behavior of the investigated spin-system becomes more complicate with frequency increasing. The values of the minimum embedding dimension are almost constant. It means that a change in the input frequency does not increases or decreases the number of degrees of freedom of the spin system.

Another parameter which was calculated is maximum Lyapunov exponent λ. It characterizes the rate of exponential divergence of phase trajectories. Dissipative systems are characterized by negative Lyapunov exponent. Zero exponents correspond to periodical motion. For the case of deterministic chaos maximum Lyapunov exponent is positive.

Table 1. Dependence of the fractal dimension, minimum embedding dimension, Lyapunov exponent, and modulation frequency from input frequency.

	Input signal frequencies (MHz)								
	2246.0	2246.2	2246.3	2246.4	2246.5	2246.6	2246.7	2247.3	2247.7
D_f	5.9	5.9	6	5.9	5.9	5.8	5.8	6.1	6.2
d_{em}	21	22	22	22	22	21	22	22	22
λ	0.08	0.073	0.087	0.08	0.09	0.081	0.086	0.083	0.091
Δf_{mod} (MHz)	2.9	2.95	2.93	2.98	2.95	2.95	2.98	2.9	2.92

For calculation of the Lyapunov exponents an algorithm introduced by Rosenstein and by Kanz was used [41]. The algorithm allows to calculate the average divergence of trajectories that starts from close placed points. For calculation the TISEAN package [42] was used.

Using the TISEAN values of maximum Lyapunov exponents λ for different frequencies of the input signal were determined. In the table 1 the dependence of the Lyapunov exponent on input frequency is shown. Positive values of λ confirm that observed solitonic signals demonstrates chaotic dynamics. Similar to fractal dimension an increase in the input signal frequency leads to weak increase of λ.

Calculations of the parameters shows that the chaotic soliton motion of the spin system depends weakly on the frequency of the input signal.

4. Conclusion

In this work the chaotic dynamics of soliton trains was studied. The results confirm deterministic nature of the observed chaotic soliton sequences.

Acknowledgement

The work at SPbETU was supported in part by the Russian Foundation for Basic Research, Grant 15-32-20357 mol_a_ved and by the Ministry of Education and Science of Russian Federation (Project "Goszadanie"). The work at ITMO University was supported by the Government of the Russian Federation (Grant 074-U01).

References

[1] Karpman V I 1975 *Nonlinear waves in dispersive media* (London: Pergamon Press)
[2] Zakharov V E and Ostrovsky L A 2009 *Physica D* **238** 540
[3] Kalinikos B A, Kovshikov N G and Slavin A N 1984 *Sov. Tech. Phys. Lett.* **10** 392
[4] Wu M, Kalinikos B A and Patton C E 2004 *Phys. Rev. Lett.* **93** 157207
[5] Wu M and Kalinikos B A 2008 *Phys. Rev. Lett.* **101** 027206
[6] Bolton S R and Acton M R 2000 *Phys. Rev. A* **62** 063803
[7] Akhmediev N, Soto-Crespo J M and Town G 2001 *Phys. Rev. E* **63** 056602
[8] Akhmediev N and Soto-Crespo J M 2005 *Phys. Rev. Lett.* **95** 024101
[9] Seghete V, Menyuk C and Marks B S 2007 *Phys. Rev. A* **76** 043803
[10] Chacon R and Martinez P J 2007 *Phys. Rev. Lett.* **98** 224102
[11] Zhu Q, Hai W and Rong S 2009 *Phys. Rev. E* **80** 016203
[12] Mirsa A P and Shukla P K 2009 *Phys. Rev. E* **79** 056401
[13] Filatrella G, Malomed B A and Salasnich L 2009 *Phys. Rev. A* **79** 045602
[14] Akhmediev N and Ankiewicz A 2008 *Dissipative solitons: From optics to biology and medicine* (Berlin: Springer)
[15] Grelu Ph and Akhmediev N 2012 *Nature Photonics* **6** 84
[16] Stavroulakis P 2006 *Chaos applications in telecommunications* (Boca Raton: SRC Press)

Journal of Physics: Conference Series **661** (2015) 012059 doi:10.1088/1742-6596/661/1/012059

[17] Uchida A 2012 *Optical Communication with Chaotic Lasers* (Weinheim: Wiley-VHS Verlag&Co. KGaA)
[18] Wu M, Hagerstrom A M, Eykholt R, Kondrashov A and Kalinikos B A 2009 *Phys. Rev. Lett.* **102** 237203
[19] Ustinov A B, Demidov V E, Kondrashov A V, Kalinikos B.A., S.O. Demokritov, Phys. Rev. Lett. **106**, 017201 (2011).
[20] Wang Z, Hagerstrom A, Anderson J Q, Tong W, Wu M, Carr L D, Eykholt R and Kalinikos B A 2011 *Phys. Rev. Lett.* **107** 114102
[21] Fujioka J, Cortes E, Perez-Pascual R, Rodrıguez R F, Espinosa A and Malomed B A 2011 *Chaos* **21** 033120
[22] Grishin S V, Sharaevskii Y P, Nikitov S A, Beginin E N and Sheshukova S E 2011 *IEEE Trans. Mag.* **47** 4382
[23] Lecaplain C, Grelu Ph, Soto-Crespo J M and Akhmediev N 2012 *Phys. Rev. Lett* **108** 233901
[24] Cartes C, Cisternas J, Descalzi O and Brand H R 2012 *Phys. Rev. Lett.* **109** 178303
[25] Latas S C V and Ferreira M F S 2012 *Optics Letters* **37** 3897
[26] Urzagasti D, Laroze D, Clerc M G, Coulibaly S and Pleiner H 2012 *J.Appl. Phys.* **111** 07D111
[27] Batalov S V and Shagalov A G 2013 *Physics Letters A* **377** 964
[28] Batalov S V and Shagalov A G 2013 *The Physics of Metals and Metallography,* ***114*** 826
[29] Clerc M G, Garcia-Nustes M A, Zarate Y and Coulibaly S 2013 *Phys. Rev. E* **87** 052915
[30] Urzagasti D, Laroze D, Clerc M G and Pleiner H 2013 *Europhys. Lett.* **104** 40001
[31] Grishin S V, Sharaevskii Y P, Nikitov S A and Romanenko D V 2013 *IEEE Trans. Mag.* **49** 1047
[32] Urzagasti D, Laroze D and Pleiner H 2014 *Eur. Phys. J. Special Topics* **223** 141
[33] Anderson J Q, Ryan R A, Wu M and Carr L D 2014 *New Journal of Physics* **16** 023025
[34] Gurevich A G and Melkov G A 1996 *Magnetic oscillations and waves* (SRC Press)
[35] Soohoo R F 1965 *Magnetic Thin Films* (Harper&Row)
[36] Kalinikos B A and Slavin A N 1986 *J. Phys. C* **19** 7013
[37] Takens F 1981 *Dynamical Systems and Turbulence, Lecture Notes in Mathematics* **898** 366
[38] Grassberger P and Procaccia I 1983 *Phys. Rev. Lett.* **50** 346
[39] Grassberger P and Procaccia I 1983 *Physica D* **9** 189
[40] Schuster H 1984 *Deterministic Chaos: An Introduction* (Weimheim: Physic-Verlag)
[41] Kantz H and Schreiber T 2004 *Nonlinear Time Series Analysis - 2nd edition* (Cambridge: University press)
[42] Hegger R, Kantz H and Schreiber T. 1999 *CHAOS* **9** 413

UV Sensor Based on Layered Ferrite-Diamond Structure

P Y Belyavskiy[1], A V Kondrashov[1,2], A A Nikitin[1,2], V V Vitko[1,2], B A Kalinikos[1], A A Semenov[1], A B Ustinov[1] and J E Butler[3]

[1]St. Petersburg Electrotechnical University, St. Petersburg, 197376, Russia
[2]Institute of Applied Physics of the Russian Academy of Science, Nizhny Novgorod, 603950, Russia
[3]Cubic Carbon Ceramics, Huntingtown, MD, 20639, USA

E-mail: pbeliavskiy@gmail.com

Abstract. Design of a UV sensor based on an yttrium-iron garnet/diamond layered cavity is presented. The operation principle of the sensor is based on the change in the spin-wave dispersion law and, consequently, in the frequency and Q-factor of eigenmodes of the cavity under study upon appearance of free carriers generated by UV light.

One of ways to improve the resolution of the photolithographic process is to use far-UV light with wavelengths of 157 to 248 nm [1]. Improving the characteristics of fast and sensitive far-UV sensors will enable wider mastering of this spectral range.

One of ways to develop a sensor for far-UV emission is to use a material with an energy gap width comparable with the energy of a UV photon. One of materials of this kind is single-crystal diamond whose energy gap width is about 5.5 eV. Recent advances in the fabrication technology of single-crystal films and layers of diamond with varied doping level open up prospects for their use as sensors for light with wavelength shorter than 250 nm. For example, emission sensors constituted by single-crystal diamond with metallic electrodes were suggested in [2--5].

In this communication, we suggest a simple UV sensor design based on an yttrium-iron garnet/diamond layered structure. The operation principle of microwave devices based on single crystal yttrium-iron garnet (YIG) films magnetized to saturation is underlain by specific features of the propagation in these devices of a particular kind of waves: magnetization waves or long-wavelength spin waves [6]. The operation principle of the UV sensor we suggest is based on the effect of free carriers generated in the semiconductor on the dispersion law of spin waves (SWs) in the YIG film.

The effect of conducting screens on the dispersion characteristics of SWs has been studied in sufficient detail. In particular, a dispersion equation for the surface SW in a film with a metallic screen was derived in [7]. The structure under study is shown schematically in Fig. 1a. In [7], the finiteness of the metal conductivity was disregarded, and the dependence of the microwave parameters of the system on the distance to the screen was analyzed. For example, it was shown that the dispersion equation for the structure described has the form:

$$\left(1+2\frac{\omega_H}{\omega_M}+2s\frac{\omega}{\omega_M}\right)\frac{\omega_M+(\omega_H-s\omega)(1+\tanh(kt))}{\omega_M+(\omega_H+s\omega)(1-\tanh(kt))}=\mathrm{e}^{-2kd}. \tag{1}$$

Here, d is the film thickness, t is the distance to the metal, and s is a parameter characterizing the propagation direction of the surface SW. Figure 1b shows the dispersion characteristics calculated by formula (1) at various distances to the metal. It can be seen in the figures that the SW dispersion substantially changes as the conducting screen approaches the film. In particular, the frequency sharply grows at small values of *k*.

(a) (b)

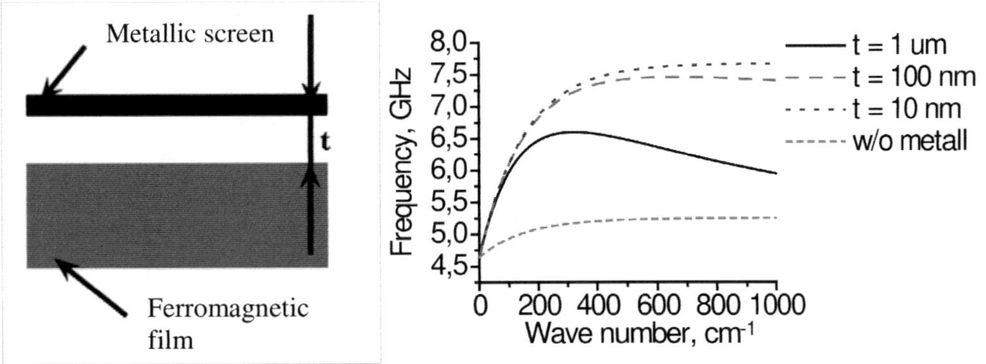

Figure 1. (a) Structure of a screened film, and (b) dispersion characteristics for various distances between the film and the metal.

The dispersion and dissipation parameters of SWs in ferrite-semiconductor structures were calculated in [8] with consideration for the finite value of the conductivity. A dispersion equation for surface SWs propagating in a planar ferrite-semiconductor waveguide was obtained in [8] in terms of the hydrodynamic model for a ferrite and semiconductor in the magnetostatic approximation, without consideration for the exchange interaction. This dispersion equation has the form

$$e^{-2kd} = \frac{(\mu^- + k)(\delta\mu^+ + k)\tanh(k_s t) + (\mu^+ + k)\sqrt{\delta}}{(\mu^+ - k)(\delta\mu^- - k)\tanh(k_s t) + (\mu^- - k)\sqrt{\delta}}. \quad (2)$$

Here, $\mu^{\pm} = k(\mu \pm \mu_a)$, $k_s^2 = \delta k^2$, $k = k_1 + ik_2$; d and t are the thicknesses of the ferrite and semiconductor layers, respectively; and μ and μ_a are the diagonal and nondiagonal elements of the ferrite permeability tensor. The value of δ determines the specific features of the SW interaction with the semiconductor layer, and $\delta = f(\omega)$. Thus, the expression obtained in [7] makes it possible to simulate both the dispersion characteristics and the variation of the SW dissipation parameter for different semiconductor wafers. In the case of $\delta = 1$, expression (2) becomes the Damon-Eschbach equation [9]. Figure 2a plots the dispersion dependences of surface SWs for various carrier concentrations. It can be seen in the figure that the SW dispersion changes under the action of the carrier concentration in the semiconductor, beginning at the concentration $N = 10^{15}$ cm^{-3}. With increasing N, the curves shift toward the curve corresponding to the metallized film. It should be noted that, because of the finite value of the conductivity of the semiconductor film, its effect is equivalent to the influence of a metallic layer moved away to a certain distance. This effect can be attributed to the existence of a skin layer in the conducting film.

Figure 2b shows calculated dependences of the loss in the ferrite-semiconductor structure at various carrier concentrations in the semiconductor. It can be seen that, with increasing *N*, the loss at large SW wave numbers decreases.

It is physically clear that measuring the SW dispersion and dissipation parameters in ferrite-semiconductor structures makes it possible to estimate the parameters of the semiconductor structures, such as conductivity and distance to the doping region. Measurements of the resonance characteristics of a cavity based on the YIG/diamond layered structure will provide information about the free carrier concentration and, consequently, about the emission intensity.

Figure 3 shows schematically the experimental setup making it possible to experimentally study the UV detector. The detector was fabricated from a YIG/diamond layered structure. The YIG film was grown on a gallium-gadolinium garnet (GGG) substrate by liquid-phase epitaxy. The thickness of the YIG film was 6 μm. A film cavity with a plane size of 3×3 mm² was cut from this film. On top of the cavity was attached a 3.5x3.5 mm² single-crystal diamond film. The resulting structure was placed in a dc magnetic field directed tangentially to the plane of the films. The field strength was chosen so that the YIG cavity was magnetized to saturation.

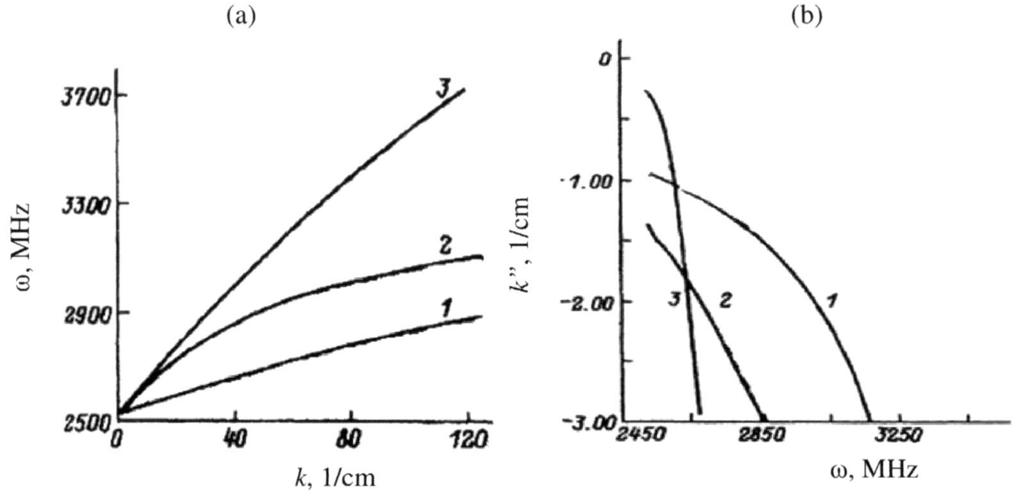

(1) $N = 10^{13}$, (2) $N = 10^{15}$, and (3) $N = 10^{17}$

Figure 2. Dependences of the microwave parameters of ferrite-semiconductor structures at various carrier concentrations in the semiconductor. (a) Dispersion characteristic and (b) frequency dependence of the loss.

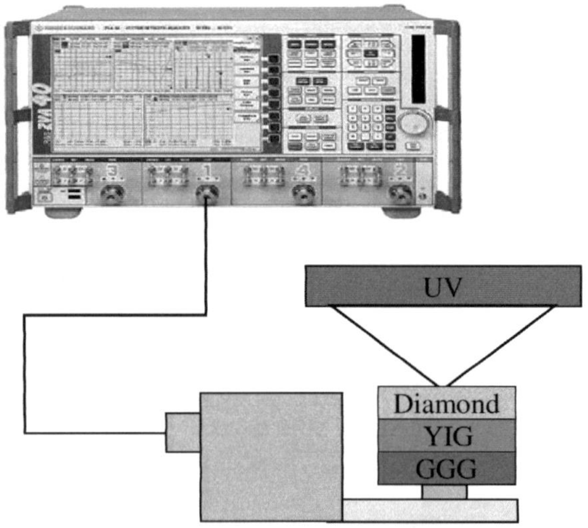

Figure 3. Schematic of the setup for studying the UV sensor based on the diamond/YIG structure.

The microstrip antenna was used to excite magnetization oscillations in the layered cavity. The frequency of the resonance modes was determined by the characteristics of the structures from which the cavity was fabricated. It should be recalled that the carrier concentration in the diamond film will affect the frequency and Q-factor of the resonance peaks.

To evaluate this effect, it is necessary to calculate the resonance frequencies of the YIG/diamond cavity on the assumption that diamond is an insulator. These frequencies will be determined by the SW dispersion in the tangentially magnetized film and by the cavity dimensions (width w and length l). The relationship between the frequency and wave number can be found using the dispersion equation of the form $f_{res} = D(k_{res})$, where $k_{res} = \sqrt{(k_{\parallel})^2 + (k_{\perp})^2}$ is the resonance wave number, $k_{\parallel} = \frac{\pi m}{w}$, $k_{\perp} = \frac{\pi n}{w}$ are the longitudinal and transverse wave numbers, and m and n are the numbers determining the distributions of the magnetostatic potential along and across the magnetization direction.

For the cavity described above, w = 3 mm and l = 3 mm. In the course of the simulation, we calculated the dispersion characteristic of the tangentially magnetized layered structure. The simulation results are presented in Fig. 4a. Choosing the values of m and n, we can determine the corresponding resonance frequencies.

(a) (b)

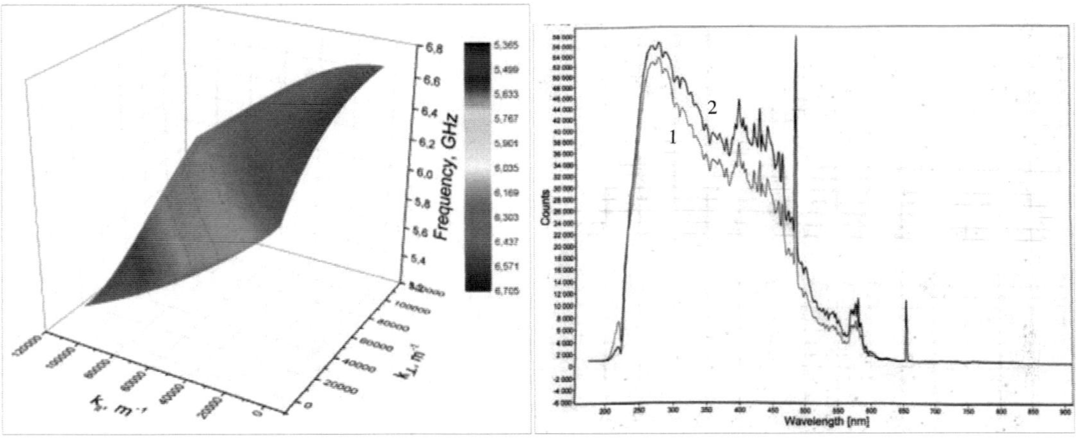

Figure 4. (a) Dispersion characteristic of the SW in a tangentially magnetized YIG film and (b) dependence of the intensity of a reflected wave on the wavelength for the cases of only a mirror (light line) and a mirror with diamond (dark line).

In order to investigate absorption of experimental sample in UV range in the second stage of the study, we measured the optical properties of the single-crystal diamond. Experimental procedure was carried out as follows: initially light in a wide range of wavelengths (200-600 nm) emitted by deuterium lamp was applied to a mirror and power of reflected signal was measured (line 1 in the figure 4b). Then an experimental sample was placed on the mirror. The mirror-sample system was exposed to light and the power of reflected light was measured again (line 2 in the figure 4b). It can be seen that diamond can reduce the reflected signal in nearly the whole range of wavelengths. However, the light line lies above the dark line at 200-300 nm. This can be attributed to the absorption of light via transitions of electrons from the valence band to the conduction band.

The absorption was measured for two samples. Figure 5 shows normalized reflectance plots. These plots were obtained by subtracting from the level corresponding to the reflection from the mirror that corresponding to the reflection from the mirror with diamond. Figure 5a corresponds to the case of diamond with low impurity content. Figure 5b shows a similar dependence for a diamond sample with high concentration of the nitrogen impurity. The presence of such an impurity resulted in that a yellow tint appeared.

It can be seen from the plots in Fig. 5a that the maximum absorption is observed at a wavelength of 220 nm, which corresponds to the energy gap width of 5.64 eV. This value is in good agreement with reference data, which points to a high purity of the diamond film and to the possibility of using this film as the conducting layer in a UV sensor.

(a) (b)

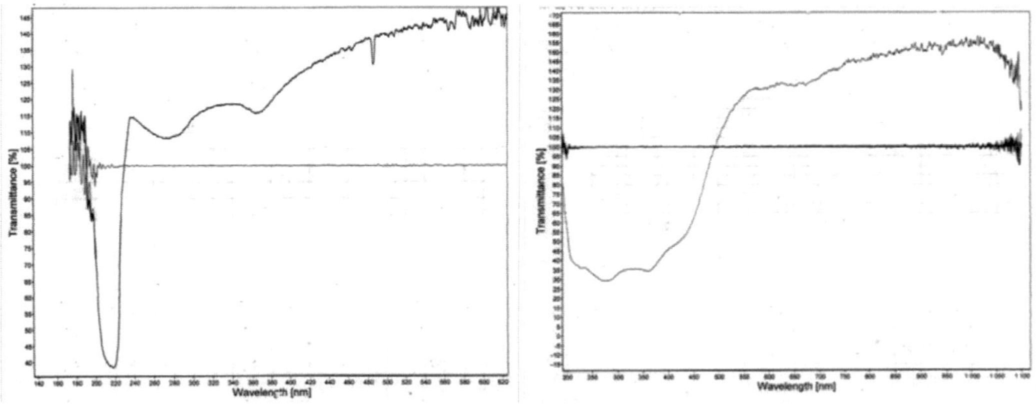

Figure 5. Normalized reflection level for (a) "pure" diamond sample and (b) impurity diamond sample.

Figure 5b shows the characteristic of the "yellow" impurity diamond. The absorption begins at 450-500 nm. This confirms the assumption that there exists a nitrogen impurity that gives rise to additional impurity sublevels in the energy gap of diamond. Electrons are excited to the conduction band from these additional levels at a lower energy, compared with pure diamond.

Thus, the design of a UV sensor based on a YIG/diamond microwave cavity was presented. The advantages of the suggested design are the following: simplicity of fabrication, fast operation speed, and high sensitivity. Its experimental characteristics will be reported elsewhere.

The work at SPbETU was supported in part by the Russian Foundation for Basic Research, grant of President of Russian Federation for young scientists and PhDs MK-6229.2015.8, and by Ministry of Education and Science of Russian Federation (Project "Goszadanie"). The work at IAP RAS was supported by Act 220 of the Russian Government (Agreement No. 14.B25.31.0021).

References

[1] Gwyn C W, Stulen R, Sweeney D and Attwood D 1998 *J. Vac. Sc. & Tech. B.* **16** 3142
[2] Kania D R, Landsrass M I, Plano M A, Pan L S, & Han S *Diamond and Related Mater.* **2** 1012
[3] Prasad R R 2006 *Diamond Radiation Detectors* (San Leandro: Alameda Applied Science Corporation)
[4] Kagan H *Nuclear Instruments and Methods in Physics Research Section A: Accelerators, Spectrometers, Detectors and Associated Equipment* **546** 222
[5] Teraji T, Yoshizaki S, Wada H, Hamada M, Ito T 2004 *Diamond and Related Mater* **13** 858
[6] Gurevich A S 1994 *Magnetic Oscillations and Waves* (Moscow: Fizmatlit)
[7] Bongianni W L 1972 *J. Appl. Phys.* **43** 2541
[8] Kindyak A S 1994 *Tech. Phys.* **64** 99
[9] Damon R W, Eschbach J R 1961 J. *Phys. Chem. Sol.* **19** 308

Theoretical analysis of phase noise spectrum of active ring microwave oscillators based on spin-wave delay lines

A V Drozdovskii, A B Ustinov and B A Kalinikos

Department of physical electronics and technology, Saint Petersburg Electrotechnical University "LETI", Saint Petersburg, 197376, Russia

E-mail: drozdovskyiav@gmail.com

Abstract. A phase noise of the microwave oscillators based on spin-wave delay lines is theoretically investigated. Theoretical results show that spin-wave damping in the delay line limits a phase reduction with increasing time delay. It is possible to find an optimum value of the time delay for each particular value of damping decrement in order to provide minimum phase noise of the generated microwave signal.

1. Introduction

Microwave signal oscillators are widely used in microwave electronics. The simplest oscillator with a carrier frequency in the microwave band typically consists of a microwave amplifier, directional coupler, and a passive circuit with feedback frequency control element. Resonators are usually used for the frequency control. Short-term stability of the generated microwave signal for dielectric resonator oscillators is usually determined from the analysis of phase noise spectrum and by the design of the generator [1]. The frequency tuning of oscillators is performed by tuning the resonance frequency of the resonator [2]. Generation occurs as a result of excitation of the system by noise existing in the components of the oscillator. The main disadvantage of dielectric resonators is their limited tuning range. Any tuning of the resonator adversely affects its quality factor even for phase-locked loop oscillator. Thus, tuning and phase noise are competing parameters. Typically, generators with low phase noise demonstrate relatively small frequency tuning and vice versa.

Another approach to design oscillators is based on the use of delay lines. The delay line operates as a frequency selective element. In this case whole active ring acts as a resonator. Active ring oscillators based on acoustic [3-5] and optical [6,7] delay lines are known for a long time. Their main disadvantages are the difficulty of the fabrication and complexity of analogue tuning over a wide frequency range. Additionally, optical oscillators have large size.

Microwave active ring oscillators based on the spin-wave delay lines are free from these disadvantages [8]. The spin-wave delay line has small sizes (from millimetres to several centimetres) and can be tuned by changing the external magnetic field. A typical range of frequency tuning is 1-20 GHz. Note also that the spin-wave devices can be easily made with conventional photolithography and surface mounted electronic components. It is possible to create miniature tunable low-noise microwave signal generators based on the spin-wave technology. In particular, the possibility for creation of tunable spin-wave delay-line oscillators with low phase noise (on the order of –130 dBc/Hz at 10 kHz offset) was demonstrated [9-13]. Although the first working models of spin-wave microwave generators were built and demonstrated a long time ago, an interest for study and development of microwave oscillators is unabated so far [14].

The aim of this work is detailed study of the influence of spin-wave oscillator design on the phase noise spectrum of generated signal.

2. Theoretical investigation

Figure 1 presents the simplest circuitry of the spin-wave generator which consists of a microwave amplifier (1), a spin-wave delay line (2), and a directional coupler (3). The coupler serves for coupling a small part of microwave signal out of the ring. This ring circuit allows one to generate pulsed and monochromatic microwave signals with different parameters [15]. Generation begins at the resonant frequencies of the active ring. The theoretical description of the resonance properties for more general case of multiferroic delay line is described in details in ref. [16].

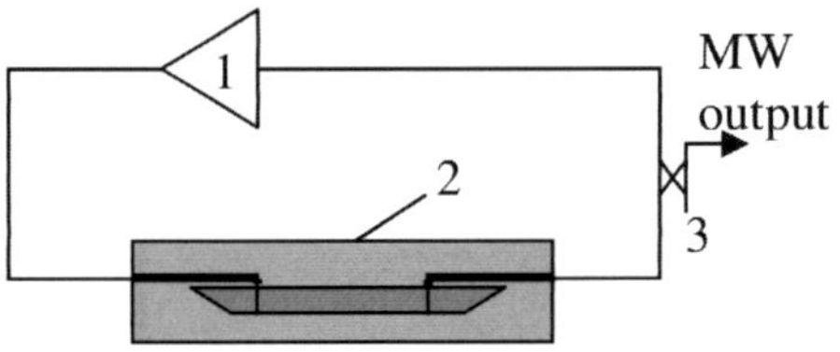

Figure 1. The block diagram of the generator.

Consider briefly the main factors determining a phase noise spectrum of the generator. The simplest spin-wave generator have a ring circuitry consisting of the spin-wave delay line and the microwave amplifier. An equation describing the spectrum of phase noise in single sideband (SSB) above the carrier frequency, taking into account the thermal noise can be written as [17]

$$\mathrm{L}(f_m)=10\lg\left(\frac{GFkT}{2P}\left[\frac{1}{4\pi^2\tau_d^2}\left(\frac{f_a}{f_m^3}+\frac{1}{f_m^2}\right)+\frac{f_a}{f_m}+1\right]\right), \tag{1}$$

where f_m is an the offset frequency from the carrier frequency; G is a gain of the amplifier; F is a noise figure of the amplifier; k is Boltzmann constant; T is temperature; P is signal power; f_0 is a carrier frequency; τ_d is delay time of the microwave signal in the active ring (i.e. this is the time of one circulation of the signal in the ring); f_a is an offset frequency at which the flicker noise of the active element (amplifier) emerges. It is seen from equation (1), that the minimum value of the phase noise is determined by delay time of microwave signal in the active ring, temperature of the device and the frequency at which the flicker noise of the active element occurs.

The term $GFkT/2P$ in equation (1) determines the threshold of the thermal noise of the generated signal and theoretically can be reduced with reduction of the required gain and noise figure of the amplifier [18]. In addition, the amplifier as any active element generates its own noise on the order of $1/f_m$ which could increase the overall phase noise at frequencies of less than detuning frequency f_a, i.e. flicker noise frequency.

Figure 2 shows influence of f_a on phase noise spectrum of the spin-wave delay-line oscillator. The calculations were done for τ_d = 100 ns. As can be seen from the figure, this influence is relatively small. For example, if f_a increases from 0 Hz (the ideal case) up to 100 kHz, the phase noise at offset 10 kHz increases from –105 dBc/Hz to –95 dBc/Hz. The main contribution to the phase noise spectrum is introduced with the time delay of microwave signal travelling around the active ring. The phase noise produced by the mechanism of the time delay is on the order of $1/f_m^2$. As is seen from figure 3, an increase in the delay time τ_d up to 10 μs reduces significantly the level of the phase noise.

Journal of Physics: Conference Series **661** (2015) 012061 doi:10.1088/1742-6596/661/1/012061

Thus, the time delay of the active ring is the most effective tool to control the phase noise spectrum of the ring oscillators.

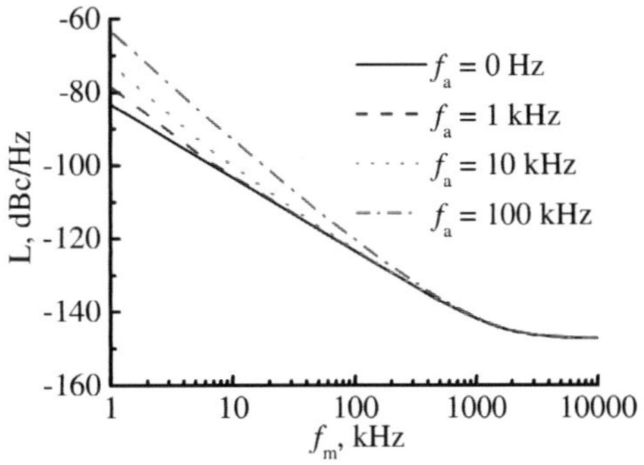

Figure 2. The phase noise spectra calculated for different values of the flicker noise frequency f_a.

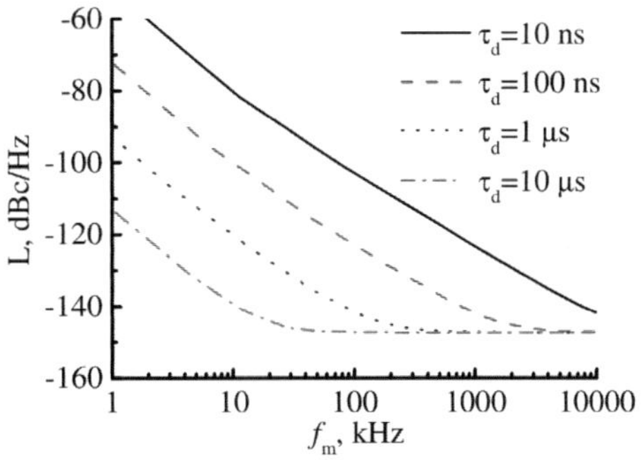

Figure 3. The phase noise spectra calculated for different values of the delay time τ_d.

Figure 4 shows influence of the product of the amplifier parameters F and G on the phase noise spectrum of the spin-wave delay-line oscillator. It is seen that the $F{\cdot}G$ determined the thermal noise floor in the operating frequency range of the amplifier. The increase of the thermal noise floor leads to an increase in the phase noise spectra close to the carrier. Therefore, the lowest phase noise occurs for minimum possible value of the $F{\cdot}G$ at the threshold of microwave signal generation.

Figure 5 shows the phase noise at 10 kHz offset as a function of τ_d for ideal case of lossless delay line. The calculation is based on the equation (1). The product $F{\cdot}G$ assumed to be constant for all values of τ_d. It is evident theoretically that an unlimited increase in τ_d allows to decrease the phase noise level down to very small values.

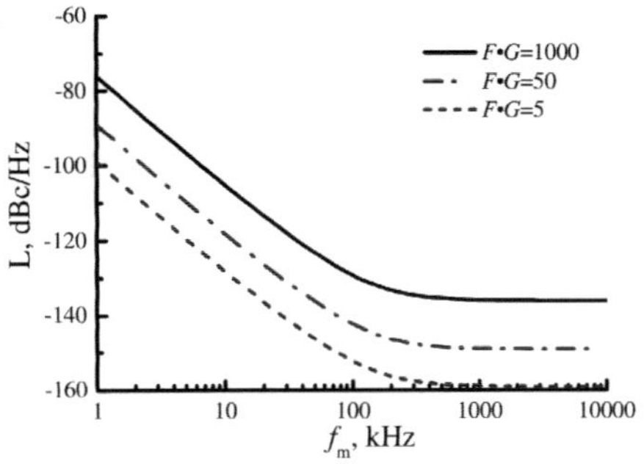

Figure 4. The phase noise spectra calculated for different values of the term GF.

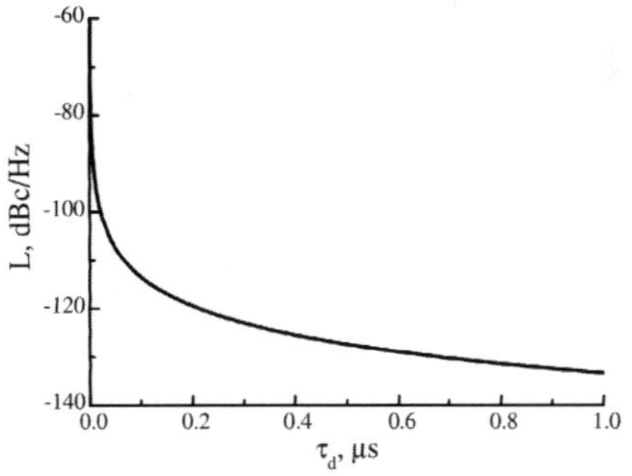

Figure 5. Phase noise at 10 kHz offset as a function of τ_d.

The equation (1) does not take into account losses which take place in passive components of the ring. For our calculations we took into account losses inserted only by the spin-wave delay line. Insertion losses of other passive components are usually relatively small. They can be neglected. Insertion loss of the spin-wave delay line is defined as $\exp(-\omega_r\tau_d)$, where $\omega_r=2\pi|\gamma|\Delta H$, ΔH is a ferromagnetic resonance linewidth, $|\gamma|$ is a module of gyromagnetic ratio. Increasing the spin-wave propagation path with increasing a distance between excitation and reception antennas one can increase delay time τ_d. At the same time it will also result in an increase in the spin-wave loss. In turn, it will require to increase the amplifier gain G. Finally, it will lead to an increase in the phase noise. Thus, we come to the following contradiction. From one hand, an increase in τ_d leads to the phase noise reduction as $1/\tau_d$. From other hand, an increase in τ_d causes an increase of the amplifier gain which is proportional to $\exp(\omega_r\tau_d)$,.

We have studied behaviour of the phase noise of the spin-wave delay-line oscillator taking into account both processes described above. Figure 6 shows the phase noise at 10 kHz offset as a function of τ_d. The curves are plotted for different values of ΔH. It is seen that for any nonzero value ΔH the phase noise reduces first, then in rises. Therefore, for each particular value of ΔH the optimum value of τ_d exists. This value provides minimum phase noise. For example, a microwave oscillator based on the spin-wave delay line made with an yttrium iron garnet (YIG) film having a thickness of 13 μm and ΔH=0.5 Oe is possible to demonstrate minimum phase noise of –110 dBc/Hz at 10 kHz offset for τ_d=0.23 μs.

The optimum delay time τ_{d_opt} for some value ΔH can be determined from the transcendental equation

$$\frac{\omega_r \exp(\omega_r \tau_{d\_opt})}{(\tau_{d\_opt})^2} - \frac{2\exp(\omega_r \tau_{d\_opt})}{(\tau_{d\_opt})^3} - \frac{2M}{(\tau_{d\_opt})^3} = 0, \tag{2}$$

where M is a total loss caused by excitation and reception of spin waves and attenuation of microwave signal in passive components. Figure 7 shows the optimum delay time as a function of ΔH (solid line). The dotted line on the graph shows the minimum possible value of phase noise at 10 kHz offset as a function of ΔH. It is seen that the losses in the active ring oscillator determined mainly by spin-wave damping in the delay line increases the minimum possible value of the phase noise L_{min}. Thus, the phase noise reduction is possible with the use of high-quality YIG films with extremely small value of ΔH. It is evident that an unlimited increase in τ_d is theoretically unlimited decrease in the level of phase noise.

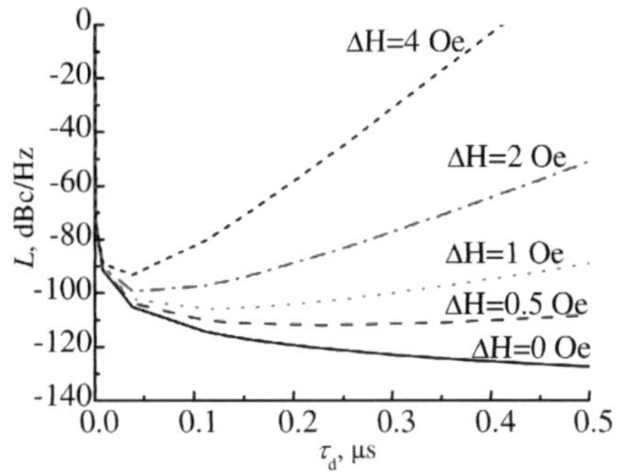

Figure 6. Phase noise at 10 kHz offset as a function of τ_d for different values of ΔH.

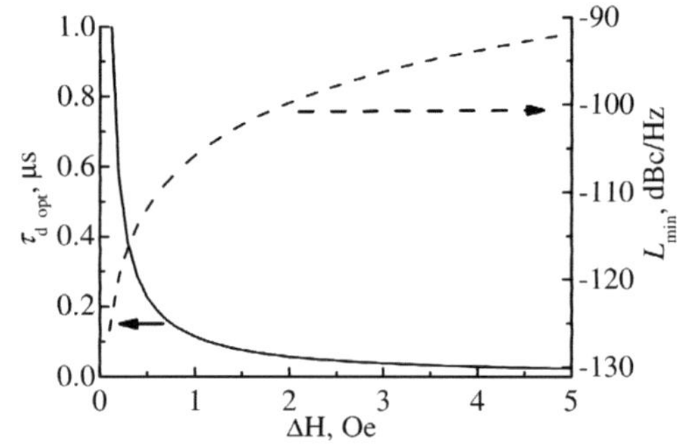

Figure 7. The optimum value of τ_d (solid line) and the corresponding value of the phase noise (dashed line) as a function of ΔH.

3. Conclusions

We analyzed in this work phase noise characteristics of spin-wave delay-line oscillators. The investigation shows that the phase noise is a function of both delay time and damping of spin waves which demonstrate minimum for particular parameters of the oscillator structure. Extremely small value of damping parameter of ferrite materials is necessary to fabricate the low-noise tunable microwave generators.

Acknowledgments

This work was supported in part by the Russian Foundation for Basic Research, Grant of President of the Russian Federation, and by Ministry of Education and Science of the Russian Federation (Project "Goszadanie").

References

[1] Westra J, Verhoeven C, Van Roermund A 2000 *Oscillators and Oscillator Systems* (Kluwer)
[2] Pozar D M 2009 *Microwave engineering* (John Wiley & Sons)
[3] Henaff J. 1979 *1979 Ultrasonics Symposium (IEEE)* 855-860
[4] Eichinger L, Fleischmann B, Russer P 1988 *IEEE Transactions on Microwave Theory and Techniques* **36(12)** 1677-1684
[5] Grajal J 2008 *Microwave and Optical Technology Letters* **50(11)** 2967-2970
[6] Yao X S, Maleki L 1996 *IEEE Journal of Quantum Electronics* **32(7)** 1141-1149
[7] Capmany J, Novak D 2007 *Nature Photonics* **1(6)** 319-330
[8] Stancil D D, Prabhakar A 2009 *Spin waves* (Springer)
[9] Castera J P 1978 *IEEE Transactions on Magnetics* **14** 826-828
[10] Adam J, Daniel M R 1981 *IEEE Transactions on Magnetics* **17(6)** 2951-2956
[11] Ishak W 1983 *Electronics Letters* **19(22)** 930-931
[12] Chen C L, Mahoney L J 1989 *Electronics Letters* **25(3)** 196-197
[13] Ishak W S, Chang K W 1986 IEEE *Transactions on Microwave Theory and Techniques* **34(12)** 1383-1393
[14] Slavin A, Tiberkevich V 2009 *IEEE Transactions on Magnetics* **45(4)** 1875-1918
[15] Kondrashov A V, Ustinov A B, Kalinikos B A and Benner H 2008 *Tech. Phys. Lett.* **34** 492
[16] Nikitin A A et al. 2012 *Technical Physics* **57(7)** 994-997
[17] Rubiola E 2008 *Phase noise and frequency stability in oscillators* (Cambridge University Press)
[18] Boudot R, Rubiola E 2012 *IEEE Transactions on Ferroelectrics and Frequency Control* **59(12)** 2613-2624

Phase noise management of spin-wave delay-line oscillators

A V Drozdovskii and A B Ustinov
Department of physical electronics and technology, Saint Petersburg Electrotechnical University "LETI", Saint Petersburg, 197376, Russia

E-mail: drozdovskyiav@gmail.com

Abstract. A phase noise of microwave oscillators having an active ring circuitry with a spin-wave delay line is theoretically and experimentally investigated. The delay line was made with yttrium iron garnet (YIG) film epitaxially grown on gadolinium gallium garnet substrate. Obtained results demonstrate a management of the oscillator phase noise with a variation of the distance between antennas used for excitation and reception of spin waves in the YIG film.

1. Introduction

It is well known, that active ring circuits containing a spin-wave delay line based on yttrium iron garnet (YIG) film and a microwave amplifier can generate microwave signals of various types. Thus, it is possible to generate highly monochromatic signals [1-3], periodic sequence of nonlinear pulses, in particular, solitons [4-8], and chaotic signals [9-13]. Carrier frequency of the generated stationary signals can be tuned by magnetic field. Phase noise of delay-line oscillators can be reduced with an increase in delay time or, in other words, with an increase in slope of phase-frequency characteristic of the delay line. This is one of the advantages of the YIG-film delay-line oscillators in comparison with oscillators based on YIG-sphere resonators. Therefore, delay-line oscillators have potentially lower phase noise in comparison with resonator oscillators.

Physically clear that, an increase in time delay of spin-wave delay line increases its insertion loss also. It requires to increase the ring gain G required for beginning a self-generation. It could increase the phase noise of the oscillator. The aim of this work was is experimental and theoretical investigation of the influence of the time delay on the phase noise spectrum of the spin-wave delay-line oscillator.

2. Design of the spin-wave delay-line oscillator

A schematic diagram of the experimental oscillator is shown in figure 1. Meander-type microstrip antennas were used for excitation and reception of the spin waves in the YIG film. The width of the microstrip and the distance between the antenna elements are 50 μm and 260 μm, respectively. The distance between the antennas d was able to be varied in the range of 4-10 mm. The YIG-film waveguide was positioned on the microstrip antennas. The film was grown on a gadolinium gallium garnet substrate by a method of liquid phase epitaxy. The film had a thickness of 13 μm and a saturation magnetization of 1950 Gs. A half of ferromagnetic resonance line-width ΔH measured for frequency of 5300 MHz was 0.5 Oe. The film demonstrated unpinned surface spins. To avoid possible influence of reflected waves on the experimental results, the ends of the film were acute. The spin-wave delay line was placed between the poles of an electromagnet created a uniform magnetic field of

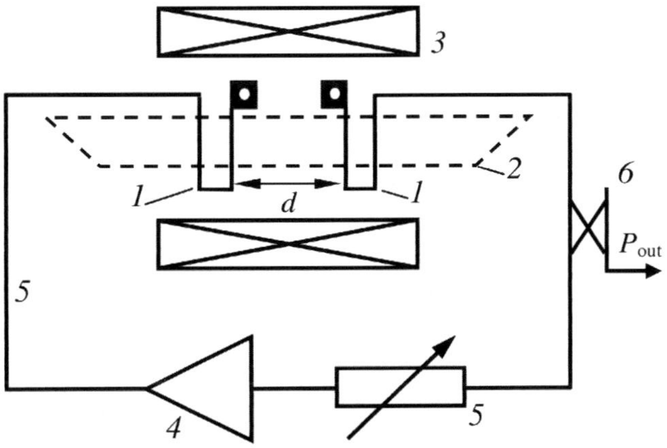

Figure 1. Schematic diagram of the oscillator. The units are designated as follows: (*1*) meander-type antennas; (*2*) YIG film; (*3*) electromagnet; (*4*) microwave amplifier; (*5*) variable attenuator; (*6*) directional coupler.

1150 Oe. Such mutual orientation of the field and the propagation direction of spin waves corresponds to the excitation of surface spin waves in the YIG film.

The experimental oscillator also included the microwave amplifier Mini-Circuits ZX60-8080E-S+. The total length of the ring circuit was less than 1 m. Variable attenuator was used to control the net gain of the ring. The directional coupler led out small portion of self-generated microwave signal for analysis. Spectrum analyzer Tectronix RSA6114B was utilized for measurements of the spectrum and phase noise of the generated continuous wave signal.

3. Results and discussion

It is well known, that microwave generation begins in the active ring if two following conditions are satisfied. The first condition is for the ring gain *G*:

$$G \geq \exp(\omega_r \tau_d) + M \,, \tag{1}$$

where $\omega_r = 2\pi|\gamma|\Delta H$, ΔH is a ferromagnetic resonance linewidth, $|\gamma|$ is a module of gyromagnetic ratio, τ_d is a time of one circulation of the signal in the ring (we will call it "a ring delay time"), *M* is a total loss caused by excitation and reception of spin waves and attenuation of microwave signal in passive components. The second condition is for phase:

$$\varphi_e(\omega) + \varphi_{sw}(\omega) = 2\pi n \,, \tag{2}$$

where φ_e is a phase shift accumulated in electrical part of the ring, φ_{sw} is a phase shift accumulated by spin waves in the delay line, n is any integer.

An example of experimentally obtained transmission characteristic of the spin-wave delay line for d = 4 mm is shown in figure 2. The arrow on the graph shows the frequency at which the above described conditions were satisfied and continuous wave (CW) generation was observed. This frequency equals 5390 MHz.

The ring delay time τ_d depends mainly on the delay time of the spin-wave delay line. Delay time of the spin-wave delay line is determined by a distance between the excitation and reception antennas as $\tau_d = d/V_{gr}(\omega)$, where $V_{gr}(\omega)$ is a group velocity of the spin waves. The experimental dependence of delay time versus spin wave propagation distance for carrier frequency f_0 = 5390 MHz is shown in figure 3. As the distance d increases the losses in the spin-wave delay line increases exponentially as the $\exp(\omega_r \tau_d)$. The spin wave losses as a function of the distance d for f_0 = 5390 MHz is shown in figure 4. To compensate the increase of insertion loss the variable attenuator was used.

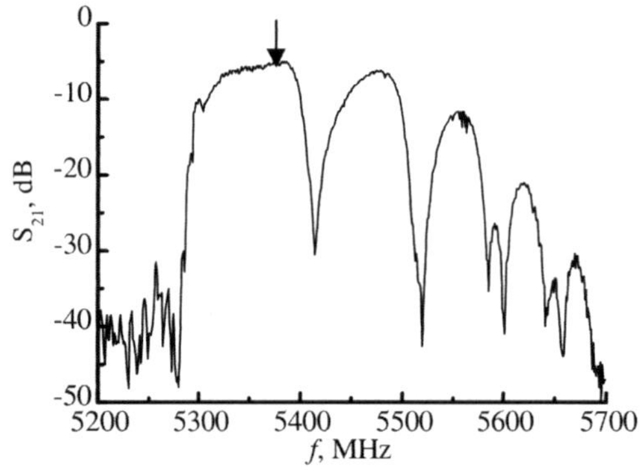

Figure 2. The spin-wave delay line insertion loss.

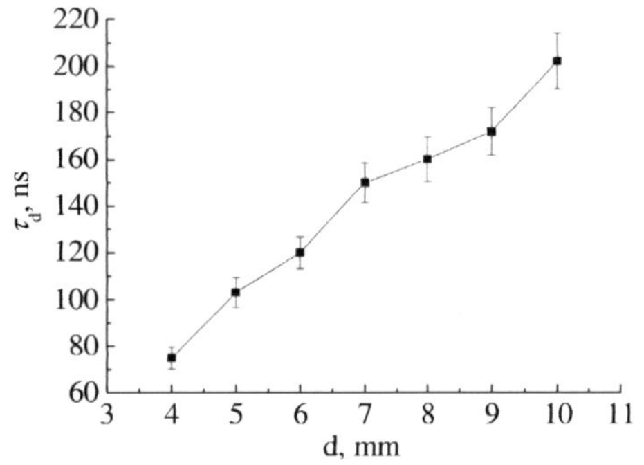

Figure 3. Delay time as a function of distance for f_0 = 5390 MHz.

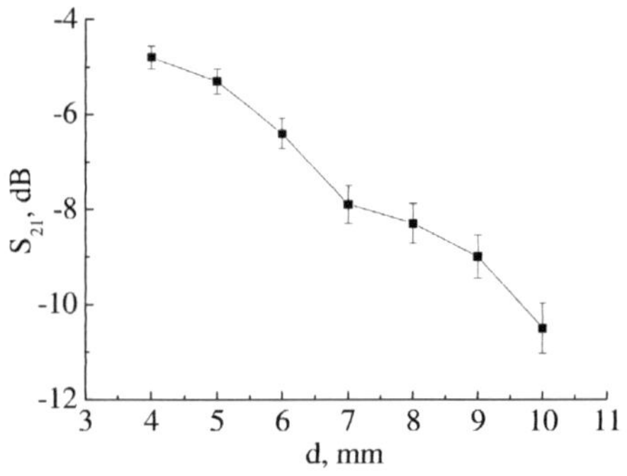

Figure 4. Insertion loss as a function of distance for f_0 = 5390 MHz.

The measurements of the phase noise were carried out for three values of d of 4 mm, 7 mm, and 10 mm corresponding to the delay time τ_d of 80 ns, 140 ns, and 200 ns. Phase noise spectrum of the generated CW microwave signal is shown in figure 5. The measured value of the phase noise at 10 kHz offset was of -100 dBc/Hz.

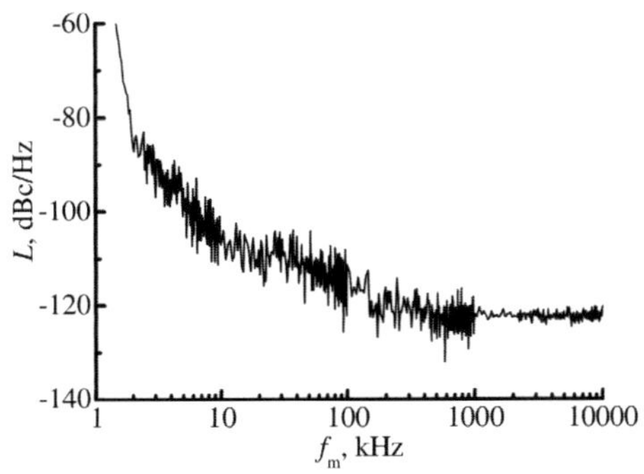

Figure 5. The experimental spectrum of phase noise.

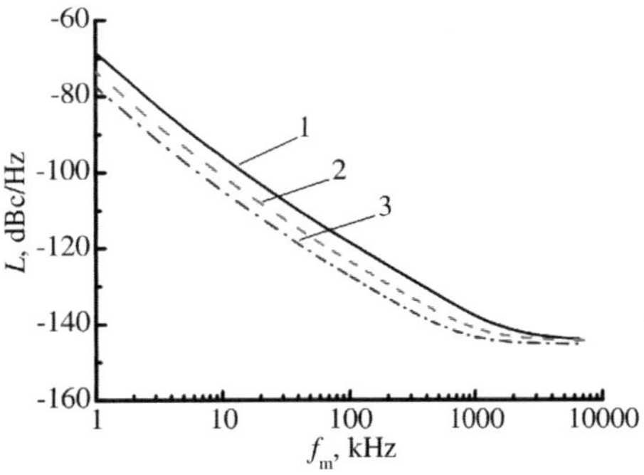

Figure 6. The theoretical spectrum of phase noise.

Consider briefly factors influencing on the phase noise spectrum of the oscillator. An equation describing the spectrum of phase noise in single sideband above the carrier frequency and taking into account the thermal noise can be written as [14]

$$L(f_m) = 10\lg\left(\frac{GFkT}{2P}\left[\frac{1}{4\pi^2\tau_d^2}\left(\frac{f_a}{f_m^3}+\frac{1}{f_m^2}\right)+\frac{f_a}{f_m}+1\right]\right), \tag{3}$$

where f_m is an the offset frequency from the carrier frequency; G is a gain of the amplifier; F is a noise figure of the amplifier; k is Boltzmann constant; T is temperature; P is signal power; f_0 is a carrier frequency; τ_d is delay time of the microwave signal in the active ring; f_a is an offset frequency at which the flicker noise of the active element (amplifier) emerges. It is seen from equation (3), that the minimum value of the phase noise is determined by delay time of microwave signal in the active ring, temperature of the device and the frequency at which the flicker noise of the active element occurs. The term *GFkT/2P* in equation (3) determines the threshold of the thermal noise of the generated signal and theoretically can be reduced with reduction of the required gain and noise figure of the amplifier [15].

Numerical simulation of phase noise spectra was carried out for parameters of the experimental oscillator. The results are shown in figure 6. The spectra were calculated for three experimental values of *d*. They were in a good agreement with measured spectra. Thermal noise level obtained in the calculations was much less than experimental one. Relatively large experimental thermal noise of -120 dBc/Hz was due to intrinsic noise of spectrum analyzer. Theoretical analysis predicted a phase noise reduction with an increase in *d*. It was confirmed by analysis of the experimental spectra (see figure 7). The experimental data demonstrate that an increase in *d* from 4 mm to 10 mm reduces the phase noise at 10 kHz offset from -100 to -104.5 dBc/Hz.

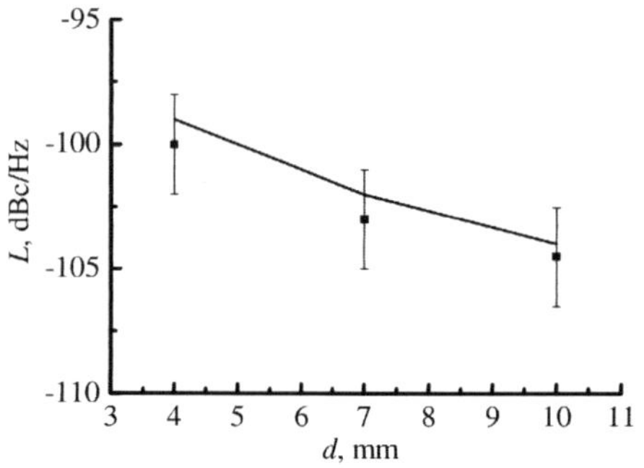

Figure 7. The theoretical (solid curve) and experimental (dots) value of phase noise at 10 kHz offset.

4. Conclusions

The influence of the time delay of the spin-wave delay line on the phase noise spectrum of the microwave oscillator is investigated. It is demonstrated that an increase in the distance between the excitation and reception spin-wave antennas leads to a reduction of the phase noise of the oscillator. A good agreement between theory and experiment is shown.

Acknowledgments

This work was supported in part by the Russian Foundation for Basic Research, Grant of President of the Russian Federation, and by Ministry of Education and Science of the Russian Federation (Project "Goszadanie").

References

[1] Castera J P 1978 *IEEE Transactions on Magnetics* **14** 826-828
[2] Carter R L, Owens J M 1981 *1981 IEEE Microwave Symposium Digest* 383-385
[3] Ishak W 1983 *Electronics Letters* **19(22)** 930-931
[4] Fetisov Y K, Kabos P, Patton C E Bistable 1996 *Electronics Letters* **32(20)** 1894-1895
[5] Kalinikos B A, Kovshikov N G, Patton C E 1998 *Physical review letters* **80(19)** 4301
[6] Kalinikos B A, Scott M M, Patton C E 2000 *Physical review letters* **84(20)** 4697
[7] Scott M M, Kalinikos B A, Patton C E 2001 *Applied Physics Letters* **78(7)** 970-972
[8] Grishin S V, Beginin E N, Morozova M A, Sharaevskii Y P, Nikitov S A 2014 *Journal of Applied Physics* **115(5)** 053908
[9] Demidov V E, Kovshikov N G 1997 *Journal of Experimental and Theoretical Physics Letters* **66(4)** 261-265
[10] Mingzhong Wu, Kalinikos B A, Patton C E 2005 *Physical review letters* **95** 237202
[11] Kondrashov A V, Ustinov A B, Kalinikos B A, Benner H 2008 *Technical Physics Letters* **34** 492
[12] Wang Zihui, Aaron H, Anderson J Q, Wei Tong, Mingzhong Wu, Carr L D, Eykholt R, Kalinikos B A 2011 *Physical review letters* **107** 114102
[13] Grishin S V, Sharaevskii Y P, Nikitov S A, Romanenko D V 2013 IEEE *Transactions on Magnetics* **49(3)** 1047-1054
[14] Boudot R, Rubiola E 2012 *IEEE Transactions on Ferroelectrics and Frequency Control* **59(12)** 2613-2624
[15] Rubiola E 2008 *Phase noise and frequency stability in oscillators* (Cambridge University Press)

Theoretical analysis of microwave magnonic crystals of finite length

A V Drozdovskii[1], A A Nikitin[2] and A B Ustinov[1]

[1]St. Petersburg Electrotechnical University, St. Petersburg, 197376 Russia
[2]ITMO University, St. Petersburg, 197101 Russia

E-mail: drozdovskyiav@gmail.com

Abstract. Propagation of spin waves in a one-dimensional magnonic crystal to a finite distance have been studied theoretically. It was shown that, in contrast to the ideal spatially unlimited magnonic crystals with zero loss, there are no band gaps in the real magnonic crystals. The effective spectrum of spin waves at frequencies close to the Bragg resonances strongly depends on the magnonic crystal length.

1. Introduction

Propagation of waves in physical systems and media with periodically varying parameters has been studied during many years (see, e.g., [1-3] and references therein). The interest in studies of this kind is due to the possibility of creating waveguide media with advanced dispersion properties, which could be further used for development of new devices.

In recent years, an increased interest in studies of magnetic spatially periodic structures has been observed [4,5]. After the studies [6,7] were published, these objects are frequently named "magnonic crystals" (MCs) or "magnetic superlattices". Originally, these structures were studied in the late 1970s and early 1980s [8-15]. The reason for the present-day interest in magnonic crystals is that, on the one hand, media of this kind are a convenient experimental model for studies of linear and nonlinear wave effects and methods of signal generation and conversion and, on the other hand, these crystals can be used to develop a new generation of microwave devices [15-22].

Studies of spin-wave phenomena in spatially periodic structures are based on the dispersion laws. A theory of the spectrum is commonly developed in terms of the model of an infinite periodic structure with zero loss. As a result, the eigenwave spectrum $\omega(k)$ has band gaps at certain frequencies. The propagation of spin waves (SWs) at these frequencies is prohibited and the dispersion characteristic has discontinuities. However, magnonic crystals of finite length with nonzero loss are used in practice. Therefore, there are, instead of band gaps, bands in which SWs propagate, even though their decay is strong. The frequency dependence of the insertion loss (amplitude-frequency characteristic) shows alternation of pass bands with small and large losses. The latter bands are hereinafter called “stop bands”. The dispersion characteristic has no discontinuities observed in the case of an infinite structure, but there are strong deviations in the dispersion law near the stop band boundaries. These specific features of the dispersion can affect, e.g., the manner in which SW envelope Bragg solitons are formed [22-25].

It should be emphasized that, despite the variety of the existing approaches to description of properties of magnonic crystals, the dispersion characteristics (spectra) of SWs in magnonic crystals

Journal of Physics: Conference Series **661** (2015) 012063 doi:10.1088/1742-6596/661/1/012063

have been theoretically studied in detail only for the case of infinite periodic magnetic structures [26,27].

The goal of our study is theoretical investigation of insertion loss and dispersion law of magnonic crystals of finite length.

2. Theoretical investigation

The structure under study is shown schematically in figure 1. The structure is a single-crystal iron-yttrium garnet (YIG) film waveguide having grooves on the film surface. It is important to note that the position of the stop bands in the frequency spectrum depends on the configuration of the periodic structure, materials from which it is fabricated, and magnitude and direction of the external magnetic bias field. The parameters of the studied structures were chosen so that several stop bands due to the Bragg reflection of SWs were situated in that part of the dispersion characteristic which is determined by the dipole-dipole mechanism.

For numerical simulation of SW spectra in magnonic crystals we used the method of wave transfer matrices [28, 29]. This approach makes it possible to take into account the finite dimensions and the decay of waves and to study their influence on SW spectrum. Let us briefly consider the main relations used in the present study.

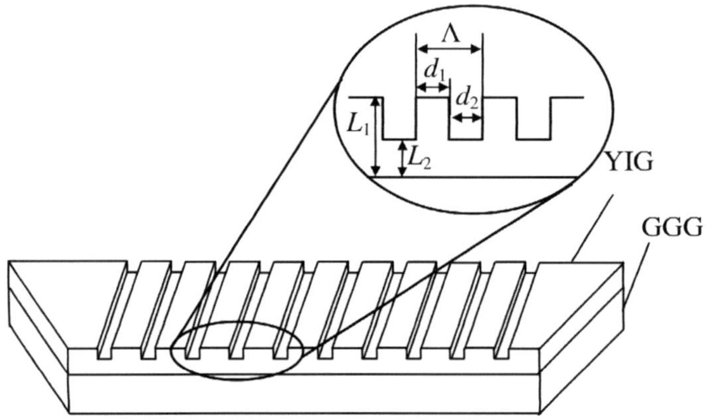

Figure 1. Schematic representation the magnonic crystal.

The structure had a period Λ = 400 μm, groove length d_2 = 50 μm, and groove depth $\Delta L=L_1-L_2=$ =2μm. The initial film thickness L_1 was 12 μm. The saturation magnetization of the ferrite was 1750 Gs. We assumed in the calculation that the grooves are oriented perpendicularl to the SW propagation direction. The external magnetic bias field with strength of 1205 Oe was assumed to be directed within the film plane along the grooves. This configuration corresponds to propagation of surface SWs in a the magnonic crystal. The calculation was done for the ideal case in which the ferromagnetic resonance linewidth and, consequently, the SW damping were taken to be zero.

Figure 2 shows the results of a numerical simulation of the influence of a magnonic crystal on its insertion loss and dispersion characteristics. The insets of figure 2 show scaled-up parts of the insertion loss and dispersion characteristics, which correspond to the rejection bands of the magnonic crystal. Curves 1 in figure 2 corresponds to the case of a structure with a single period. It can be seen in the figure that the damping introduced in the SW propagation at all frequencies is nearly zero and the dispersion characteristic is similar to that of a homogeneous YIG film. Curves 2 and 3 are plotted for the structures comprising 15 and 1000 periods, respectively.

The run of these dependences can be explained as follows. Outside the rejection bands, when the phase accumulation on a period of the structure is not a multiple of π, there is no effective reflection of SWs from the periodic structure. The resulting phase accumulation is proportional to the distance travelled by the wave. A transmission band is observed in the MC spectrum, and the dispersion characteristic of the MC is similar to that of a homogeneous film.

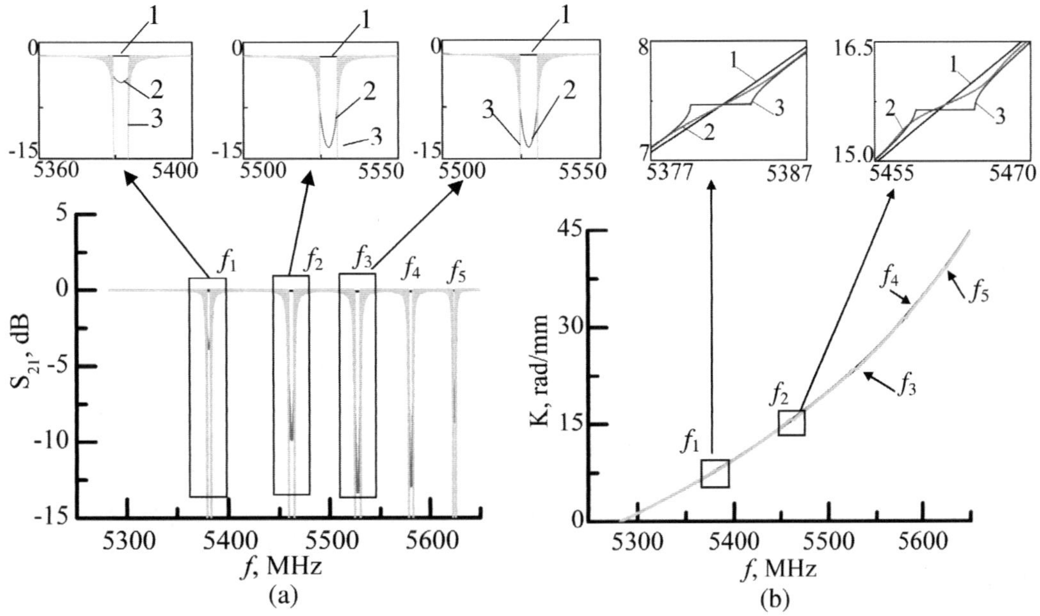

Figure 2. The insertion loss (a) and the dispersion characteristic (b) of magnonic crystal.

In the case when an SW propagates with a carrier frequency close to the stop band, a certain role is played, together with the dispersion of the medium, by the periodicity-caused dispersion. When the Bragg reflection condition is approximately satisfied, the wave begins to experience reflection and thereby decays, with its energy transferred to back-reflected waves. This is equivalent to the SW wave number becoming complex. This fact must be manifested in the frequency transmission coefficient of an infinite magnonic crystal as band gaps in which propagation of waves is impossible. However, it can be seen in figure 2(a) that, in the case of an MC with finite length, the transmission characteristic has, instead of band gaps, bands with comparatively large, but finite insertion loss, i.e., stop bands. With increasing length of a periodic magnetic structure, the loss within the band gap also grows as $\exp(-k''d)$, where k'' is the damping decrement within the stop band, and d is the length of the periodic structure. The dependences of the loss introduced within the stop band on the structure length are shown on the logarithmic scale for the first four rejection bands in figure 3(a).

The waves re-reflected by the periodic structure make a substantial contribution to the formation of phase-frequency and, consequently, dispersion characteristics of a magnonic crystal. In the dispersion characteristic, the stop bands appear as "bending" bands (Figure 2(b)). With increasing number of periods, these bands change their slope ratio $\Delta f/\Delta K$ (see figure 3(b)). For example, at n = 1, the slope ratio of the dispersion characteristic within the rejection band corresponds to that for the regular YIG film and $\Delta f/\Delta K \to 8$ mm/(rad s). At n = 1000, the plot of $\Delta f/\Delta K$ is nearly vertical and has a value on the order of 1000 mm/(rad s), and the dispersion characteristic is almost similar to the dispersion characteristics corresponding to infinite periodic structures reported, e.g., in [6, 7, 24].

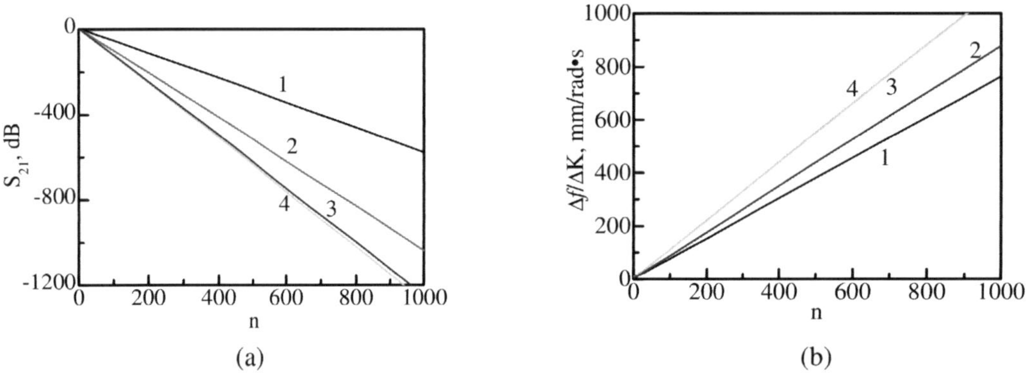

Figure 3. (a) The insertion loss at frequencies of stop bands MC. (b) The slope of dispersion characteristic of the magnonic crystal at frequencies of stop bands MC. Digits is the number of Brillouins band gap.

3. Conclusions

The method of wave transfer matrices was used for numerical analysis of the wave spectrum in a spatially periodic magnetic structure. An influence of the length in the formation of the dispersion and transmission characteristics of a one-dimensional magnonic crystal was studied. It was shown that an increase in the length of the periodic magnetic structure results in a rise in the insertion loss and in the slope of the dispersion characteristic at the Bragg resonance frequencies. A decrease in the length of the structure make smaller the contribution of waves re-reflected by the periodic structure to the formation of the magnonic crystal characteristics. The data obtained can be used for studies of linear and nonlinear spin wave processes in periodic magnetic structures.

Acknowledgments

This work at SPbETU was supported in part by the Russian Foundation for Basic Research, Grant of President of the Russian Federation, and by Ministry of Education and Science of the Russian Federation (Project "Goszadanie"). The work at ITMO University was supported by the Government of the Russian Federation (Grant 074-U01).

References

[1] Elachi C 1976 *Proc. of the IEEE* **64(12)** 1666-1698
[2] Yariv A, Yeh P 1984 *Optical waves in crystals* (Wiley, New York,).
[3] Brillouin L 2003 *Wave propagation in periodic structures* (Dover Publications, USA).
[4] Serga A A, Chumak A V, Hillebrands B 2010 *J. Phys. D: Appl. Phys.* **43** 264002
[5] Kruglyak V V, Demokritov S O, Grundler D 2010 *J. Phys. D: Appl. Phys.* **43** 264001
[6] Nikitov S A, Tailhades Ph, Tsai C S 2001 *JMMM* **236** 320-330
[7] Gulyaev Y V, Nikitov S A, Zhivotovskii L V, Klimov A A, Tailhades P, Presmanes L, Filimonov Y A 2003 *JETP Lett.* **77** 567-570
[8] Sykes C G, Adam J D, Collins J H 1976 *Appl. Phys. Lett.* **29** 388-391
[9] Owens J M, Smith C V, Lee S N 1978 *IEEE Trans. on magn.* **MAG-14** 820-825
[10] Seshadri S R 1979 *IEEE Trans. MTT* **MTT-27** 199-204
[11] Gulyaev Y V, Nikitov S A, Plesskii V P 1980 *Sov. Phys. Solid State* **22** 1651
[12] Kostylev M P, Stashkevich A A 2010 *Phys. Rev. B.* **81(5)** 054418
[13] Voronenko A V, Gerus S V 1984 *Pisma v zhurnal teknicheskoi fiziki* **10** 746-748
[14] Tsai C S, Young D, Nikitov S A 1998 *J. Appl. Phys.* **84** 1670

[15] Gouzerha J, Stashkevich A A, Kovshikovb N G, Matyushevb V V, Desvignesa J M 1991 *JMMM* **101(1)** 189-190
[16] Ustinov A B, Drozdovskii A V, Kalinikos B A 2010 *Appl. Phys. Lett.* **96** 142513
[17] Grishin S V, Beginin E N, Dulin Y V, Nikitov S A, Sharaevskii Y P 2012 *Tech. Phys.Lett.* **38(7)** 638-641
[18] Zhu Y, Chi K H, Tsai C S 2014 *Appl. Phys. Lett.* **105** 022411
[19] Grishin S V, Beginin E N, Sharaevskii Yu P, Nikitov S A 2013 *Appl. Phys. Lett.* **103** 022408
[20] Grishin S V, Beginin E N, Morozova M A, Sharaevskii Yu P, Nikitov S A J. 2014 *Appl. Phys. Lett.* **115** 053908
[21] Inoue M, Baryshev A, Takagi H, Lim P B, Hatafuku K 2011 *Appl. Phys. Lett.* **98** 132511
[22] Kryshtal R G, Medved A V 2012 *Appl. Phys. Lett.* **100** 192410
[23] Chen Niu-Niu, Slavin A N, Cottam M G 1993 *Phys. Rev. B* **47** 8667
[24] Ustinov A B, Kalinikos B A, Demidov V E, Demokritov S O 2010 *Phys. Rev. B* **81** 180406(R)
[25] Drozdovskii A V, Cherkasskii M A, Ustinov A B, Kovshikov N G, Kalinikos B A 2010 *JETP Letters* **91(1)** 16
[26] Kolodin P A, Hillebrands B 1996 *JMMM* **161** 199-202
[27] Krawczyk M, Puszkarski H 2008 *Phys. Rev. B* **77** 054437
[28] Seshadri S R 1986 *J. Appl. Phys.* **60** 1758
[29] Chumak A V, Serga A A, Wolff S, Hillebrands B, Kostylev M P 2009 *Appl. Phys. Lett.* **94** 172511

Journal of Physics: Conference Series **661** (2015) 012064 doi:10.1088/1742-6596/661/1/012064

Dual control of spin-electromagnetic wave chaos in active ring oscillators based on artificial multiferroics

A V Kondrashov[1], A B Ustinov[1], E Lähderanta[2,3], O V Pakhomov[2], A A Nikitin[2,3] and B A Kalinikos[1]

[1]St. Petersburg Electrotechnical University, St. Petersburg, 197376 Russia
[2]ITMO University, St. Petersburg, 197101 Russia
[3]Lappeenranta University of Technology, Lappeenranta, 53850 Finland

E-mail: kondrashov_av@yahoo.com

Abstract. Properties of spin-electromagnetic wave chaos developed in active ring oscillators have been investigated. A multiferroic structure composed of yttrium iron garnet film and barium strontium titanate (BST) slab served as a nonlinear dispersive medium of the oscillator. Dual control of the fractal dimension of the chaotic signal attractor was realized by variation of the ring gain and dielectric permittivity of the BST slab.

1. Introduction

For many years, investigations of nonlinear dynamics have attracted a great research interest. Such fascinating nonlinear wave phenomena as solitons and self-modulation instability have received perhaps the most attention. They were studied for waves of different nature (see. e.g. [1-10]). In recent years, chaotic phenomena have become important due to their great potential for application in telecommunications [11-13]. In this connection, such fundamental questions as a route from stationary excitations to chaos [14-20], chaos control [18,21], synchronization of chaotic generators [22], and chaotic dynamics of solitons [19,20,23-26] are under extensive investigations especially for optical fiber ring lasers [12] and ferrite-film active rings [27].

At present, an increased research interest in magnetoelectric interactions in different kind artificial multiferroic structures is evident [28-30]. Recently, nonlinear properties of spin-electromagnetic waves were begun to study [31-34]. In particular, experimental observation of dark envelope solitons and chaos was reported in [34].

The aim of this work is analysis of spin-electromagnetic wave chaos developed in active ring oscillators based on multiferroic (ferrite-ferroelectric) structure.

2. Experimental set-up

Figure 1 shows the experimental structure of the feedback active ring auto-oscillator. For the present experiments we used an artificial multiferroic waveguide made in the form of bilayer composed of a magnetic yttrium iron garnet (YIG) film strip and a ferroelectric barium strontium titanate (BST) ceramic slab. The YIG strip had lateral dimensions of 2 by 40 mm and a thickness of 5.7 μm. The BST slab had lateral dimensions of 5 by 5 mm and a thickness of 500 μm. Two microstrip antennas were placed over the YIG strip at a distance of 6.7 mm. The output antenna was connected to the input antenna through an active feedback loop consisting of a wideband microwave amplifier and a variable precision attenuator controlling the net gain coefficient G in the ring. The bilayer was magnetized in its plane and transverse to the wave propagation direction. This orientation of the magnetic field provided the propagation of quasi-surface spin-electromagnetic waves in the multiferroic waveguide.

The microwave excitations were propagating in the ring as spin waves, hybrid spin-electromagnetic waves, and electromagnetic waves. Thus, the electromagnetic wave travelling in the feedback loop transforms to a surface spin wave by the input antenna. Then, the excited spin wave propagates short distance in the YIG film and enters the multiferroic structure. This is the region where the YIG film is in contact with the BST slab. At the border of this region, the surface spin wave transforms into the quasi-surface spin-electromagnetic wave, which then propagates in the YIG-BST multiferroic structure. The spin-electromagnetic wave propagating through the structure reaches the YIG film where it transforms back to the spin wave and is received by the output microstrip antenna to be transformed to electromagnetic wave microwave signal. This signal is amplified in the feedback loop (see figure 1).

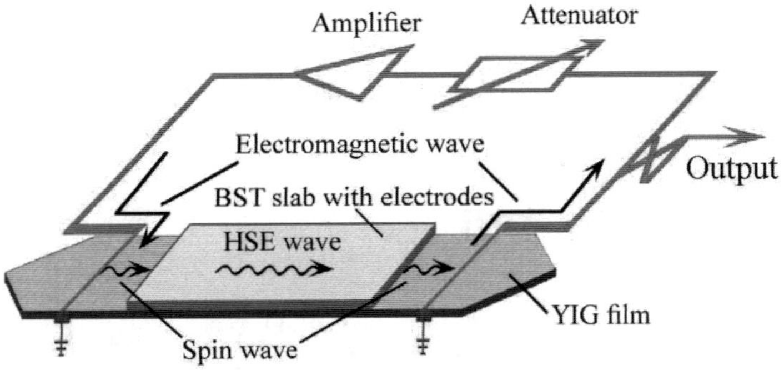

Figure 1. Block diagram of the experimental oscillator.

3. Experimental results and discussion

Increase of the ring gain coefficient up to G = 0.3 dB led to change of the active ring self-generation regime. Increased power of the microwave signal circulating in the ring led to appearance of a nonlinear interaction of hybrid waves in ferrite-ferroelectric structure. Due to the interaction the additional regularly spaced modes appeared in the frequency spectrum of the signal. In a time domain periodical amplitude modulation of signal was observed. On the figure 2(a,b) spectrum and waveform are presented. A carrier frequency was f_0 = 7.005 GHz. Multicomb frequency spectrum has a separation between Fourier-harmonics of Δf = 3.56 MHz. A repetition period of the corresponding nonlinear pulses was measured to be 273 ns.

For further analysis multy-dimensional phase portrait was reconstructed. Phase trajectory of corresponding regime was obtained using waveform envelope by the standard time-delay method [35] as follows:

$$\boldsymbol{x}(t) = \{U(t), U(t+\tau), ..., U(t+(d-1)\tau)\}, \tag{1}$$

where τ is time delay, d is a phase space dimension and U(t) is the amplitude of analyzed signal. In our calculation maximum dimension of phase space was set to 25. Phase portrait of the periodic regime in two-dimensional phase space is represented on figure 2(c). The form of the phase portrait was determined by an attractor, which attracts all phase trajectories in the phase space. In the case of periodic signal the trajectory represented a loop. It apparently corresponded to the attractor in the form of limit cycle. According to the work [35] this phase portrait could be used for calculation of attractor numerical parameters such as fractal dimension, Lyapunov exponents, Kolmogorov entropy etc. Fractal dimension was calculated to prove the attractor topology. Calculation was done by the algorithm of Grassberger-Procaccia [36,37]. First correlation sums for a number of phase space dimensions were calculated according to next formula:

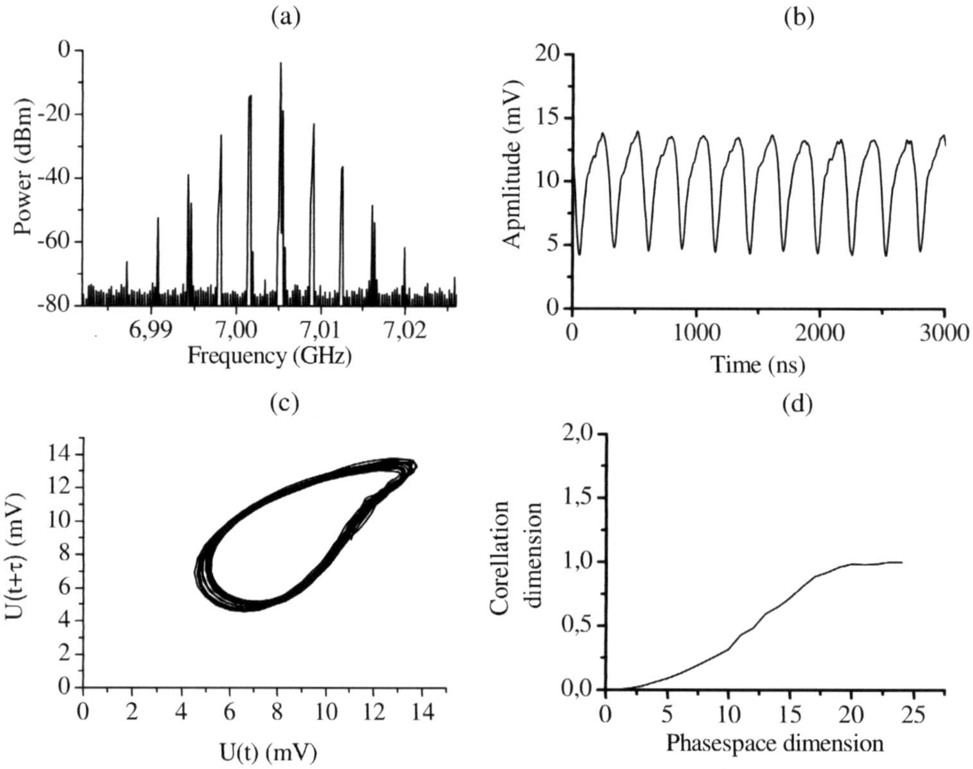

Figure 2. Periodic signal generation regime characteristics: (a) spectrum, (b) envelope waveform, (c) 2D phase portrait, (d) dependence of correlation dimension on phase space dimension (G = 0.35 dB, E = 0 V/μm).

$$C(\varepsilon) = \frac{2}{N(N-1)} \sum_{i=1}^{N} \sum_{j=1}^{N} \theta(\varepsilon - \|\mathbf{x}_i - \mathbf{x}_j\|), \quad (2)$$

where **x** is the coordinate of the phase trajectory point, N is a number of point, ε is a characterization distance between points, and θ is the Heaviside step function. Logarithmic dependence of the correlation sum on the characterization distance has a linear part. Thus

$$C(\varepsilon) = \varepsilon^{D_c}, \quad (3)$$

where D_c is so-called correlation dimension of the attractor embedded into the phase space with certain dimension d. On the figure 2(d) dependence of correlation dimension on phase space dimension calculated by described method is represented. Increase of the phase space dimension led to increase of the correlation dimension. The dependence has a saturation level. The saturation shows that the *whole* attractor is embedded into the phase space. This level is called fractal dimension D_f. Fractal dimension value characterizes attractor topology. In the case of periodical signal fractal dimension was found to be D_f= 1. It proves that while self-generation of periodical modulated signal in the phase space appeared limit cycle.

Further increase of ring gain coefficient up to 0.82 dB led to enrichment of spectrum and complicacy of amplitude modulation. Calculation of fractal dimension shown that the observed regime is quasiperiodical regime with fractal dimension D_f = 2. In this regime in the phase space attractor in the form of 2D-torus was observed.

Values of $G > 1.1$ dB corresponded to a self-generation threshold of the signal with broadband spectrum. Each self-generated harmonic were broadened. In time domain nonperiodic signal was observed. Further increase of gain coefficient made broadened parts of the spectrum wider until broadband basement appeared in the frequency spectrum. During this process waveform modulation became more complicated. On the figure 3(a,b) spectrum and waveform corresponding to G = 1.3 dB are shown. The broadband spectrum is 100 MHz wide. According to described above methods phase portrait and fractal dimension were investigated. On the figure 3(c) 2D phase portrait is shown. One can see that phase trajectories are mixed. It could be caused by appearance of strange attractor, but also by influence of a noise. To investigate whether the nature of dynamic is determined or not, fractal dimension was calculated. On the figure 3(d) saturation of correlation dimension could be seen. It means that the dimension of attractor is finite, and observed dynamics is a dynamical chaos. The value of the fractal dimension for G = 1.3 dB were found to be D_f = 8.5. Characterization of a noise signal would not demonstrate saturation of correlation dimension and fractal dimension would be infinite.

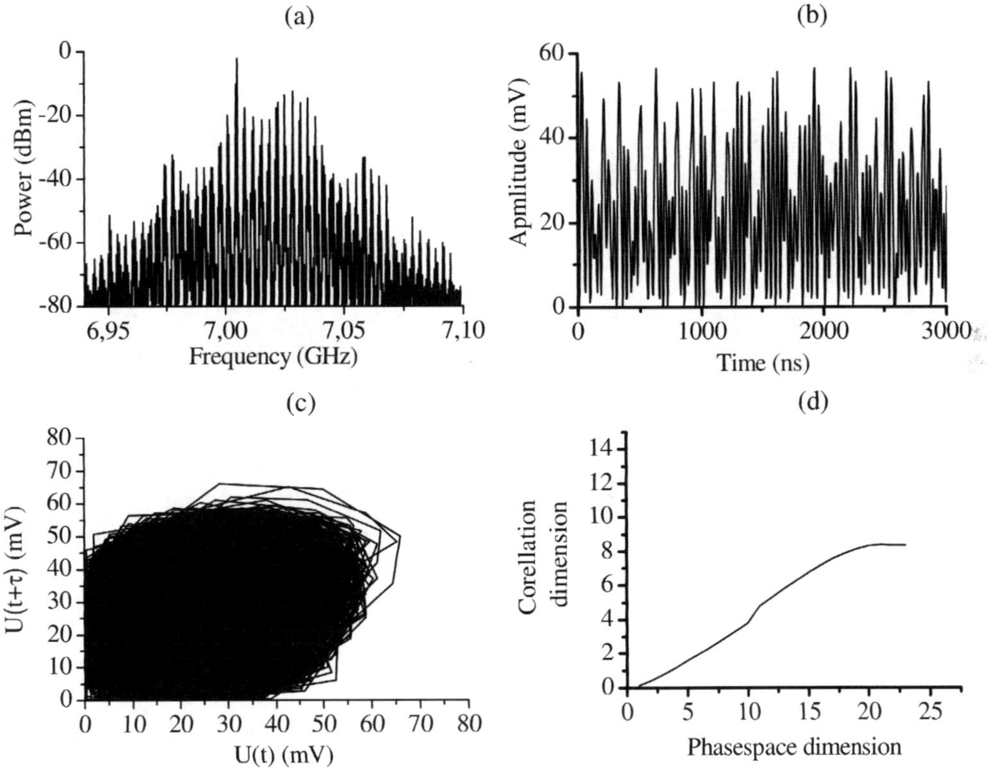

Figure 3. Dynamical chaos generation regime characteristics (a) spectrum, (b) – envelop waveform, (c) – phase portrait, (d) - dependence of correlation dimension on phase space dimension (G = 1.3 dB, E = 0 V/μm).

For investigation of chaotic signal tunability fractal dimensions for a range of ring gain coefficients were calculated. Figure 4 represents the dependence of fractal dimension for different self-generation regimes in the range of the ring gain coefficient from 0 dB up to 2.2 dB. One can see that increase of the ring gain coefficient leads to increase of the fractal dimension. For regular dynamic fractal dimension changed stepwise. Each step corresponded increase of fractal dimension by one. It is so-called Hopf bifurcation. Transition to dynamical chaos through sequence of Hopf bifurcation is the Ruelle-Takens scenario. In chaotic regime Increase of fractal dimension led to gradual fractal dimension increase from 4.5 till 11.5. Above G = 2 dB weak change of fractal dimension was observed. Constant fractal dimension indicates that complicacy of chaotic dynamic is also constant, and topology of attractor does not change.

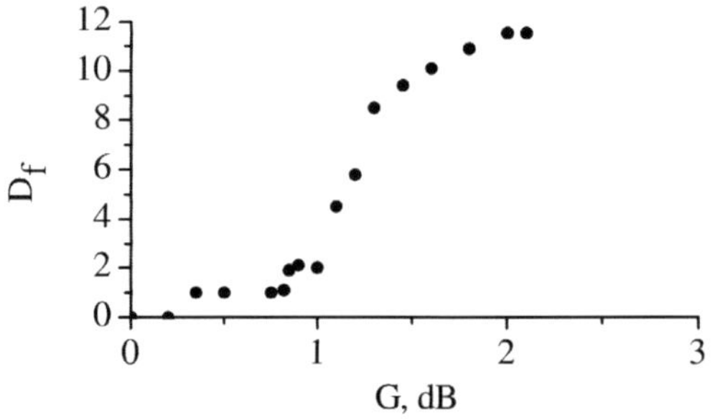

Figure 4. Dependence of fractal dimension on ring gain coefficient for E = 0 V/μm.

For the value of ring gain coefficient corresponding to maximum fractal dimension influence of electric bias field on chaotic signal was estimated. In the range of 0-1.6 V/μm fractal dimensions were calculated. The dependence of the fractal dimension on the bias electric field is shown on figure 5. One can see that increase of bias field leads to increase of fractal dimension. In comparison to the gain tunability influence of electric field is weaker. Fractal dimension changed from 11.5 up to 12.9

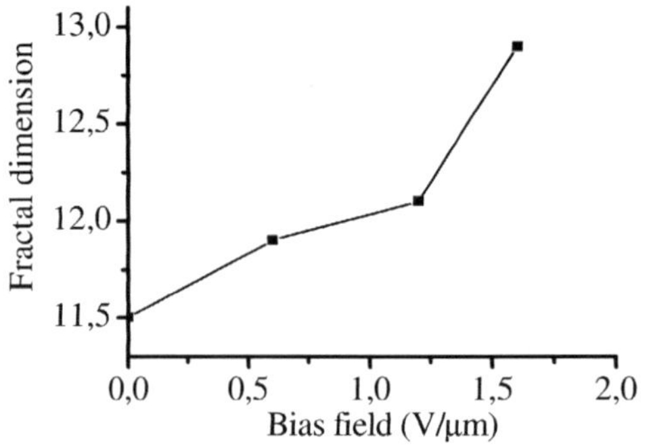

Figure 5. Dependence of fractal dimension on bias electric field (G = 2.1 dB).

Therefore in this work we observed and studied a dual control of the fractal dimension. In wide range the control could be done by varying of ring gain coefficient. In a narrow range the control could be done by variation of electric bias field.

Acknowledgement
The work at SPbETU was supported in part by the Russian Science Foundation. The work at ITMO University was supported by the Government of the Russian Federation (Grant 074-U01).

References

[1] Remoissenet M 1999 *Waves Called Solitons: Concepts and experiments* (Berlin: Springer-Verlag).
[2] Agraval G 2013 *Nonlinear fiber optics* (Amsterdam: Academic Press)
[3] Grelu Ph and Akhmediev N 2012 *Nature Photonics* **6** 84
[4] Stancil D D and Prabhakar A 2009 *Spin waves: Theory and applications* (New York: Springer)
[5] Zakharov V E and Ostrovsky L A 2009 *Physika D* **238** 540
[6] Mollenauer L F, Stolen R H and Gordon J P 1980 *Phys. Rev. Lett.* **45** 1095
[7] Kalinikos B A, Kovshikov N G and Slavin A N 1983 *JETP Lett.* **38** 132
[8] Kalinikos B A, Kovshikov N G and Slavin A N 1984 *Sov. Tech. Phys. Lett.* **10** 392
[9] Tai K, Hasegawa A and Tomita A 1986 *Phys. Rev. Lett.* **56** 135
[10] Khaykovich L et. al. 2002 *Science* **296** 1290
[11] Stavroulakis P 2006 *Chaos applications in telecommunications* (Boca Raton: SRC Press)
[12] Uchida A 2012 *Optical Communication with Chaotic Lasers* (Weinheim: Wiley-VHS Verlag&Co. KGaA)
[13] Akhmediev N and Ankiewicz A 2008 *Dissipative solitons: From optics to biology and medicine* (Berlin: Springer)
[14] Soto-Crespo J M and Akhmediev N 2005 *Phys. Rev. Lett.* **95** 024101
[15] Seghete V, Menyuk C and Marks B S 2007 *Phys. Rev. A* **76** 043803
[16] Zhu Q, Hai W and Rong S 2009 *Phys. Rev. E* **80** 016203
[17] Wu M, Hagerstrom A M, Eykholt R Kondrashov A and Kalinikos B A 2009 *Phys. Rev. Lett.* **102** 237203
[18] Kondrashov A V, Ustinov A B and Kalinikos B A 2010 *Tech. Phys. Lett.* **36** 224
[19] Ustinov A B, Demidov V E, Kondrashov A V, Kalinikos B A and Demokritov S O 2011 *Phys. Rev. Lett.* **106** 017201
[20] Wang Z, Hagerstrom A, Anderson J Q, Tong W, Wu M, Carr L D, Eykholt R and Kalinikos B A 2011 *Phys. Rev. Lett.* **107** 114102
[21] Hagerstrom A, Wu M, Eykholt R and Kalinikos B A 2011 *Phys. Rev. B* **83** 104402
[22] Dmitriev B S, Hramov A E, Koronovskii A A, Starodubov A V, Trubetskov D I and Zharkov Y D 2009 *Phys. Rev. Lett.* **102** 074101
[23] Grishin S V, Sharaevskii Y P, Nikitov S A and et al. 2011 *IEEE Trans. Mag.* **47** 3716
[24] Fujioka J, Cortes E, Perez-Pascual R, Rodriguez R F, Espinosa A and Malomed B A 2011 *Chaos* **21** 033120
[25] Lecaplain C, Grelu Ph, Soto-Crespo J M and Akhmediev N 2012 *Phys. Rev. Lett.* **108** 233901
[26] Cartes C, Cisternas J, Descalzi O and Brand H R 2012 *Phys. Rev. Lett.* **109** 178303
[27] Wu M 2011 Nonlinear Spin Waves in Magnetic Film Feedback Rings *Solid State Physics* **62** 163
[28] Ozgur U, Alivov Ya, and Morkoc H 2009 *J.Mater Sci: Mater Electron* **20** 911
[29] Sun N X and Srinivasan G 2012 *SPIN* **2** 1240004
[30] Vaz C A F 2012 *J. Phys.: Condens. Matter* 24 333201
[31] Cherkasskii M A, Kalinikos B A 2013 *Tech. Phys. Lett.* **39** 182
[32] Cherkasskii M A, Kalinikos B A 2013 *JETP Letters* **97**, 611
[33] Ustinov A B, Kalinikos B A and Srinivasan G 2014 *Appl. Phys. Lett.* **104** 052911
[34] Ustinov A B, Kondrashov A V, Nikitin A A and Kalinikos B A 2014 *Appl. Phys. Lett.* **104** 234101
[35] Takens F 1981 *Lecture Notes in Mathematics* **898** 366
[36] Grassberger P and Procaccia I 1983 *Physica D* **9** 189
[37] Grassberger P and Procaccia I 1983 *Phys. Rev. Lett.* **50** 346

Spin wave chaos in active resonant rings based on ferromagnetic films

A V Kondrashov, A B Ustinov, B A Kalinikos
Department of Physical Electronics and Technology, St.Petersburg Electrotechnical University, 5 Popov Street, St.Petersburg 197376, Russia

E-mail: kondrashov_av@yahoo.com

Abstract. Development of a microwave dynamical chaos in an active ring oscillator based on a ferromagnetic film upon a gradual increase of the ring gain has been studied. It was demonstrated that the transition of the system from steady-state to chaotic dynamics is in good agreement with the Ruelle-Takens scenario. The attractors of the dynamic system were reconstructed from the experimentally measured waveforms, and a numerical analysis of their parameters was made.

1. Introduction

Recently, a steadily increasing attention has been given to the dynamical chaos. The keen interest in this phenomenon is due to a number of its inherent properties, such as the communication confidentiality, larger information capacity, possible self-synchronization of a transmitter and receiver, etc. In this context, a study of new methods for generation of the dynamical chaos and of the process in which it appears becomes topical.

It has been demonstrated that the dynamical chaos can be generated by using bulk ferrite samples [1, 2], active resonant rings based on ferromagnetic films [3-8] or ferrite-ferroelectric multiferroic structures [9]. We stress, that transition to dynamical chaos is caused by appearance and development of nonlinear parametrical processes. For ferrite-ferroelectric structures nonlinearity of hybrid spin-electromagnetic waves have dual electric and magnetic nature [9]. In contrast to the multiferroics, nonlinear parametric processes in ferrite films are caused by only magnetic nonlinearity. In papers [1-8], the nonlinear processes developed due to either three-wave or four-wave spin wave interaction. Transition to dynamical chaos in multiferroic and ferrite structures may have similar scenario but different underlying physics. Therefore ferrite-film-based active resonant rings with single nonlinearity are of particular interest.

The goal of our study was to examine the development of chaos in active resonant rings based on ferromagnetic films under the conditions of a nonlinear only four-wave scattering of backward volume spin waves.

2. Experimental set-up

To perform an experiment, we assembled a model generator as an active ring. The block diagram of the generator is shown in figure 1. It comprised series-connected ferromagnetic delay line (1), microwave amplifier (2), adjustable attenuator (3), and directional coupler (4). The delay line was the frequency-setting element of the circuit, and the amplifier served to compensate for the loss of the microwave signal in the delay line and attenuator. The attenuator was used to vary the ring gain. The microwave signal being generated was extracted from the ring with a directional coupler.

The delay line was fabricated from a single-crystal yttrium-iron garnet (YIG) film (5) grown by epitaxy on a gadolinium-gallium garnet substrate. The film was placed on spin-wave transducers (6) fabricated as segments of microstrip transmission lines. The 50-μm-wide and 2-mm-long transducers were short-circuited at ends. The microwave power was fed into the input transducer and extracted from the output transducer with microstrip lines (7) having a wave impedance of 50 Ohm. The YIG was 5.2 μm thick and its saturation magnetization was 1750 Gs. The film was magnetized to saturation with a dc magnetic field directed tangentially to the film plane and in parallel to the spin wave propagation direction. This orientation of the field provided excitation of the so-called backward volume spin waves in the film. The magnetization field strength of 1300 Oe was chosen so that the nonlinearity of the spin waves was only determined by four-wave interaction processes. The model generator is a resonator whose spectrum can be obtained if the attenuation is made stronger and a weak signal is introduced [10].

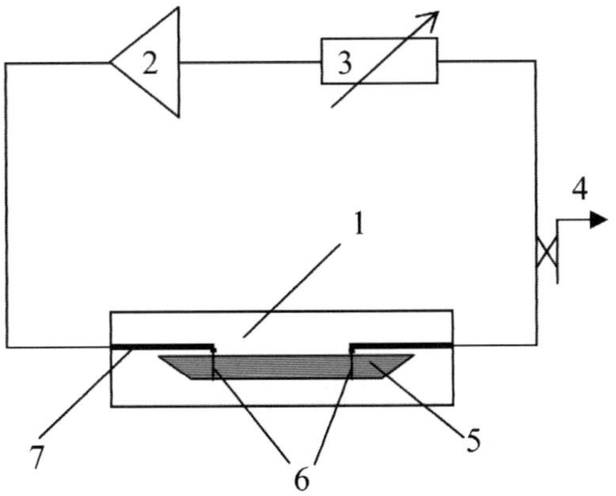

Figure 1. Block diagram of the experimental model generator.

3. Experimental results and discussion

In the course of the experiment, we gradually raised the ring gain G. At its certain value, the gain fully compensated for the attenuation and a self-generation of monochromatic spin waves appeared in the ring. This value of the gain was taken to be zero. Further increase in the gain resulted in a change of generation regimes. In the frequency domain, these regimes can be conditionally divided into three types; generation of a monochromatic signal, generation of a frequency comb, and generation of a dynamical chaos.

Figure 2 shows spectral characteristics (right-hand panel) and waveforms for four ring generation regimes. Figure 2(a) corresponds to the monochromatic generation mode. A single generated frequency of 5.653 GHz can be seen in the spectrum. Further increase in the gain results in that this frequency becomes unstable and breaks down into a frequency comb. In this case, generation of a sequence of nonlinear pulses, envelope solitons, was observed in the time domain. The pulses had a repetition period of 186 ns, width of 54 ns, and a carrier frequency of 5.652 GHz. At $G > 0.35$ dB, the generation of solitons was disrupted and a periodic stationary sequence of nonlinear pulses with a nonsoliton shape was generated (see figure 2(c)). With the gain raised to an even greater extent to $G = 0.9$ dB, the signal being generated became stochastized.

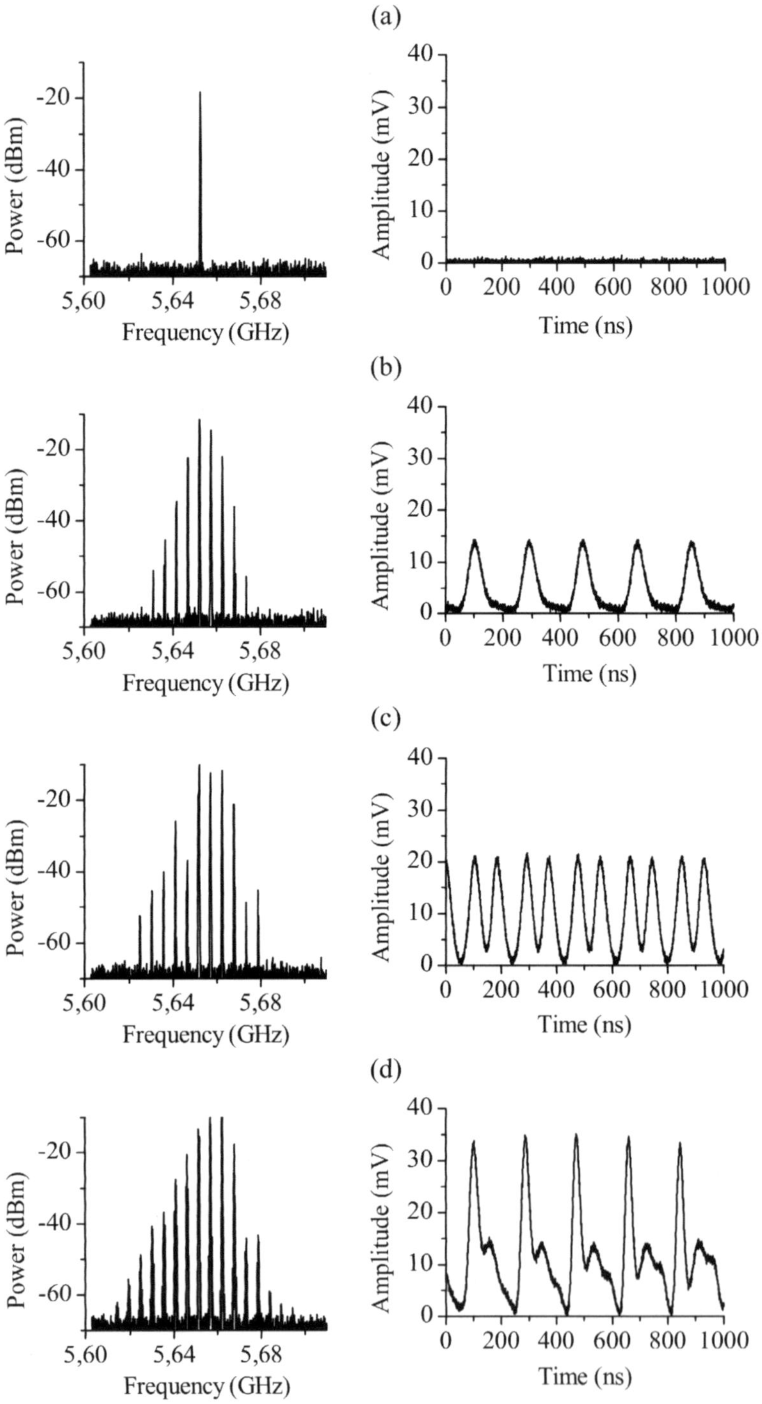

Figure 2. Spectral characteristics and waveforms for different gains G (dB): (a) 0, (b) 0.2, (c) 0.35, and (d) 1.2.

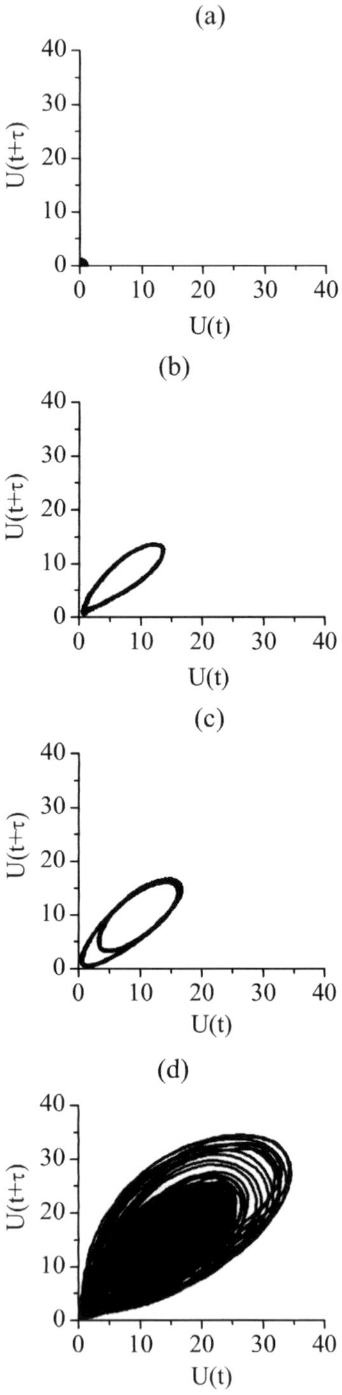

Figure 3. Attractor projections: (a) monochromatic signal, (b) envelope solitons, (c) periodic stationary pulses, and (d) dynamical chaos.

As is seen from figure 2(a) formation of bright *spin wave* solitons took place. It is due to attractive nonlinearity of backward volume spin waves propagating in tangentially magnetized YIG film. This situation is different from active rings based on multiferroic structures. The ring structure reported in [9] generates dark *spin-electromagnetic wave* solitons. Nonlinearity of the spin-electromagnetic waves (and consequently soliton properties) is controlled by electric field, that confirms different nature of nonlinear excitations.

It can be seen in figure 2 that the transition from the monochromatic generation to the chaotic type occurs via a sequence of bifurcations in which the spectrum of the microwave signal being generated is enriched in new frequencies. To further analyze the results obtained, we constructed phase portraits for each generation regime. The attractors were reconstructed by the standard time-delay method [11]. In this case, the components of the d-dimensional vector $\boldsymbol{x}(t)$ in the phase space were obtained by the time-shift of the signal $U(t)$ by a certain delay τ:

$$\boldsymbol{x}(t) = \{U(t), U(t+\tau), ..., U(t+(d-1)\tau)\}. \tag{1}$$

Figure 3 shows 2D phase portraits of the four generation modes described above. To the monochromatic signal generation mode corresponds a regular attractor, a stable point (see figure 3a). At $G > 0.2$ dB, there occurs a Hopf bifurcation corresponding to generation of a limit cycle from the stable point (see figure 3(b)). With the gain increasing further, a period-doubling bifurcation occurs (see figure 3(c)). At $G > 1.2$ dB, a strange attractor was formed (see figure 3(d)). This transition corresponds to the Ruelle-Takens scenario.

The phase portraits were used to calculate the quantitative characteristics of the signals: correlation integrals, fractal dimensions, and embedding dimensions. The calculation was carried out with the standard Grassberger-Procaccia algorithm [12]. The value of the correlation integral shows the number of pairs of points spaced by a distance l. Figure 4 demonstrates the dependence of the correlation integral on l for different attractor embedding dimensions.

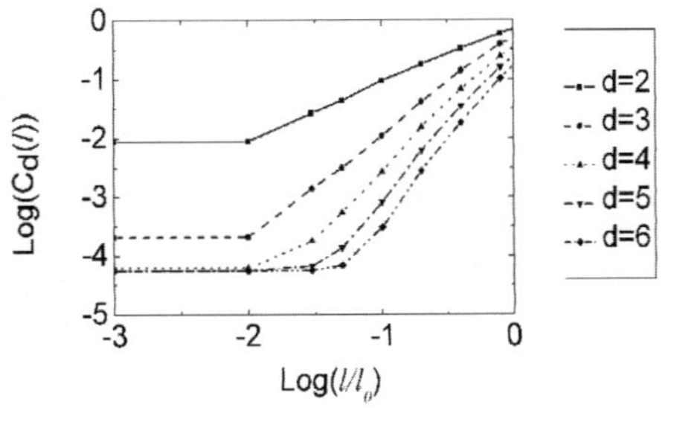

Figure 4. Curves describing the dependence of the logarithm of the correlation integral, $\log(C_d(l))$, on the logarithm $\log(l/l_0)$ for different embedding dimensions; l_0 is the a normalization constant.

Figure 5 shows how the correlation dimension depends on the attractor embedding dimension. It can be seen from the figures that, in all cases, this dimension tends to approach a constant value ("saturates") at a certain attractor embedding dimension. This value is the attractor embedding dimension and characterizes the number of degrees of freedom of the given system. At the same time, the value of the correlation dimension D_c is the fractal dimension of the attractor of the dynamic system.

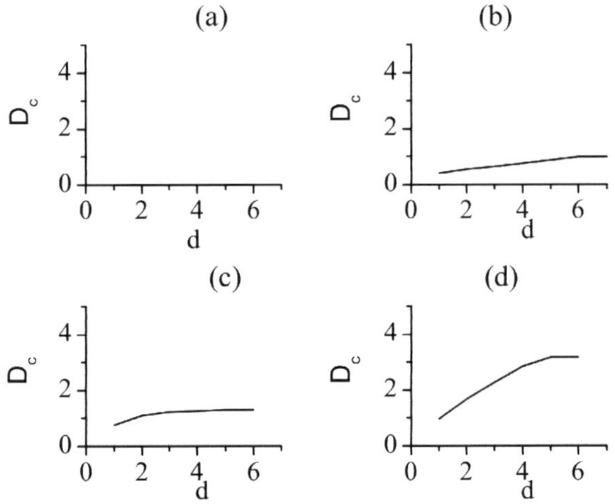

Figure 5. Correlation dimension plotted against the embedding dimension: (a) monochromatic signal, (b) envelope solitons, (c) periodic pulses, and (d) dynamical chaos.

It can be seen in figure 5(a) that the fractal dimension is 0 for the monochromatic generation mode and this value is reached already at the embedding dimension of 1. For the case of soliton generation, the fractal dimension and the embedding dimension are 1 and 6, respectively (see figure 5b). This means that the attractor is a limit cycle (see figure 3(b)).

Figure 6 shows how the fractal dimension depends on the ring gain. It can be seen in the figure that the fractal dimension grows with increasing gain. In the chaotic generation regime, the fractal dimension was 3.2. These results suggest that a dynamical chaos starts to develop in the ring at a gain $G = 1.2$ dB.

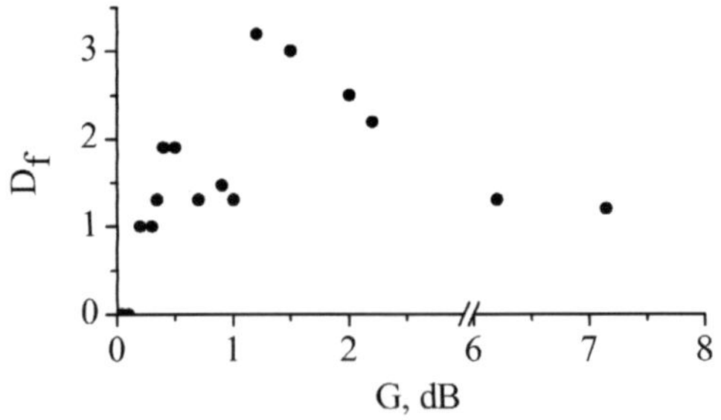

Figure 6. Dependence of the fractal dimension on the ring gain.

It is notable that, as the gain increases to more than 6, the oscillation spectrum becomes depleted. In the time domain, periodic oscillations are again observed and the fractal dimension decreases to 1.3. This transition from chaotic oscillations to those of the periodic type invites a further study.

Acknowledgement
The work was supported in part by the Russian Foundation for Basic Research and by Ministry of Education and Science of Russian Federation (Project "Goszadanie").

References
[1] Yamazaki H 1988 *J.Appl.Phys.* **64** 5391
[2] Gibson G and Jeffries C 1984 *Phys. Rev. A* **29** 811

[3] Kondrashov A V, Ustinov A B, Kalinikos B A and Benner H 2008 *Tech. Phys. Lett.* **34** 492
[4] Kondrashov A V, Ustinov A B and Kalinikos B A 2010 *Tech. Phys. Lett.* **36** 224
[5] Wang Z, Hagerstrom A, Anderson J, Tong W, Wu M, Carr L D, Eykholt R and Kalinikos B A 2011 *Phys. Rev. Lett.* **107** 114102
[6] Grishin S V, Sharaevskii Y P, Nikitov S A and Romanenko D V 2013 *IEEE Trans. on Magn.* **49** 1047
[7] Grishin S V, Golova T M, Morozova M A, Romanenko D V, Seleznev E P, Sysoev I V, Sharaevskii Y P 2015 *JETP* **121** 714
[8] Demidov V E, Kovshikov N G 1999 Technical Physics **44** 960
[9] Ustinov A B, Kondrashov A V, Nikitin A A and Kalinikos B A 2014 Appl. Phys. Lett. 104 234101
[10] Kalinikos B A, Scott M M, and Patton C E 2000 *Phys. Rev. Lett.* **84** 4697
[11] Takens F 1981 *Dynamical Systems and Turbulence, Lecture Notes in Mathematics* **898** 366
[12] Kantz H and Schreiber T 2004 *Nonlinear Time Series Analysis - 2nd edition* (Cambridge: University press)

GaAs- A_3B_5 heterostructures for high-speed power diodes manufacturing

F Yu Soldatenkov[1,2], V A Kozlov[1,2], I L Shulpina[1], V I Ivanovskiy[3]

[1]Ioffe Physical-Technical Institute, Politekhnicheskaya 26, St. Petersburg, 194021 Russia
[2]Power Semiconductors Ltd., Gzhatskaya 27, St. Petersburg, 195220 Russia
[3]Saint Petersburg State Electrotechnical University (LETI), ul. Professora Popova 5, St. Petersburg, 197376 Russia

E-mail: f.soldatenkov@mail.ioffe.ru

Abstract. Physical basis of using the controlled defect formation in InGaAs heteroepitaxial layers to vary the carrier lifetime is considered. It is shown that the lifetime of nonequilibrium carriers in the base layers of a diode can be controllably varied from several to hundreds of nanoseconds. The results obtained in a study of (i) structural defects and their rearrangement related to the lattice mismatch between the heteroepitaxial InGaAs layer and the GaAs substrate and (ii) influence exerted by these defects on the carrier lifetime and on the voltage blocked by diode structures are reported. Dynamic switching characteristics of high-speed power heteroepitaxial diodes at different temperatures are presented.

1. Introduction

One of the main tendencies in the development of modern power electronics and pulse technology consists in raising the conversion efficiency of electrical signals and reducing the mass and size of technical devices used in electricity converters. Further advances in this field primarily require that high-frequency fast power semiconductor devices with improved set of static and dynamic characteristics should be used in converter devices. As applied to the class of bipolar diodes, this requires that devices with short switching times should be developed, with the minimum possible forward voltage drops in the conducting state, low leakage current under reverse biases, and extremely high working temperatures of the diode crystal preserved. It is nearly impossible to simultaneously satisfy all these requirements because of the opposite correlation of the static and dynamic characteristics of a diode on parameters of its semiconductor structure. Use of GaAs-A_3B_5 heterostructures instead of the conventional Si-based devices opens up new opportunities for development of fast power diodes and improvement of the whole set of their frequency and static characteristics [1, 2].

One of the most important dynamic characteristics of high-power pulse bipolar diodes, which limit frequency parameters, is the reverse recovery time t_{rr} (the time in which the blocking properties of a p-n junction are restored after the diode is switched from the conducting to blocking state). When a diode is switched off, two phases in which the reverse resistance is restored are distinguished: the stage of a high reverse conductivity in which the p-n junction remains forward-biased despite the change of the polarity of the applied external voltage and the stage of recovery of the high reverse resistance in which a space-charge layer is formed near the p-n junction and, while extending, this layer blocks the voltage applied to the diode. The switch-off parameters of a diode can be controlled by varying the conditions of minority carrier accumulation and resorption and, in the first place, by changing the values and distribution profile of minority carrier lifetimes in the lightly doped base layers of the high voltage diode. In the present work, the main results on investigation and elaboration of methods for decreasing the values of charge carrier lifetime by means of application of a technique for generation of additional intrinsic structural defects arising in crystals during epitaxial growth of mismatched heterostructures are presented.

2. Sample fabrication

We fabricated the structures of high-voltage gradual p-i-n junctions by liquid-phase epitaxy (LPE) technique the fundamentals of which had been developed at the Ioffe Physical-Technical Institute [3]. This technique is simple in implementation, economically efficient, and, in our opinion, is the only presently possible method for obtaining high-voltage GaAs p-n junctions capable of blocking voltages on the order of 1000 V and more [4]. Epitaxial layers of GaAs and $In_xGa_{1-x}As$ solid solutions with InAs content *x* of up to 6% were grown on p^+-GaAs (111) and (100) substrates doped with zinc to (2-10)·10^{18} cm^{-3}. The growth equipment used in the study made it possible to growth epitaxial layers on 2- and 3-inch GaAs substrates. The epitaxial growth of lightly doped gradual p-n junction layers was carried out from a confined Ga-As or In-Ga-As melt in the hydrogen atmosphere in quartz boat from 900ºC to room temperature. Heavily doped (with Te or Sn) n^+-GaAs emitter layers of the diode structures were grown in piston graphite cassette.

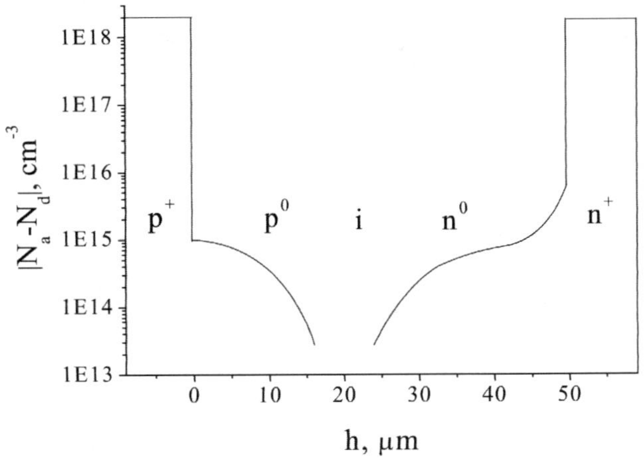

Figure 1. Typical free carrier distribution across the thickness of GaAs or InGaAs high voltage p^0-i-n^0 layers grown on p^+-GaAs substrates with a later-grown n^+-GaAs emitter.

The content of electrically active defects in epitaxial layers grown by this method is controlled by the content of residual (background) impurities in the melt and growth system, temperature and duration of a preliminary annealing of the melt, and flow rate and humidity of hydrogen, as well as by the growth velocity of the epitaxial film under programmed cooling of the system. In this mode of LPE growth of p-i-n structures, it is possible to obtain thick lightly doped i-type layers of GaAs or InGaAs with a free carrier concentration of ~10^{13} cm^{-3}, which makes it possible to reach U_b of up to 2000 V. Typical free carrier distribution across the thickness of diode structures with GaAs or InGaAs base layers are shown in figure 1.

The relative amounts of the components of the In-Ga-As liquid phase used to obtain $In_xGa_{1-x}As$ layers of required composition was found by calculations in terms of the quasi-regular approximation model, with consideration for the elastic stresses appearing because of the lattice mismatch between the layer and substrate for elastically strained layers [1,5,6] and with elastic stresses disregarded for layers in which these stresses are partly relaxed (under relatively large lattice mismatch in heterostructure). It has been found previously [7] that, when layers for which the combination of the lattice mismatch and thicknesses is such that elastic stresses are partly relaxed and mismatch dislocations are thereby formed (see below), the composition of the solid solution of a layer tends to be the closest to that in equilibrium. This was confirmed by X-ray microanalysis on a Camebax installation for $In_xGa_{1-x}As$ layers in which elastic stresses are partly relaxed via formation of a network of linear mismatch dislocations revealed by X-ray topography [7].

Diode chip samples had a form of «mesa» with round base which was obtained by chemical etching of active layers up to substrate around contact pads. The double-layer lithography with lift-off photoresists (LOR) was used to form thick (up to 2-3 μm) contact pads during single vacuum evaporation process [8, 9]. The Cr-Ni-Au or Cr-Ni-Au contact systems was used to make ohmic contacts to p^+-GaAs substrate, and AuGe-Ni-Au, AuGe-Ni-Ag, or Ti-Ge-Pt-Au [10] – to n^+-GaAs emitter layers.

3. Study of the crystalline structure of $In_xGa_{1-x}As$ heteroepitaxial layers. Relationship between the layer crystalline structure and the device characteristics of diodes

The real crystalline structure of lightly doped layers in $In_xGa_{1-x}As$-GaAs heterostructures was studied with $Cu_{K\alpha}$ radiation and 422, 533, and 331 asymmetric reflections in the back reflection X-ray diffraction topography at different information layer depths. This procedure for studying the crystalline structure of the layers was described in detail in [11].

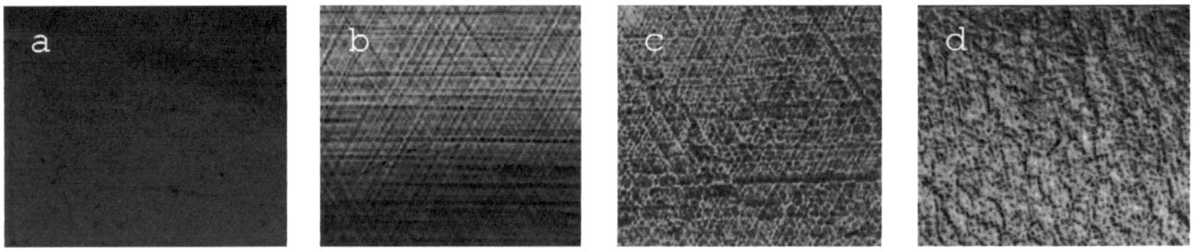

Figure 2. Topograms of the homostructure GaAs (a) and heterostructures $In_xGa_{1-x}As$-GaAs (b-d) with equal thicknesses (50-65μm) and different x of the base layers: $x = 0$ (a); $x \sim 2\%$ (b); $x \sim 3.5\%$ (c); $x \sim 5\%$ (d). Bragg method, CuKα-radiation; reflection: 422 (a); 422 (b); 511 (c); 331 (d).

Figure 2 shows how the crystalline structure of a heteroepitaxial layer changes with increasing lattice mismatch (increasing x) at the same base layer thicknesses. After the critical elastic strain relaxation thicknesses of $In_xGa_{1-x}As$ layers were reached, there appeared networks of mismatch dislocations (figure 2, b), with the type of these networks determined by the solid solution concentration x, layer thicknesses, and crystallographic orientation of wafers. In [7], the range of compositions x at which a network of linear mismatch dislocations is characteristically formed in $In_xGa_{1-x}As$ (111) layers with thicknesses of 50 to 65 μm was experimentally found to be x = 1.5-3.3 mol %. At these thicknesses and compositions of the $In_xGa_{1-x}As$ solid solutions, a quasi-uniform network of linear mismatch dislocations is formed in the epitaxial layers (figure 2, b).

With x increasing at the same thickness of epitaxial layers, the average density of mismatch dislocations in the layers became higher and the shape of cells of the dislocation network for $In_xGa_{1-x}As$ (111) layers changed from triangular to rounded (figure 2, c), being transformed at $x \approx 4$ mol % to a cellular dislocation structure with dense cell boundaries (figure 2, d). In this case, the dislocation density at dislocation cell boundaries increased to about 10^6 cm^{-2} and more.

In a study of samples with high isovalent doping levels ($x > 4$ mol %), optical methods also revealed characteristic growth figures on the surface of the epitaxial layers. As a rule, these figures combine several dislocation cells at once. These growth figures are also well identified in X-ray topographs in reflections with a small X-ray beam incidence angle (331 in comparison with 422 or 533). Their presence is indicative of the high content of structural defects in the $In_xGa_{1-x}As$ epitaxial layer.

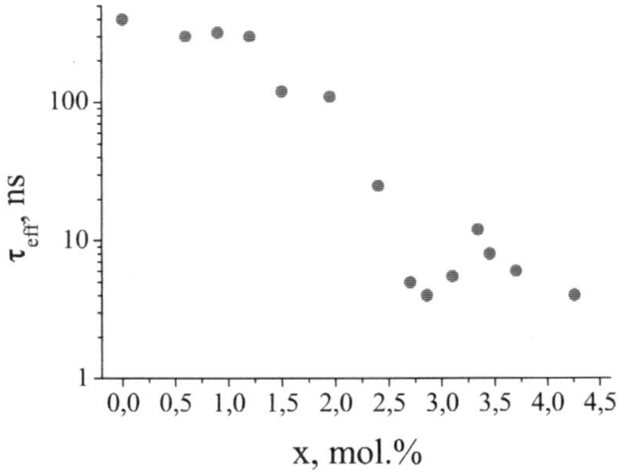

Figure 3. Effective nonequilibrium charge carrier lifetime τ_{eff} in $In_xGa_{1-x}As$ layers grown on GaAs substrates depending on x.

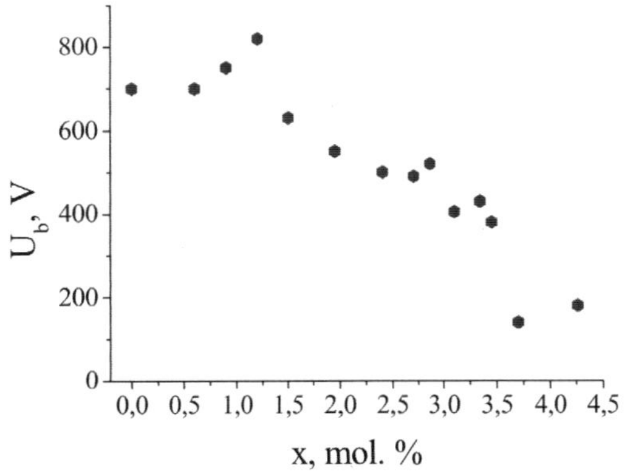

Figure 4. Maximum blocking voltages U_b of $In_xGa_{1-x}As$-GaAs heterostructures of thickness up to 65μm depending on x.

The process in which the dislocation of heteroepitaxial layers changes with increasing x is accompanied by a substantial decrease in the lifetime of nonequilibrium carriers near the p-n junction from hundreds to several nanoseconds (figure 3) and also by a gradual decrease in voltages blocked by diodes (figure 4), and at $x > 3.5$ mol %, by a sharp fall of these voltages. We attribute the sharp decrease in U_b for diode heterostructures with higher content of InAs (figure 4, $x > 3.5$ mol %) to a fundamental change of the dislocation structure of $In_xGa_{1-x}As$ layers, i.e., to a major transition from a uniform network of rectilinear dislocations with symmetry in the (111) plane to a "cellular" dislocation structure with very high dislocation density at boundaries of cellular networks and formation of growth figures on the surface of an epitaxial layer. This conclusion about the effect of dense dislocation walls on the decrease in the blocking capacity of p-n junctions is well correlated, both qualitatively and quantitatively, with the results of a study of dislocations in silicon epitaxial structures of power devices [12].

The forward voltage drops U_f for InGaAs-GaAs diodes with nanosecond times τ_{eff} and $U_b < 600$ V had values of U_f not exceeding 2 V at a working current density $j_f = 150$ A/cm^2. At increased current densities, the forward voltage drop U_f on diodes of this kind has quite acceptable values $U_f < 3.0$ V at $j_f = 1000$ A/cm^2. It should be noted that this combination of small values of U_f and a nanosecond switching-off time of a diode cannot be reached in silicon-based devices. The electrical capacitance C of GaAs diodes weakly depends on voltage and has a value on the order of $C_0 \approx (1\text{-}2)\ 10^3$ pF/cm^2 under zero bias and $C_{300V} \approx (2\text{-}4)\ 10^2$ pF/cm^2 under a reverse bias $U_r = 300$ V.

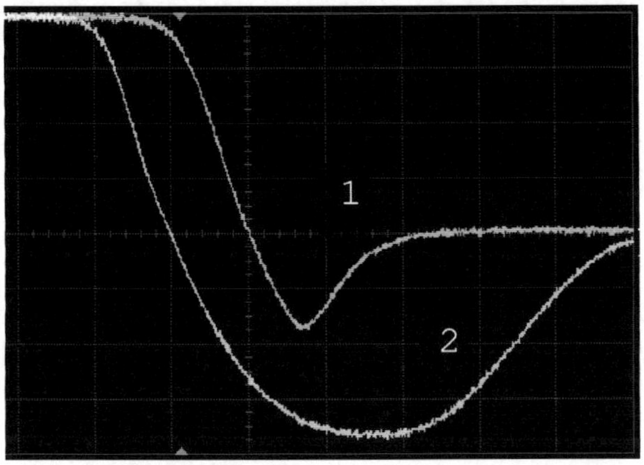

Figure 5. Waveforms of 4A current turn-off switching for InGaAs-GaAs heterojunction diode (blue curve 1) and GaAs homojunction diode (green curve 2). Vertical scanning – 1A/div and horizontal one – 10 ns/div; $T = 25$ °C.

Figure 5 shows for comparison oscillograms of the switching-off dynamics of diodes with different base layers: heteroepitaxial InGaAs base layer (curve 1 in figure 5; t_{rr} ~ 16 ns) and homoepitaxial GaAs base layers (curve 2 in figure 5; t_{rr} ~ 60 ns). Comparative tests of device structures of this kind demonstrated that the heterostructure growth technology with controlled defect formation makes it possible to fabricate diodes having the lowest amount of accumulated charge, shortest switching-off time, and minimum switch-on and switch-off losses among Si and GaAs bipolar diodes.

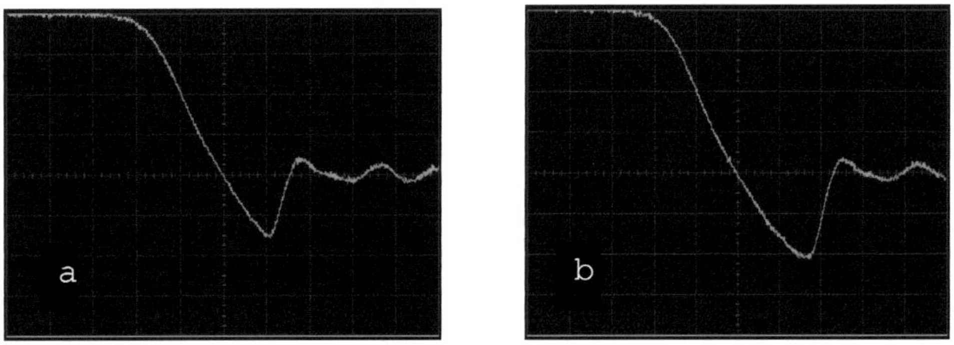

Figure 6. Waveforms of 1A current turn-off switching for InGaAs-GaAs heterojunction diode under different temperature measurements: under 25 °C (a) and 175 °C (b). Vertical scanning – 0.25A/div and horizontal one – 5 ns/div.

As regards the whole set of their dynamic characteristics, InGaAs-GaAs diodes are nearly not inferior to SiC Schottky diodes, especially at high current densities [7]. InGaAs-GaAs diodes are actually capable of operating at frequencies on the order of several megahertz at crystal temperatures T of up to 250°C (provided that the diode chip surface is passivated and protected by, e.g., insulators; without protection of the lateral surface of a chip, power GaAs diodes are properly functioning up to 190-200°C). As the temperature of the crystal of heteroepitaxial diodes increases from room temperature to 175°C, the time and charge of reverse recovery of InGaAs-GaAs diodes grow by a factor of ~1.5 (see figure 6), whereas when Si bipolar diodes are heated to 175°C, the recovery times commonly become more than three times longer.

4. Summary

The principal result of the study is that the carrier lifetime in base layers of power InGaAs-GaAs diodes can be controlled in the range from several to hundreds of nanoseconds by using the method of intrinsic defect formation in growth of mismatched heteroepitaxial base layers. This growth technique can be successfully used in fabrication of high-voltage high-temperature power diodes with working voltages of 300-800 V and reverse recovery times on the order of 10 ns and shorter at an acceptable loss in the conducting state.

Acknowledgments

The work has been supported by the Russian Science Foundation, agreement No. 14-29-00178.

5. References

[1] Soldatenkov F Yu, Danilchenko V G and Korolkov V I 2007 *Semiconductors* **41** 211-4
[2] Danilchenko V G, Korolkov V I and Soldatenkov F Yu 2009 *Semiconductors* **43** 1055-7
[3] Alferov Zh I, Korolkov V I, Nikitin V G, Stepanova M N and Tretiakov D N 1976 *Pis'ma v Zhurnal Tekhnicheskoi Fiziki (in Russian)* **2** 201-4
[4] Danilchenko V G, Korolkov V I, Ponomarev S I and Soldatenkov F Yu 2011 *Semiconductors* **45** 515-8

[5] Soldatenkov F Yu, Ulin V P, Yakovenko A A, Fedorova O M, Konnikov S G and Korolkov V I 1999 *Technical Physics Letters* **41** 852-4

[6] Soldatenkov F Yu 2007 Russian Federation patent No 2297690

[7] Kozlov V A, Soldatenkov F Yu, Danilchenko V G, Korolkov V I and Shulpina I L 2014 Defect engineering for carrier lifetime control in high voltage GaAs power diodes *Proc. 25th Advanced Semiconductor Manufacturing Conf. (ASMC-2014) (Saratoga Springs, USA, 19-21 May 2014)* pp. 139-144

[8] Soldatenkov F Y, Sorokina S V, Timoshina N Kh, Khvostikov V P, Zadiranov Y M, Rastegaeva M G and Usikova A A 2011 *Semiconductors* **45** 1219-26

[9] Andreev V M, Kalyuzhniy N A, Lantratov V M, Soldatenkov F Yu and Usikova A A 2010 Russian Federation patent No 2428766

[10] Andreev V M, Soldatenkov F Yu, Ilinskaya N D and Usikova A A 2010 Russian Federation patent No 2391741

[11] Shulpina I L, Ratnikov V V, Kozlov V A, Soldatenkov F Yu and Voitovich V E 2014 *Technical Physics* **59** 1566-9

[12] Shulpina I L and Kozlov V A 2012 X–ray topography study of defects in silicon multilayer epitaxial power devices *Proc. 11-th biennial conf. on high resolution X-ray diffraction and imaging (X-TOP 2012) (St.Petersnurg, Russia, 15-20 Sept. 2012)* pp. 182-3

Journal of Physics: Conference Series **661** (2015) 012067 doi:10.1088/1742-6596/661/1/012067

Anisotropic thermal expansion in crystals of different categories

A Zamkovskaya and E Maksimova
V. Vernadsky Crimea Federal University, Simferopol, Russian Federation

E-mail: lenamax112@rambler.ru

Abstract. The three-dimensional (3D) indicatory surfaces of thermal expansion in crystals of different categories were constructed in program MathCad. Indicatory surface of thermal expansion is a sphere, spheroid, ellipsoid or surface of multiple parts, depending on the category of the crystal symmetry. The symmetry elements of thermal expansion include the symmetry elements of the point group of the crystal according Neumann's Principle.

1. Introduction

Natural and synthetic crystals are widely used in the modern instrument-making. A characteristic feature of the physical properties of single crystals is their anisotropy - different properties in different directions. We investigated the anisotropy of thermal expansion of crystals, because this property has a large practical value: different crystals are exposed thermal expansion to both at instrumental processing and at functioning in technical devices. For the evident graphic image of anisotropy of thermal expansion it is convenient to use indicatory surfaces. Radius-vector of this surface is proportional to the size of the relative lengthening of crystal in this direction at heating, [1].
The purpose of work: investigation of conformity between symmetry of surface thermal expansion and crystal symmetry.

2. Research method

Quantitatively, the thermal expansion is described by the coefficient of linear thermal expansion α_n, which equals to the ratio of the relative change in body length along some particular direction n at heating to the temperature increment:

$$\alpha_n = \frac{\Delta l/l}{\Delta T} \tag{1}$$

Thermal expansion of any crystal is described by nine coefficients α_{ij}, formative the second-order symmetric tensor.
For a definite choice of the coordinate system (the choice is determined by installation rules and associated to the elements of the symmetry of the crystal) number of independent coefficients α_{ij} decreases. The higher symmetry of the crystals, the smaller of coefficients describe the thermal expansion of crystal, table 1. Coefficient of thermal expansion in any direction, [2]:

$$\alpha_n=\sum\alpha_{ij}\cdot n_i\cdot n_j, \tag{2}$$

where:
α_{ij} - coefficients of thermal expansion;
n_i,n_j - components of the vector of unit length determining an some direction in the crystal.

Table 1. The equations of indicatory surfaces of thermal expansion.

Category of symmetry	View of the tensor of thermal expansion	The number of independent parameters	The equations of indicatory surfaces
Highest	$\begin{pmatrix} \alpha_{xx} & 0 & 0 \\ 0 & \alpha_{xx} & 0 \\ 0 & 0 & \alpha_{xx} \end{pmatrix}$	1	$\alpha_n = \alpha_{xx}$
Middle	$\begin{pmatrix} \alpha_{xx} & 0 & 0 \\ 0 & \alpha_{xx} & 0 \\ 0 & 0 & \alpha_{zz} \end{pmatrix}$	2	$\alpha_n = \alpha_{xx} \cdot (n_x^2 + n_y^2) + \alpha_{zz} \cdot n_z^2$
Lowest	$\begin{pmatrix} \alpha_{xx} & 0 & 0 \\ 0 & \alpha_{yy} & 0 \\ 0 & 0 & \alpha_{zz} \end{pmatrix}$	3	$\alpha_n = \alpha_{xx} \cdot n_x^2 + \alpha_{yy} \cdot n_y^2 + \alpha_{zz} \cdot n_z^2$
	$\begin{pmatrix} \alpha_{xx} & 0 & \alpha_{xz} \\ 0 & \alpha_{yy} & 0 \\ \alpha_{zx} & 0 & \alpha_{zz} \end{pmatrix}$	4	$\alpha_n = \alpha_{xx} \cdot n_x^2 + \alpha_{yy} \cdot n_y^2 + \alpha_{zz} \cdot n_z^2 + 2 \cdot \alpha_{xz} \cdot n_x \cdot n_z$
	$\begin{pmatrix} \alpha_{xx} & \alpha_{xy} & \alpha_{xz} \\ \alpha_{yx} & \alpha_{yy} & \alpha_{yz} \\ \alpha_{zx} & \alpha_{zy} & \alpha_{zz} \end{pmatrix}$	6	$\alpha_n = \alpha_{xx} \cdot n_x^2 + \alpha_{yy} \cdot n_y^2 + \alpha_{zz} \cdot n_z^2 + 2 \cdot \alpha_{xy} \cdot n_x \cdot n_y + 2 \cdot \alpha_{xz} \cdot n_x \cdot n_z + 2 \cdot \alpha_{yz} \cdot n_y \cdot n_z$

For the construction of 3D revolving models of indicatory surfaces would be convenient to use application package MathCad. Great mathematical capabilities of this package make it a convenient tool for physical research [3, 4].

Within this package the program was created, that allows to construct of indicatory surfaces of thermal expansion. Analysis of the received indicatory surfaces allows defining the symmetry and the anisotropy of the properties, and, if necessary, establishing the directions of its extreme value.

3. Result and discussion

3.1 Highest category

Thermal expansion of crystals of the highest category symmetry (cubic crystal system) is described by a single coefficient α_{xx}. For any cubic crystal, indicatory surface has a form of a sphere, figure 1. Thus, the crystals of higher category evenly broaden on all directions, saving the primary form.

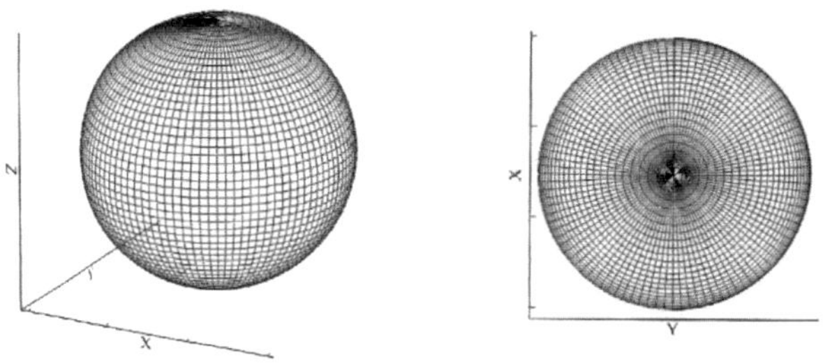

Figure 1. The indicatory surface of thermal expansion of the cubic crystal and its projection on the plane (XOY).

3.2 Midlle category

For the crystals of middle category (crystals of this category have an third, fourth or sixth order axis of rotation; the axis Z is directed along its) there are number of interesting cases:

a) all the coefficients of thermal expansion are positive, and $\alpha_{xx} < \alpha_{zz}$. For example, for crystal of zinc: α_{xx}= $14 \cdot 10^{-6}$, °C^{-1}, α_{zz}= $55 \cdot 10^{-6}$ °C^{-1} [5], figure 2.

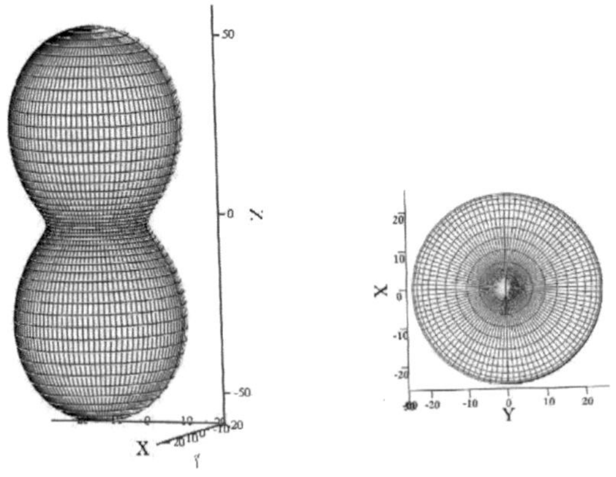

Figure 2. The indicatory surface of thermal expansion of zink and its projection on the plane (XOY).

For the zinc the indicatory surface is elongated along the axis Z. This axis coincides with the axis of the sixth order. With a uniform heating of zinc, in this direction crystal lengthened stronger than in the perpendicular direction.

b) all the coefficients of thermal expansion are positive, and $\alpha_{zz} < \alpha_{xx}$. As, for example, for α-quartz: α_{xx}=$14 \cdot 10^{-6}$ °C^{-1}, α_{zz}= $9 \cdot 10^{-6}$ °C^{-1}, figure 3.

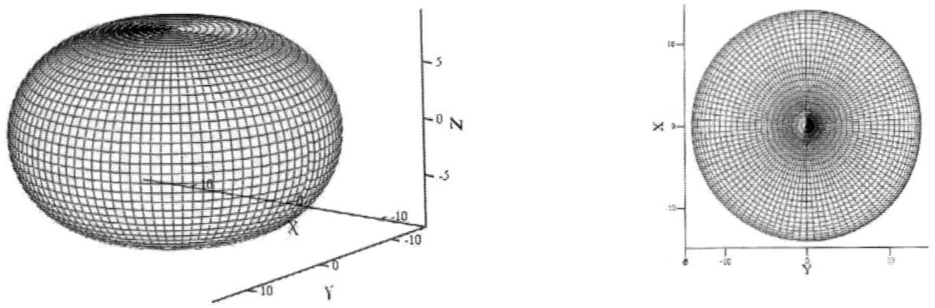

Figure 3. The indicatory surface of thermal expansion of α-quartz and its projection on the plane (XOY).

It is seen that surface of thermal expansion in this case is the spheroid, flattened along the axis of Z. This direction of minimum increase of quartz at heating. And maximally - in perpendicular direction.

c) especially highlight case when one of the coefficients of thermal expansion is less than zero. As an example, the indicatory surface of thermal expansion of calcite $CaCO_3$, which is widely used in an optical instrument production. Its coefficients of thermal expansion: α_{xx}= -5,2·10^{-6} °C^{-1}, α_{zz}= 22,6·10^{-6} °C^{-1}. The indicatory surface of thermal expansion of calcite - surface of multiple parts with positive (light) and negative (dark) areas of thermal expansion, figure 4. It is seen, that along the axis Z, the thermal expansion of calcite is maximal. Perpendicular to this axis is the region of negative thermal expansion (compression). Thus, at heating, calcite is lengthened in one direction, and in other is shortened. Also near the axis of Z there is a cone of directions with half-angle 75°56′, along which the expansion (compression) is zero. In these directions a crystal does not change at heating.

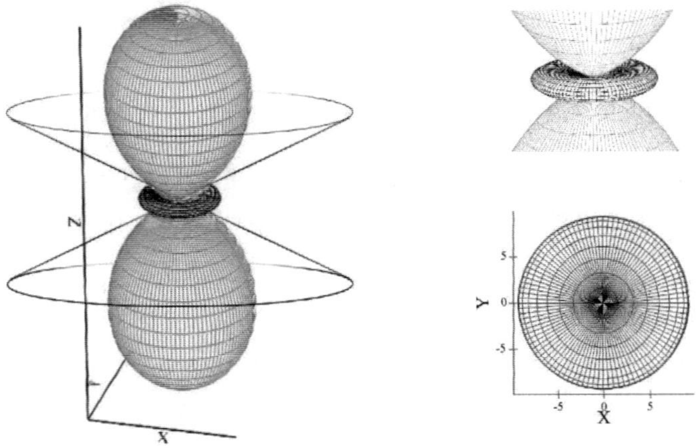

Figure 4. The indicatory surface of thermal expansion of calcite with increased fragment of the negative part and its projection on the plane (XOY).

Thus indicatory surfaces of thermal expansion of crystals middle category are spheroids. The axis of rotation is the main axis of symmetry (axis Z).

3.3 Lowest category

For crystals of the lowest category, depending on the symmetry and, accordingly, the number of independent coefficients of thermal expansion, there are also a number of cases:

a) three independent coefficients of thermal expansion for the orthorhombic crystal system, characterized by the presence of three mutually perpendicular two-order axes of symmetry.

For example, for a crystal of aragonite: $\alpha_{xx}= 35\cdot10^{-6}$ °C^{-1}, $\alpha_{yy}= 17\cdot10^{-6}$ °C^{-1}, $\alpha_{zz}=10\cdot10^{-6}$ °C^{-1}, The indicatory surface thermal expansion of aragonite is shown in figure 5.

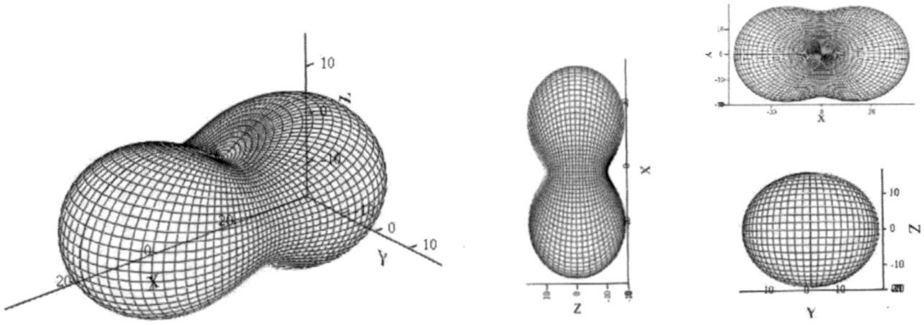

Figure 5. The indicatory surface of thermal expansion of aragonite and its projection on the planes (XOY), (XOZ) and (ZOY).

It is seen that all the projections are ellipses. Indicatory surface of such crystals is a triaxial ellipsoid with axes coinciding with the coordinate axes, which coincide with the second order axis of symmetry.

b) in less symmetrical crystals in this category (the monoclinic crystal system) the indicatory surface is described by four thermal coefficients. For example, for a crystal of potassium tartrate: $\alpha_{xx}= 12\cdot10^{-6}$ °C^{-1},$\alpha_{yy}= 44{,}8\cdot10^{-6}$°C^{-1}, $\alpha_{zz}=32\cdot10^{-6}$ °C^{-1}, $\alpha_{xz}= -12\cdot10^{-6}$ °C^{-1}. The indicatory surface of such crystals in projection has one ellipse only - perpendicular to the axis Y, that coincides with the only axis of symmetry in these crystals, figure 6.

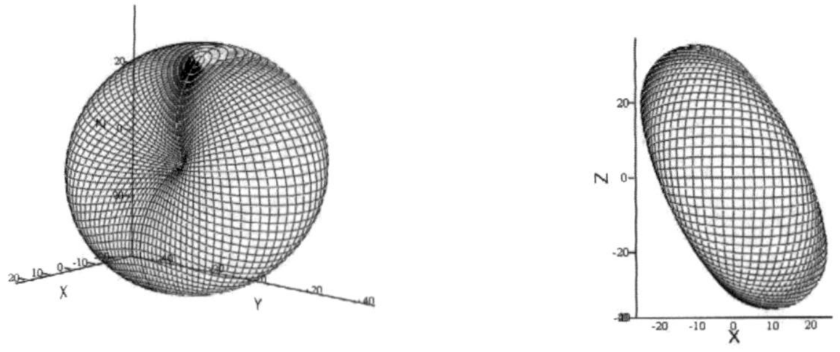

Figure 6. The indicatory surface of thermal expansion of potassium tartrate and its projection on the plane (XOZ).

4. Conclusion

The form and orientation of the indicatory surfaces of thermal expansion of the crystals is related to their symmetry according Neumann's Principle: the symmetry elements of any physical property of a crystal must include the symmetry elements of the point group of the crystal.

Journal of Physics: Conference Series **661** (2015) 012067 doi:10.1088/1742-6596/661/1/012067

As the result of investigation of anisotropy of thermal crystal expansion has been shown:

- Thermal expansion of crystals of the highest category is isotropic.
- When heated, the crystals of middle and lowest category can elongate in all directions or in some directions to shorten and elongated in other ones.
- An indicatory surface of thermal expansion of such crystals is an ellipsoid if all α_{ij} -positive or surface with positive and negative parts, if some α_{ij} -negative.
- In last case there are certain directions in a crystal along which thermal expansion is equal to zero.

References

[1] Wooster W A 1973 Tensors and group theory for the physical properties of crystals (Oxford: Claredon Press)
[2] Sirotin Y I, Shaskol'skaya M P 1979 Fundamentals of crystal physics (Moscow: Science)
[3] Polulyakh S N 1998 Application packages in physics (Simferopol: Pyramid Crimea)
[4] Maksimova E M 2002 Tensor properties of crystals (Simferopol: TNU)
[5] Perelomova N V, Tagiyeva M M 1982 Book of problems in crystal physics (Moscow: Science)